Second Edition

THE SCIENCE OF NUTRITION and its APPLICATION in CLINICAL DENTISTRY

ABRAHAM E. NIZEL, D.M.D., M.S.D., F.A.C.D.

Assistant Clinical Professor of Periodontology (Nutrition), Tufts University, School of Dental Medicine; Research Associate in Nutrition and Food Science, Massachusetts Institute of Technology

W. B. SAUNDERS COMPANY   Philadelphia and London

W. B. Saunders Company   West Washington Square
Philadelphia, Pa. 19105

12 Dyott Street
London W.C.1

The Science of Nutrition and its Application in Clinical Dentistry

SBN 0-7216-6806-2

Print No.:   9   8   7   6

WITH AFFECTION TO
my wife, Jeannette,
and my sons, David and Jonathan

and
with appreciation to
my esteemed teacher,
scholarly collaborator and cherished friend

ROBERT S. HARRIS, Ph.D.
PROFESSOR OF
NUTRITIONAL BIOCHEMISTRY
MASSACHUSETTS INSTITUTE
OF TECHNOLOGY

for inspiring, encouraging and guiding me
in the art and science of nutrition.

Contributors

WERNER ASCOLI, M.D., M.P.H.
Chief, Applied Nutrition Branch, Institute of Nutrition of Central America and Panama, Carretera Roosevelt, Zona 11, Guatemala, Central America.

BASIL G. BIBBY, B.D.S., Ph.D., D.M.D., D.Odont. (Hon.)
Director, Eastman Dental Center; Clinical Professor of Dentistry, University of Rochester, Rochester, New York.

FINN BRUDEVOLD, D.D.S., M.S., A.M. (Hon.), Dr. Odontologiae honoris causa (Univ. Oslo)
Professor of Dentistry, Forsyth Dental Center, and Harvard School of Dental Medicine, Boston, Massachusetts.

HOWARD H. CHAUNCEY, Ph.D., D.M.D.
Research Professor, Department of Oral Pathology, Tufts University School of Dental Medicine, Boston; Chief, Research in Oral Diseases, Research Service, Department of Medicine and Surgery, Veterans Administration Central Office, Washington, D. C.

GEORGE CHRISTAKIS, M.D., M.P.H.
Adjunct Assistant Professor in Clinical Nutrition at Institute of Nutrition Sciences, Columbia University; Research Associate, Department of Medicine, St. Luke's Hospital, New York.

SAMUEL DREIZEN, D.D.S., M.D.
Associate Professor of Nutrition and Metabolism, Northwestern University Medical School, Chicago, Illinois; Former Assistant Director, Spies Nutrition Clinic, Hillman Hospital, Birmingham, Alabama.

WILLIAM J. DARBY, M.D., Ph.D.
Professor of Nutrition and Chairman, Department of Biochemistry, Vanderbilt University School of Medicine, Nashville, Tennessee.

STANLEY N. GERSHOFF, Ph.D.
Associate Professor, Department of Nutrition, Harvard School of Public Health, Boston, Massachusetts.

SAMUEL A. GOLDBLITH, Ph.D.
Professor of Food Science and Executive Officer, Department of Nutrition and Food Science, Massachusetts Institute of Technology, Cambridge, Massachusetts.

RONALD J. GIBBONS, Ph.D.
Associate Staff Member, Forsyth Dental Center, and Assistant Professor of Bacteriology, Harvard School of Dental Medicine, Boston, Massachusetts.

BERNARD S. GOULD, Ph.D.
Associate Professor of Biochemistry, Department of Biology, Massachusetts Institute of Technology, Cambridge Massachusetts.

v

ROBERT S. HARRIS, Ph.D.
Professor of Nutritional Biochemistry, and Director of Oral Science Program, Department of Nutrition and Food Science, Massachusetts Institute of Technology, Cambridge, Massachusetts.

KRISHAN K. KAPUR, B.D.S.; D.M.D., M.S.
Professor and Chairman, Department of Oral Biology and Director of Research and Graduate Studies, University of Detroit, School of Dentistry, Detroit, Michigan.

WALTER J. LOESCHE, D.M.D.
Associate in Nutrition, Massachusetts Institute of Technology, Cambridge; Staff Associate in Periodontology, Forsyth Dental Center, Boston, Massachusetts.

HAROLD G. McCANN, M.S.
Chief of Analytical Chemistry, Forsyth Dental Center, Boston, Massachusetts.

SANFORD A. MILLER, Ph.D.
Associate Professor of Nutritional Biochemistry, Department of Nutrition and Food Science, Massachusetts Institute of Technology, Cambridge, Massachusetts.

HOWARD M. MYERS, D.D.S., Ph.D.
Professor of Oral Biology and Lecturer in Biochemistry, University of California, San Francisco Medical Center, San Francisco, California.

ABRAHAM E. NIZEL, D.M.D., M.S.D., F.A.C.D.
Assistant Clinical Professor of Periodontology (Nutrition), Tufts University School of Dental Medicine, Boston; Research Associate, Department of Nutrition and Food Science, Massachusetts Institute of Technology, Cambridge, Massachusetts.

NEVIN S. SCRIMSHAW, M.D., Ph.D.
Professor of Nutrition, and Head, Department of Nutrition and Food Science, Massachusetts Institute of Technology, Cambridge, Massachusetts; Consulting Director, Institute of Nutrition of Central America and Panama.

I. M. SHARON, D.D.S.
Research Associate, Poultry Husbandry Experimental Station, University of California, Berkeley; Senior Research Member, Institute of Medical Science, Presbyterian Medical Center, San Francisco, California.

GERALD N. WOGAN, Ph.D.
Associate Professor of Food Toxicology, Department of Nutrition and Food Science, Massachusetts Institute of Technology, Cambridge, Massachusetts.

THEODORE B. VAN ITALLIE, M.D.
Clinical Professor of Medicine, Columbia University College of Physicians and Surgeons; Director of Medicine and Attending Physician, St. Luke's Hospital, New York.

DAVID WEISBERGER, M.D., D.M.D.
Professor of Dental Medicine, Harvard School of Dental Medicine; Oral Surgeon and Chief of Dental Medicine Service, Massachusetts General Hospital, Boston, Massachusetts.

CHARLOTTE M. YOUNG, Ph.D.
Professor Medical Nutrition and Secretary, Graduate School of Nutrition, Cornell University, Ithaca; Medical Nutritionist, Sage Hospital and Gannet Medical Clinic, Ithaca, New York.

ISADORE ZIPKIN, Ph.D., F.A.C.D.
Assistant Chief of Laboratory of Biochemistry, National Institute of Dental Research, National Institutes of Health, Bethesda, Maryland.

Preface

This book is more than just a revision and updating of the first edition. Rather it is a completely new text which deals not only with the art of nutrition in dentistry but also with the science of nutrition and its effect on the oral and para-oral structures. The dynamic growth and change in the last five years in both nutrition and oral biology dictated this more sophisticated emphasis and approach. Depth and breadth had to be infused into the basic science section so that a better understanding could be reached concerning cellular as well as tissue level changes in the oral cavity that can be traced to nutrients.

In addition, this book reflects a newly increased cooperative and sympathetic attitude toward helping dentistry resolve its problems by our scholarly colleagues in the basic sciences. We deem ourselves fortunate and foresee much good emanating from this partnership. The extensive contribution of these scientists to this edition bespeaks their genuine concern and intent to help the dental profession meet adequately its major challenge, prevention of disease.

The clinical aspects of applied nutrition in dental practice have been developed more fully so that the practical would strike a balance with the theoretical, for after all, it is through application and implementation that the contribution of scientific knowledge can be objectively evaluated and appreciated. A formal knowledge of the principles of nutrition is not enough. The student must learn how to apply these principles in practice and be able to offer individual and personalized counseling to patients with dental problems that have resulted from faulty food and nutrition practices.

Twenty-four contributors who are recognized leaders and experts in such disciplines as biochemistry, bacteriology, nutrition, food science, medicine, oral science and clinical dentistry have provided authoritativeness to this text. It is interesting to note that even those contributors who do not have formal dental training were most empathic with our oral problems and have written their chapters with this type of bias.

The contents of this book were designed to conform with the recommendations of the Conference on Nutrition Teaching in Dental Schools held in March, 1965. (The proceedings are published in the March 1966 issue of the Journal of Dental Education.) The book deals with five general areas: The Nutrients and Their Oral Relevance; Food; Clinical Nutrition; Nutrition and Oral Biology; and Applied Nutrition in Dental Practice.

In Sections One and Two an attempt is made to present basic concepts and the newest in scientific knowledge concerning nutrients and food. The metabolic fate of the different nutrients as well as their functional role in life's processes, such as calcification and collagen biosynthesis, are discussed. Also, the effects of each of the nutrients on the oral and para-oral structures is thoroughly documented. (The material on oral relevance of the nutrients presented in Chapters Four, Five, Seven, Eleven and Thirteen was written by me and added as a supplement to the respective chapters with the permission of each author.) Translating nutrients into foods, factors involved in selection and nutritional requirements under different conditions are described so that the dentist will be armed with facts to combat misinformation and misconceptions that are often unwittingly acquired these days from the various mass communication media.

In Section Three an acquaintance with some major nutritional problems in medicine is presented.

New advances in oral biology–nutrition interrelationships and an appreciation that there is much information about mechanisms that we can glean from the laboratory are the reasons that Section Four was included in this text.

Section Five details step-by-step techniques in the actual management of dietary and nutritional problems for the patient with high caries susceptibility, or with periodontal disease, or who is going to undergo oral surgery, or the geriatric patient with new dentures. The latter chapters are the "bread and butter" section of the text. As in the previous edition an attempt is made to present nutrition as an objective discipline and as a specific individualized service that can be rendered in any dental office.

I am grateful to each of my distinguished contributors and colleagues for their unhesitant cooperation in spite of their own full and busy personal schedules. An especial thanks goes to Professor Sanford Miller, of the Massachusetts Institute of Technology, who not only became personally involved by contributing chapters but who also gave much moral support and wise counsel on the format of this text.

The typing of the manuscript was done by my able secretary, Mrs. Violette Biggins, and by Miss Susan Doughty, and for their devotion and skill I am most grateful.

As always, the W. B. Saunders Company has been most helpful with handling ably the myriad details concerned with the publishing of this text. For this I thank them sincerely.

Last, but not least, my lovely, charming wife and understanding children deserve my appreciation for humoring and encouraging me during periods of minor frustrations and for being unselfish about the time that I took away from them and gave to the assembling, editing and writing of this book.

ABRAHAM E. NIZEL

Contents

SECTION TWO
FOOD

SECTION FIVE
APPLIED NUTRITION IN DENTAL PRACTICE

Contents

Section One

The Nutrients and
Their Oral Relevance

Chapter One

The Science of Nutrition, a Discipline in Oral Biology and a Parameter in Preventive Dentistry

Abraham E. Nizel, D.M.D.

PROLOGUE

"The lack of nutrition makes the teeth weak, thin and brittle. An excess of nutrition excites a kind of inflammation similar to that of the soft parts. A deficiency of nourishment not only causes the tooth to die away but also enlarges the cavities," said Galen in 131 A.D.

Tomes in 1873 concluded that "caries is the effect of external causes . . . that it is due to the solvent action of acids which have been generated by fermentation going on in the mouth . . . and when once the disintegration is established in some congenitally defective point, the accumulation of food and secretions in the oral cavity will intensify the mischief by furnishing new supplies of acids" (Hanke, 1931).

These are only two quotations of the many that could be cited which attest to the recognition that a cause and effect relationship exists between food and/or nutrition and dental health. It is significant that one of these statements dates back to the beginning of recorded history, some 2000 years ago. Yet today we, the dental profession, are still timid and wary about utilizing the art and science of nutrition in a practical and objective fashion in our practice of clinical dentistry. It is to dispel this negative attitude and replace it with a positive approach toward nutrition–oral health interrelationships resulting from newer knowledge that this book was written. Hope-fully, this book will also serve to excite inquiry and research as well as strengthen the concept that the science of nutrition is truly an indispensable discipline in oral biology and a practical parameter in preventive dentistry.

WHAT IS NUTRITION?

Nutrition is a life science.

Nutrition cuts across a number of basic disciplines, making it a multidisciplinary integrated science. For its knowledge it draws on such basic general sciences as chemistry, physics, biology as well as their subdivisions, physical chemistry, biophysics and molecular biology. It deals with growth and development, reproduction and the maintenance of life by the provision of energy, building of tissue and preservation of a relatively constant environment. It answers such questions as: "What are the substances required by an organism for growth, maintenance and reproduction and in what quantities? Which foods will enable the organism to meet these requirements and in what quantities are these foods required? What are the results of failure to meet these requirements? What is the physiological role of each of these nutrients? How does failure of these physiologic functions lead to overt signs of deficiency?" (Harper, 1966).

3

In a sentence, nutrition may be defined as the science concerned with food and the dynamic interrelationships of the nutrients so that health can be attained and maintained.

WHAT IS SCIENCE?

Science is systematized knowledge based on keen and alert observations. It is, in a sense, a search for truths to satisfy curiosity and inquisitiveness. Science is the desire to know—in fact, it is derived from the Latin word *scientia,* the present participle of the verb *scire* which means "to know."

The primary aim of science is to describe the facts as exactly, as simply and as completely as possible. The scientist is made aware of certain facts that interest him, and by careful and critical observation he attempts to verify them and communicate them to others. The facts are assembled and then arranged in an orderly fashion in order to find their common denominator, to discover the condition of their occurrence, to describe them and, finally, to sum them up in a general formula or law.

The scientific method consists of systematic thinking based on observing certain logical processes and deriving certain inferences. There are three modes of inference: (a) from particular to particular (analogical reasoning), (b) from particular to general (inductive reasoning) and (c) from general to particular (deductive reasoning). Actually, induction occasionally interlaced with deduction is the major type of reasoning used in science.

The following are the steps to be taken when applying the scientific method.

Collection of Data. This implies clear, impartial observations of that which is or ought to be the facts, not vague impressions of experiences or secondhand evidence that suits the experiment.

Measurement of Data. Accurate and precise measurement and registration of the facts, which may be laborious and tedious to collect, is an absolute necessity.

Arrangement of Data. In order to understand the facts better one must try to resolve and describe them in simple and understandable general terms.

Analysis of Data. A scientific hypothesis is a tentative solution which must be tested. The scientist makes intellectual keys and then tries to see if they fit the lock. If one hypothesis does not fit, it is rejected and another is made. No matter how a hypothesis is conceived, be it by induction from many particulars or deduction from previous conclusions, it has to be tried and tested under controlled conditions before it can become a theory.

Formulation of Conclusion. What has been proved must be summed up in clear, terse terms. A theory is stated, a formula is derived, and a new set of facts is tested by an old law.

To sum up, science involves the collection of observations, organization of them into an orderly array and derivation from them of a principle or axiom.

THE SCIENCE OF NUTRITION, A DISCIPLINE IN ORAL BIOLOGY

Accumulation of scientific knowledge involves establishing the facts, and this requires the use of investigative tools. For example, the science of nutrition has developed much of its knowledge by using four reliable, time-tested research procedures, namely, epidemiological surveys, clinical trials, microbiological assays and animal experiments. It is interesting that the development of knowledge in oral biology has also come about as a result of the application of some of these same methods of research, particularly the use of animal experimentation.

Indeed, the use of the experimental animal has been shown to be so important in providing clues to the possible mechanism of action of specific nutrients and foods on dental and periodontal tissues that it seems appropriate at this juncture to discuss the proper design of animal feeding experiments. It will be seen that the steps in the scientific method just described are applicable here. If the reader is informed as to what constitutes a proper design, then he will develop a critical scientific attitude and be able to judge knowledgeably the reliability of scientific literature.

PRACTICAL FACTORS IN PLANNING AND DESIGNING ANIMAL FEEDING EXPERIMENTS

The use of experimental animals has been, by far, the most popular method of nutritional investigation used to date. This is understandable because as a research tool it offers such advantages as the ability to control more readily the experimental parameters, genetics, diet and environment; the ability to detect systemic effects of nutrients within weeks rather than years; and a comparatively small upkeep and cost.

Before designing a protocol for any experiment there are three overriding questions that must be answered:

1. What is the specific problem to be solved?
2. What parameters and measurements are going to be used for the solution of the problem?
3. Will the experimental design provide the kind of data that can be interpreted as a reasonable answer to the original question?

Once these questions are answered satisfactorily, the practical factors—the animal itself, its diet, the length of the experimental period and the treatment of the data—should be considered.

Selection of Animals. The selection of the species for the research is a prime consideration since the species selected must respond to the nutrient in a fashion similar to man if we intend to relate the results to man. If, for example, one intends to do dental caries research, the rat rather than the cat is selected because the rat is much more susceptible to dental decay and his dentition responds to the same nutrient challenge as man. A cat is not caries prone. On the other hand, if some fundamental rather than applied information is being sought, it is quite feasible to select a species whose organ or structure is more easily studied. For instance, if one is interested in elucidating a basic biochemical mechanism involved in the heart beat, a simpler species like the frog might be selected, even though the basic information obtained might not be directly applicable to man.

The number of animals in each experimental group is the second consideration. Even though inbred animals can provide considerable uniformity in response to a treatment and, therefore, are preferably used in nutrition experiments, there still will be considerable intra-individual variation. This can be offset by having a sufficient number of animals. The number of animals used in any experiment depends on how clear-cut a response is expected; e.g., is life or death to be the criterion? If it is, a minimal number of animals is adequate. On the other hand, if the response is going to involve a subtle measurement such as the comparative size of carious lesions, then larger numbers are needed. In other words, the number depends on the amplitude of variation expected in response to treatment—the greater the difference, the smaller the number, and vice versa. There are, in fact, statistical methods of determining adequacy of a sample prior to setting up the experiment. There are also rule of thumb procedures based on experience for arriving at adequate sample size. In general, 10 experimental animals per group for a feeding experiment is a required minimum.

The third consideration is assigning the animals to their proper groups. There should be a control and an experimental group or groups. The animal population should be equally represented in each of these groups, and this is done by randomizing the sample. Randomizing by weight is the simplest procedure. All animals are weighed and are then assigned numbers according to their weight, the heaviest animal being number 1, the next heaviest number 2, etc. Then, if, for example, the experiment consists of four groups, animals number 1, 5, 9, 13 are assigned to group A and animals number 2, 6, 10, etc., to group B. From a practical standpoint, unless there are very large numbers of dams and young, it is very difficult to actually randomize each group according to sex, litter and weight. The simplest method is to distribute the animals according to weight, using the same sex for the whole experiment and determining the genetic effect by statistical analysis.

Randomizing can also be done by using a published table of random numbers. Randomness is gained for the sampling by assigning numbers to the lot to be sampled, and then assigning them into groups according to the numbers that are drawn from the book of random numbers (Snedecor, 1956).

The husbandry of all animals participating in the experiment must be optimal and constant at all times. This means that the controllable environmental factors such as housing, lighting, feeding procedure, temperature and handling (which is the art in animal experimentation) must be exactly alike for all animals throughout the experimental period.

Diet. The nutrition of the animals is the next important consideration in the design of the experiment. The diet should be one that is acceptable in its physical form, palatable and conducive to maintaining the animal in good health (i.e., promoting growth, development, reproduction, and maintenance of good health) unless, of course, a purposeful nutritional deficiency is the experimental variable. A control diet for the rat usually consists of 18 to 22 per cent high quality protein (egg protein or casein supplemented with methionine); 55 to 65 per cent carbohydrate (starch or mixed carbohydrate consisting of starch, dextrin, sucrose and dextrose); 10 to 20 per cent fat (corn oil or lard or a mixture of both); 1 or 2 per cent all essential vitamins; 2 to 4 per cent all essential minerals; 3 to 10 per cent roughage, which may or may not be required (agar, celluflour), and water (ad lib or mixed in the diet to form a gel).

Experimental Period. An experiment should be run long enough for a reasonable response to occur. Extending it for longer periods is wasteful in time, energy and money, and sometimes the sensitivity of the effect is lost. Termination of the experiment before measurable data can be obtained obviously makes it valueless.

Significance of the Data. Finally, the data should be treated by statistical analysis, which is simply determining the probability or odds that this is a fact. To interpret the response to the treatment under investigation there should be recorded and reported the number of animals, the range of observed values, the mean of the observed values and the standard deviation or the standard error of the mean. From these data it is then possible to determine the significance of the observation,

which should have a mean value that is greater than the variation within the group to be significant. Otherwise, it should be considered a trend.

A further step is to calculate the significance of the difference between the means of the control and experimental groups or between the experimental groups themselves; for this Fischer's "T test" is useful. Actually, the statistical significance is expressed on the basis of probability. In other words, it is possible to determine whether the difference in the result of a deficiency feeding is due to chance or to a real difference in the experimental treatment. This chance or probability is expressed in per cent. In biological experiments a probability of 1 per cent or less is considered significant; a 5 per cent probability is considered a trend.

THE CONTRIBUTION OF THE SCIENCE OF NUTRITION TO DENTAL KNOWLEDGE

As mentioned previously, animal experimentation is only one of the nutritional investigative methods that have been used in dental research. The others, epidemiological surveys, clinical trials and in vitro experiments have been used to good advantage to re-enforce the validity of scientific observations. It is interesting to point out briefly how the use of these tools of the science of nutrition has influenced the development of knowledge about dental caries, periodontal disease and tooth morphology.

DENTAL CARIES

Historically, modern knowledge on the effect of nutrients on oral tissues began with the use of only the experimental animal. Mellanby (1918) was the first to describe the effect of what was then termed fat-soluble vitamin A (now known as vitamin D) on tooth calcification and tooth eruption. This was followed by histological studies on the effect of scorbutic diets and vitamin A deficient diets on guinea pig teeth (Wolbach, 1925a; 1925b).

Since dental caries was and is the most

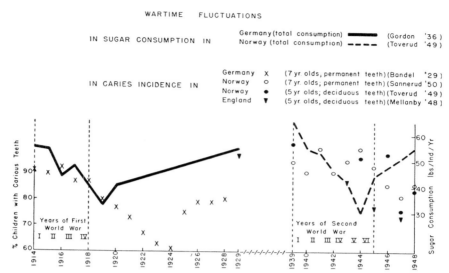

FIGURE 1-1. *Factors influencing caries in developing teeth. Is the susceptibility to dental caries influenced by factors operating during the period of tooth development? (From Sognnaes, R. F.: Sugar and Dental Caries Symposium, J. Calif. Dent. Assn., 26:47, 1950.)*

prevalent chronic disease in man, it can be readily understood that the maximal effort in dental research has been directed and expended here. It was the first dental disease that was associated in a scientific fashion with food or nutrients. This was accomplished through epidemiological surveys and clinical trials. Furthermore, the etiology and control of dental caries have been explained on the basis of animal feeding experiments as well as on studies dealing with the nutrition-oral flora interrelationships. A brief résumé of a few of the significant classic experiments is given here; these are later described in greater detail.

Epidemiological Proof. Survey data on the incidence of dental caries in African Masai tribesmen (Oranje et al., 1935) and primitive Greenlanders (Waugh, 1930) demonstrated that there was a direct correlation between increased dental decay and increased use of sophisticated foods, in general, and refined sugars, in particular. This concept has been confirmed by many others, notably Toverud (1949), who concluded from his data that lowering of refined carbohydrate intake and simultaneous increase in the consumption of the more natural protective foods during the food rationing period of World War II were the factors responsible for decreasing

caries incidence in children (Fig. 1-1). Recently, independent studies in Tristan da Cunha, Hungary, Ghana, Ireland, Australia and Ethiopia (Lossee, 1965) have also implicated sugars as being cariogenic.

Clinical Proof. There are numerous clinical observations (Koehne et al., 1934; Collins et al., 1942) which show that fermentable carbohydrates promote the initiation and progression of caries, the classic one being the five-year study at a mental institution in Vipeholm, Sweden (Gustafsson et al., 1954). A basal diet was fed to a control group. Nine other experimental groups were fed the basal diet supplemented by a sugar-containing foodstuff. The results indicated that the caries incidence did not parallel the amount of sugar consumed but rather correlated with the physical nature of the sweets and the frequency with which they were eaten. (See Chapter Four, p. 33.)

Experimental Animal Proof. Although McCollum et al. (1922) was perhaps the first to relate nutrition to dental caries in rodents, the classic work of Hoppert, Webber and Caniff (1931) is regarded as the beginning of controlled animal experimentation in dental caries research. They used a nutritionally adequate natural food diet consisting in the main of corn of coarse particle size to produce what was described as a fracture type caries. Since then, purified diets of fine powdery consistency high in sucrose, in addition to other chemically defined constituents, have been shown to produce caries in rodents in the absence of mechanical fracture of molar cusps (Sognnaes, 1948). The importance of carbohydrates in the diet to caries development was demonstrated by Shaw (1954) by alternate feeding of carbohydrate-free and carbohydrate-rich diets to rats. In other experimental groups sialadectomizing the animals, a recognized technique of causing an increase in caries, was found to be without effect if the animals were fed a carbohydrate-free diet. There are also data to show that the severity of caries varies with the type of carbohydrate consumed; it is increased markedly by sucrose, moderately by glucose and only slightly by raw wheat starch (Grenby, 1963).

Microbiological Proof. W. D. Miller (1890) theorized from his test tube experiments that the acid production from microbial metabolism of carbohydrates was the prime causative mechanism for the initiation of dental caries in man. Only a little over 10 years ago the presence of bacteria was shown to be indispensable to the occurrences of dental caries according to the classic study of Orland et al. (1954). More recently, it has been demonstrated that plaque bacteria can synthesize and store intracellular polysaccharides of the glycogen-amylopectin type from glucose and carbohydrates and that the number of organisms capable of storing large quantities of polysaccharides is greater in caries-active than caries-inactive people (Gibbons, 1964).

This same methodological approach has been applied to the development of the fluorine–dental caries hypothesis. It began with the classic epidemiological studies of Dean (1941; 1942), who demonstrated an inverse relationship between the fluoride levels of domestic water supplies and the prevalence of dental caries in children of 21 cities in 14 states. This prompted the fluoridation of communal water supplies with 1 ppm of sodium fluoride to prove Dean's hypothesis (Ast, 1962). There are also animal experiments which confirm this fluorine–dental caries relationship by showing that occlusal fissure and smooth surface dental caries can be in-

hibited by supplementing an otherwise cariogenic diet of rats with 50 ppm sodium fluoride (McClure, 1941). Finally, there is bacteriological confirmation that ingested fluorides lower lactobacilli counts (Hodge, 1946).

Thus, it is seen from these citations of the literature that these four methods of nutritional investigation have contributed to the development of dental knowledge.

GINGIVAL AND PERIODONTAL DISEASES

In the field of relating foods and nutrients to gingival and periodontal health in man and animals, the state of our knowledge is very limited. In fact, it is at the stage today that dental caries research was 35 years ago. In man, our data are sparse and unconfirmed. Epidemiologically, Russell (1963) attempted to relate periodontal disease to the nutritional status of 21,559 persons in eight different areas and could not find a clear-cut association. A dental health survey of Tristan da Cunha islanders seemed to implicate increased consumption of soft sticky fermentable sweets with more extensive deterioration of periodontal health (Holloway, 1963). Through the use of the experimental animal, particularly the rice rat, we are beginning to make some progress. For example, Auskaps et al. (1957) showed in an initial experiment that the carbohydrate component of the diet was probably responsible for the soft tissue periodontal syndrome in the rice rat. When more than half the sucrose in a periodontal syndrome-producing diet was replaced by lard, there was significant reduction in the extent of destruction of gingival soft tissues and alveolar bone resorption.

TOOTH MORPHOLOGY

In a third area which involves relating nutritional factors to growth and development of teeth and jaws, Paynter and Grainger (1956) have demonstrated that the morphology of the occlusal cusps and intercuspal areas as well as the sizes of the rat molars can be influenced by dietary challenges such as supplemented fluorides, vitamin A deficiencies and marked imbalances in Ca:P ratio. The provocative implication of this research is that if nutritional demands during the period of tooth development are not met adequately, the genetic pattern and constitution might be altered.

In summary, then, the historical development of our present dental knowledge resulting from nutritional research is in a more reliable and authentic state with respect to dental caries than with any other oral disease. The correlation between nutrition and periodontal disease is only on the threshold of discovery, and with respect to its effect on dental and oral tetralogy, it is only in the gestation period. The dental researcher has employed the sound scientific method and the standard nutritional investigative tools, epidemiological surveys, clinical trials, experimental animal feeding trials, and microbiological study with some significant success in resolving, at least partially, some of his problems. It can be concluded that dentistry, through the science of nutrition, is ridding itself of empiricism and clinical impressions; it is placing increased emphasis on the scientific approach to establish its truths.

THE SCIENCE OF NUTRITION, A PARAMETER IN PREVENTIVE DENTISTRY

Today, concepts of the practice of dentistry are beginning to change. For the past 50 or more years our primary task has been restoring, fabricating and rehabilitating teeth and mouths. In spite of our laudable contribution, the need keeps outpacing the fulfillment. Because of the increase in population, and the corresponding rise in dental needs, we appreciate more and more each day that we are fighting a losing battle by mere emphasis on therapy; this is obviously neither adequate nor progressive. Prevention, not therapy, is today's challenge and, hopefully, tomorrow's achievement. Only by better use of our present knowledge of which foods can contribute to dental health and by prescription of individualized diets for our patients are we going to reach the goal of all health professions—prevention of disease.

Since both the initiation and the continuation of our major problems in dental practice, dental caries and periodontal disease are, in part, nutrient and food in origin, it is logical and necessary that our management be geared to dealing with these causes. Research has proved that partial inhibition or arrest of these diseases can be realized by dealing directly with the food factor. We can negate the disease effect by either a decrease in or a restriction of the use of deleterious food factors or by promotion of the use of beneficial foods.

Dietary counseling is to date one of the few practical biological techniques that is available for use in a dental office. It is a practical objective method of implementing some of the newer knowledge that we have acquired concerning the effect of food and nutrients on biological systems. Furthermore, it has been shown to be effective (Howe, 1942). We need no longer focus all our teaching, learning and practice on technical skills and repair. We need not limit ourselves to talking about the biological approach; we can now put it into action.

The art of nutrition is another area that should be given serious consideration because it can fill the emotional needs of the patient and help immeasurably in attaining our ultimate goal, optimal physiological and psychological health for our patient. Making eating a joyful and pleasurable experience is as important as providing the proper amounts and kinds of vitamins, minerals, protein, etc. The art of nutrition allows for the practice of understanding and human relations. Food and food habits are very personal considerations, and respecting them and dealing with them artfully and humanistically are important.

We sometimes become so involved with the science and the mechanics of our service that we forget to act as a fellow human, with empathy and compassion for our patients who are beset with personal problems and difficulties. We must not let the "milk of human kindness be curdled by molecular biology." We should combine and interrelate the science of dentistry with the art of human relations. Thus, our proper and deserved place on the health team will be assured because we will be practicing the best possible kind of dentistry.

REFERENCES

Ast, D. B., and Fitzgerald, B. A.: Effectiveness of water fluoridation. J. Amer. Dent. Assn., 65:581, Nov., 1962.

Auskaps, A. M., Gupta, O. P., and Shaw, J. H.: Periodontal disease in the rice rat. 3. Survey of dietary influences. J. Nutr., 63:325, 1957.

Collins, R. C., Jansen, A. L., and Becks, H. J.: Study of caries free individuals. II. Is an optimum diet or a reduced carbohydrate intake required to arrest dental caries? J. Amer. Dent. Assn., 29:1169, 1942.

Dean, H. T., Jay, P., Arnold, F. A., Jr., and Elvove, E.: Domestic water and dental caries. II. A study of 2832 white children, aged 12-14 years, of eight suburban Chicago communities, including *Lactobacillus acidophilus* studies of 1761 children. Pub. Health Rep., 56:761, 1941.

Dean, H. T., Arnold, F. A., Jr., and Elvove, E.: Domestic waters and dental caries. V. Additional studies of the relation of fluoride domestic waters to dental caries experience in 4425 white children age 12-14 years, of thirteen cities in four states. Pub. Health Rep., 57:1155, 1942.

Gibbons, R. J.: Bacteriology of dental caries. J. Dent. Res., 43:1021, 1964.

Grenby, T. H.: The effects of some carbohydrates on experimental dental caries in the rat. Arch. Oral Biol., 8:27, 1963.

Gustafsson, B. E., Quensel, C., Lanke, L., Lundqvist, C., Grahnen, H., Bonow, B. E., and Krasse, B.: The Vipeholm dental caries study: The effect of different levels of carbohydrate intake on caries activity in 436 individuals observed for five years. Acta Odont. Scand. 11:232, 1954.

Hanke, M.: Role of diet in the cause, prevention and cure of dental disease. J. Nutr., 3:433, 1931.

Harper, A. E.: Why and how should nutrition be taught in dental schools? J. Dent. Educ., 30:69, 1966.

Hodge, H. C., and Sognnaes, R. F.: Experimental caries and a discussion of the mechanism of caries inhibition by fluorine. *In* Moulton, R. F. (ed.): Dental Caries and Fluorine. Washington, D.C., Am. Assoc. for Adv. of Science, 1946.

Holloway, P. J., James, P. M. C., and Slack, G. L.: Dental disease in Tristan da Cunha. Brit. Dent. J., 115:19, July, 1963.

Hoppert, C. A., Webber, P. A., and Canniff, T. L.: The production of dental caries in rats fed an adequate diet. Science, 74:77, 1931.

Howe, P. R., White, R. L., and Elliott, M. D.: The influence of nutritional supervision on dental caries. J. Am. Dent. Assn., 29:38, 1942.

Koehne, M., Bunting, R. W., and Hadley, F. P.: Review of recent studies of cause of dental caries. J. Amer. Diet. Assn., 9:445, 1934.

Lossee, F. L.: Enamel caries research: 1962-1964. J. Amer. Dent. Assn., 70:1428, June, 1965.

McClure, F. J.: Observations on induced caries in rats. III. Effect of fluoride on rat caries and on the composition of rat's teeth. J. Nutr., 22:391, 1941.

McCollum, V., Sunnends, N., Kenney, E. M., and Grievs, C. J.: The relation of nutrition to tooth development and tooth preservation. Johns Hopkins Hosp. Bull., 33:202, 1922.

Mellanby, M.: An experimental study of the influence of diet on teeth formation. Lancet, 2:767, 1918.

Miller, W. D.: The Microorganisms of the Human Mouth. Philadelphia, S. S. White Dental Mfg. Co., 1890.

Oranje, P., Noriskin, J. N., and Osborn, T. W. B.: The effect of diet upon dental caries in South African Bantus. S. African J. Med. Sci., 1:57, 1935.

Orland, F. J., Blayney, J. R., Harrison, R. W., Reyniers, J. A., Trexler, P. C., Wagner, M., Gordon, H. A., and Luckey, T. D.: Use of germfree animal technic in the study of experimental dental caries. I. Basic observations on rats reared free of all microorganisms. J. Dent. Res., 33:147, 1954.

Paynter, K. J., and Grainger, R. M.: The relation of nutrition to the morphology and size of rat molar teeth. J. Canad. Dent. Assn., 22:519, 1956.

Russell, A. L.: International nutrition surveys: A summary preliminary finding. J. Dent. Res., 42:233, 1963.

Shaw, J. H.: Effect of carbohydrate-free and carbohydrate-low diets on the incidence of dental caries in white rats. J. Nutr., 53:151, 1954.

Snedecor, G. W.: Statistical Methods. 5th ed. Ames, Iowa State College Press, 1956, pp 9-14.

Sognnaes, R. F.: Caries conducive effect of a purified diet when fed to rodents during development. J. Amer. Dent. Assn., 37:676, 1948.

Toverud, G.: Dental caries in Norwegian children during and after the last World War: A preliminary report. Proc. Roy. Soc. Med., 42:249, 1949.

Waugh, L. M.: Health of the Labrador Eskimo with special reference to mouth and teeth. J. Dent. Res., 10:387, 1930.

Wolbach, S. B., and Howe, P. R.: Tissue changes following deprivation of fat-soluble A vitamin. J. Exp. Med., 42:753, 1925a.

Wolbach, S. B., and Howe, P. R.: The effect of the scorbutic state upon the production and maintenance of intercellular substances. Proc. Soc. Exp. Biol. Med., 22:400, 1925b.

Chapter Two

Highlights in the History
of the Science of Nutrition

SAMUEL A. GOLDBLITH, PH.D.

APOLOGIA
Brevis esse labõro obscũrus fio

Sydney Smith, writing in the Edinburgh Review in 1825, said: "One great use of a Review, indeed, is to make men wise in ten pages who have no appetite for a hundred pages; to condense nourishment, to work with pulp and essence, and to guard the stomach from unmeaningful bulk."

Certainly in attempting to write a critical review of the history of nutrition, following through what McCollum (1957) has called "the sequence of ideas" and attempting to do it within reasonable bounds of space, one is faced with the imponderable problem of selecting and choosing, of attempting to be brief and yet not obscure. It is hoped that the "condensed nourishment" which follows in this chapter is "complete" nourishment and contains but little "unmeaningful bulk."

FROM PREHISTORIC MAN TO THE GREEKS AND THE ROMANS

Undoubtedly, prehistoric man carried on nutritional experiments through the ages as he began to supplement his vegetarian diet with eggs, reptiles and small mammals. We learn from archeological explorations that a million years ago, in China, "homo sapiens" ate animal meat, including the marrow of the leg bone (which supplied vitamins, iron and phospholipids) (Furnas and Furnas, 1937). Through the millennia of the ages, man became more skillful as a hunter, a food seeker, and then as a food cultivator. As food became more plentiful, he relied on appetite as the key to selection of his food supply. Appetite must have stimulated early man's senses to distinguish new and desirable odors and thus tempted him to experiment and taste unfamiliar foods. Those that delighted his taste buds, he accepted; those that repulsed his gastronomical senses, he refused, much as animals do when grazing in the fields. Thus, flavor and odor must have had much to do with early man's nutritional status, for it is a cardinal point that in order to be nutritious a food must first be ingested. Satiety also was used as a yardstick of nutritional value, for the function of food as an energy producer must have been recognized millennia ago.

THE FIRST NUTRITIONAL EXPERIMENT

Our first gleanings of actual nutritional experiments may be found in the Holy Scriptures in the Book of Daniel, 1:3-16, written about 2100 years ago.

And the king spoke unto Ashpenaz his chief officer, that he should bring in certain of the children of Israel, and of the seed royal, and of the nobles, youths in whom was no blemish, but fair to look on, and skillful in all wisdom, and skillful in knowledge, and discerning in thought, and such as had ability to stand in the king's palace; and that he should teach them the learning and the tongue of the Chaldeans. And the king appointed for them a daily portion of the king's food, and of the wine which he drank, and that

11

they should be nourished three years; that at the end thereof they might stand before the king. Now among these were, of the children of Judah, Daniel, Hananiah, Mishael, and Azariah. And the chief of the officers gave names unto them: unto Daniel he gave the name of Belteshazzar; and to Hananiah, of Shadrach; and to Mishael, of Meshach; and to Azariah, of Abed-nego.

But Daniel purposed in his heart that he would not defile himself with the king's food, nor with the wine which he drank; therefore he requested of the chief of the officers that he might not defile himself. And God granted Daniel mercy and compassion in the sight of the chief of the officers. And the chief of the officers said unto Daniel: 'I fear my lord the king, who hath appointed your food and your drink; for why should he see your faces sad in comparison with the youths that are of your own age? so would ye endanger my head with the king.' Then said Daniel to the steward, whom the chief of the officers had appointed over Daniel, Hananiah, Mishael, and Azariah: 'Try thy servants, I beseech thee, ten days; and let them give us pulse to eat, and water to drink. Then let our countenances be looked upon before thee, and the countenance of the youths that eat of the king's food; and as thou seest, deal with thy servants.' So he hearkened unto them in this matter, and tried them ten days. And at the end of ten days their countenances appeared fairer, and they were fatter in flesh, than all the youths that did eat of the king's food. So the steward took away their food, and the wine that they should drink, and gave them pulse.

It was natural for pulses (derived from Latin "puls," meaning pottage) to be the subject of the first written experimental material being tested for nutritional value inasmuch as legumes were among the earliest food crops to be cultivated by man. Archeologists have traced them, as cultivated crops, back to neolithic times, when man was transforming from the hunting and food gathering into the food producing stage of development and beginning to live in farming villages, the precursor of urban civilization.

Pulses are mentioned not only in the Book of Daniel but also in Ezekiel (Chapter 4:9), in Genesis (Chapter 25) in the story of Jacob and Esau, and the victory of Shammah in the field of lentils (II Samuel 23:11), and the present of Shobi to David (II Samuel 17:28) (Aykroyd and Doughty, 1964).

Students of the history of medicine have stated that Hippocrates (460-364 B.C.), the Father of Medicine, paid close attention to the diet of his patients as a feature of his therapeutic regime (Furnas and Furnas, 1937). Yet as early as the time of Hippocrates, the belief existed that a universal food sub-

stance or nutrient(s) existed in all foods which could be acted upon by the digestive tract. This was called an "aliment" (Mendel, 1923). This idea really persisted until even the early 1800's, as Mendel cites a quotation from a popular textbook published in 1813 (the American edition of Richerand's *Elements of Physiology*):

> By aliment is meant whatever substance affords nutrition, or whatever is capable of being acted upon by the organs of digestion. Substances which resist the digestive action, those which the gastric juice cannot sheathe, whose asperities it cannot soften down, whose nature it cannot change, possess, to a certain degree, the power of disturbing the action of the digestive tube, which revolts from whatever it cannot overcome. . . . However various our aliments may be, the action of our organs always separates from them the same nutritious principles; in fact, whether we live exclusively on animal or vegetable substances, the internal composition of our organs does not alter; an evident proof, that the substance which we obtain from aliments, to incorporate with our own, is always the same, and this affords an explanation of a saying of the father of physic. "There is but one food, but there exist several forms of food."

We note also that the great physician from Troy, Galen (131-200 A.D.), who practiced in Rome some 600 years after Socrates, was unable to add anything to the ancient doctrines taught by the Greeks (Lusk, 1922). Socrates (470-399 B.C.) had felt that the object of food was to "replace the loss of water from the skin and the loss of ponderable heat" (Lusk, 1922). Galen really had little to add to the knowledge of nutrition that Socrates and Hippocrates and their followers had not already known.

THE MIDDLE AGES

The great Hebrew scholar and celebrated physician Moses Maimonides (1135-1204 A.D.) wrote a treatise on asthma consisting of some 13 chapters (Muntner, 1963). Six of the chapters deal with food, drink and dietary measures that he recommended for asthmatics. While Maimonides, perhaps one of the greatest physicians of the Middle Ages, knew of the role of proper nutrition in the treatment of disease, he did not understand its importance. In fact, for 1300 years after the time of Galen, our knowledge of nutrition as a science advanced very little. As McCollum succinctly put it in the opening sentence of the first edition of his *Newer Knowledge of Nutrition* in 1919: "Our knowledge of nutrition has progressed hand in hand with the development of the science of Chemistry." The alchemists were busy trying to make gold from the baser metals and trying to create medical panaceas. However, until knowledge of the chemistry of living things was developed, there could really be no knowledge of the function of food. The Middle Ages were really the Dark Ages of nutritional science.

THE INFLUENCE OF THE CHEMICAL REVOLUTION ON NUTRITION

An old proverb states that "it is better to kindle a light than to curse the darkness." The light of the science of nutrition was kindled and began to emerge from the Dark Ages in the eighteenth century, through the chemical revolution.

Many scientists participated in studies which led to the birth of modern chemistry and, with this, the nascency of nutritional science. In this short narrative there is room to mention but a few whose work has been of major significance to the early development of nutritional science, which was really the science of metabolism. These include Priestley (1733-1804), one of the discoverers of oxygen (who to the end of his days still believed in the phlogiston theory of combustion); Scheele (1742-1786), a Swedish apothecary who also discovered oxygen and yet believed in the phlogiston theory; and, finally, Lavoisier (1743-1794), who overthrew the phlogiston theory in 1783 and founded modern chemistry. Lavoisier's contributions were many, but particularly germane to this work, in addition to the above, is the fact that Lavoisier was the first to really stress accuracy in measurements. He carried on respiration studies with animals; he used the analytical balance and the thermometer to study life itself and laid the basis of our modern chemical physiology. Lavoisier was not satisfied to work just on animals and, in 1789, together with Seguin, studied the respiration of humans—the subject being his co-worker Seguin (Lusk, 1922).

Until the middle of the nineteenth century,

nutritional scientific research was, in the main, in the fields of calorimetry and metabolism. Yet chemistry was beginning to make its impact. William Prout (1785-1850) classified foodstuffs in a way which we can recognize today. He divided "organized matter" into several classes of substances, e.g., the "saccharine group," "oleaginous bodies," the "gelatine" portion of animal bodies (water soluble) and the "albumen" portion (water insoluble). He classified "fibrin" and curd (of milk) as modifications of the "albuminous" principle. Another "albuminous" material was "gluten": "This substance though most abundant in vegetables, so far resembles the fleshy parts of animals, as to be, in like manner, capable of separation into two portions, analogous to gelatine and albumen" (Mendel, 1923).

François Magendie (1783-1855) was the first to begin to distinguish clearly between the nitrogenous and non-nitrogenous groups of foods. Magendie also pointed out that vegetables also contained nitrogenous substances (Mendel, 1923). Magendie used dogs to study the nutritive value of various incomplete diets (McCollum, 1957).

In 1839, the Dutch physiological chemist G. J. Mulder (1802-1880) suggested use of the term "protein." He pointed out the preeminent role of protein in life processes (Mendel, 1923; McCollum, 1957).

Jean Baptiste Boussingault (1802-1887) was a French mining engineer who, after 10 years of working at this profession in South America, returned to France and carried on important research on the nutritive value of various animal feeds. Whereas Mulder and Liebig carried out chemical tests on foods and speculated on the chemical nature of protein, Boussingault carried out animal experimentation on a large scale, using cows and horses, and recognized the importance of the non-nitrogenous as well as the nitrogenous portions of the diet. He drew up a table on the nutritive value of different feedstuffs (Mendel, 1923; McCollum, 1957). Moreover, in 1839, Boussingault devised a method which "was prophetic of future metabolism studies" (Lusk, 1928).

Justus von Liebig (1803-1873), the famous German chemist, was the father of organic analysis. His research made possible organic, physiological and agricultural chemistry. Mulder's work inspired him to consider proteins. He created organic chemistry and applied it to biological problems. He emphasized the nutritive value of proteins and suggested that the nitrogen in the urine might be a measure of protein destruction in the body. Lusk (1922) quotes his own teacher, the famous Carl von Voit, in describing Liebig's services: "All these chemical discoveries, to which Liebig so largely contributed, gave him his fruitful conceptions concerning the processes in the animal body. Before him the observations were single building-stones without interrelation, and it required a mind like his to bring them into ordered relation. It is a service which the physiologists of our own day do not sufficiently recognize. In order to appreciate this, one has only to read physiological papers written before the publication of his books and afterward in order to witness how his writings changed the mental attitude toward the processes in the organism."

If Liebig had a fault, it was in overemphasizing the role of proteins. Liebig's reputation was so great that many of his demonstrably wrong conceptions were fostered unduly long, such as his classification of foods into nitrogenous and non-nitrogenous sources of energy; or that fat or starch does not serve nutrition but merely facilitates the respiratory processes. Voit pointed out in 1881 that the "one" universal and unchangeable aliment pre-existent in the food, as postulated by Hippocrates, was replaced by protein, which became endowed with nutritive powers and into which every nutritious substance had to be converted (Mendel, 1923). Nevertheless, as McCollum (1957) has stated in speaking of Mulder and Liebig: "The great merit of these two men of genius was their attempts at constructive thought, and their strong motivation to test the truth or falsity of their working hypotheses. . . . They broke new ground and offered logical conclusions from their limited data. Their assertions stimulated the imagination and thought of many of the best minds in science in their generation, and fostered the development of new methods for chemical analysis and animal experimenting with foods. . . . Liebig and Mulder deserve unstinted admiration of later nutrition investigators."

The metabolism studies of Max von Pettenkofer (1818-1901) and Carl von Voit (1831-1908) at Munich were more critical than those done prior to their time. With the use of respiration apparatus large enough to accommodate a man, they determined the respiratory quotients of protein, fat and carbohydrate when metabolized in the body.

Voit established the so-called Munich School, and included among his pupils were Max Rubner, who discovered the specific dynamic action of foods and who solved the problem, first initiated by Lavoisier, of demonstrating that the law of the conservation of energy held true for the animal organism; and Graham Lusk, the eminent physiologist at Cornell University Medical College, researcher and student of the history of metabolism.

Life is a process of combustion and, in this, energetics play an important role. Thus, metabolic studies continued into the early part of the twentieth century with the researches of Atwater, Benedict, and others. These studies illustrate but some of the many early milestones in the growth of nutritional science as an outgrowth of the development of modern chemistry.

THE BIOLOGICAL METHOD OF ANALYSIS OF FOODS

In 1907 Dr. Elmer V. McCollum, then at the University of Wisconsin, proposed the use of rats as a means of evaluating the nutritive value of feedstuffs (McCollum, 1964). The first use of rats in nutrition studies is purported to have been by Savory in 1863 in London (McCollum, 1957). There are several advantages to using these animals. First of all, their gestation period is short and their life span is also short. Moreover, they consume relatively little food and this makes it possible to feed purified feedstuffs. McCollum at first met with resistance from those people who felt that direct studies made on large animals were the only way to conduct experiments. Moreover, to the farmer, a rat was a pest, and Dean Henry L. Russell, a noted food microbiologist, felt that this would be difficult to explain to the state legislature when it was necessary to account for funds used to purchase rats and to perform research with them.

At that time, Dr. Stephen M. Babcock was perhaps the single most influential man on the Wisconsin campus. He had developed the simple Babcock test for fat in milk, which made it possible for every farmer to weed out his herd rapidly and to retain the high fat producers, an economic advantage inasmuch as milk was sold on the basis of fat content. Dr. Babcock was friendly to McCollum and persuaded the university to support the establishment of a rat colony as McCollum wished.

In January 1908, McCollum established the first rat colony in this country. With the exception of the enthusiastic support of Professor Babcock, it was at first tolerated rather than supported. McCollum had studied Maly's *Jahresberichte* and learned of the efforts to nourish small animals, usually mice. He was "struck by the fact that in every instance in which small animals had been restricted to such 'purified' diets they promptly failed in health, rapidly deteriorated physically, and lived only a few weeks" (McCollum, 1964). McCollum felt that with these small animals one could make definitive studies. Moreover, from Babcock he had quickly learned that chemical analyses, per se, meant little until the results had been confirmed by animal studies. As Babcock had pointed out, certain soft coals, when analyzed by food-analysis procedures, yielded results that indicated that coal was a good, well-balanced foodstuff!

McCollum's rat colony became the foundation of the "biological method for analysis of foods." This technique was vital in the subsequent elucidation of the various vitamins, a task which could not have been done without McCollum's method. McCollum, with the help of his assistant, Miss Marguerite Davis, originally developed the method for the purpose of discovering the nature of the deficiencies of cereal grains. They fed incomplete foods with single and multiple purified supplements and measured the physiological responses of the animals to known substances, thus obtaining information on the nutrient requirements of the animals.

McCollum's biological method for the analysis of foods, when used together with the advances of Liebig and others in the chemical field, made possible the era of study of the essential nutrients, which commenced in the early 1900's.

EXPERIMENTS WITH PURIFIED DIETS

Dr. Thomas B. Osborne (1859-1929) of the Connecticut Experiment Station and Dr. Lafayette B. Mendel (1872-1935) began their classic studies on the feeding of purified proteins in 1909. These were published as monographs of the Carnegie Institution (1911). In the research work leading to this publication they also discovered that coprophagy was a source of protein. They prepared "protein-free milk" (sugar, milk salts, and unidentifiable materials later shown to be the vitamins). With this, Osborne and Mendel were able to show the varying nutritional qualities of proteins from different sources.

Hopkins, one of the leaders in biochemistry in the early part of the twentieth century, not only isolated and described tryptophan but, in 1906, together with Willcock, first demonstrated the indispensability of an amino acid (Willcock and Hopkins, 1906-1907).

William C. Rose (1887-   ), a student of Mendel at Yale, continued Mendel's work while at Illinois on the isolation and identification of the amino acids, and his own isolation and identification of threonine in 1935 made possible for the first time the feeding of diets consisting solely of chemically purified amino acids, thus permitting the establishment of the essential amino acid requirements of various species of animals (McCoy, Meyer and Rose, 1935-1936; Meyer and Rose, 1936).

THE VITAMIN HYPOTHESIS

Although the twentieth century has been considered the era of the vitamins, in actuality, vitamins were almost discovered in 1881 by Nicholas Lunin, who fed purified diets to mice, diets that contained carefully purified protein, fat and carbohydrate, plus the whole ash of milk. Lunin found that the mice fed on this diet, which he considered identical to the control diets of whole milk upon which mice flourished, still did not grow and survive. He concluded that animals needed some unknown substances other than fat, protein, carbohydrate and minerals to survive (Lunin, 1881).

Eijkman, in 1897, observed that beriberi was prevalent among those who ate polished rice but not among those who consumed unmilled rice (Eijkman, 1897). Beriberi was a disease that had been known for centuries in several Far Eastern countries where rice was the staple food.

Jansen, who crystallized thiamine in 1926, has stated that the "modern science of nutrition originated from or at least was greatly stimulated by the beriberi research" (Jansen, 1956). It was really the research for the cause of beriberi that put an end to the acceptance of the classic theory of nutrition.

At the turn of the century, the world had been witnessing some remarkable revolutions in medicine. Pasteur, Koch, Lister and others had demonstrated the causes of many diseases to be positive etiological agents, microbes. It was difficult to believe that some diseases could be caused by the lack of small quantities of nutrients. Only the eminence of Hopkins of Cambridge, perhaps the leading biochemist of his era, could convince scientists that deficiency factors could cause disease. His paper in 1912 illustrated the importance of accessory factors in normal dietaries.

Funk (1912) discussed the deficiency diseases—beriberi, scurvy, pellagra, and others—and pointed out that they presented certain general characteristics "which justified their inclusion in one group, called deficiency diseases." He proposed the term "vitamine" (vital amine), as the name for the accessory dietary factors. The term vitamine is a most fortuitous one. Originally, Funk felt that the accessory food factors were organic bases, hence his suggestion of "vitamine" to describe them. Mendel later suggested the term "hormone" and Lusk, the word "food hormone" in 1917. The word vitamin came into universal use shortly thereafter.

The first fat-soluble vitamin (later named vitamin A) was discovered by McCollum and Davis (1913). Drummond proposed the alphabetical nomenclature of the vitamins in 1920 in order to get around some of the difficulties which existed for naming compounds whose chemical structure was not known (Drummond, 1920).

McCollum and his associates (1922) demonstrated the existence of "a fourth vitamin (vitamin D) whose specific property, as far as we can tell at present, is to regulate the metabolism of the bones." The work by McCollum in this field was important inasmuch as it also showed the important role that pathologists play in the study of nutritional diseases (McCollum, 1964). In subsequent work, Dr. E. A. Park played an important part in the elucidation of vitamin D and its healing properties in rickets. Mellanby (1918 a,b) was impressed with the power of butter fat to cure rickets, although he wrongly concluded that the substance which stimulated bone calcification was the same as McCollum's "fat-soluble A."

The 1930's saw the identification and synthesis of thiamine and other vitamins of the B-complex, and the unfolding of the multiple nature of the B-complex, which became apparent as more animal species began to be studied and as the use of more sophisticated methods of isolation and purification were developed (Cline et al., 1937).

In the era from the 1920's to the 1940's the saga of the isolation and identification of the antiscorbutic vitamin was unfolded by Szent-Györgyi (1928), Svirbely and Szent-Györgyi (1932), Waugh and King (1932), Tillmans, Hirsch, and Hirsch (1932), Zilva (1932), and others. This, too, culminated many years of work on the dread disease scurvy, which had been so aptly described by Lind two centuries earlier (Stewart and Guthrie, 1953).

Once the vitamins were isolated and identified, their role in metabolism could be elucidated. In this, the 1930's and 1940's were exciting years. With Barker's work on the vitamin B_{12} co-enzyme in 1958 and 1960, an era of nutrition was ending—that of the B-complex.

Not all of the important milestones in the vitamin saga have been covered in this brief review, nor indeed have even all of the vitamins been discussed. Later chapters in this book will discuss these and their implications for the nutrition and well-being of man.

In view of the fact that this book is designed primarily for the dental profession and inasmuch as a great deal of space will be devoted to discussion of mineral metabolism and nutrition, it is sufficient here to note that the role of minerals in nutrition had been studied for many years by a number of researchers, including the famous Professor G. von Bunge, the Russian-Swiss physiologist who started N. Lunin on his now famous experiments. In the field of mineral metabolism, just as in the field of vitaminology, the agricultural chemists had an important role in furthering knowledge of nutrition. As an example, Hart and his group at Wisconsin discovered the role of iron in nutrition (Hart et al., 1925).

EPILOGUE AND PROLOGUE

Lepkovsky aptly pointed out in 1959 that "we are witnessing the end of an era in nutrition—the era of essential nutrients" (Lepkovsky, 1959). Nutritional science today is concerned with discovering the interrelationships among the nutrients and other metabolites, their role in disease at all stages of life—infectious diseases, metabolic diseases, etc. Men are laboring toward utilizing the knowledge of nutritional science "pro bono publico" —in public health work, by trying to solve many of the nutritional problems of expanding populations, and in the problems of the overfed as well as the underfed. Among the problems of the future is determination of the role of nutrition in appetite, satiety, behavior, etc.

Brillat-Savarin (1825) wrote a classic book on the physiology of taste, and in this he recognized the role of flavor, odor, texture, etc., in the art of living. Since that time, various workers have been concerned with the role of flavor and odor of food in experimental rations for animal diets. However, it was not until recent years, with the emergence of new knowledge and techniques in biochemistry and physiology, primarily neurophysiology, that recognition has been given to the basic mechanisms affecting hunger and satiety. The organoleptic characteristics of foods appear today to be as elusive as were the accessory food factors in 1906 to Hopkins, McCollum, Funk and other pioneers. Yet with the tools now available in terms of newer knowledge, equipment, techniques, etc., perhaps in the next two decades the role of these organoleptic factors which are so important to dental science will also be elucidated in terms of optimal nutrition. Perhaps as Woolley (1964) has succinctly put it, through studies of the hypothalamus and those other parts of the primitive brain which are the seat of emotion and fundamental behavior, the so-called "biogenic amines," i.e., serotonin, epinephrine, and acetylcholine, will be encountered, thus passing "from the frontier of the 'vital amines' (the vitamins) as Funk called them, to that of the biogenic amines which somehow control behavior."

Will this not be the exciting future of nutritional science, which is so closely interwoven with the role of the mouth and its associated apparatus, i.e., the teeth, tongue, salivary glands, etc.? Thus the future developments of nutrition may well be intimately bound to dental science, and the next 50 years may prove to be more exciting in this field than the first half of this century when classical nutrition was transformed.

REFERENCES

Aykroyd, W. K., and Doughty, J.: Legumes in Human Nutrition. F. A. O. Studies No. 19, Food and Agriculture Organization of the United Nations, Rome, 1964.

Barker, H. A., Smyth, R. D., Weissbach, H., Toohey, J. I., Ladd, J. N., and Volcani, B. E.: Isolation and properties of crystalline cobamide coenzymes containing benzimidazole or 5,6-dimethylbenzimidazole. J. Biol. Chem., 235:480, 1960.

Brillat-Savarin, J. A.: Physiologie du Goût. Just Tessier, Paris, 1825.

Chittenden, R. H.: The Nutrition of Man. New York, Frederick A. Stokes Publishing Co., 1907.

Cline, J. K., Williams, R. R., and Finkelstein, J.: Studies of crystalline vitamin B₁. XVII. Synthesis of vitamin B₁. J. Amer. Chem. Soc., 59:1052, 1937.

Drummond, J. C.: LIX. The nomenclature of the so-called accessory food factors (vitamins). Biochem. J., 14:660, 1920.

Eijkman, C.: Ein Versuch zur bekämpfung der Beriberi. Virch. Arch. für Pathologische Anatomie, 149: 187, 1897.

Funk, C.: The etiology of the deficiency diseases. J. State Med., 20:341, 1912.

Funk, C.: The Vitamines. Authorized translation from second German edition by H. E. Dubin. Baltimore, The Williams & Wilkins Co., 1922.

Furnas, C. C., and Furnas, S. M.: Man, Bread and Destiny. New York, Reynal & Hitchcock, 1937.

Goldblith, S. A., and Joslyn, M. A. (ed.): Milestones in Nutrition. Westport, Connecticut, The Avi Publishing Co., 1964.

Hart, E. B., Steenbock, H., Elvehjem, C. A., and Waddell, J.: Iron in nutrition. I. Nutritional anemia on whole milk diets and the utilization of inorganic iron in hemoglobin building. J. Biol. Chem., 65:67, 1925.

Hopkins, F. G.: Feeding experiments illustrating the importance of accessory factors in normal dietaries. J. Physiol., 44:425, 1912.

Jansen, B. C. P.: Early nutritional researches on beriberi leading to the discovery of vitamin B$_1$. Nutr. Abstr. Rev., 26:1, 1956.

Kleiber, M.: The Fire of Life: An Introduction to Animal Energetics. New York, John Wiley & Sons, Inc., 1961.

Lepkovsky, S.: Potential pathways in nutritional progress. Food Tech., 13:421, 1959.

Lunin, N., von: Ueber die Bedeutung der anorganischen Salze fur die Ernahrung des Thieres. Ztschr. Physiol. Chem., 5:31, 1881.

Lusk, G.: A history of metabolism. In Endocrinology and Metabolism. New York, D. Appleton & Co., 1922.

Lusk, G.: The Science of Nutrition. 4th ed. Philadelphia, W. B. Saunders Co., 1928.

McCollum, E. V.: The Newer Knowledge of Nutrition. New York. The Macmillan Co., 1919.

McCollum, E. V.: A History of Nutrition. Boston, Houghton Mifflin Co., 1957.

McCollum, E. V.: From Kansas Farm Boy to Scientist. Lawrence, Kansas, University of Kansas Press, 1964.

McCollum, E. V., and Davis, M.: The necessity of certain lipins in the diet during growth. J. Biol. Chem., 15:167, 1913.

McCollum, E. V., Simmonds, N., and Becker, J. E.: Studies on experimental rickets. XXI. An experimental demonstration of the existence of a vitamin which promotes calcium deposition. J. Biol. Chem., 53:293, 1922.

McCoy, R. H., Meyer, C. E., and Rose, W. C.: Feeding experiments with mixtures of highly purified amino acids. VIII. Isolation and identification of a new essential amino acid. J. Biol. Chem., 112:283, 1935-1936.

Mellanby, E.: The part played by an "accessory factor" in the production of experimental rickets. Proc. Physiol. Soc., xi-xii, Jan. 26, 1918a.

Mellanby, E.: A further demonstration of the part played by accessory factors in the aetiology of rickets. Proc. Physiol. Soc., liii-liv, Dec. 14, 1918b.

Mendel, L. B.: Nutrition: The Chemistry of Life. New Haven, Connecticut, Yale University Press, 1923.

Meyer, C. E., and Rose, W. C.: The spatial configuration of α-amino β-hydroxy-n-butyric acid. J. Biol. Chem., 115:721, 1936.

Muntner, S. (ed.): The Medical Writings of Moses Maimonides: Treatise on Asthma. Philadelphia, J. B. Lippincott Co., 1963.

Osborne, T. B., and Mendel, L. B.: Feeding Experiments with Isolated Food Substances. Parts I and II. Carnegie Institution of Washington, Publication No. 156, 1911.

Stewart, C. P., and Guthrie, D.: Lind's Treatise on Scurvy. Edinburgh, Scotland, The University of Edinburgh Press, 1953.

Svirbely, J. L., and Szent-Györgyi, A.: CV. The chemical nature of vitamin C. Biochem. J. 26:865, 1932.

Szent-Györgyi, A.: CLXXIII. Observations on the functions of peroxidase systems and the chemistry of the adrenal cortex: Description of a new carbohydrate derivative. Biochem. J. 22:1387, 1928.

Tillmans, J., Hirsch, P., and Hirsch, W.: Das Reduktionsvermögen pflanzlicher Lebensmittel und seine Beziekung zum Vitamin C. I. Der reduzierende Stoff des Citronensaftes. Ztschr. Untersuch. Lebensmittel, 63:1, 1932.

Waugh, W. A., and King, C. G.: Isolation and identification of vitamin C. J. Biol. Chem., 97:325, 1932.

Willcock, E. G., and Hopkins, F. G.: The importance of individual amino acids in metabolism. Observations on the effect of adding tryptophane to a dietary in which zein is the sole nitrogenous constituent. J. Physiol., 35:88, 1906-1907.

Woolley, D. W.: Dedicatory foreword. In Goldblith, S. A., and Joslyn, M. A., ed.: Milestones in Nutrition. Westport, Connecticut, The Avi Publishing Co., Inc., 1964.

Zilva, S. S.: CXCIII. The non-specificity of the phenolindophenol reducing capacity of lemon juice and its fractions as a measure of their antiscorbutic activity. Biochem. J., 26:1624, 1932.

Chapter Three

Calories and Energy Metabolism

SANFORD A. MILLER, PH.D.

The fundamental concepts of the life sciences, as in the physical sciences, are those concerned with energy, its sources, conversion and utilization. In the broadest sense, life is energy. To survive, all living things must be assured of an adequate supply of materials which can be used to produce energy; the more complex the living organism, the less flexible its needs. In terms of diet, more food is necessary to provide for the energy requirement than for all other purposes combined. This is not to imply that other metabolic needs are not as essential for life but rather that if the energy need is satisfied from a variety of food sources, it is probable that other nutrient requirements will be satisfied.

HISTORY

The central and obvious need for energy resulted in its being the first of the major dietary needs to be investigated. Just before the French Revolution, Lavoisier investigated the heat produced by an animal and was able to demonstrate that there was a relationship between the heat production, oxygen utilization and carbon dioxide production. The series of experiments are considered today to be the foundation of modern nutrition.

In the years following this monumental work, many talented men devoted much effort to elucidation of the principal factors inherent in the energy exchanges of organisms. Pettenkofer, Voit, Rubner, Zuntz and Johansson were among those who contributed much during the period from 1860 to 1890 toward our understanding of these processes. In 1892, W. O. Atwater, an American pupil of Voit, returned to the United States and initiated the significant series of studies which established the essential quantitative physiological knowledge upon which all assessments of the energy needs of man are based. Atwater and his co-workers built and utilized calorimeters which were capable of measuring heat production in man with an accuracy of 0.1 per cent, a truly remarkable technical achievement for that day.

When, in 1906, scientists began to discover the presence of "accessory food factors" in the diet, studies in energy metabolism became less fashionable and less significant in nutritional research. The great period of vitamin research was beginning. This was followed by studies of mineral needs, and next came the investigations of protein requirements and metabolism. Since World War II, however, a renaissance in research on energy and its conversion has emerged, but is now concerned with energy conversions at the molecular level in the cell. This research will ultimately lead to a firmer foundation for our understanding and knowledge of life and its processes.

BASIC PRINCIPLES

In essence, life is concerned with five forms of energy: solar, chemical, mechanical, thermal and electrical. In all living organisms all forms of energy are interchangeable, and thus energy metabolism must follow the first law of thermodynamics, which deals with the principle of conservation of energy. Plants can utilize energy from the sun directly to synthesize new molecules, but animals cannot

19

and must obtain their energy through relatively complex processes in which the energy must be supplied in a chemical form.

For the higher animals and man, this energy is contained in the complex organic molecules of the diet—carbohydrates, fats and proteins. It is the function of metabolism to extract energy from the chemical bonds of these materials, to convert it to forms useful to the body, and to use it for synthesis of new molecules and, thus, new tissues. To understand energy needs, therefore, it is necessary to understand the metabolism of nutrients and the ways in which an organism can use them to produce energy.

In general, the organism obtains its energy by the oxidation of the food which it ingests. Over-all, this process is similar to the oxidation of fuels in engines and furnaces. It differs, however, in detail and in the steps by which the process occurs. Energy in the living organism is expressed in terms of the heat unit or calorie, which, by convention, is defined as the amount of heat required to raise the temperature of 1 gm. of water 1°C. from 14.5 to 15.5°C. In nutrition we often utilize the kilogram calorie (Calorie), a unit which is 1000 times that of the calorie and is defined as the amount of heat required to raise the temperature of 1 kg. of water 1°C. from 14.5°C. to 15.5°C.

In a machine, fuel is oxidized in a single or at most a limited number of steps to provide energy for heat or work. The living organism differs in that it obtains its energy in a series of steps so that the energy is obtained not in just one large burst but rather in limited amounts over a period of time. In this way the living organism can utilize more efficiently the energy it obtains and thus is able to control the process. These steps, or intermediary metabolism, provide the framework of energy conversion in the organism.

INTERMEDIARY METABOLISM

The dynamic state of body constituents provides a pool of carbohydrate, protein and fat. This pool is also contributed to by the diet. From this pool, molecules undergo specific reactions which result ultimately in their conversion to carbon dioxide, water and energy. Carbohydrates are phosphorylated and anaerobically oxidized to pyruvate or lactate. Fatty acids are oxidized to acetate, and the amino acids are then deaminated to other intermediates. All, however, share a common aerobic pathway known as the Krebs or tricarboxylic acid cycle in which the various intermediates of the specific nutrients are converted to energy, carbon dioxide and water. It becomes obvious, therefore, that one function of metabolism is to convert the various nutrients to a form suitable for entrance into this cycle of metabolic events.

During each step of the metabolism of the molecule, energy is being released. To transport and to provide a common basis for the utilization of this energy, the organism utilizes a group of compounds containing a high-energy phosphate bond. Of these the most significant is adenosine triphosphate or ATP. This compound, called the common currency of energy metabolism, is unique in the predominant role it plays in the process by which energy is generated, stored and transferred. Most processes in the organism which require energy for their function utilize ATP; muscle contraction, synthesis of new molecules, nerve impulse propagation all at one point or another require ATP. This is not to say that other compounds do not exist to perform similar functions. There is no question, however, that the vast bulk of phenomena involving energy utilize the ATP molecule to some extent. It is possible, therefore, to discuss energy metabolism in terms of ATP generation and utilization.

The measurement of the utilization of ATP is a relatively complex procedure and not suitable for use in the whole animal. For this purpose, measurements of oxygen utilization, carbon dioxide production or heat production are more suitable. Oxygen utilization and carbon dioxide production are obviously related to energy metabolism since all energy-producing reactions are oxidations. Techniques involving the measurement of these parameters are known as indirect calorimetric measurements. Heat production is also a useful parameter for the estimation of energy requirements. This measurement, known as direct calorimetry, cannot, however, give the total energy expenditure since it will not measure energy that is utilized for mechanical or electrical work or for the synthesis of new compounds.

ENERGY CONTENT OF NUTRIENTS IN FOODS

If a food is burned under conditions in which it is completely consumed and converted to carbon dioxide and water, the heat derived can be measured by a process known as bomb calorimetry. For foods containing only carbon, hydrogen and oxygen, such as carbohydrates and fats, the energy derived from this process is approximately the same as that obtained by oxidation in the body (Fig. 3-1). On the other hand, nutrients containing other atoms, such as the nitrogen in protein, yield smaller caloric values when oxidized in the body. This is because the nitrogen is not oxidized but is excreted as urea.

The oxidation of different compounds of the same class results in different caloric values. Glucose yields 3.7 calories per gram while starch and sucrose yield 4.1 and 4.0 calories per gram, respectively. For fats and proteins,

these differences are somewhat smaller. Most fats average about 9.0 calories per gram, whereas most proteins average 4.1 calories per gram. Atwater, over 50 years ago, determined the absorption of nutrients and applied these values to the caloric contents of foods. From these data, he derived a series of caloric factors which are used today. These factors, called "Atwater factors," are 4.0 calories per gram for protein and carbohydrate and 9.0 calories per gram for fat. It is important to remember that these values take into account calories lost in urine and through digestion and do not represent the total energy value of any compound. Furthermore, it is also important to remember that they are averages and approximations, useful only in estimating the caloric content of diets from nutrient analysis. For more accurate information, it is necessary to know the exact nutrient composition rather than proximate analysis. The most accurate determination can be made only by using data obtained by bomb calorimetry and correcting for losses in urine and feces. However, since the differences in results obtained by the most and least accurate methods are generally smaller than the variation in different samples of the same food, the use of Atwater factors and proximate analysis under clinical conditions can

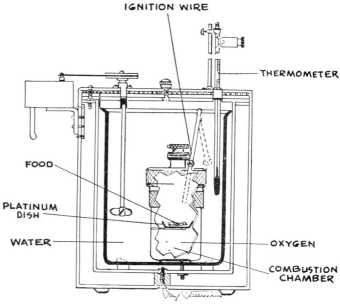

FIGURE 3-1. *The bomb calorimeter. (Langley, L. L., and Cheraskin, E.: Physiology of Man. McGraw-Hill Book Company, Inc., 1954. Used by permission.)*

be useful in estimating the caloric value of diets for practical purposes. For research, the most accurate methods are required.

In the organism, both direct and indirect calorimetric methods have been used to estimate energy production for specific foodstuffs. Theoretically, direct methods appear to be the easiest. In practice, however, the execution of this technique is more difficult. For a variety of reasons, the direct measurement of heat production is a long and arduous activity with many variables to be considered, and indirect calorimetric measurements have been more popular. Generally, in this method, oxygen consumption is measured using either the rate of disappearance of oxygen from a reservoir or the direct measurement of the difference in oxygen entering or leaving the test chamber or face mask. It is the first of these that is more commonly used in clinical applications.

RESPIRATORY QUOTIENT

By using indirect techniques it is possible to determine, under normal conditions, which of the major sources of energy are being utilized at any particular time. This is done by determining the relative amounts of carbon dioxide produced and oxygen consumed during the time of the experiment. This ratio of the volumes of carbon dioxide to oxygen is called the "respiratory quotient" or RQ.

This measurement is based upon the following considerations. When carbohydrates are oxidized, the simplified reaction is as follows:

$$C_6H_{12}O_6 + 6\ O_2 \longrightarrow 6\ CO_2 + 6\ H_2O$$
$$RQ = \frac{6\ CO_2}{6\ O_2} = 1.0$$

Since fats contain less oxygen than carbohydrates, more oxygen must be supplied in order to complete their oxidation.

$$2\ C_{51}H_{98}O_6 + 145\ O_2 \longrightarrow 102\ CO_2 + 98\ H_2O$$
$$RQ = \frac{102\ CO_2}{145\ O_2} = 0.7$$

Thus, there is a big difference in the RQ of fat and of carbohydrate, which allows an estimate of the relative proportions of each being oxidized at any one time. If this is to be done, however, the measured RQ must be corrected for the amount of protein being oxidized and thus contributing to the RQ; this is done by measuring urinary nitrogen excretion. It has been determined that 4.76 liters of carbon dioxide and 5.94 liters of oxygen are required to oxidize sufficient protein to produce 1 gram of urinary nitrogen. Using these factors and the total urinary nitrogen loss, one can then estimate the carbon dioxide and oxygen correction for the RQ. The corrected RQ can then be used to estimate the proportion of fat and carbohydrate being utilized.

CALORIC REQUIREMENTS

The caloric requirement is composed of three factors: the amount of energy required to maintain the organism in its minimal state (basal metabolism); the amount of energy required for the specific dynamic action of foods (SDA); and the energy required for growth, repair and physical activity.

The basal metabolism represents essentially the base level of energy needs. It includes the energy required for temperature regulation, to keep the heart and lungs operating, as well as the energy needs of the resting tissues. In general, it is defined as the energy need 12 to 18 hours after the ingestion of food and measured under conditions in which the organism is awake, at a comfortable temperature, but at complete rest. Since this value is dependent upon the amount of active tissue in the organism, it is conventionally related to some function of actively metabolizing tissue, usually the surface area. Generally, it is expressed as the basal metabolic rate (BMR) and is defined as the number of calories given off by the organism per square meter of body surface per hour. In the human, the BMR varies from 36 to 42 calories per square meter per hour (cal./m.²/hr.) in the male and from 34 to 36 cal./m.²/hr. in the female. Clinically, the BMR is often expressed as a deviation, plus or minus, from the average. A number of factors can influence the BMR; these include endocrine disorders, such as those of the thyroid gland, emotional state, age, climate and environmental temperature, and diet. All these factors must be considered when evaluating BMR data.

The specific dynamic action (SDA) of foods is still an unknown entity in nutrition. When food is ingested, there is a rise in the heat output not associated directly with the oxidation of the food. This increase varies for each of the nutrients, ranging from about 5 per cent when carbohydrate alone is fed to as high as 20 to 30 per cent when protein alone is consumed. In mixed diets, however, these values are somewhat lower. In any case, it is a factor to be considered in evaluating energy needs but is probably of less practical importance than may be inferred from many experimental situations.

ENERGY REQUIREMENTS FOR OTHER ACTIVITIES

For physical activity the energy requirements have been estimated empirically. Some of these are presented in Table 3-1. These range from a low of 15 cal./hr. for sitting to as high as 960 cal./hr. for climbing. Mental activity, as exhausting as it might appear, is usually ignored since it has been shown to account for a metabolic increase of only 3 to 4 per cent. For convenience in estimating energy needs, the activities of an individual may be classified as sedentary, moderate,

heavy or very heavy. For each of these classifications, an energy increment above the basal requirement may be estimated and ranges from 225 cal./day for sedentary occupations to as high as 2500 cal./day for very heavy work.

An additional increment must be allowed for growth, pregnancy and lactation, since each of these activities requires additional energy. For growth, the energy need depends upon the rate of growth and the age and varies from 15 to 20 cal./day/kg. of body weight. The relatively greater activity of children also results in an increase in the caloric requirement, which may be as much as 25 per cent greater than that of the adult.

The requirements for pregnancy and lactation are obvious, but are not nearly so large as many women seem to assume, and this may become a major problem in the management of pregnancy.

TOTAL CALORIC REQUIREMENTS

To enable diets of practical importance to be developed, a variety of estimates of caloric requirements have been produced. These include those of the FAO, the Food and Nutrition Board of the National Academy of Sciences–National Research Council in the United States, and the Canadian Dietary Standards (see Chapter Fourteen). Each group has attempted to estimate caloric needs of the populations they serve and then to come

TABLE 3-1.   APPROXIMATE ENERGY EXPENDITURE WITH DIFFERENT ACTIVITIES*

Activity	Cal./hr.	Activity	Cal./hr.
Dressing and undressing	33	Mental work	7–8
Sitting at rest	15	Mixed carpentry work	180
Standing relaxed	20	Sawing wood	420
Standing stiffly	20–30	Rapid typing	16–40
Walking	130–200	Coal mining (average)	320
Running	500–930	House painting	160
Singing	37	Tailoring	44
Reading aloud	20	Bookbinding	51
Dishwashing	59	Riveting	276
Ironing	59	Cycling	180–300
Sewing	25–30	Swimming	200–700
Sweeping	110	Climbing	200–960
Writing	10–20	Rowing	120–600
Knitting	31	Wrestling	980

* Average 70 kg. man or 58 kg. woman.
(Adapted from The Heinz Handbook of Nutrition. McGraw-Hill Book Co., New York, 1959.)

to some practical conclusions concerning the calories required for optimum health. The bases for these conclusions are presented in the various reports from each of the organizations. As useful as these standards are, they suffer from the disadvantages of any averaged value.

Food intake, in general, is affected by many factors other than those having to do with physiological function. For this reason it is important that the energy need of the individual be appreciated and a rational and proper dietary plan designed just for him.

REFERENCES

General

Kennedy, E. P.: Energetics and metabolic function. *In* Bourne, G. H., and Kidder, G. W. (ed.): Biochemistry and Physiology of Nutrition. Vol. 2. New York, Academic Press, Inc., 1953, pp. 197-229.

Kleiber, M.: The Fire of Life. New York, John Wiley & Sons, Inc., 1961.

Swift, R. W., and Fisher, K. H.: Energy metabolism. *In* Beaton, G. H., and McHenry, E. W. (ed.): Nutrition: A Comprehensive Treatise. New York, Academic Press, Inc., 1964, pp. 181-260.

Widdowson, E. M.: Assessment of the energy value of human foods. Proc. Nutr. Soc., *14*:142, 1955.

Chapter Four

Carbohydrates in Nutrition

SANFORD A. MILLER, PH.D.

Carbohydrates have occupied a significant place in the diet from early times when man became a cultivator rather than a hunter of food. Today carbohydrates still supply a major portion of the energy of the diet of man, ranging from about 90 per cent in the diets of people of underdeveloped areas to 50 per cent in diets of countries such as the United States.

Carbohydrates are principally a source of dietary energy, but these compounds have additional functions which should be recognized. For example, they can act as starting materials for the synthesis of other compounds in the body, such as fatty acids and amino acids. In addition, they play a role as part of the structure of other biologically important materials, such as glycolipids, glycoproteins, nucleic acids and heparin. It is important, therefore, not to lose sight of these additional functions and to consider the carbohydrates only as an energy source.

CHEMISTRY

The carbohydrates are structurally related ketonic and aldehydic compounds containing only carbon, hydrogen and oxygen, the hydrogen and oxygen being present in the same proportion as in water.

The carbohydrates may be classified in terms of the number of saccharide units of which they are composed. A *monosaccharide* is defined as a carbohydrate which cannot be broken down to simpler sugar by acid hydrolysis. *Disaccharides* are those which are formed by the condensation of two monosaccharides, while *polysaccharides* are the polymers of many monosaccharides. From the viewpoint of nutrition, the most important of these are given in Table 4-1.

Glucose is the most common of the carbohydrates. It is the only hexose known to exist in the free form in the fasting human. Most glucose occurs in a combined form and is thus found in nearly all foods. Sucrose is, of course, the most familiar sugar in domestic use, whereas lactose is unique to mammals. Starch is the energy stored in most plants and seeds. Upon digestion, starch is broken down to form maltose and, ultimately, glucose. Glycogen is the animal equivalent of starch. In terms of amount, however, it does not represent a significant source of carbohydrate; its importance lies in its metabolic function. Cellulose is generally not digestible by man or by many animals, except the herbivorous animals; its role in nutrition is confined to its action as roughage in maintaining the tone of the gastrointestinal tract.

In addition to the carbohydrates discussed previously, there are a number of hexose derivatives which play an increasingly important role in man's diet. These include sorbitol, mannitol and dulcitol. Each is the alcohol of its respective hexose, sorbitol being derived from glucose, and mannitol and dulcitol being derived from mannose and galactose, respectively. Their importance rests on their slow absorption from the intestine, which prevents any increase in the blood sugar level. Since they do possess a sweet flavor, they are generally used as a sugar substitute in proprietary preparations of foods for use by diabetics and others requiring a limited sugar intake. It must be remembered, however, that although these compounds do not raise blood

25

TABLE 4-1. NUTRITIONALLY IMPORTANT CARBOHYDRATES

Name	Classification	Monosaccharide Components	Principal Source
Glucose	Monosaccharide		Most foods in trace amounts More in grapes
Fructose	Monosaccharide		Fruits, honey
Galactose	Monosaccharide		Generally only in bound form
Sucrose	Disaccharide	Glucose and fructose	Sugar cane Sugar beets
Lactose	Disaccharide	Glucose and galactose	Milk
Mannose	Disaccharide	Glucose	Malting barley
Starch	Polysaccharide	Glucose	Plants
Cellulose	Polysaccharide	Glucose	Plants
Glycogen	Polysaccharide	Glucose	Animal liver, shellfish

sugar levels, they do contribute to the caloric load and should be considered when calculating energy contents of a diet.

DIGESTION AND ABSORPTION

Digestion of the utilizable carbohydrates begins in the mouth, where the salivary amylase, ptyalin, starts the hydrolysis of starch. This action continues to some extent in the stomach. When a carbohydrate reaches the intestines, pancreatic amylase continues the process until all the starch has been converted to maltose. This maltose and any other disaccharides present are then hydrolyzed to their component hexoses by the action of specific carbohydrases. Ultimately, only the simple hexoses remain.

All of the monosaccharides are practically completely absorbed in the small intestine. Prior to absorption, the solutions containing the carbohydrates are brought to approximate isotonicity. The major portion of the monosaccharides is then transported via the portal circulation to the liver. There is, however, a difference in the relative rates of absorption of the various carbohydrates. In decreasing order, galactose is absorbed most rapidly, followed by glucose, fructose, mannose and the various pentoses.

There is some evidence that absorption is a two-component process, involving both active transport and passive diffusion. The active transport component depends upon phosphorylation of the molecule in the cells of the intestinal mucosa and, thus, the expenditure of energy. The passive phase depends upon a concentration gradient between the lumen of the intestine and cells of the mucosa. Since glucose, galactose and fructose are apparently absorbed in large measure during the active transport phase, their rates of absorption are thus higher than those of the other simple sugars.

A number of factors are required for proper hexose absorption. These include some of the B vitamins, particularly thiamine. For this reason, glucose absorption is reduced in thiamine deficiency.

PHYSIOLOGICAL REGULATION OF THE CARBOHYDRATES

The hexoses, after being absorbed into the portal circulation and reaching the liver, enter into one of the most finely controlled physiological systems in the body. The function of this system is to enable the animal to control the level of blood sugar at any time and under any condition. At the same time, it provides for the maintenance of this system and also makes certain a sufficient supply of glucose for the tissues. This system is, in essence, the result of the balance between the withdrawal of glucose from the blood and the release of glucose to the blood by the liver. The mechanisms involved in regulating the withdrawal of glucose include uptake by the liver of hexoses other than glucose for conversion to glucose, storage of glucose in the form of glycogen, utilization of glucose for energy, and utilization of glucose for synthesis of other molecules, such as fatty acids. On the other hand, the mechanisms involved in glucose release to the circulation include conversion of glycogen to glucose, synthesis of glucose from amino acids, and the demand by other tissues for glucose.

Once in the general circulation, the glucose can be utilized by the extrahepatic tissue. Much of the circulating glucose enters the skeletal muscle where it provides the energy needed for the contraction of muscle. This reaction is both aerobic and anaerobic. If, however, insufficient oxygen is available, lactic acid accumulates. This lactic acid then leaves the muscle cell and is transported to the liver where it is converted to glycogen. This sequence of liver glycogen to blood glucose to muscle glycogen to blood lactate back to liver glycogen is known as the Cori cycle, after the investigators who first described it. An insufficient amount of oxygen may be available in a variety of circumstances, such as extreme muscular activity or occlusion of the arteries of the leg. Lactate will also accumulate when a thiamine deficiency is present.

The nervous tissue also is an important user of glucose. In contrast to other tissue, nervous tissue obtains almost all its energy from glucose. Furthermore, nervous tissue apparently utilizes glucose directly from the blood and, thus, these tissues have no reserve stores of glycogen. It may be for this reason, among others, that blood glucose is so closely regulated.

INTERMEDIARY METABOLISM

By far the greatest proportion of glucose is utilized by the tissues for energy. The process by which glucose is oxidized to carbon dioxide, water and energy can be conveniently divided into two parts: the first, an anaerobic process, and the second, aerobic in nature.

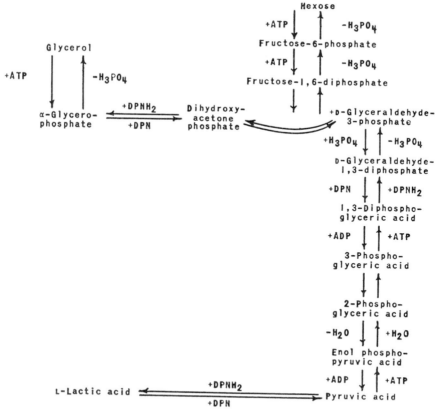

FIGURE 4-1. *Meyerhof-Embden-Parns scheme of glycolysis.* (*Sourkes, T. L.* In *Bourne, G. H., and Kidder, G. W.: Biochemistry and Physiology of Nutrition. Vol. 1. New York, Academic Press, Inc., 1953.*)

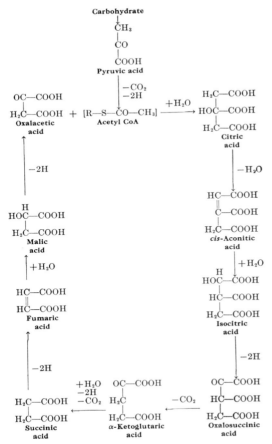

FIGURE 4-2. *Aerobic carbohydrate metabolism.* (*Krebs' or tricarboxylic acid or citric acid cycle.*) (*Hawk, P. B., Oser, B. L., and Summerson, W. H.: Practical Physiological Chemistry, 13th ed. New York, The Blakiston Co., Inc.*)

The sequence of reactions involved in these systems is shown in Figures 4-1 and 4-2.

It is the function of the anaerobic phase of carbohydrate metabolism to convert glucose to a form suitable for entrance into the aerobic phase during which most of the energy is released. This first phase, the Embden-Meyerhof scheme of glycolysis, involves the phosphorylation of glucose followed by conversion to fructose phosphates, which, in turn, are split into triose phosphates. The triose phosphates are then converted to pyruvate, which can, under proper conditions, be converted to form acetyl coenzyme A (CoA).

Although some energy is released during this phase of metabolism, most is released during the next, the aerobic, phase. This sequence of reactions is called the Krebs or tricarboxylic acid cycle. In these reactions, the acetyl CoA is combined with oxalacetate, which then passes through the various reactions of the cycle, releasing carbon dioxide, water and energy.

In addition to the major pathways of carbohydrate oxidation, there is a supplementary pathway called the hexose monophosphate shunt. One of its important functions is to supply reduced triphosphopyridine nucleotide (TPNH), a cofactor required in the biosynthesis of many molecules, including fatty acids.

As in all areas of the living system, the metabolism of glucose is closely interrelated with the metabolism of other nutrients. When, for example, carbohydrate is consumed in excess of the need of the tissue and in excess of the limited ability of the liver and muscle tissues to synthesize glycogen, the extra amount is converted to fat and is stored in adipose tissues. On the other hand, when insufficient carbohydrate is supplied in the diet, the organism calls first on its adipose tissue and then on its protein to make up the deficiency. It is the latter problem which must be prevented. If it continues for too long a time, the protein is drawn from the metabolic pool and structural components of the body. The ability of glucose to prevent the loss of protein is known as its "protein-sparing action."

The major nutrients are also related in that they all can enter, to a greater extent, the Krebs' cycle. Fats can enter via acetyl CoA, while certain of the amino acids enter through a number of points in this cycle. For this reason one cannot discuss carbohydrate need by itself. It must be related to the needs of other nutrients as well.

RELATION OF CARBOHYDRATES TO DISEASE

There is evidence that sugar and sugar-containing food contribute to several diseases, obesity, dental caries, diabetes mellitus and cardio-vascular disease. Yudkin (1957) has pointed out that there is a better correlation between consumption of sugar and myocardial infarction than with the consumption of total fat or of any particular sort of fat. Some of these diseases are discussed in detail elsewhere in this text (dental caries in Chapter Twenty-six and obesity and cardiovascular disease in Chapter Twenty-one).

Certain problems associated with the carbohydrates are due to metabolic abnormalities, as in the case of diabetes mellitus. This syndrome is characterized by the failure of the liver and muscle to utilize glucose properly. This may result from either the lack of insulin or the inactivation of available insulin. The result of this abnormal condition is an increase in lipid (see Chapter Five) and amino acid utilization to correct for the energy deficiency. In consequence, there is an accumulation of acetyl CoA which, in turn, recombines to form acetoacetic acid. This latter compound can then either be reduced to beta hydroxy butyric acid or oxidized to acetone. These three compounds are known as the ketone bodies. Accumulation of ketone bodies can lead to a wide variety of deleterious events which can ultimately result in death.

Diet plays an important role in the control of diabetes. Design of diets for such patients must take into account the interrelationships of the nutrients as well as the specific needs of the patient.

A number of other abnormalities are associated with the carbohydrates. These include glycogen storage disease (von Gierke's disease),

fructosuria, galactosemia and pentosuria. In each case, as in diabetes, diet plays an important role in controlling the disease and maintaining the patient in a reasonably normal physiological state.

CARBOHYDRATES IN THE DIET

The multipoint interrelationships of the nutrients give to the living organism a flexibility which allows it to adapt to a wide variety of environmental conditions. It is evident, however, that a living system requires carbohydrate, particularly glucose, in one form or another. The amount of carbohydrate that should be included in the diet is questionable. As indicated earlier, people survive in an adequate manner on diets containing as much as 90 per cent carbohydrate. On the other hand, the Eskimo diet contains only about 20 per cent. The proper amount probably lies somewhere in between.

The pattern and trend of carbohydrate consumption are influenced by the ease and cheapness of production of foodstuffs that are rich in carbohydrates and the general affluence of the society. Actually, one finds that with an increase in income there is a moderate increase in the consumption of calories and protein, considerable increase in the consumption of fat and little change in total carbohydrate intake. However, with a higher

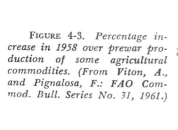

FIGURE 4-3. *Percentage increase in 1958 over prewar production of some agricultural commodities. (From Viton, A., and Pignalosa, F.: FAO Commod. Bull. Series No. 31, 1961.)*

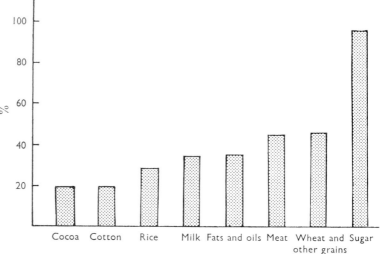

TABLE 4-2.   YEARLY WORLD CONSUMPTION OF SUGAR*

Year	Consumption	
	kg/head	Index
1899	5·5	100
1909	7·5	135
1924	10·2	185
1929	12·4	225
1938	12·0	218
1949	11·8	215
1957	15·5	282

* Viton, A., and Pignalosa, F.: FAO Commod. Bull. Series No. 31, 1961.

income there is a corresponding increase in consumption of refined sugar. In the wealthiest countries sugar contributes nearly 20 per cent of the total calories consumed. Compared to other food commodities, the world production of sugar is increasing most rapidly (Fig. 4-3). The average consumption of sugar has increased nearly threefold since the beginning of the century (Table 4-2). The changing sugar consumption in some countries is shown in Figure 4-4.

There is no evidence that one particular source of carbohydrate is better than another. Glucose derived from starch in potatoes or

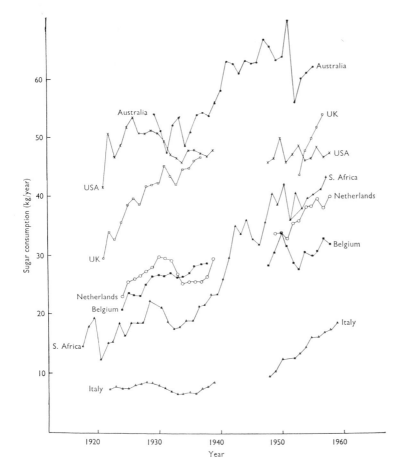

FIGURE 4-4. *Changing sugar consumption in some selected countries. (From Viton, A., and Pignalosa, F.: FAO Commod. Bull. Series No. 31, 1961.)*

from lactose in milk or from fructose in honey is utilized in exactly the same way as glucose produced by the chemist. Insofar as the diet is concerned, the amount of carbohydrates included in the diet is the important fact.

It has been stated that the ability of man to create civilization is the result of his ability to cultivate and store food. If this be true, then his ability to utilize a variety of sources of carbohydrate has contributed to this success.

THE ORAL RELEVANCE OF CARBOHYDRATES

Six aldose sugars, galactose, glucose, mannose, fucose, "rhamnose," and xylose, have been recovered from the hydrolysates of the organic matter of bovine and human tooth enamel (Burgess et al., 1960). Stack (1956) has found that 10 times as much carbohydrate is found in the outermost layer of enamel as in the inner. Furthermore, in pooled "chalky" enamel, which has 3 to 4 times as much organic matter as normal enamel, he has recovered 12 times as much carbohydrate as in normal enamel.

In sound dentin Armstrong (1960) found 0.4 per cent "glucose" units of unrefined carbohydrate material and 10 times this amount in carious dentin. Furthermore, the fractions of carious dentin resistant to collagenase action or autoclaving contain 30 times as much carbohydrate material as sound dentin. This same investigator further hypothesized that the reaction of the dentin matrix with carbohydrates or carbohydrate derivatives probably occurs during the carious process (Armstrong, 1964).

DENTAL CARIES

The historical development of the correlation between carbohydrates and dental caries is detailed in Chapter Twenty-six; but there are a few additional historical notes which are informative and interesting and therefore worthy of being included here.

The Arab physicians of the eighth and ninth centuries A.D. gave an important place to sugar in their pharmacopeia, which may have accounted for the increased cultivation of sugar cane. Since sugar had to be imported from India and Arabia, it was a rare and expensive commodity and considered a luxury. When new trade routes were opened in the fifteenth century by the Portuguese, the price of sugar was lowered and its use for preservation of food became more widespread. The effect of this increased usage of sugar was reflected in Englishmen's bad teeth which "were fairly common at that time among those who could afford to indulge a taste for sweetmeats." In particular, Paul Hentzner observed Queen Elizabeth's black teeth—"a defect the English seem subject to, from their too great use of sugar." (Anonymous [Pertinax] 1964).

Microbiological Experiments. From a scientific standpoint, perhaps, the classic test tube experiments of W. D. Miller in the last decade of the nineteenth century (1890) can be pointed to as the first of the ongoing investigations these last 75 years on the mechanism of dental caries production. He established that a carbohydrate substrate was necessary for oral bacterial action, and that anaerobic glycolysis by microorganisms provided the acids which decalcify the mineral portions of teeth.

Fosdick and Burrill (1943) pointed out that the only available substrates from which acids can be formed in the mouth are carbohydrates, particularly sucrose and glucose. Furthermore, rapid acid formation giving a pH as low as 4.5 has been shown to occur when sugars are placed on bacterial plaques which adhere to human teeth (Stephan and Miller, 1943).

The evidence that pure sucrose is a strong decalcifying agent was brought out by Jenkins et al. (1959) when they incubated calcium phosphate as well as whole teeth with saliva and sucrose. They found that far more calcium phosphate was dissolved when pure sucrose was present than with treacle or cane sugar, partly because the last two substances are heavily buffered.

Animal Experiments. The mandatory presence of carbohydrates in the diet (Shaw, 1954) and its local contact with the tooth (Kite et al., 1950) are now proved. In Shaw's experiment it was demonstrated that not only did rats have to be fed some carbohydrate in the diet (more than 5 per cent) to develop decay, but even sialodectomized rodents that usually develop extreme caries when fed a cariogenic diet were caries-resistant when carbohydrates were omitted from their diets. In the experiment of Kite and his associates, caries did not occur in even 1 of 13 caries-susceptible animals fed a high-sugar diet by stomach tube; whereas all their littermates, which were fed the same type of cariogenic diet by the usual oral route, developed some caries.

The chemistry of the carbohydrate, mono-, di- or polysaccharide, is also an essential consideration with respect to the degree of its cariogenic action. For example, sucrose or glucose supplements are very active cariogenic agents but starches and dextrins are not (Schweigert et al., 1945; Shafer, 1949). The reason that starch may not be very cariogenic is that the large starch molecule does not penetrate and diffuse through the dental plaque so readily as the smaller di- and monosaccharides. Starch is more likely to remain on the outer surface of the plaque where it can be washed away and eliminated. On the other hand, the sucrose that has diffused through the plaque will degrade to acid beneath the plaque and react with the tooth surface. In rats fed a sugar supplement, caries was a consistent finding (Koenig and Grenby, 1965), but when wheat starch, white or whole kernel flour, and white flour replaced by brown or fine offals were fed as dietary supplements instead of sugar, caries was practically negligible.

Another important observation that has been made in experimental animals with regard to dietary carbohydrate and dental caries is that the cariogenic effect of sucrose is accentuated when its feeding is initiated during early tooth formation, development and calcification (Sognnaes, 1948; Volker, 1951; Steinman and Haley, 1957). The authors postulate that sucrose is replacing a protective mineralizing factor present in trace amounts, perhaps, in natural foods which are more beneficial to the tooth during its maturation. A significant imbalance in the phosphate:carbonate ratio created by decreased phosphate and increased carbonate was found in the enamel of molars of weanling rats that received sugar solutions intra-peritoneally during the latter part of the suckling period (Luoma, 1961). This lowered phosphate:carbonate ratio has been associated with increased caries susceptibility (Sobel et al., 1960).

The level of sucrose and glucose in the diet has also been found to be a significant consideration in the amount of caries produced. For example, Mitchell et al. (1951) found that animals fed a ration with 45 per cent sucrose had more caries than those fed rations with only 15 or 30 per cent. Likewise, Keyes and Likens (1946) found that animals fed sucrose at 40 and 60 per cent levels showed more caries than those fed sucrose at a 20 per cent level.

The physical form of the carbohydrate is another variable which influences the extent of caries production. Sugar in solid form has been shown to be more cariogenic than liquid sucrose. Of 13 rats that were fed granulated sugar, only 2 remained caries-free. However, 7 of 13 littermate rats were without a single carious lesion when a sucrose solution was ingested (Haldi et al., 1953).

Clinical Observations. When a primitive, native-type food intake pattern is replaced by a civilized sophisticated one, dental caries will increase. This has been shown with the Bantus (Oranje et al., 1935) and Eskimos (Waugh, 1930) and more recently with the natives of Tristan da Cunha (Holloway et al., 1963). The diet of the Tristan da Cunha islander in 1938 consisted mainly of two staples, potatoes and fish. The average potato consumption per person was about 4 times higher, and the average fish consumption was about 3 times greater than that of the English. The natives also ate young sea birds, red crowberries, apples and pumpkins, and sparingly of beef, mutton, milk and eggs. In the last decade, this dietary pattern has changed radically because imported foodstuffs like flour, sugar, condensed milk, jam, dried and canned fruit, canned fish and sweets were made available. They now consume an average of 1 pound of sugar per week per person as compared to 1938 when they had no sugar. The change in dietary pattern has been accompanied by a simultaneous change in dental health. In 1938 Sognnaes found not a single carious first permanent molar even on radiographic examination in any of the young people under 20 years of age. On the other hand, in 1962 when Holloway et al. made dental examinations in a comparable age group, they found that 50 per cent of the teeth were carious.

In still another large scale general population study the influence of caries proneness of the kinds of foodstuffs eaten prior to tooth eruption has been noted. The incidence of postwar dental caries and the consumption of sugar during the war showed a high degree of correlation (Sognnaes, 1948; Toverud, 1949). Molars of Scandinavian children were made caries resistant during the war period when less refined foods were available. Three to 5 years later when the teeth were exposed to a more cariogenic diet, the innate resistance was sufficient to ward off the local attacking forces of increased sweets. Likewise, the incidence of decay in first permanent molars was found lowest in German children who 5 years previously had been on a low sugar ration.

In addition, there have been clinical trials in orphanages, homes, etc., that have added more evidence of the inverse correlation between sugar and caries. Some of these studies have been explained on a developmental or systemic basis; others have given more credit to the local environmental mechanism. The answer probably lies somewhere between both these philosophies or explanations. For example, in the majority of children ill with diabetes mellitus who ate more liberal amounts of milk, eggs, meat, vegetables and fruit than most children, there was little or no extension of caries for years (Boyd, 1943). The small number who failed to observe their dietary regime experienced a much higher rate of caries. The children who were put on this low sugar diet early had less caries than those who started the diet later. Similarly, Howe et al. (1942) showed that children who followed dietary instructions and ate a diet of high nutritional quality with simultaneous lowering of carbohydrate content had significantly less tooth decay than comparable children who were not so counseled. On the other hand, Koehne and Bunting (1934) added large amounts of carbohydrates in the form of sugar or candy to the diets of children and found an increase in caries. This emphasized a local extrinsic factor as the predominating mechanism.

Not only has clinical caries research established and confirmed the animal findings with respect to the hypothesis that carbohydrates are essential for caries production, but there is some very good evidence in humans that the type of carbohydrate, starch versus sugar, as well as the physical form and frequency of usage are each important variables in the magnitude of caries experience to be expected. Individuals who suffer from hereditary fructose intolerance, which is an inborn error of metabolism, have a deficiency in the enzyme fructose-1-phosphate aldolase. They can tolerate starch but not sucrose and fructose; therefore, they avoid sugars and eat plentifully of foods like bread, noodles and potatoes instead. These patients have nearly caries-free mouths in spite of the large amounts of starchy foods that they eat (Froesch et al., 1963).

The well-controlled Vipeholm study is probably the classic illustration of the importance of the effect of form and frequency of carbohydrate ingestion on caries development in humans (Gustafsson et al., 1954). A long-term, 5-year, nutritional study was carried out on 436 mental patients with a mean age of 32 who were confined, practically permanently, in an institution in Vipeholm, Sweden. Diets were carefully supervised as to preparation, and nurses were able to insure cooperation of the patients in following the experimental prescriptions.

The first year constituted an adjustment period in which a base line caries index was established and the patients consumed a diet rich in vitamins and other protective foods four times a day with no candy or chocolate. The next four years consisted of the carbohydrate study period in which ten groups were fed the same basal diet, but they differed from each other in that some groups had increased amounts of sugar *at mealtime* and others had increased amounts of sugar *between meals*. There were four main groups:
1. Basal diet
2. Basal diet and additional sugar in solution at meals
3. Basal diet and additional sugar in bread consumed at meals
4. Basal diet and additional sugar in the form of sweets consumed between meals

Those who were on the basal diet (the con-

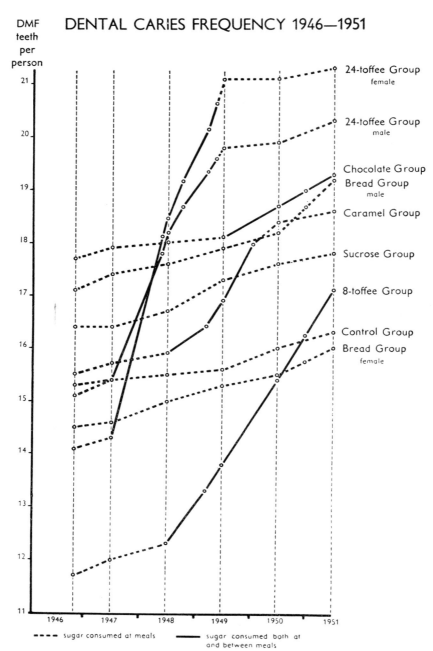

FIGURE 4-5. *Relation of carbohydrate intake to dental caries. (From Gustafsson, B., Quensel, C., Lanke, L., Lundqvist, C., Grahnen, H., Bonow, B., Krasse, B.: The Vipeholm Dental Caries Study: The effect of different levels of carbohydrates intake on caries activity in 436 individuals observed for five years, Acta Odont. Scand.* **11:232, 1954.)**

trol group throughout the study) had a low caries activity. (See Fig. 4-5.) In those groups who had as much as 300 grams of sucrose added to the meal in liquid form as a beverage or in food preparation, the caries activity was only slightly increased. This is about twice the mean sugar consumption of most Western countries. It is interesting to note that the same slight caries activity increase was noted in the group who were given bread *at meals*, which contained only 50 grams of refined sugar.

In all groups other than those on the basal diet, there was a very significant increase in dental caries. Even if there was a small amount of additional sugar added, such as those who ate candy between meals, there was a marked increase in caries activity, indicating that quantity of sugar was not the all important factor to account for this result. Furthermore, when the sweets were withdrawn from between-meal periods, the caries activity decreased to the level of the initial preparatory period.

The important conclusion from this experiment is that if sugar with only a slight tendency to be retained, such as sucrose solution, was ingested *at meals* or if sugar-rich bread which has a strong tendency to be retained was consumed *at meals,* the risk of increasing caries activity was least. However, when sugar with a strong tendency to be retained in the mouth was eaten *between meals* frequently, the risk of increasing caries activity was greatest.

Lundqvist measured the time sugar could be detected in the saliva of the participants of the Vipeholm study. In Figure 4-6 it can be seen that in those groups who ingested sugar at meals, regardless of whether it was the control group, or the sucrose group, or the bread group, only four peaks of sugar in the saliva were noted, corresponding to the four meals. Of special interest is that the sucrose groups, who ate twice as much sugar (but at meals) as the control group, had an identical salivary glucose level.

In short, dental caries activity increased in connection with consumption of sugar in sticky form between meals, but it decreased when the consumption was interrupted. Furthermore, when sugar was consumed in solution at meals, in amounts twice the Swedish average consumption, no increase in dental caries was observed.

The results of Gustafsson's and Lundqvist's studies prove several points: (1) sugar exerts its caries-promoting effect locally in the mouth, (2) starchy foods like bread are not so cariogenic as the disaccharide sucrose, (3) the amount of sugar is not of paramount importance, (4) the form and composition of the sweets is critical (retentive worse than nonretentive), and (5) the frequency of usage is a prime factor in caries activity. These conclusions, particularly about the importance of *at meal* or *between meal* timing of sugar intake, are confirmed by the results of Mack (1949), King et al. (1955), Jay (1947) and Potgieter et al. (1956). The first two investigators reported that when extra sugar was eaten at meals no increase in caries was noted. In the last two studies sugar was given between meals and caries increased.

In summary, all these in vitro animal and human data certainly point to the adverse effect of carbohydrates, particularly sucrose, on dental caries. But it is recognized that food habits and ingrained cultural practices are difficult to change, particularly when dealing with a disease like dental caries, which is not a matter of life or death. Although some of our attention is being directed to understanding the psychological and social aspects that influence a patient's dietary pattern, our actual management will be concerned in the future with substituting for carbohydrate, foods with less cariogenic potential than sucrose. This is discussed in detail in Chapter Twenty-six.

PERIODONTIUM

There is some preliminary evidence that carbohydrate foods, particularly those that are readily retained and easily fermented, play an important role in the etiology of periodontal disease. The incidence of the periodontal syndrome was markedly increased in rice rats (which are constitutionally prone to this disease) when fed highly cariogenic diets containing 67 per cent sucrose. Those animals fed a carbohydrate-free ration experienced a major reduction in periodontal soft tissue lesions and moderate reduction in bone

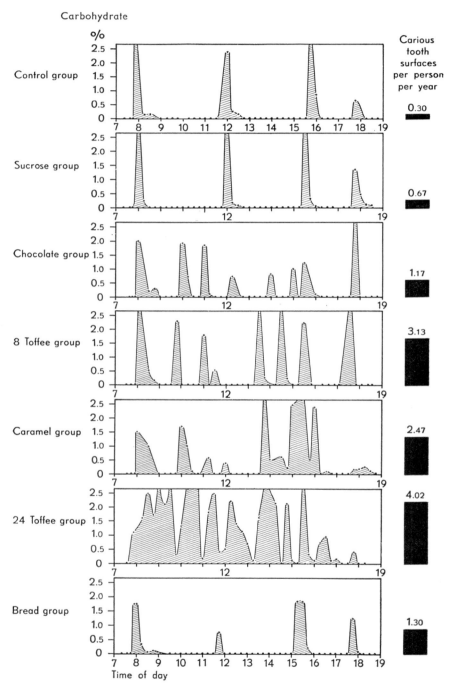

FIGURE 4-6. *Sugar clearance in day-series in different study groups. One individual series in agreement with each group's average value is chosen. The caries activity is expressed as an average value for each group within which clearance determination has been performed. (Lundqvist, C.: Odontologisk Revy, Vol. 3, Suppl. 1, 1952.)*

loss. Furthermore, animals that were fed a diet with reduced dietary carbohydrate and increased lard content showed a somewhat lower reduction of periodontal lesions when compared to those fed high sucrose diets. It is possible that the periodontal syndrome produced by the dietary regimes may be really a reflection of bacterial activity and the availability of optimal nutrition for the growth of these microorganisms (Auskaps et al., 1957).

Another bit of indirect evidence that high carbohydrate diets adversely affect gingival health and even wound healing can be found in the animal studies on protein-free (Stahl, 1962) and low-protein (Stahl, 1963) diets. Actually, these animals were fed diets with 70 to 75 per cent starch content and, in both instances, showed delay in connective tissue and bone repair after being wounded. Furthermore, there is the finding of Frandsen et al. (1953), that rats fed on diets containing no protein and 86 per cent sucrose showed severe osteoporosis and increased rate of bone resorption. Stahl's findings and those of Frandsen and his associates might be interpreted as the result of a dual dietary aberration, namely, a protein deficiency compounded by an excess carbohydrate diet. The more important of these two dietary variables in the etiology of the disease was not demonstrated.

Human Data. A deterioration in the periodontal health of Tristan da Cunha islanders from 1938 to 1962 was noted by Holloway et al. (1963). The diet of these islanders had changed from no ingestion of sugar in 1938 to 1 pound per person per week in 1962. Sognnaes reported in 1938 only 10 per cent of the adult population showed advance periodontal disease with bone loss and gingival recession. In 1962 the percentage of the population afflicted with this disease had risen to 32 per cent.

The possibility that carbohydrate metabolic patterns might be related to the integrity of the periodontium was investigated recently (Shannon and Gibson, 1964). Since the subjects of this research were healthy males with relatively mild periodontal disease, the object of the study was to use the oral glucose tolerance test as an index of susceptibility rather than a chemical finding correlated with the presence of advanced disease. They found no correlation between the oral glucose tolerance test and periodontal health.

In conclusion, our knowledge on the relation of carbohydrates to periodontal disease is still sparse, but the initial experimental results do seem to hold some promise for developing a tenable hypothesis that this nutrient has periodontal disease-producing potential.

REFERENCES

General

Pigman, W., and Goepp, R. M.: Chemistry of the Carbohydrates. New York, Academic Press, Inc., 1948.

Sinclair, H. M.: Carbohydrates and fats. *In* Beaton. G. A., and McHenry, E. W. (ed.): Nutrition: A Comprehensive Treatise. New York, Academic Press, Inc., 1964, pp. 41-114.

Soskin, S., and Levine, R.: Carbohydrate Metabolism. Chicago, University of Chicago Press, 1958.

Stanbury, J. B., Wyngaarden, J. B., and Fredrickson, D. S.: The Metabolic Basis of Inherited Disease. New York, McGraw-Hill Book Co., Inc., 1960.

Oral Relevance

Anonymous (Pertinax): Without prejudice, Brit. Med. J., *5422*:1457, 1964.

Armstrong, W. G.: Carbohydrate material in carious dentine. 1. Preliminary investigations. Arch. Oral Biol., 2:69, 1960.

Armstrong, W. G.: Modifications of the properties and composition of the dentin matrix caused by dental caries. *In* Staple, P. (ed.): Advances in Oral Biology. New York, Academic Press, 1964, pp. 309-332.

Auskaps, A. M., Gupta, O. P., and Shaw, J. H.: Periodontal disease in the rice rat. 3. Survey of dietary influences. J. Nutr., *63*:325, 1957.

Boyd, J. D.: Long term prevention of dental caries among diabetic children. Am. J. Dis. Child., *66*: 349, 1943.

Burgess, R. C., Nikeforuk, S., and MacLaren, C.: Chromatographic studies of carbohydrate components in enamel. Arch. Oral Biol., *3*:8, 1960.

Fosdick, L. S., and Burrill, D. Y.: The effect of pure sugar solutions on the hydrogen in concentration of carious lesions. Font. Rev., Chicago Dent. Soc., *6*:7, 1943.

Frandsen, A. M., Becks, H., Nelson, M. M., and Evans, H. M.: The effect of various levels of dietary protein on the periodontal tissues of young rats. J. Periodont., *24*:135, 1953.

Froesch, E. R., Wolf, H. P., Baitsch, H., Prader, A., and Labhart, A.: Hereditary fructose intolerance. An inborn defect of hepatic fructose-1-phosphate splitting aldolase. Am. J. Med., *34*:151, 1963.

Gustafsson, B. E., Quensel, C. E., Lanke, L., Lundqvist, C., Grahnen, H., Bonow, B. E., and Krasse, B.: The Vipeholm dental caries study: Effect of different levels of carbohydrate intake on caries activity in 436 individuals observed for five years. Acta Odont. Scand., 11:232, 1954.

Haldi, J., Wynn, W., Shaw, J. H., and Sognnaes, R. F.: The relative cariogenicity of sucrose when ingested in the solid form and in solution by the albino rat. J. Nutr., 49:295, 1953.

Holloway, P. J., James, P. M. C., and Slack, G. L.: Dental disease in Tristan da Cunha. Brit. Dent. J., 115:19, 1963.

Howe, P. R., White, R. L., and Elliott, M. D.: The influence of nutritional supervision on dental caries. J. Amer. Dent. Assn., 29:38, 1942.

Jay, P.: The reduction of oral lactobacillus acidophilus counts by the periodic restriction of carbohydrates. Amer. J. Orth. & Oral Surg., 33:162, 1947.

Jenkins, G. N., Forster, M. G., Speirs, R. L., and Kleinberg, I.: The influence of the refinement of carbohydrates on their cariogenicity. Brit. Dent. J., 106:195, 362, 1959.

Keyes, P. H., and Likens, R. C.: Plaque formation, periodontal disease and dental caries in Syrian hamsters. J. Dent. Res., 25:116, 1946.

King, J. D., Mellanby, M., Stones, H. H., and Green, H. N.: The effect of sugar supplements on dental caries in children. Med. Res. Couse. Spec. Rep. Ser., No. 288, London, 1955.

Kite, O. W., Shaw, J. H., and Sognnaes, R. F.: Prevention of experimental tooth decay by tube feeding. J. Nutr., 42:89, 1950.

Koehne, M., and Bunting, R. W.: Studies in the control of dental caries. II. J. Nutr. 7:657, 1934.

Koenig, K. G., and Grenby, T. H.: The effect of wheat grain fractions and sucrose mixtures on rat caries developing in two strains of rats maintained on different regimes and evaluated by two different methods. Arch. Oral Biol., 10:143, 1965.

Luoma, H.: The effect of injected monosaccharides upon the mineralization of rat molars. Arch. Oral Biol., 3:271, 1961.

Mack, P. B.: A study of institutional children with particular reference to the caloric value as well as other factors of the dietary. Washington, D.C., Soc. Res. in Child Development, 13:62, 1949.

Mitchell, D. F., Chernausek, D. S., and Helman, E. Z.: Hamster caries. The effects of three different dietary sugar levels and an evaluation of scoring procedures. J. Dent. Res., 30:778, 1951.

Muller, W. D.: The Microorganisms of the Human Mouth. Philadelphia, S. S. White Dental Mfg. Co., 1890.

Oranje, P., Noriskin, J. N., and Osborn, T. W. B.: The effect of diet upon dental caries in South African Bantus. S. African J. Med. Sci., 1:57, 1935.

Potgieter, M., Morse, E. H., Erlenbach, F. M., and Dall, R.: The food habits and dental status of some Connecticut children. J. Dent. Res., 35:638, 1956.

Schweigert, B. S., Shaw, J. A., Phillips, P. H., and Elvehjem, C. A.: Dental caries in the cotton rat. 3. Effect of different dietary carbohydrates on the incidence and extent of dental caries. J. Nutr., 29:405, 1945.

Shafer, W. G.: The caries-producing capacity of starch, glucose, and sucrose diets in the Syrian hamster. Science, 110:143, 1949.

Shannon, I. L., and Gibson, W. A.: Oral glucose tolerance responses in healthy young adult males classified as to caries experience and periodontal status. Periodontus, 2:292, 1964.

Shaw, J. H.: Effect of carbohydrate-free and carbohydrate-low diets on incidence of dental caries in white rats. J. Nutr., 53:151, 1954.

Sobel, A. E., Shaw, J. H., Harok, A., and Nobel, S.: Calcification. XXVI. Caries susceptibility in relation to composition of teeth and diet. J. Dent. Res., 39:462, 1960.

Sognnaes, R. F.: Caries-conducive effect of a purified diet when fed to rodents during tooth development. J. Amer. Dent. Assn., 37:676, 1948.

Stack, M. V.: The carbohydrate content of human enamel. J. Dent. Res., 35:966, 1956.

Stahl, S. S.: The effect of a protein-free diet on the healing of gingival wounds in rats. Arch. Oral Biol., 7:551, 1962.

Stahl, S. S.: Healing of gingival wounds in female rats fed a low protein diet. J. Dent. Res., 42:1511, 1963.

Steinman, R. R., and Haley, M. J.: The biological effect of various carbohydrates ingested during the calcification of the teeth. J. Dent. Child., 24:211, 1957.

Stephan, R. M., and Miller, B. F.: A quantitative method for evaluating physical and chemical agents which modify production of acids in bacterial plaques on human teeth. J. Dent. Res., 22:45, 1943.

Toverud, G.: Dental caries in Norwegian children during and after the last World War: A preliminary report. Proc. Roy. Soc. Med., 42:249, 1949.

Volker, J. F.: Some observations concerning dental caries production in the Syrian hamster. (Abst.) J. Dent. Res., 30:484, 1951.

Waugh, L. M.: Health of the Labrador Eskimo with special reference to the mouth and teeth. J. Dent. Res., 10:387, 1930.

Chapter Five

Fats in Nutrition

Sanford A. Miller, Ph.D.

INTRODUCTION

In contrast to carbohydrates, the amount of fat in the diet appears to vary directly with the economic well-being of the individual. Ranging from a low of about 8 per cent in diets in Japan to as high as 41 per cent in those in the United States, the quantity of dietary fat appears to be a hallmark of a high standard of living. Unfortunately, however, the increase in dietary fat appears also to be associated with an increase in body fat. Obesity is an endemic disease in countries such as the United States and may well represent the major health hazard of our time.

The importance of fat in the diet is the result of a number of interrelated factors. Fat is not only an excellent dietary source of energy, supplying 9 calories per gram, but it also has a number of other desirable characteristics. For example, it adds to the palatability of the diet not only as a result of its flavor but also because of its lubricating and textural characteristics. The absorption of fat occurs at a slower rate than does the absorption of other nutrients; thus, fat contributes for a longer period of time to the feeling of "fullness" and satiety generally experienced at the end of a meal. Under normal conditions, the ingestion of fat soluble vitamins is also associated with an adequate fat intake. In addition, Deuel (1951-1957) has also indicated a number of other advantages, including the possibility of the fats being able to increase the capacity of the individual for work, the length of survival during fasting and the resistance of the individual to certain kinds of stress. Finally, the capacity of dietary fat for sparing protein and thiamine is well known. For these reasons, re-ducing the fat content of a diet simply to reduce its caloric content is not rational. The development of an adequate diet for weight reduction requires a full understanding of all the attributes of the fats.

CHEMISTRY

For the layman, the word "fat" has a particular connotation and relates to foods such as butter and lard. For the biochemist, the word has another meaning. To distinguish between these conceptions it is customary to refer to this class of compounds as the lipids, including in this all those compounds covered by the housewife's designation of "fat."

The lipids constitute a heterogeneous group of compounds that have the following characteristics:

1. They are insoluble in water but soluble in certain organic solvents such as ether, chloroform and benzine.

2. They contain or are combined with one or more fatty acids in their natural state.

3. For the nutritionist, they include only those compounds utilizable by or occurring in biological systems.

The classification of the lipids is also relatively complex. A number of compounds of varying molecular structure are included in this group. Table 5-1 lists some of the categories of lipids important in nutrition.

The most commonly occurring form of fat in nature is the triglyceride. This type of compound is the ester of glycerol and three fatty acids, usually long chain and of even number. Since there are over 40 different types of fatty acids found in nature, the possible combina-

TABLE 5-1.   CLASSIFICATION OF LIPIDS

Primary Group	Secondary Groups	Constituent Groups
Simple	Fats	Esters of fatty acids and glycerol
	Waxes	Alcohols and alcohol esters other than those of glycerol
Compound	Phospholipids	Esters of fatty acids and glycerol which are combined with phosphoric acid and a nitrogenous base
	Lipoproteins	Combination of proteins with lipids
Derived	Fatty acids Alcohols Sterol alcohols Bases	Compounds derived from above groups by hydrolysis
Compounds associated with lipids in nature	Carotenoids Tocopherols K vitamins Steroids	

tions provide an important flexibility in the structure of triglycerides. Thus, the fats generally consist of mixtures of these triglycerides. By convention, if the fat is solid at room temperature, it is called a fat, while if it is liquid, it is called an oil.

The melting point of fat is dependent upon its constituent fatty acids. Thus, hard fats contain mostly saturated fatty acids, and oils contain principally unsaturated fatty acids. Examples of the hard fats include butter, lard and cheese. Cottonseed, corn, peanut and olive oil are examples of oils. By a process known as hydrogenation, which basically involves the addition of hydrogen to double bonds, the soft fats or oils are converted into the harder, more saturated type.

The phospholipids constitute the next large group of lipids in the body, there being three types that are of biological importance. These are lecithins, cephalins and sphingomyelins. The first two consist of glycerol, two fatty acids, phosphoric acid and one nitrogenous base. The sphingomyelins contain one fatty acid, phosphoric acid and two nitrogenous bases. The phospholipids appear to be an active component of all living cells. They are part of the cell wall and mitochondria and also function in the transport of fatty acids.

As indicated earlier, there are a wide variety of naturally occurring fatty acids. In general, they can be divided into either saturated or unsaturated forms, depending upon whether they contain one or more double bonds. Examples of unsaturated fatty acids of nutritional importance include acetic acid from vinegar, butyric acid from butter, and stearic acid found in animal and vegetable fats. Fatty acids with one double bond (monoethenoid) include oleic acid; with two double bonds (diethenoid), linoleic acid; with three double bonds (triethenoid), linolenic acid; with four or more double bonds (polyethenoid), arachidonic acid.

Three of these fatty acids (linoleic, linolenic, and arachidonic) are considered to be necessary for growth and maintenance since they cannot be synthesized by the animal and must be supplied in the diet; because of this they are called essential fatty acids (EFA). They have recently regained importance as a result of the suggestion that relative deficiencies in these compounds may be involved in certain degenerative diseases.

Among the waxes, the one that has recently received most attention is cholesterol. This compound is widely distributed in animals and occurs either in a free form or esterified with fatty acids. Cholesterol serves as a precursor of bile salts, vitamin D in the skin and hormones of the adrenal and genital organs. It is also important in facilitating the absorption of fatty acids. Although much cholesterol is consumed in the diet, man is not dependent upon his diet for his supply since the body can synthesize all it needs.

DIGESTION, ABSORPTION AND TRANSPORT

Most digestion of fats occurs in the intestines since the emulsification necessary for the function of the lipid enzymes does not occur in the stomach. In the intestines, bile salts, small quantities of fatty acids and monoglycerides all help to produce a stable emulsion suitable for attack by the pancreatic lipase. The action of this lipase is to release the outer fatty acids of the triglycerides, which are then absorbed. Other lipases reduce the remaining monoglycerides to glycerol and fatty acid. The fatty acids and other lipids are then absorbed into the cells of the intestinal mucosa. Here, the fatty acids of 12 or more carbons are used in the resynthesis of triglyceride, although without the original glycerol. The shorter fatty acids are transported directly to the portal vein. Cholesterol, after being de-esterified in the lumen, is re-esterified in the mucosa. The phospholipids are partially hydrolyzed in the lumen and resynthesized in the mucosa, generally as lecithins; some, however, are absorbed unchanged.

Ultimately, the lipid appears in the lacteals as a milky emulsion called chyle, which passes through the lymphatics to the thoracic duct and enters the venous blood. About 60 to 70 per cent of the fat passes this way, the remainder entering the portal circulation directly.

Apparently, the most important factor influencing digestibility of fats is the melting point. Most of the edible fats such as butter, lard and vegetable oils melt below 50°C. and are relatively easily and completely digestible. Those which are still hard at this temperature, e.g., hydrogenated peanut oil, are apparently less well utilized.

The neutral lipids, the glycerides, reach the circulation in the form of emulsified and finely divided particles called chylomicrons. The level of chylomicrons in the blood reaches a peak about 4 to 6 hours after ingestion. It is interesting to note that the coagulability of the blood also increases during this period of hyperlipemia. Some of the fat in the blood enters the adipose tissue directly, and the remainder enters the liver where it is metabolized.

Although it is apparent that dietary levels of lipid can influence the blood lipid level, the nature of the dietary lipid can also modify this parameter. When a large amount of saturated fatty acid is ingested, blood cholesterol levels tend to increase. On the other hand, if the proportion of unsaturated fatty acids is high, then blood cholesterol decreases. The importance of these changes in cardiovascular disease is under debate.

METABOLISM

The glycerol obtained by hydrolysis of the lipids is phosphorylated in the liver and ultimately converted to glucose. The fatty acids follow a separate pathway of metabolism. A number of theories have been advanced to explain the oxidation of fatty acids. The most accepted view is the beta oxidation theory of Knoop. According to Knoop, the fatty acids are first esterified with coenzyme A (CoA), which is then oxidized between the beta and gamma carbons to yield a double bond. The oxidized ester is hydroxylated at the beta carbon, the hydroxyl group then oxidized to a keto group and the compound split between the beta and gamma carbons to yield a two-carbon fragment and a shorter fatty acid ester; this process is repeated until all the fatty acid has been converted to acetate. The two-carbon fragments, as acetyl CoA, can then enter the Krebs cycle by condensation with oxalacetate and ultimately be oxidized to carbon dioxide, water and energy. It is apparent, therefore, that under conditions in which oxalacetate is reduced, the acetyl CoA formed from fatty acids will accumulate. This occurs when glucose becomes limiting, as in diabetes. Under these conditions, the acetyl CoA then forms the ketone bodies (see preceding chapter), the accumulation of which results in the characteristic ketosis of diabetes. This condition can also result when the supply of fatty acids is in excess of the capacity of the body to metabolize them, such as when the fat-carbohydrate ratio is too high or under conditions of starvation.

The synthesis of fats can occur in many tissues, including the liver, mammary glands and adipose tissue. In general, this occurs in two possible ways. In the first, called the mitochondrial system, acetate is added to fatty acids of eight carbons or more to build up fatty acids of varying chain length. However, most synthesis occurs via the second, or malonyl CoA, system. In this system, acetyl CoA is first carboxylated to form malonyl CoA. A varying number of molecules of malonyl CoA are then combined. The long chain resulting from these combinations of malonyl CoA is combined with acetyl CoA or propionyl CoA to form the final fatty acid.

Fats, as well as carbohydrates and protein, ingested in excess of need are stored in the adipose tissue, particularly under the skin, in the peritoneal cavity and around the ovaries and kidneys. The carbohydrates and proteins (amino acids) are first converted to fat before being stored. On the other hand, these stores represent a labile energy reserve and are reduced relatively rapidly when energy needs are in excess of caloric intake. It is apparent, therefore, that weight reduction can take place only by reduction of caloric intake in comparison to need.

NUTRITIONAL IMPORTANCE OF FATS

As in all areas of life, the level of fat in the diet must be in balance, not only in terms of its relationship to other nutrients, but also in the relationship of various types of fats to each other. If fat is ingested in excess of need, it will be deposited in the adipose tissue and obesity will result. If the relationship of fat to carbohydrate is impaired, then ketosis is a consequence. If the proper balance among the fatty acids is not maintained, cardiovascular disease may develop. It is important to determine what the optimum quantity and quality of dietary fats should be. Unfortunately, this question has not been settled. In years past, the recommendations of both the Food and Nutrition Board of the National Research Council and the Canadian Dietary Standards had been that fats supply 20 to 25 per cent of the total dietary calories. More recently, however, it has been felt that the questions of the level of fat and the balance among the various kinds of fat in the diet have not been fully resolved. For this reason, these standards do not presently recommend any definite amount or type of fat for the diet. It is probably safe to recommend a diet containing 20 to 25 per cent of the daily calories as fat and to suggest that this fat be supplied from both animal and vegetable sources. The importance of the lipids in nutrition cannot be ignored. Not only do they add to the enjoyment of food, but their essential physiological roles demand that they be included in any well-balanced diet.

DEFICIENCY STATES

It is possible to describe two types of fatty acid deficiency. In the first, the ingestion of a fat-free diet produces an essential fatty acid (EFA) deficiency. This is a pure deficiency state. Under these conditions in the rat, growth ceases, skin becomes scaly, reproduction is impaired and kidney damage is observed. These changes can be alleviated by feeding small amounts of either linoleic or arachidonic acid.

The second type of deficiency, called a relative deficiency, is more difficult to describe. This syndrome is said to result when a diet has a low ratio of EFA to non-EFA. As a consequence to the ingestion of such a diet, there is a fall of linoleate and arachidonate in body lipids and a rise in serum cholesterol. As indicated earlier, the relationship of these changes to cardiovascular disease is an important question in nutrition today.

THE ORAL RELEVANCE OF LIPIDS

It is only recently, since the development of sophisticated laboratory tools such as chromatography techniques and use of microsilicic acid columns, that accurate lipid analyses of teeth could be made. However, it was speculated 35 years ago that lipid is normally present in teeth and that a fatty metamorphosis of dentin may be a histological explanation for decay (Bodecker, 1931).

A very small amount of lipids, primarily phospholipid and cholesterol, has been found in the organic portion of enamel and dentin (Hess, 1956). Dirksen has applied special techniques for the qualitative (1963) and quantitative (1964) analysis of the lipid content of dentin. In general, he was able to demonstrate the presence of cholesterol esters, cholesterol, triglycerides, diglycerides, monoglycerides, fatty acids and various phospholipids. Sound dentin was found to contain inositol phosphatide, sphingomyelin, lecithin, phosphatidylethanolamine, lysocephalin and three unidentified phosphatides. In decayed dentin, he was able to identify, in addition, phosphatidyl serine and other phosphatides which had undergone various degrees of degradation.

The function of lipids in teeth has been investigated with the aid of a Sudan dye. Sudanophilic inclusions which are interpreted to be lipid granules have been found in ameloblasts and odontoblasts, and it is felt that these lipids may be sites for calcification of the teeth (Irving, 1958; 1963).

DENTAL CARIES

When larger than usual amounts of fat have been added to experimental animal diets at the expense of sucrose, a caries decrease has been reported (Rosebury and Karshan, 1939; Schweigert et al., 1946; Granados et al., 1948; Gustafsson et al., 1955). The logical explanation for this is that there is less fermentable carbohydrate available for bacterial degradation and acid formation. Another hypothesis for the action of fats in protecting against caries is that they coat the food particles or the teeth with an oily film which prevents their direct contact with either bacteria or acid. On the other hand, when a 5 per cent supplement of hydrogenated fat was added to a cariogenic diet at the expense of its corn component, dental caries increased. In this instance it was theorized that the plastic fat acts as a vehicle for retaining the sucrose in the pits and fissures, thus providing the major cariogenic factor with an opportunity for extended local contact with the tooth surface (Harris et al., 1965). In short, it appears from the animal experimental evidence that, from a nutritional standpoint, fat enrichment of the diet at the expense of carbohydrate can prove to be cariostatic, but, in terms of caries development, the physical nature of the fat will determine whether it acts in a positive or negative fashion.

In a human clinical study, Boyd (1944) attempted to correlate dietary fat intake with dental caries incidence. Significant reductions in caries were not found in association with a high fat diet. In this study, a group of children were fed a high fat diet consisting of a protein:carbohydrate:fat ratio of 7:9:21. Later, a lower fat diet (protein:carbohydrate: fat ratio of 7:15:11) was fed to the same group. Even though the caries experience was found to be slightly lower during the period the children were on the high fat diet, Boyd concluded that this result was due not to the high fat:low carbohydrate intake but rather to better supervision during the period when the high fat diets were given.

PERIODONTIUM

Knowledge of the effect of fats on the periodontium is very meager. The comparative effects of fat-depleted diets, high fat diets (30 per cent) and a normal diet (5 per cent) on the histology of the epithelial attachment, gingival corium, interdental papillae, periodontal membrane, cementum and bone were studied in the rat (Rao, 1965). It was noted that inflammatory changes occurred in the fat-depleted group but that tissue changes among the high fat group were mostly of a degenerative type. The gingivae showed very thin epithelial covering compared to those of the controls. Loss of tissue detail, irregular fibrosis, bone resorption and the presence of proliferative tissue replacing some of the cementum and bone were some of the chief microscopic details that were noted. This experiment indicates the need for further exploration into the dietary fat-periodontium relationship, not only in the experimental animal but in man himself.

REFERENCES

General

Deuel, H. J., Jr.: The Lipids, Their Chemistry and Biochemistry. 3 vol. New York, John Wiley & Sons, (Interscience) Inc., 1951-1957.

Frazer, A. C.: Fat absorption and its disorders. Brit. Med. Bull., *14*:212, 1958.

Hannahan, D. J.: Lipide Chemistry. New York, John Wiley & Sons, Inc., 1960.

Lynen, F.: Biosynthesis of saturated fatty acids. Fed. Proc., *20*:941, 1961.

Olson, R. E., and Vester, J. W.: Nutrition-endocrine interrelationships in the control of fat transport in man. Physiol. Rev., *40*:677, 1961.

Sinclair, H. M.: Animal and vegetable fats as human foods. Proc. Nutr. Soc., *20*:149, 1961.

Oral Relevance

Bodecker, C. F.: The lipid content of dental tissues in relation to decay. J. Dent. Res., *11*:277, 1931.

Boyd, J. D.: Dental caries as influenced by fat versus carbohydrate in the diet. Am. J. Dis. Child., *67*:278, 1944.

Dirksen, T. R.: Lipid components of sound and carious dentin. J. Dent. Res., *42*:128, 1963.

Dirksen, T. R., and Ikels, K. G.: Quantitative determination of some constituent lipids in human dentin. J. Dent. Res., *43*:246, 1964.

Granados, H., Glavind, J., and Dam, H.: Observations on experimental dental caries; effect of purified rations with and without dietary fat. Acta Path. Microbiol. Scand. *25*:453, 1948.

Gustafsson, G., Stelling, E., Abramson, E., and Brunius, E.: Experiments with various fats in a cariogenic diet. Experimental dental caries in golden hamsters. Acta Odont. Scand., *13*:75, 1955.

Harris, R. S., Walsh, N. B., and Nizel, A. E.: Cariostatic action of fat-embedded potassium phosphate when fed in the diet of rats. Arch. Oral Biol., *10*: 477, 1965.

Hess, W. C., Lee, C. Y., and Peckham, S. C.: The lipid content of enamel and dentin. J. Dent. Res., *35*:273, 1956.

Irving, J. T.: Sudanophil inclusions in ameloblasts, odontoblasts and cells of the oral epithelium. Nature (Lond.), *181*:569, 1958.

Irving, J. T.: The sudanophil material at sites of calcification. Arch. Oral Biol., *8*:735, 1963.

Rao, S. S., Shourie, K. L., and Shankwalkar, G. B.: Effect of dietary fat variations on the periodontium. An experimental study on rats. Periodontics, *3*:66, 1965.

Rosebury, T., and Karshan, M.: Susceptibility to dental caries in rat. Further studies of influence of vitamin D and of fatty oils. J. Dent. Res., *18*:189, 1939.

Schweigert, B. S., Potts, E., Shaw, J. H., Zepplin, M., and Phillips, P. H.: Dental caries in the cotton rat. VIII. Further studies on the dietary effects of carbohydrate, protein, and fat on the incidence and extent of carious lesions. J. Nutr., *32*:405, 1946.

Chapter Six

Proteins in Nutrition

SANFORD A. MILLER, PH.D., AND ABRAHAM E. NIZEL, D.M.D

INTRODUCTION

Chemists of the eighteenth century noticed that the white of egg (albumen), a substance in milk (casein), and a component of blood (globulin) changed from the liquid to the solid state when heated. These substances were classified as "albuminous" since they reacted in a manner similar to albumen in egg white. In 1839 the Dutch chemist Mulder worked out a root formula for albuminous substances and called them "protein" (derived from the Greek word *protos* meaning "of first importance"). Presumably, the term was meant merely to signify that this root formula was of importance in determining the structure of albuminous material but, as it turned out, the word protein was very apt for the substances themselves because they have proved to be the fiber of life and of first importance to life itself. In fact, von Liebig established that proteins were more essential for life than carbohydrates or fats.

Proteins constitute the major part of such fundamental organic materials as cytoplasm and nucleus. They may be conjugated with complex compounds such as carbohydrates to form mucin, with nucleic acid to form nucleoproteins, or with heme, an active iron pyrrole group, to form hemoglobin. They also serve as essential components of enzymes and hormones. In fact, they determine the specificity of the enzymes.

Fibrous proteins in the form of keratin, myosin, collagen, elastin, ossein and dentin contribute to the structural framework of the body. A globular type of protein is found in animals which is exemplified by caseinogen in milk, albumen in egg white and plasma globulins. Glutens are proteins found in plants such as wheat.

CHEMISTRY

Proteins are large complex colloidal molecules, ranging in molecular weight from thousands to millions, and are formed from simple units, the amino acids. Each amino acid has an amine group (NH_2) and a carboxyl group (COOH) attached to the same carbon atom. The rest of the molecule varies with the particular amino acid.

Amino acids are structurally represented by the following typical formula

$$R - \overset{\overset{\displaystyle H}{|}}{\underset{\underset{\displaystyle NH_2}{|}}{C}} \text{—— COOH}$$

in which the R radical may be hydrogen or an aliphatic group (carbon atoms which form straight-branched chains) or an aromatic group (carbon atoms which form closed rings) or a heterocyclic group (ring compounds which contain other atoms besides carbon atoms).

The essential chemical difference between proteins, carbohydrates and fats is that proteins contain nitrogen and various combinations of other elements such as sulfur, phosphorus, iron, iodine, copper, manganese and zinc, in addition to carbon, hydrogen and oxygen. Furthermore, when carbohydrates and

45

fats are oxidized in the organism, their end products are carbon dioxide and water, but protein end products always include, in addition, nitrogen in the form of urea, uric acid or ammonia.

BIOSYNTHESIS OF PROTEIN

There are about 22 different amino acids and, in essence, proteins are polymers of these amino acids. Amino acids are linked together in the protein molecule by peptide linkage, and the number of amino acids that are so linked will determine whether the compound is a dipeptide, tripeptide, proteose or peptone. The important point is that these amino acids are not polymerized in any random fashion; rather, they are put together in a very specific pattern which provides the essential specificity of the protein molecule. Thus, proteins give to each species its specific genetic and immunological characteristics and also provide individuality within the species.

How does each cell maintain this specificity through each generation? What is the mechanism for the manufacture of each protein molecule so that it conforms to a specific design? This specificity, this factor that controls heredity, this basic blueprint of life is a function of the deoxyribonucleic acid (DNA) molecule. DNA is found in every living cell and consists of two intertwining strands of polymers which are in the shape of helical coils.

The subunit molecules of DNA are called nucleotides. They are composed of purines or pyrimidines bound to phosphorylated sugar. In the DNA molecule, adenine (a purine) is always joined to thymine of the other strand. A similar relationship exists for guanine (a purine) and cytosine.

"Messenger RNA" (ribonucleic acid), which is derived from DNA, carries the blueprint for protein construction from the nucleus to the ribosomes in the cytoplasm. RNA differs chemically from DNA in that it contains ribose instead of deoxyribose as the sugar and uracil instead of thymine as the pyrimidine base.

DNA also gives rise to smaller single strand units called "transfer RNA" which, with the help of adenosine triphosphate (ATP) and enzymes bring together the amino acids in the proper sequences to build on the ribosome the protein chain ordered, in the first place, by "messenger RNA." The protein then leaves the ribosome and is ready for use by the organism. Thus, the precise predetermined pattern has been passed on from DNA to "messenger RNA" and with "transfer RNA" ultimately to the protein.

PROPERTIES AND METHOD OF MEASURING PROTEIN CONCENTRATION IN FOOD

In general, proteins, because they are macromolecules, cannot traverse semipermeable membranes. They are coagulated by heat and are precipitated by heavy metals. Moderate heat, ultraviolet light, alcohol, or mild acids or alkalies will cause denaturation.

The protein concentration of a food is generally estimated by measuring its total nitrogen content. Since a total nitrogen analysis measures nitrates, nitrites and other nitrogenous substances that are not necessarily part of the protein molecule, the assumption that the total nitrogen value measures protein alone is not completely valid. It is, however, sufficiently accurate for most purposes. Based on the assumption that the protein of food contains 16 per cent nitrogen, a factor of 6.25 gm. of protein per gm. of nitrogen can be used to calculate the protein content of food from its nitrogen value. Thus, the Kjeldahl nitrogen value is multiplied by 6.25 to give "total protein." (If the protein in food contains an amount of nitrogen different from this percentage, the factor for converting nitrogen to protein is different.)

DIGESTION

Since proteins are large molecules, they must undergo hydrolysis into amino acids before being absorbed by the intestinal mucosa. This is accomplished by the proteolytic enzymes pepsin, trypsin and erepsin.

The "apparent digestion" of protein is simply the difference between the amount eaten as measured by the nitrogen content of the food and the amount of nitrogen in the feces.

In normal individuals, under ordinary conditions, the fecal nitrogen approximates 10 per cent of the intake, which means that 90 per cent has been digested; there is, however, evidence that this relation may vary. Since gastrointestinal secretions and the sloughing off of cells of the mucosa contribute to the fecal nitrogen, too, the amount of nitrogen lost in the feces may not necessarily be an accurate measure of the food nitrogen actually absorbed.

The variation in digestibility of proteins is relatively small and, therefore, under normal conditions the nutritive value of most proteins is not significantly affected by this parameter.

ABSORPTION AND TRANSPORT

After the proteins are digested to amino acids, the latter are absorbed, transported to the tissues and enter the cells where they are utilized. It has been demonstrated that amino acids are actively absorbed and do not depend on the presence of a concentration gradient.

A number of factors are involved in the absorption of amino acids. For example, amino acid uptake into cells is enhanced by the presence of adequate amounts of vitamin B_6, suggesting that the latter may be a carrier molecule. In addition, hormones like insulin, corticosteroids and sex hormones have been shown to influence and even to control the transport of amino acids.

METABOLISM

Originally, as suggested by Folin, it was believed that protein metabolism simply had an endogenous and an exogenous component (Fig. 6-1). The metabolism of amino acids from foods was exogenous and that from body protein was endogenous. Actually, it was believed that the tissues slowly wasted away as a result of "wear and tear" and that constant replenishment was necessary to replace the lost protein. The synthesis of new additional protein was believed to nearly cease at adulthood. Nitrogen entering into cells was considered to be on a "one-way street." Thus, the nitrogen requirement of the adult was considered to be that required to replace the constant loss due to the endogenous metabolism.

This concept was challenged, however, when Schoenheimer (1942) and his associates, as a result of experiments with stable isotopes, showed that proteins are in a continuous dynamic state (Fig. 6-2). Amino acids are deposited and withdrawn from a "metabolic pool" to which both food and tissue proteins contribute. This allows for an interchange between "endogenous" and "exogenous" sources of protein. Although it is convenient to speak of an "amino acid metabolic pool," this is not to say that there is a single specifically defined component. It is probable that there are many "pools" or compartments, some serving as more direct precursors of specific proteins or nitrogenous components than others and each pool in equilibrium with the others. This dynamic state allows the body to adapt to different levels of protein intake. It has been estimated that the "metabolic pool" in man is approximately 0.5 gm. of nitrogen per kg., while the total synthesis of protein is about 0.3 gm. per kg. per day.

Some of the amino acids derived from either food or body protein enter into energy catabo-

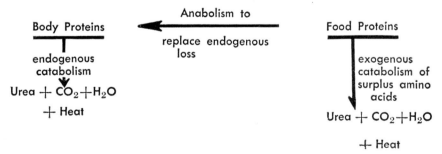

FIGURE 6-1. *Protein metabolism, according to Folin.*

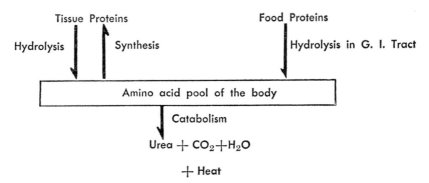

FIGURE 6-2. *Protein metabolism, according to Schoenheimer.*

lism, leading to formation of simpler products, and are excreted as urea, uric acid, ammonia and creatinine.

In summary (Fig. 6-3), body protein is constantly being broken down to its constituent amino acids. These amino acids then enter into a "metabolic pool" where they are joined by amino acids derived from the proteins of the diet. By means of the anabolic metabolic pathway, proteins are synthesized from these amino acids. However, by means of the catabolic metabolic pathway some amino acids from this mixed pool are converted to carbohydrate intermediates and oxidized via the Krebs cycle to carbon dioxide and water (the glucogenic amino acids). Still others are metabolized to lipid intermediates and are metabolized through fat pathways (ketogenic amino acids). Despite the ability of the organism to adapt to

lower levels of protein intake, it is apparent that nitrogen equilibrium can be readily upset in view of the dynamic properties and multiple uses of protein by the organism.

PROTEIN RESERVE

Does one have to continually supply amino acids to the organism? Is the amino acid pool a storage depot? Do we have a protein reserve? This is a matter of current debate. There is no organ in the body that contains a static reserve of protein similar to glycogen in the muscle or fat in the adipose tissue. However, there is labile protein. When the body is deprived of protein, certain types of protein whose functions are not absolutely vital for the existence of the body are broken down readily. These include certain enzymes and

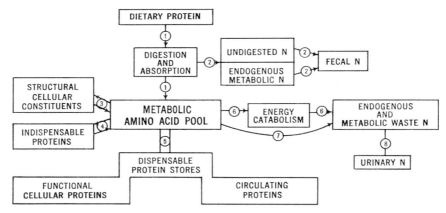

FIGURE 6-3. *Schematic outline showing the general pathways of protein metabolism. (Longenecker: Newer Methods of Nutritional Biochemistry. Courtesy of Academic Press.)*

certain structural nonessential cytoplasmic molecules. These molecules break down rapidly and tend to supply amino acids to the pool when the synthesis of labile protein molecules is reduced. The amino acids in the pool are available for the synthesis of vital protein. These protein molecules then are used, in a sense, as storage and are called "labile protein molecules" or labile protein reserves.

ESSENTIAL AND NONESSENTIAL AMINO ACIDS

All amino acids are equally important in terms of protein synthesis and all must be supplied in one way or another at the site of protein synthesis. They are needed for growth of new tissue, maintenance of established tissue and metabolic processes. They are supplied either by the organism synthesizing them or they are obtained, preformed, in the diet. Since animals have limited powers of converting one amino acid into another, they must obtain some of the amino acids they require from the diet. These are called "essential" or "indispensable" amino acids. A definition of an essential amino acid, therefore, is one that the organism cannot synthesize itself in quantities great enough to meet its needs. On the other hand, the so-called "nonessential" or "dispensable" amino acid is one that can be synthesized by the organism, providing a sufficient amount of nitrogen is present. The number of essential amino acids varies from species to species because each species has different capabilities for synthesis. The amount of each needed may depend on the quantity of other amino acids in the diet. For example, when cysteine is present in abundance, less methionine is needed and, similarly, when plenty of tyrosine is present, the organism can get along with less phenylalanine. Also, the rate of synthesis and relative needs for various amino acids vary with the physiological state.

The amino acid requirement of the human was first determined by Rose (1949) and his associates. In his work with various amino acid mixtures, each of which was limited with respect to one amino acid, and the ability of the subject to maintain nitrogen balance was determined. A dietary supply of eight amino acids was found to be essential for the human.

They are lysine, tryptophan, phenylalanine, leucine, isoleucine, threonine, methionine and valine. For the rat, histidine is required for growth and maintenance.

Certain of the essential amino acids are not only important for protein synthesis but they have other physiological roles. For example, tryptophan is a precursor of niacin and methionine acts as a methyl donor.

METHODS OF ESTIMATING NUTRITIVE VALUE OF PROTEINS

There have been several methods devised for determining the nutritive value of proteins. The best known is biological value (BV), which is defined as the percentage of the absorbed nitrogen retained in the body. For all practical purposes, retention of protein is equal to the ability of the protein to make tissue since the body has little storage ability with respect to protein. Thus, a protein that has a biologic value of 100 per cent is able to replace, gram for gram, daily body losses of protein.

The protein of whole egg has a biologic value of practically 100 per cent. This means that the egg contains the correct ratio and type of essential amino acids for the anabolic body needs and that, at relatively low dietary protein levels, very little if any is left for energy catabolism. In short, egg protein is very efficiently utilized and its amino acid composition is ideal. The relative biological value of other proteins compared to that of eggs is as follows: animal protein, 70 to 100; casein, 70 to 75; wheat, 60; and vegetable protein, 40 to 65. In other words, in comparison with egg protein, only about half a gram of body protein can be synthesized, for instance, from a gram of vegetable protein.

Any method for evaluating the quality of a protein must be based upon the ability of the protein to satisfy the needs of the body. This can be done in two general ways. The first method involves direct observations of how well the animal utilized the protein for growth, maintenance and reproduction; these are biological methods. The second method attempts to compare the pattern of essential amino acids in the protein to a standard pattern which has been empirically derived from observation in animals; these are chemical methods.

There are several biological methods for evaluating dietary proteins (Hegsted, 1964).

Growth Rate. The simplest method is to compare growth rates of animals fed equivalent levels of protein or nitrogen. This method, however, has not been used very much since decreased food intake associated with the feeding of protein of poor quality made accurate comparisons difficult. To correct for this variation, attempts have been made to express the results on a standard basis.

Protein Efficiency Ratio. The protein efficiency ratio (PER) is the gain in body weight per gram of protein consumed. Since the PER is not only a function of the protein sources, it can have meaning only when performed under standard conditions. In this assay, weanling animals are fed a diet containing 10 per cent of a particular protein. Since animals fed poor quality proteins do not eat so much food, it is commonly believed that the PER is a better method of evaluating protein than weight gain alone since it is assumed that a correction for the level of food intake is made.

Carcass Analysis. One major problem of the techniques based on weight gain is that one must assume that nitrogen content of the tissue gained is constant. Since this is not always true, a more accurate measurement would involve the direct determination of nitrogen gain in the carcass. In this method, the carcass is analyzed for the quantity of nitrogen that is laid down. The relationship between the protein (nitrogen) eaten and the increase in carcass nitrogen is termed the net protein utilization (NPU).

Depletion-Repletion. Adult animals instead of weanlings may be fed a standard depletion diet, low in or devoid of protein, until a certain amount of weight is lost. Test proteins are then compared by the weight gain during the repletion period. This allows the same animal to be reutilized, but this raises an objection since, after one or two tests, the nutritional background of the test animals may be dissimilar.

Nitrogen Balance. Since absorption and retention of nitrogen affect growth, measurement of these parameters can provide a basis for estimating the ability of an animal to utilize a protein. Furthermore, in contrast to measurement of growth, direct determination of the absorption and retention allows a more accurate estimation of the quality of a protein. In other words, these methods can indicate whether the problem with a protein source is in its absorption and digestion or in its pattern of amino acids. Evaluations based upon these measurements are known as nitrogen balance methods; the assumption is that the quality of a protein is inversely related to the excretion of nitrogen in the urine and feces.

The most widely used of these techniques is the Biological Value (BV) of Thomas and Mitchell. For this calculation it is necessary to know the amount of nitrogen fed during the test (D); the urinary nitrogen excretion (U); the urinary nitrogen loss when no nitrogen is fed (U_o); the fecal nitrogen loss (F); and the fecal nitrogen loss when no nitrogen is fed (F_o). With this information, one can calculate the absorbed nitrogen (AN) as follows:

$$AN = D - (F - F_o)$$

Nitrogen retained (NR) can be calculated as follows:

$$NR = AN - (U - U_o) =$$
$$D - (F - F_o) - (U - U_o)$$

From these, biological values (BV) can be calculated:

$$BV = \frac{NR}{AN} \times 100 =$$
$$\frac{D - (F - F_o) - (U - U_o)}{D - (F - F_o)} \times 100$$

The formulas given above indicate a number of important points. For one thing, it is apparent that there is a continual loss of nitrogen in the feces and urine. The origin of this so-called endogenous loss was discussed earlier. Thus, in order to determine what part of the excretory loss was due to the protein fed, it is necessary to correct for the losses. This leads to the assumption that these losses are constant. There are, in fact, data which suggest that this endogenous loss is variable and may change with the protein fed.

To obviate the need for calculation of U_o and F_o, collectively NE_o, Allison has suggested that the slope of the line relating nitrogen balance (NB) and absorbed nitrogen be used as an estimate of protein utilization. This slope (k), known as the nitrogen balance index, is equivalent to the biological value and can be calculated as follows:

$$\frac{NB - NE_o}{AN} = k$$

While this equation is, in essence, identical to that of the biological value, it is possible, since it is a linear function, to obtain k by graphic methods without calculation of NE_o; thus, the problems associated with NE_o is avoided.

DIETETIC CONSIDERATIONS WITH RESPECT TO PROTEIN AND AMINO ACID

MUTUAL SUPPLEMENTATION

The proteins of fish, meat and poultry are of high biological value while those of nuts and vegetable proteins are lower. In order to provide good nutrition, plant proteins of lower quality must be supplemented with other foods which will supply the missing or inadequate amino acids. For efficient use in tissue synthesis, the essential amino acids must be present at the same time in a definite pattern or proportion. On the basis of this concept, mutual supplementation of proteins has been practiced instinctively by man for ages. Even though a protein may be of low biological value, it may be useful if it is supplemented with another protein which provides the missing constituent (for example, corn eaten with beans, or bread or cereal with milk, or macaroni with cheese). The important factor to be re-emphasized is that these foods must be eaten at the same meal because amino acids are not stored for any appreciable length of time. Therefore, for the fabrication of tissue protein, all the amino acids must be available at the site of synthesis at the proper time. In fact, the amount of tissue protein that is produced is limited by the essential amino acid which is present in the smallest amount in relation to the need of the organism. This is called the most limiting amino acid.

It should be emphasized that it is possible to be well nourished while consuming only proteins of plant origin as long as the diet contains plant proteins of such divergent composition that they will provide the entire spectrum of essential amino acids. A good example of this is Incaparina, developed at the Institute of Nutrition of Central America and Panama to combat protein deficiency. This product contains only proteins of plant origin, namely, corn, sorghum, cottonseed meal and torula yeast, and has a biological value similar to that of casein.

AMINO ACID BALANCE

In order to realize an amino acid balance, essential or indispensable amino acids are required by the body in certain definite proportions. Proteins are considered nutritionally balanced if they provide amino acids in the same proportions as are required by the body. Furthermore, a balanced protein is required in relatively small quantities because it is used so efficiently.

When the indispensable amino acids in a protein are low and the proportions deviate from the ideal, the protein is unbalanced. If the proteins are not too greatly out of balance, an increased intake of protein will compensate for the deficiency and permit maximal performance. However, if there is a disproportionate deficiency of a single amino acid, this imbalance can be rectified only by supplementation with the limiting deficient amino acid. Amino acid imbalances can cause reduced growth, accumulation of liver fat and decrease of liver enzyme activities.

STATUS OF PROTEIN NUTRITION IN THE UNITED STATES

In the United States there has been a gradual increase over the last 45 years in the consumption of dairy products, eggs and meat, with a concomitant decrease in consumption of grain products (Table 6-1). According to the U.S. Department of Agriculture, there was available in the 1957 food supply 96 gm. of protein per capita per day along with 3200 calories. In general, many in the population

TABLE 6-1.   SOURCES OF PROTEIN IN U. S.
FOOD SUPPLY IN 1957

Foods	Percentage of Protein from Specified Sources
Dairy products	25.4
Eggs	6.6
Meat, poultry, fish, fats, oils	35.0
Grain products	19.5
Legumes, nuts	5.5
Other vegetable sources	8.0
Total	100.0
Total animal sources	67.0
Total vegetable sources	33.0
Protein per person per day (grams)	96

Source: Supplements for 1956 and 1957 to Consumption of Food in the United States 1909-1952. Washington, D. C., U. S. Department of Agriculture, Agriculture Handbook 62.

of the United States have diets adequate in protein quality and quantity. However, there are exceptions within families, some members not sharing in the general protein supplies in accordance with their nutritional needs. For example, on the basis of recommended allowances, women have been found to have diets poorer in protein than those of men. Older people are likely to have diets low in protein. The dietary protein intake of younger children is relatively better than that of older children, and boys do better than girls. Diets of adolescent girls, especially, are likely to be low in protein. The amino acids that are most likely to be limited in the United States diets are the sulfur-containing amino acids and tryptophan.

PROTEIN REQUIREMENTS AND RECOMMENDED ALLOWANCES

The minimum protein need is closely related to the basal metabolism of the animal species; in man this has been calculated to amount to 12.5 mg. protein per basal calorie per day. This estimate is based on the evidence that minimal endogenous urinary nitrogen is about 2 mg. N/Cal. of basal metabolism, which is multiplied by 6.25 (the factor one uses to calculate protein from nitrogen value). In addition to this basal requirement, one must add 0.4 mg. of nitrogen for fecal loss and 0.8 mg. of nitrogen for loss through

skin and integumental growth. When these last two values are multiplied by the 6.25 factor and added to the original 12.5 mg. the total becomes 20 mg. of protein per basal calorie. For an adult with a basal metabolic rate of 1500 calories per day, the protein requirement would be the product of 1500 × 20 mg., or 30 gm. per day using a protein with a biological value of 100. If a dietary protein of less than 100 biological value is used, the minimum need would be increased accordingly. For example, if a vegetable or cereal protein with a biological value of 50 were used, the minimum need of protein would be 60 gm. per day.

The minimum protein requirements for children are relatively higher than those for adults because the growth factor must also be considered. The maintenance requirement for protein is also 20 mg. of protein per basal calorie as it is for adults. The gain in weight per day multiplied by 0.18 gives the protein deposited. The sum of these values is the amount of absorbed protein needed.

There are several problems in translating information on minimal needs into dietary recommendations: (1) Is the method of determining requirements sufficiently sensitive? (2) Has the factor of individual variation been determined? (3) Are the experimental conditions sufficiently different from the "natural" conditions to be important? (4) How large should the allowance for safety be?

In Chapter Fourteen there is detailed discussion of the recommended allowance for protein for different age groups under different physiological conditions and using standards adopted by different countries. In the United States the allowance has been set at 1 gm. of protein per kg. of body weight per day. This allowance was based upon data collected in more than 100 nitrogen balance experiments in the course of 25 independent investigations, including work in several countries and with both sexes. The average determined in these studies was increased by 50 per cent to cover individual differences in need as well as variations in the nutritional efficiency of the protein of different diets. The Food and Nutrition Board in the United States has retained this recommendation even though the Food and Agriculture Organization of the United Nations has suggested that the safe practical allowance ranges from 0.66 to 0.84 gm. per kg.

PROTEIN DEFICIENCY

In young animals, protein deficiency will produce growth retardation and decrease in food intake. There are a rarefaction of bone, failure of collagen formation, atrophy of skin and muscle, loss of hair, decrease in hemoglobin production and other changes in the liver, pancreas, reproductive organs and endocrines.

In man, as was seen during the Second World War, hunger edema associated with starvation may result from simple protein deficiency. Since the albumin level is affected by protein deficiency, and since osmotic pressure depends in part on albumin and plasma proteins, edema may result if the intravascular and extravascular fluid volumes are not maintained during periods of decreased plasma protein concentration.

Protein-calorie malnutrition in infants and young children is probably the major nutritional disease in the world today. There are two variants, marasmus and kwashiorkor. Marasmus approaches starvation and is characterized by wasting, nearly complete loss of body fat and greatly retarded development. Kwashiorkor is a protein deficiency characterized by hypoalbuminemia, edematous fatty liver, dermatosis, gastrointestinal disturbances and psychic changes and occurs when diets low in protein but high or adequate in calories are fed (Scrimshaw and Béhar, 1961) (Fig. 6-4). These protein-deficient children are more susceptible and more seriously affected by common childhood infections than normal children (Scrimshaw et al., 1959). Infection not only decreases food intake but increases metabolic losses. The treatment of these diseases is to provide adequate calories and 3 to 5 gm. of protein per kg. of body weight as well as to correct any potassium deficiency. Infection must be controlled. Prevention is possible if adequate diets high in protein are provided.

A more detailed discussion of protein deficiency and infection is given in Chapter Twenty.

Since the discovery of the niacin-tryptophan relationship, pellagra has been considered to

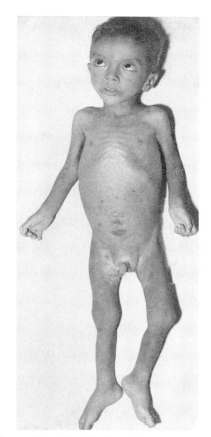

FIGURE 6-4. *Kwashiorkor. (Courtesy of Dr. N. S. Scrimshaw.)*

result from a deficiency in both these nutrients. Therefore, pellagra is a human syndrome in which protein deficiency plays an important role.

In this country, nutritional liver disease is a common syndrome of protein deficiency. It is seen most often in alcoholics, who maintain an adequate calorie intake but ingest inadequate amounts of protein. They develop fatty liver disease, which can be helped by a high protein diet with or without choline supplements.

Besides these deficiencies of protein that arise from exogenous sources, there are also some endogenous protein deficiency states, for example, inborn errors of metabolism such as phenylketonuria. There are also hypoproteinemias that are associated with the reproductive syndromes, chronic blood loss, long standing infections and chronic diarrhea.

As yet, there is no evidence that excessive protein intake is deleterious under normal circumstances.

THE ORAL RELEVANCE OF PROTEIN

TEETH

Protein Composition. Human dental enamel is made up of 0.2 to 0.4 per cent organic matter which, like other tissues of ectodermal origin (hair, nails, epidermis), contains an insoluble protein, eukeratin. Since the amino acid content of mature enamel contains relatively large amounts of glycine, alanine, hydroxyproline and proline (the precursors of collagen), and relatively small amounts of cystine, a characteristic constituent of hard keratins, the question arises whether enamel truly fits into the classification of being a eukeratin. Actually, it has been suggested that this protein material be called a "soft keratin" (pseudokeratin) (Eastoe, 1963), or a delta keratin (Perdock, 1961) rather than eukeratin, which is commonly associated with nails.

The protein found in the developing enamel of unerupted teeth differs from mature enamel in that it contains exceptionally large amounts of proline, glutamic acid and histidine. It is considered to be the true matrix in which apatite crystals are laid down by epitaxy, the oriented growth of one crystalline substance on a different crystalline substance (Glimcher, 1959).

Normal human dentin, which is 21 to 22 per cent organic matter, is primarily protein in nature (only 1 or 2 per cent is not protein). The principal amino acids in dentin are glycine, alanine, proline and hydroxyproline, which makes it chemically similar to collagen. Carious human dentin has less arginine, proline and hydroxyproline than sound dentin, which may be due to a contamination of the matrix by noncollagenous protein or peptide material as well as to some proteolytic degradation of the matrix (Armstrong, 1964).

Effect of Protein Deficiency

Tooth and Jaw Structure. Compared to bone, teeth are very much less sensitive to protein or amino acid deficiency. However, under very stringent deficiencies of lysine (Irving, 1956) or tryptophan (Bavetta, 1954), formation of poorly calcified dentinal matrix characterized by a number of interglobular areas can take place. Another characteristic is an irregular predentin layer.

In children who suffer from protein-calorie malnutrition there has been noted a marked crowding and even rotation of teeth. This might be interpreted as being the result of an inadequate development or retarded growth of the jaw bone matrix due to the extreme sensitivity of newly forming osteoid tissue to protein deprivation. Normal sized teeth in an undersized jaw will naturally produce a crowded arch and give an appearance of a mouth full of jumbled teeth (Trowell, 1954).

Tooth Size, Morphology and Eruption. The ability of nutrition to influence genetic pattern and heredity is indeed a challenging concept which has been provoked by some very excellent experiments in which a protein deficiency was purposely created in rodents. Offspring that are bred from a dam that is fed a protein-deficient diet during pregnancy and lactation will have smaller teeth than expected (Paynter, 1956; Shaw, 1963), the cuspal pattern of the molars will be altered, and the eruption of third molars will be delayed by as much as five days beyond the usual eruption date (Holloway, 1961).

Caries Susceptibility. There is suggestive evidence that protein factors are involved in dental caries production and prevention, at least in experimental animals. On the other hand, there is no firm evidence that this occurs in man; in fact, there are findings from an epidemiological nutritional and dental survey of relatively primitive population groups like Eskimo Scouts (Russell, 1961) as well as surveys of personnel in the armed forces (Shannon, 1964) that show that there is no correlation between dental caries incidence and serum protein levels.

McClure and Folk (1953) noted that smooth surface caries increased in white rats when they were fed diets which included heat processed skim milk powder in which the lysine had been destroyed. Further, they demonstrated that this cariogenic diet can be made cariostatic by the addition of L-lysine (McClure, 1957; Bavetta and McClure, 1957). A confirmation of this type of finding was made by a group of Russian scientists, who found that supplementing a cariogenic diet with 100 to 200 mg. DL-lysine monohydrochloride per 100 gm. of diet was an effective method of reducing caries by 50 per cent (Sarpenak et al., 1963). Although Dodds (1964) under her experimental conditions did not find lysine to be as cariostatic as McClure, she did find that increasing the total protein level of the diet with dried egg white or casein was effective in decreasing the incidence of caries in rats.

PERIODONTIUM

Protein deficiency or deprivation affects the activity of matrix-forming cells, namely fibroblasts, osteoblasts and cementoblasts. Therefore, microscopic findings like atrophic and degenerative changes in the connective tissue of the gingiva and periodontal ligament, increased size of cancellous bone spaces, less than normal amounts of osteoid tissue (Fig. 6-5), as well as retardation in deposition of cementum are to be expected. These tissue changes were produced even without a challenge of a local irritating factor. There are other studies in which protein deficiencies have been compounded by either a soft diet (Ruben, 1962), a local irritant (Stahl et al., 1955) or an induced wound (Stahl, 1965), re-

sulting in more extensive inflammation and more rapid dystrophic changes. The resorption of the alveolar crest, downgrowth of the epithelial attachment and increase of the general inflammatory infiltrate are accentuated when a foreign body is purposely implanted between the teeth (Stahl et al., 1958). However, this deficiency effect could be reversed if a normal diet complete in protein was fed. In fact, the connective tissue repair of a gingival wound in animals under a low protein nutritional stress will be noticeably delayed, and there will also be evidence of sequestration of the alveolar bone. This latter finding was reported in rats fed a low protein diet consisting of 8 per cent casein when compared to control littermates fed a diet with 27 per cent casein, the animal's normal protein requirement (Stahl, 1963).

Four mediating substances (histamine, peptides, proteins and inactivators of vasoconstrictor amines) released from tissue cells are capable of inducing vascular alterations resulting in increased permeability. These mediators are liberated as a result of proteolytic activity in injured tissue along with disturbances of connective tissue ground substance (Schultz-Haudt and Lundqvist, 1962).

Amino acids and plasma proteins make up a part of the contents of the fluid of the gingival sulcus. Specifically, albumin and alpha, beta and gamma globulin have been identi-

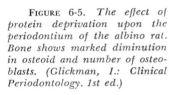

FIGURE 6-5. *The effect of protein deprivation upon the periodontium of the albino rat. Bone shows marked diminution in osteoid and number of osteoblasts. (Glickman, I.: Clinical Periodontology. 1st ed.)*

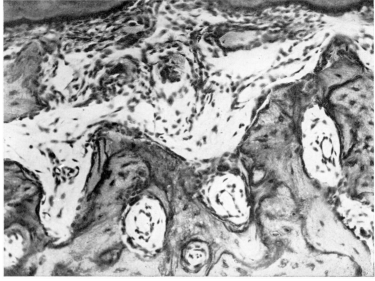

fied in the protein fractions of gingival fluid (Carraro et al., 1964).

SALIVA

The total protein of unstimulated saliva has been found to range from 140 to 640 mg. The proteins present in the highest concentration are glycoproteins, proteins conjugated with a carbohydrate moiety, found primarily in the sublingual glands. They are the mucus-forming proteins of tissues and secretions often referred to as mucin or mucoids. In addition, there have been found protein-containing substances such as enzymes (saliva amylase) secreted particularly in the parotid gland serum proteins (albumin and globulin), group specific substances and antibodies (Afonsky, 1961; Stoffer et al., 1962).

There are 20 amino acids present in saliva (Afonsky, 1961), which could conceivably make it a suitable culture medium for bacterial growth; but as yet no correlation between the protein content of saliva and caries activity has been proved, even though some preliminary evidence had been presented correlating tryptophan levels with caries incidence (Turner, 1947). In fact, no significant differences were found in the amount of total protein, protein-bound hexosamine fucose, hexose and sialic acid between the parotid saliva of caries-immune and caries-active adults (Mandel et al., 1965).

Other nitrogenous constituents found in saliva are urea, creatine, uric acid and ammonia. Because caries-free individuals have been shown to have more ammonia in their saliva than caries-prone individuals, this agent has been suggested for use as an anticaries agent (Grove and Grove, 1935). However, there is no evidence to support this hypothesis.

REFERENCES

General

Allison, J. B.: Evaluation of protein. Phys. Rev., *35*: 664, 1955.

Block, R. J., Weiss, K. W., Almquist, H. J., Carroll, D. B., Gordon, W. G., and Saperstein, S.: Amino Acid Handbook. Springfield, Ill. Charles C Thomas, 1956.

Committee on Amino Acids: Evaluation of Protein Nutrition. Washington, D. C., National Academy of Sciences–National Research Council, Publication No. 711, 1959.

Hegsted, D. M.: Proteins. *In* Beaton, G. H., and McHenry, E. W. (ed.): Nutrition. New York, Academic Press, Inc., 1964.

Rose, W. C.: Amino acid requirements of man. Fed. Proc., *8*:546, 1949.

Schoenheimer, R.: Dynamic State of Body Constituents. Cambridge, Harvard University Press, 1942.

Scrimshaw, N. S., and Béhar, M.: Protein malnutrition in young children. Science, *133*:2039, 1961.

Scrimshaw, N. S., Taylor, C. E., and Gordon, J. E.: Interaction of nutrition and infection. Amer. J. Med. Sci., *237*:367-403, 1959.

Oral Relevance

Afonsky, D.: Saliva and Its Relation to Oral Health. Montgomery, University of Alabama Press, 1961, pp. 101-144.

Armstrong, W. G.: Modifications of the properties and composition of the dentin matrix caused by dental caries. *In* Staple, P. (ed.): Advances in Oral Biology. Vol. 1. New York, Academic Press, Inc., 1964.

Bavetta, L. A., Bernick, S., Geiger, E., and Berger, W.: The effect of tryptophane deficiency on the jaws of rats. J. Dent. Res., *33*:309, 1954.

Bavetta, L. A., and McClure, F. J.: Protein factors in experimental rat caries. J. Nutr., *63*:107, 1957.

Carraro, J. J., Milstein, S., Sznajder, N., and Zdrojewski, D.: Electroforesis en agar del fluid gingival de encias normales. Rev. A. odont. Argentina, *52*: 77, 1964.

Chawla, T. N., and Glickman, I.: Protein deprivation and the periodontal structures of the albino rat. Oral Surg., *4*:578, 1951.

Dodds, M. L.: Protein and lysine factors in the cariogenicity of a cereal diet. J. Nutr., *82*:217, 1964.

Eastoe, J. E.: The amino acid composition of proteins from the oral tissues. II. The matrix proteins in dentin and enamel from developing human deciduous teeth. Arch. Oral Biol., *8*:633, 1963.

Glimcher, M. J.: The molecular biology of mineralized tissues with particular reference to bone. Rev. Mod. Phys., *31*:359, 1959.

Grove, C. T., and Grove, C. J.: Chemical study of human saliva indicating that ammonia is an immunizing factor in dental caries. J. Amer. Dent. Assn., *22*:247, 1935.

Holloway, P. J., Shaw, J. H., and Sweeney, E. A.: Effect of various sucrose:casein ratios in purified diets on the teeth and supporting structures of rats. Arch. Oral Biol., *3*:185, 1961.

Irving, J. T.: Action of the hypophysis and of dietary protein on the calcifying tissues. Nature, *178*:1231, 1956.

Mandel, J. D., Zorn, M., Ruiz, R., Thompson, R. H., and Ellison, S. A.: The proteins and protein-bound carbohydrates of parotid saliva in caries immune and caries active adults. Arch. Oral Biol., *10*:471, 1965.

McClure, F. J.: Effect of lysine provided by different

routes on cariogenicity of lysine deficient diet. Proc. Soc. Exp. Biol. Med., *96*:631, 1957.

McClure, F. J., and Folk, J. E.: Skim milk powders and experimental rat caries. Proc. Soc. Exp. Biol. Med., *83*:21, 1953.

Paynter, K. J., and Grainger, R. M.: The relation of nutrition to the morphology and size of rat molar teeth. J. Cand. Dent. Assn., *22*:519, 1956.

Perdock, W. G., and Gustafsson, G.: X-ray diffraction studies of the insoluble protein in mature human enamel. Meded. med. prodent. Res., *19*:1, 1961. (Abstract in Dental Abstracts, *8*:17, 1963.)

Rubin, M. P., McCoy, J., Person, P., and Cohen, D. W.: Effects of soft dietary consistency and protein deprivation on the periodontum of the dog. Oral Surg., *15*:1061, 1962.

Russell, A. L., Consolazio, C. F., and White, C. L.: Dental caries and nutrition of the Alaska National Guard. J. Dent. Res., *40*:594, 1961.

Sarpenak, A. E., Bobyleva, V. R., and Gorozanina, L. A.: Role of protein, lysine, some minerals, vitamins A and D in prevention of dental caries. Vop. Pitan., 1963, 22 No. 239-44 (Kaf. biohim. med. stomatol. Inst. Moscow) Russia.

Schultz-Haudt, S. D., and Lundqvist, C.: The biochemistry of periodontal disease. Int. Dent. J., *12*: 180, 1962.

Shannon, I. L., and Gibson, W. A.: Serum total protein, albumin and globulin in relation to periodontal status and caries experience. Oral Surg. *18*: 399, 1964.

Shaw, J. H., and Griffith, D.: Dental abnormalities in rats attributable to protein deficiency during reproduction. J. Nutr., *80*:123, 1963.

Stahl, S. S.: Healing of gingival wounds in female rats fed a low protein diet. J. Dent. Res., *42*:1511, 1963.

Stahl, S. S.: The healing of experimentally induced gingival wounds in rats on prolonged nutritional deprivations. J. Periodont., *36*:283, 1965.

Stahl, S. S., Miller, S. C., and Goldsmith, E. D.: Effect of various diets on the periodontal structure of hamsters. J. Periodont., *29*:7, 1958.

Stahl, S. S., Sandler, H. C., and Cahn, L. R.: The effects of protein deprivation upon the oral tissues of the rat and particularly upon periodontal structures under irritation. Oral Surg., *8*:760, 1955.

Stoffer, H. R., Kraus, F. W., and Holmes, A. C.: Immunochemical identification of salivary proteins. Proc. Soc. Exp. Biol. Med., *111*:467, 1962.

Trowell, H. C., Davies, J. N. P., and Deand, R. F. A.: "Kwashiorkor." London, E. Arnold & Co., Ltd., 1954.

Turner, N. C., and Crowell, G. E.: Dental caries and tryptophan deficiency. J. Dent. Res., *26*:99, 1947.

Chapter Seven

Water and Electrolyte Balance

G. N. WOGAN, PH.D.

INTRODUCTION

Water is the solvent in which the intracellular biochemical reactions of living cells take place. It also provides the fluid matrix which, in the form of the extracellular fluids, permits exchanges of nutrients and metabolic end products among cells and tissues and places the cells into indirect contact with the external environment.

The body fluids are intimately involved in the basic cellular processes of the living organism. These fluids consist of solutions of various inorganic and organic solutes in an aqueous medium. The principal components are water, certain inorganic electrolytes and proteins. In health the volume and the composition of the body fluids are carefully regulated within narrow limits, and the homeostatic state is maintained despite a wide range of dietary intakes. Since the principal electrolytes are sodium, potassium and chloride, this discussion will deal with these and not consider other inorganic ions present in body fluids in small amounts. The requirements for water and these major electrolytes are intimately interrelated and depend largely upon the functional state of the regulatory mechanisms governing their retention and excretion.

DISTRIBUTION, COMPOSITION AND EXCHANGE OF BODY FLUIDS

The body fluids may be classified, according to their distribution, into two principal forms, intracellular and extracellular (Gamble, 1958). The intracellular fluid consists of the water and solutes within the cells of the various tissues and is the site of the metabolic processes. The extracellular fluids surround the cells, thereby providing the internal aqueous environment as well as the pathway between the cells and the organs for exchange with the external environment. The extracellular fluids are further subdivided into the interstitial fluid and the blood plasma contained within the vascular system.

The total volume of body water (the sum of the intracellular, interstitial and plasma volumes) is determined by dilution of compounds such as deuterium oxide or tritiated water, which distribute themselves uniformly throughout all phases of the body water. As measured by this technique, the magnitude of the total body water in healthy adults is approximately 50 to 65 per cent of the body weight (Schloerb et al., 1950). This value varies inversely with the body fat content; i.e., the fatter the individual, the smaller the proportion of body weight which is water. However, the water content of the lean body mass, approximately 70 per cent, is relatively constant.

The magnitude of the extracellular fluids is determined by the dilution volume of solutes which distribute themselves freely throughout the plasma and interstitial fluids but are excluded from the intracellular space. When estimated by use of such solutes as inulin, sucrose, sodium or chloride, the extracellular fluid is found to constitute 15 to 20 per cent of the body weight.

Plasma, a subfraction of the extracellular fluid, comprises approximately 5 per cent of the body weight. This value is determined by

58

the dilution volume of solutes such as the dye Evans blue (T-1824).

The volume of the intracellular fluid cannot be determined directly by the dilution technique and is estimated by calculation of the difference between simultaneous measurements of the total body water and extracellular fluid volume. Therefore, the accuracy of this estimation depends upon the accuracy with which these two volumes have been determined. An approximate value for the intracellular fluid is 50 per cent of the body weight.

The extracellular and intracellular fluids differ sharply in their electrolyte and protein composition (Fig. 7-1). The principal ionic solutes of the intracellular fluid are potassium and magnesium as cations and protein and phosphate as anions. In the extracellular fluid, sodium is the major cation and chloride and bicarbonate are the principal anions. The small amount of protein in the extracellular fluids is essentially restricted to the plasma, which contains approximately 7 per cent protein. With regard to its other components, plasma is in equilibrium with interstitial fluid, as shown in Figure 7-1.

The exclusion of sodium and, to a certain extent, the accumulation of potassium in the intracellular fluid is an important property of intact, living cellular membranes. It is important to note that, despite the great differences in concentrations of these electrolytes, the intracellular and extracellular fluids are in os-

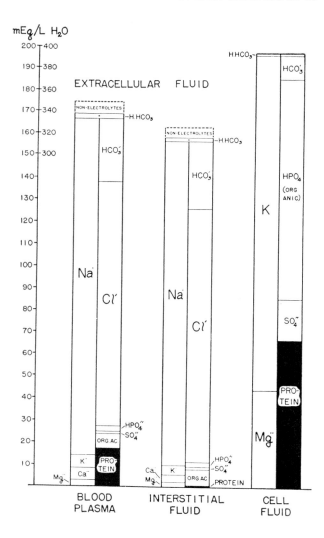

FIGURE 7-1. *The chemical composition of extracellular fluids and cell fluid. Note that the values are given as milliequivalents per liter of water contained in the fluid instead of per liter of plasma. It will be seen that the patterns of blood plasma and interstitial fluid are almost identical; the greatest single item of difference is in the amounts of protein. This makes necessary adjustment of the concentrations of the diffusible ions which will preserve the total cation-anion equivalence (Donnan equilibrium). The non-electrolyte concentration (glucose, urea, etc.) is seen to be very small in comparison with that of the electrolytes although the total quantity carried to the tissue cells and into the urine over a unit of time is several times larger. Note the predominance of potassium and the high protein content of cell fluid. (Adapted from Gamble, J. L.: Chemical Anatomy, Physiology and Pathology of Extracellular Fluid. 5th ed. Cambridge, Mass., Harvard University Press, 1950.)*

motic equilibrium. Since sodium and potassium are the major active solutes, losses or gains in one or the other of these constituents inevitably give rise to alterations in the osmotic pressure of the cells involved. Such changes lead to shifts of water to which the cellular membranes are freely permeable according to the osmotic gradient. Gross alterations in osmotic pressures result in impaired cellular function.

REGULATORY MECHANISMS FOR VOLUME AND COMPOSITION OF BODY FLUIDS

An adequate intake of water and electrolytes is essential for maintenance of the volume and composition of the body fluids within normal limits. Under conditions of normal supply of water and average diet, intake of water, sodium, chloride and potassium by healthy individuals is in excess of physiological needs, and the metabolic and regulatory mechanisms which preserve constancy operate principally through control of their excretion.

WATER

The routes by which water is lost from the body include excretion of urine and extrarenal loss through lungs, skin and stool. An approximate daily water balance of an adult man is shown in Table 7-1. The loss via the lungs and skin (insensible water loss) occurs by evaporation of water from these surfaces and amounts to 800-1200 ml./day. The energy for this process, which in the skin occurs without participation of the sweat glands, is derived from the body heat, and this mechanism accounts for about 25 per cent of the daily heat loss. This process, therefore, represents an obligatory and inevitable loss of body water, the magnitude of which varies with heat production and environmental temperature. Under conditions in which active sweating occurs, the additional water lost contributes considerably to depletion of the body water. The only other extrarenal loss of body water is the small amount contained in formed

TABLE 7-1.   NORMAL ROUTES OF INTAKE
AND OUTPUT OF WATER

Intake		Output	
Preformed water	1200 ml.	Urine	1500 ml.
Water in food	1000 ml.	Stool	100 ml.
Metabolic water	300 ml.	Insensible water	900 ml.
TOTAL	2500 ml.	TOTAL	2500 ml.

stools. In health this is a minor volume, but in disease states involving severe diarrhea, the increase in water loss can be sufficient to result in serious dehydration.

The daily urinary output of water strikes a balance between the extrarenal water losses and the intake. Urine volume, therefore, depends upon the water and solute loads remaining after extrarenal losses have occurred. The magnitude of this output is determined by the operation of receptors in the hypothalamus and posterior lobe of the pituitary gland (Dingman, 1958). Areas of the hypothalamus are sensitive to changes in osmotic pressure, and small increases in plasma osmotic pressure stimulate these receptors, which have neural connections with the posterior pituitary. As a result of this stimulation, antidiuretic hormone (ADH) is released into the general circulation and ultimately reaches the renal tubules, where it increases the reabsorption of water and thus decreases the urine volume.

The renal tubules themselves possess the inherent capacity to reabsorb approximately 80 to 90 per cent of the glomerular filtrate, which has an approximate volume of 180 L./day (Smith, 1956). This process is not under the control of ADH, which acts only upon the remaining 10 to 20 per cent of the glomerular filtrate, reducing its volume to only 1 per cent or less by enhancing water reabsorption in the collecting ducts.

By means of this mechanism, the final urine volume varies according to the state of the body fluids. Thus, deprivation of water leads to increased plasma osmotic pressure which, in turn, causes ADH release and a decrease in urine volume. On the other hand, ingestion of fluids in excess of the requirement decreases plasma osmotic pressure, suppresses ADH release and increases urine volume to re-establish fluid equilibrium.

SODIUM

Since the body is not able to produce sodium, this electrolyte is derived solely from intake. The amount of sodium (principally as NaCl) in average diets differs, and the intake usually varies from 5 to 15 gm. per day. This is far in excess of the normal requirement, and equilibrium concentration in the extracellular fluids is maintained through urinary excretion, dependent upon interactions between the adrenal cortex and the kidney.

The renal tubules possess an inherent capacity to reabsorb the bulk of sodium contained in the glomerular filtrate. As indicated previously, the usual intake of sodium is greater than the requirement, and reabsorption is sufficient to just maintain the sodium concentration (135-150 mEq./L.) in extracellular fluid; the excess is excreted in urine. However, in situations such as inadequate intake or excessive sodium loss (e.g., in profuse sweating) when conservation of the ion is required, a supplementary mechanism is called into play. This involves the secretion of aldosterone, a steroid hormone produced by the adrenal cortex, which acts directly on the renal tubules by greatly augmenting sodium reabsorption (Gaunt et al., 1955). The efficiency of this conservation system is illustrated by the fact that a patient on a sodium-free diet rapidly adapts by producing, within a few days, a virtually sodium-free urine. If extrarenal sodium losses did not occur, such a patient could continue almost indefinitely recirculating his own sodium stores without need for replenishment. Thus, unless there is adrenal cortical dysfunction, sodium depletion can be accomplished only by feeding a sodium-free diet for prolonged periods.

POTASSIUM

The potassium content of the body is also derived solely from the dietary intake, which in average diets amounts to some 4 to 8 gm. per day. Absorbed potassium enters the extracellular fluid, from which it then becomes incorporated into the intracellular compartments, where its concentration is some 15 to 20 times higher than in the extracellular fluids (Fig. 7-1). The low concentration remaining in the extracellular fluid (3.5-5.0 mEq./L.) is maintained by excretion via the urine, feces or sweat.

The extracellular levels of potassium, therefore, represent a net balance between the amount ingested, the amount incorporated into cells and the portion excreted. The relative constancy of the extracellular concentration is maintained through regulatory mechanisms which usually prevent excessive accumulation of this ion. Accumulation results in altered myocardial, neuromuscular and neurologic functions.

Again, as with water and sodium, the kidneys and hormonal agents have the principal roles in effecting excretion of this electrolyte. Potassium is actively secreted into urine by the renal tubules, this process being under the control of aldosterone. In contrast to its effect on sodium reabsorption, aldosterone stimulates the secretion of potassium. When potassium intake is reduced to zero, the renal conservation mechanism is less efficient than for sodium, and a deficiency syndrome can be created by a prolonged potassium-free diet. In the rat, potassium deficiency results in the production of renal tubular damage (Oliver et al., 1957).

The efficiency of the mechanisms which regulate the volume, distribution and composition of the body fluids is illustrated by the relative constancy of the body weight, body water, sodium, potassium and other electrolytes despite large fluctuations in the intake of water and electrolytes. However, the equilibrium is markedly altered in a variety of disease states which can result in either deficits or excesses of the body fluids and electrolytes.

DISEASE STATES LEADING TO DEFICITS OF WATER AND ELECTROLYTES

A variety of etiologically unrelated disease states result in abnormal composition or distribution of the body fluids (Elkington and

Danowski, 1955). Those leading to deficits in water or electrolytes include starvation, dehydration, vomiting, diarrhea, sweating and renal dysfunction.

Loss of appetite is a common accompaniment of many diseases. Interruption of all intake except water under these circumstances brings into play the renal mechanisms which help to conserve some body constituents. Within a few days renal excretion of sodium decreases to nearly zero, as described earlier. However, potassium excretion continues to be quantitatively important, and significant deficits of this ion can result from prolonged starvation.

Dehydration in its least complicated form can occur when the daily water intake fails to meet body requirements. This can occur, for example, in patients with obstructive lesions of the upper gastrointestinal tract, in mentally ill patients who refuse fluids or in unconscious or neglected patients. Deficits of body water under these conditions are attributable principally to the inability of the body to reduce the insensible water loss (about 1000 ml./day) in addition to the continued excretion of the obligatory urine volume (approximately 300 ml./day).

Since these losses occur through surfaces adjacent to the extracellular fluid, the volume of this fluid is the first to decline. Because, in this situation, water is lost without concomitant losses in electrolytes, the osmotic pressure in the extracellular fluid rises. This results in a flow of water from the intracellular compartments into the extracellular spaces until the gradient is dissipated. Thus, even though dehydration initially occurs in the extracellular fluid, it eventually affects the whole body.

The sequelae of profuse vomiting include all of the changes occurring in starvation and dehydration. Vomiting also results in deficits of sodium, potassium and chloride because of the significant amounts of these ions in gastric and intestinal secretions. Similar losses are incurred in severe and prolonged diarrhea, such as that associated with diseases such as cholera, dysentery or ulcerative colitis.

Sweat contains variable, though usually large, amounts of sodium and chloride. Hence, any illness characterized by fever and profuse sweating can result in deficits if replacement therapy is not employed.

The role of the kidneys in the development of water and electrolyte deficits may be correlated with functional alterations. In chronic renal disease, the kidneys may fail to conserve water to the extent that the urine becomes the chief route of water loss, exceeding the insensible losses. On the other hand, renal sodium and chloride wastage are a constant feature of adrenal cortical failure (Addison's disease). In this disease, excessive salt loss occurs because of the absence of aldosterone secretion.

Deficits of sodium chloride and potassium may also be created by prolonged use of certain diuretics, particularly organic mercurials and carbonic anhydrase inhibitors, which function as diuretics by virtue of their interference with renal tubular reabsorption of chloride and sodium.

Thus, a variety of disease states, though otherwise unrelated, may cause profound changes in the distribution, volume and composition of body fluids by creating deficits of water or elctrolytes, or both, the consequences of which manifest themselves in a variety of functional changes.

The most common symptom of water-deprivation dehydration is that of mental confusion, which has been noted in the classical descriptions of persons lost in the desert and of castaway sailors with limited water supplies. Death from dehydration is caused by respiratory failure, presumably a consequence of central nervous system damage.

Depletion of sodium and chloride results in a decrease in osmotic pressure of the extracellular fluids with consequent flow of water into the intracellular compartment. The sequelae of this shift are decreased plasma volume, hemoconcentration and increased blood viscosity. These changes, in turn, lead to decreased cardiac output, decreased venous return and hypotension. Hence, circulatory collapse is the inevitable result of excessive sodium depletion.

The chief feature of potassium deficiency is muscular weakness. This symptom is poorly characterized and is associated with microscopic necrosis of skeletal and cardiac muscle, which leads to impaired cellular integrity and function. The latter may well be the most deleterious effect of potassium deficiency. The renal lesions noted in rats fed potassium-free diets (Oliver et al., 1957) have not been observed in man.

EXCESSES OF WATER AND ELECTROLYTES AND THEIR EFFECTS

It has already been pointed out that regulatory mechanisms maintain water, sodium, potassium and chloride content of the body within rather narrow limits. It follows, then, that excesses of these constituents within the body can occur only when the intake exceeds the regulatory capacity or when the functional capacity of the regulatory mechanisms is compromised.

WATER

When renal and adrenocortical functions are normal, it is impossible to ingest enough water by mouth to produce a net positive balance. However, in some disease states such as acute tubular necrosis, when oliguria or anuria is present, a positive water balance can be established even with normal intake. "Water intoxication" results when the excess is great enough to cause significant dilution of the intracellular fluid.

Water excesses develop frequently in conjunction with decreased or increased levels of sodium and chloride in body fluids. In these cases, if the patient is already deficient in sodium and chloride, the administration of water aggravates the clinical state by superimposing water intoxication on the sodium chloride deficiency. Replacement fluids should, therefore, contain sufficient electrolytes to correct the primary electrolyte deficiency in addition to re-establishing the volume equilibrium (Brooke and Anast, 1962).

SODIUM

Disease processes or factors which involve the kidney, adrenal cortex or higher centers may interfere with regulatory mechanisms and produce retention of ingested or administered salt. In all these disease entities (such as renal disease, congestive heart failure and cirrhosis), sodium retention results from intake in excess of excretion. The positive net balance is frequently accompanied by retention of water in order to attain osmotic equilibrium. Since sodium and chloride are extracellular ions, the end result is expansion of the extracellular fluid volume and the development of edema. As edema progresses, it results in increased body weight, distention of viscera, subcutaneous swelling and fluid accumulation in body cavities.

POTASSIUM

As the principal mechanisms for regulation of body potassium levels are similar to those governing the distribution and concentration of sodium, many of the etiologic factors in potassium excess are similar to those for sodium. Thus, adrenal cortical failure, renal disease and other pathologic states in which excretion cannot keep up with intake result in potassium retention. The consequences of potassium excess are reflected in muscular weakness and cardiac dysfunction. An increase in serum potassium concentration to 6.5 mEq./L. (normal 4.5 mEq./L.) results in alterations in the electrocardiogram, and levels greater than 14 mEq./L. can cause death from cardiac arrest (Winkler et al., 1938).

THE ROLE OF THE DIET IN WATER-ELECTROLYTE BALANCE

The point has already been made that the average dietary intake provides sodium, chloride and potassium in excess of the physiologic requirements. The regulatory mechanisms discussed previously maintain equilibrium principally by altering rates of excretion of these body constituents to correspond with the intake of each. This situation tends to obscure the fact that there are real and demonstrable nutritional requirements for water, sodium and potassium.

The water requirement for adults is dependent upon body heat production, renal solute load and extrarenal loss. The magnitude of the requirement is in the order of 2200 to 3000 ml. of preformed water per day, under conditions of minimal extrarenal loss.

TABLE 7-2. SODIUM CONTENT OF
SELECTED FOODS*

Food	Description	mg./100 gm. Edible Portion
	Rich Sources	
Bread	All types	265–674
Sausages	Cold cuts and luncheon meats	740–1234
Fish	Fresh, unsalted	54–177
Cheese	Cream	250
Cheese	Cheddar	700
Butter	Salted	987
Margarine	Salted	987
Prunes	Dehydrated	329–940
Eggs	Whole, fresh	122
Eggs	Whole, dried	427
Milk	Whole, dried	405
	Moderate Sources	
Vegetables	Carrots, beets, corn, peas, spinach	
	Raw	2–71
	Canned	2–46
Ham	Cooked	65
Bacon	Cooked	65
Oatmeal	Rolled or ground	1–218
Meats	Beef, pork, lamb (cooked)	60–70
Milk	Whole, pasteurized	50
Ice cream	All types	33–63
Breakfast cereals	Corn, oats, rice, wheat (dry)	2–1267
Chocolate	Candy, sweet	33
Coffee	Dry, powder	72
	Poor Sources	
Butter	Unsalted	10
Fruit	Apples, pears, bananas	
	Fresh	1
	Dried	1–5
Nuts	Roasted, unsalted	1–5
Fruit juices	Orange, pineapple, grapefruit (fresh)	1–5
Marmalades, jams and preserves	All types	14

* Data from U. S. Department of Agriculture, Agriculture Handbook Number 8.

This quantity must obviously be increased in conditions of extreme heat, sweating, fever, etc. The recommended allowance for sodium is one additional gram of salt (NaCl) for each liter of water in excess of 4 liters daily. The requirement for potassium has not been precisely determined but has been estimated to be on the order of 1 gm. of potassium daily. It should be noted that a 1500 calorie diet from animal and vegetable sources contains approximately 2.5 gm. of potassium. (See Tables 7-2 and 7-3.) Therefore, an extremely low calorie diet would be necessary to provide less than the estimated requirement.

Although primary water-electrolyte deficiencies are uncommon, excesses or deficits of these dietary components may appear as secondary features of disease processes which impair the mechanisms regulating their distribution and excretion. In the treatment of such disorders it is an important therapeutic adjunct to adjust the dietary intake.

The dietary level of sodium chloride has been associated with the development of essential hypertension, and the use of low sodium diets in the treatment of this syndrome

TABLE 7-3. POTASSIUM CONTENT OF
SELECTED FOODS*

Food	Description	mg./100 gm. Edible Portion
Bacon	Cooked	390
Bread	All types	67–454
Breakfast cereals	Corn, oats, rice, wheat (dry)	99–947
Butter	Salted	23
Butter	Unsalted	10
Cheese	Cream	74
Cheese	Cheddar	82
Chocolate	Candy, sweet	269
Coffee	Dry, powder	3256
Eggs	Whole, fresh	129
Eggs	Whole, dried	463
Fish	Fresh, unsalted	160–525
Fruit	Apples, pears, bananas	
	Fresh	110–370
	Dried	144–1477
Fruit juices	Orange, pineapple, grapefruit (fresh)	162–940
Ham	Cooked	390
Ice cream	All types	95–181
Margarine	Salted	23
Marmalades, jams and preserves	All types	33–88
Meats	Beef, pork, lamb (cooked)	370–390
Milk	Whole, pasteurized	144
Milk	Whole, dried	1330
Nuts	Roasted, unsalted	464–773
Oatmeal	Rolled or ground	61–352
Prunes	Dehydrated	760–2170
Sausages	Cold cuts and luncheon meats	140–269
Vegetables	Carrots, beets, corn, peas, spinach	
	Raw	280–470
	Canned	96–250

* Data from U. S. Department of Agriculture, Agriculture Handbook Number 8.

is well documented (Dahl and Love, 1957). Another classic example of this type of dietary adjustment is the use of low salt diets for patients with congestive heart failure, a syndrome associated with diminished ability to excrete sodium and frequently accompanied by edema formation on unrestricted salt diets. Elkington and Danowski (1955) have formulated a useful series of such diets, which provide from 30 to 800 mg. of sodium per day. It should be pointed out that continuous and prolonged use of such diets can result in sodium depletion, since the renal conservation mechanism is not completely efficient. Production of the deficient state is also accelerated by significant extrarenal loss (e.g., through sweat) or by the simultaneous use of diuretics such as the mercurials and carbonic anhydrase inhibitors, which increase the urinary loss of both sodium and potassium. Intake of these electrolytes must, therefore, be carefully regulated to avoid sodium depletion during this type of therapeutic program.

ORAL RELEVANCE

There is some preliminary evidence that the level of sodium in the diet of experimental animals may be related directly or indirectly to dental caries incidence. However, the evidence is conflicting. For example, Dodds and Lawe (1964) have found that rats fed a cariogenic diet in which no sodium in any form was added (the constituents of the control diet provided only 0.009 per cent Na) developed 40 per cent higher caries scores than those fed the same basal diet with 0.4 per cent sodium added in the form of NaCl. The mechanism for this cariogenic action was postulated as follows: Since protein deficiency, particularly of lysine in rats, might promote caries and since sodium deficiency might cause metabolic impairment with respect to protein synthesis, it is conceivable that sodium deficiency may have been responsible in some measure for lowering the resistance of the teeth to dental caries. Unfortunately, however, one must be careful in interpreting and making generalizations based on the results of feeding experiments that use distorted, imbalanced and inadequate diets such as have been used in creating lysine deficiency in animals.

An opposite effect with regard to dietary sodium was found by Muhlemann (1964) when he added sodium chloride to a cariogenic diet like Stephan diet 580. These results, plus the finding that the sodium or potassium cation of a phosphate compound is an important consideration in determining the relative effectiveness of these compounds for caries inhibition, suggest that this area is an important one for more extensive research. In fact, there has been some speculation that differences in cariogenicity of two diets comparable in sucrose content might be the result of differences in Na:K ratios (Wynn et al., 1959). This theory needs to be supported by additional experimental evidence.

In a study of 502 young adult men, Shannon and his associates (1963) found that the dental caries experience of the group was not correlated with the sodium bicarbonate content of parotid saliva or with the rate of flow from this gland.

REFERENCES

General

Brooke, C. E., and Anast, C. S.: Oral fluids and electrolytes. J.A.M.A., *179*:792, 1962.

Dahl, L. K., and Love, R. A.: Etiological role of sodium chloride intake in essential hypertension in humans. J.A.M.A., *164*:397, 1957.

Dingman, J. F.: Hypothalamus and the endocrine control of sodium and water metabolism in man. Amer. J. Med. Sci., *235*:79, 1958.

Elkington, J. R., and Danowski, T. S.: The Body Fluids; Basic Physiology and Practical Therapeutics. Baltimore, The Williams & Wilkins Co., 1955.

Gamble, J. L.: Chemical Anatomy, Physiology and Pathology of Extracellular Fluid. Cambridge, Harvard University Press, 1958.

Gaunt, R., Renzi, A. A., and Chart, J. J.: Aldosterone: A review. J. Clin. Endocrinol. Metab., *15*:621, 1955.

Oliver, J., MacDowell, M., Welt, L. G., Holliday, M. A., Hollander, W., Winters, R. W., Williams, T. F., and Segar, W. E.: The renal lesions of electrolyte imbalance. I. The structural alterations in potassium depleted rats. J. Exp. Med., *106*:563, 1957.

Schloerb, P. R., Friis-Hansen, B. J., Edelman, I. S., Solomon, A. K., and Moore, F. D.: The measurement of total body water in the human subject by deuterium oxide dilution. J. Clin. Invest., *29*:1296, 1950.

Smith, H. W.: Principals of Renal Physiology. New York, Oxford University Press, 1956.

Winkler, A. W., Hoff, H. E., and Smith, P. K.: Electrocardiographic changes and concentration of potassium in serum following intravenous injection of potassium chloride. Amer. J. Physiol., *124*:478, 1938.

Oral Relevance

Dodds, M. L., and Lawe, R.: Mineralization of a cereal diet as it affects cariogenicity. J. Nutr., *84*: 272, 1964.

Muhlemann, M. M.: In an informal discussion at Conference on Phosphates and Dental Caries. J. Dent. Res., *43*:1163, 1964.

Shannon, I. L., Isbell, G. M., and Chauncey, H. H.: Parotid fluid and serum sodium and potassium as related to dental caries experience. J. Dent. Res., *42*:180, 1963.

Wynn, W., Haldi, J., Bentley, K. D., and Law, M. W.: Further studies on the difference in cariogenicity of two diets comparable in sucrose. J. Nutr., *67*:569, 1959.

Chapter Eight

Macro Minerals:
Calcium, Phosphorus,
Magnesium, and Calcification

ROBERT S. HARRIS, PH.D.

Inorganic elements are essential for life processes. Though essentially all elements on earth have been identified in plant and animal tissues, only a few are known to be necessary for the growth and development of animals. The human being requires calcium, phosphorus, magnesium, sodium, potassium, sulfur, chlorine, iron, copper, cobalt, iodine, manganese, zinc and fluorine in his daily diet. Molybdenum and selenium are essential for experimental animals and may yet be proved requisite for man.

The nutritional functions of minerals are usually complex and interrelated with one another as well as with vitamins, amino acids, fatty acids and carbohydrates. Calcium and phosphorus are principally involved in the formation of bones and teeth, yet each has many functions of a very different nature. Iron, copper and cobalt (as in vitamin B_{12}) are interrelated in the synthesis of hemoglobin and the formation of red cells, yet each has other activities. Sodium, potassium, chlorine, phosphorus and calcium are involved in the maintenance of body fluids; magnesium and calcium are concerned with nerve cell functions; iodine is a constituent of thyroxine, a thyroid hormone; and manganese, molybdenum, zinc and other minerals serve as activators of a number of enzymes that are concerned with a variety of metabolic reactions.

Of the thirteen or more minerals essential to man, and which are normally obtained from the diet, only three (calcium, iron, iodine) tend to be seriously deficient in the diets of specific population groups.

CALCIUM

Essentially 99 per cent of body calcium is present in bones and teeth; the remaining 1 per cent is to be found in the soft tissues and body fluids. In common with most other tissues, bone is in dynamic equilibrium with the components in blood plasma (Neuman and Riley, 1947; Thomas et al., 1952). The rate of calcium exchange is much greater than the rate of deposition of new bone. The high mineral content of bones and teeth is responsible for their strength and rigidity. Besides serving as a supporting structure, bones have a mineral reserve which is called upon in times of stress, such as lactation. The readily mobilized calcium is located in the bone trabeculae. The enamel and dentin of teeth are metabolically much less reactive than bone; as a result, little if any calcium is lost from molar teeth when diets are low in calcium content (Lund and Armstrong, 1942) or when the phosphorus turnover is low (Hevesy, 1940).

FOOD SOURCES
OF CALCIUM

In terms of mg. per 100 gm., the major food sources of calcium in the United States are:

cheeses, 201 to 1140; Irish moss, 885; molasses, 165 to 684, kale, 276; soybean flour, 199 to 265; turnip greens, 246; almonds, 235; parsley, 203; collards, 203; Brazil nuts, 186; cowpeas, 146 to 172; chick peas, 150; ice cream, 78 to 146; dandelion greens, 140; whole milk, 118; beans, 118; enriched white bread, 112 (Watt and Merrill, 1963). Thus, green vegetables, milk and milk products, nuts and legumes represent the richer sources of calcium. As mentioned elsewhere in this chapter, the availability of the calcium in some green vegetables is reduced by interfering substances, sometimes to such an extent that the calcium is not available at all (Fincke and Garrison, 1938). Breads that are fortified with milk powder or with self-rising flour (containing monocalcium phosphate) may contain as much as 265 mg. of calcium per 100 gm.

Bronner et al. (1955) fed 250 mg. quantities of calcium (labeled with 1 μc. Ca^{45}), combined in four different salts, to a group of 20 children. Feedings were made according to a Latin square design. Calcium absorption was studied by estimating total calcium and Ca^{45} in serum after $3\frac{1}{2}$ hours, and in urine and feces over a period of 5 days. Though the data resulting from the feeding of milk calcium and calcium citrate tended to show higher levels, no significant difference was found in the total absorption of calcium whether given as gluconate, lactate, citrate, carbonate or milk. Thus, it appears that any edible form of calcium (even relatively insoluble $CaCO_3$) may be used to supplement the calcium content of human foods and food products in order to raise the levels of calcium intake to adequate levels.

Dibasic calcium phosphate is often used as a supplement to diets in which milk and milk products are restricted, during pregnancy or in alleviating osteoporosis. When the physiological action of the calcium ion is desired, as in patients with tetany and inadvertent parathyroidectomy, calcium may be given orally or by injection. Calcium gluconate or calcium levulinate is less irritating than calcium lactate or calcium chloride and can be given in large doses by mouth, intravenously or intramuscularly. These salts may be used also to supply calcium during rapid growth and development. Calcium phosphates are not used to treat tetany resulting from parathyroidectomy and glomerular insufficiency because an increase in phosphorus intake seems undesirable. Calcium glycerophosphate was once considered to have therapeutic value on the ground that it is a component of body phosphatides; there is no proof that this is so (Anon., 1961).

Physiology

FUNCTIONS OF CALCIUM

The level of calcium in normal blood serum is closely regulated between 9 and 11 mg. and is generally about 10 mg. per 100 ml.; approximately 60 per cent is present in an ionized and soluble form and 40 per cent is bound to proteins. The calcium proteinate performs as a weak electrolyte, and its dissociation is governed by the mass law (Drinker et al., 1939):

$$\frac{(Ca)^{++} \times (protein)^{--}}{Ca\ proteinate} = K$$

The level of serum calcium is held nearly constant despite large variations in calcium intake and large reserves in the skeleton. Red cells contain but trace amounts of calcium.

Many of the actions of the calcium ion are more pharmacological than physiological and are often not described in textbooks of nutrition. Calcium controls the permeability of all cell membranes, in competition with both sodium and potassium. Lecithin appears to be the substance in cell membranes which binds calcium (Kimizuku and Kotetsu, 1962).

Muscle and nerve irritability is regulated by serum calcium ions through complex pathways in which the amounts and ratios of calcium, magnesium, sodium and potassium ions, as well as pH, are involved. The irritability of nervous tissue increases as the blood calcium level decreases. Low blood calcium levels cause tetany with convulsions; high levels of blood calcium depress nerve irritability. An optimum range of blood calcium is essential for maintenance of a normal pulse rate and for cardiac contraction.

Studies in hypercalcemia have helped to establish the role of calcium in kidney function. Polyuria and low osmotic pressure of the urine are related to a failure to reabsorb water and sodium. In the clotting of blood, the formation of fibrin from fibrinogen requires thrombin, and calcium is an essential part of the prothrombin complex. Calcium is an activator of such enzymes as lipases, alkaline phosphatase, choline esterase, the adenosine triphosphatase of myosin and succinic dehydrogenase (Dixon and Webb, 1958). The permeability of membranes (Reid, 1943), and probably of cells, is influenced by calcium since the integrity of the intercellular cement substance appears to depend on the presence of calcium. Recent evidence indicates that calcium is essential for the absorption of vitamin B_{12} from the intestinal tract and its absorption in the cell membrane. It is clearly evident that the levels of ionized calcium in the body fluids must be maintained under fine control.

CALCIFICATION OF BONES AND TEETH

The mass of bone mineral is in the form of hydroxyapatite $(Ca_{10}(PO_4)_6(OH)_2)$, the only Ca-P salt that is stable under physiological conditions. Bone mineral is not pure hydroxyapatite, for it contains trace amounts of carbonate, citrate, iron, magnesium, sodium and other elements. Bone is laid down initially as minute crystals which, in man, have a combined surface area of 80 to 100 square meters. These surfaces are in equilibrium with the surrounding extracellular fluid.

Neuman and Neuman (1958) have explained this equilibrium, which is instantly changeable but stable in its effect, as follows:

Because the crystals are so tiny and Ca is non-polarizable, strong electric fields are projected away from the crystal surface. These fields give rise to a bound-ion layer, which, in turn, acquires an insulating layer of water—the hydration shell. Exchange between ions in solution and those in the hydration shell is extremely rapid. . . . Exchange within the crystal is quite slow and is dependent on . . . defects and vacant lattice positions.

While admitting that any ion present in solution will penetrate the hydration shell, they suggest that a certain specificity is required of ions which penetrate the surface, and that only a few of them can enter the interior of the crystal.

Neuman and Neuman (1960) suggested that the initial step in calcification is precipitation of calcium phosphate from the supersaturated serum (pH 7.4), that the main function of bone alkaline phosphatase is to split the phosphate esters and thus increase the Ca × P product to a precipitation level and prevent the esters from inhibitory calcification. It should be noted that there are other inhibitors of precipitation, for Fleisch and Bisaz (1962) have isolated inorganic pyrophosphate which permits urine to become supersaturated with calcium and phosphorus and is hydrolyzed by pyrophosphatase in bone. Hard tissue collagen acts as a "seeding template" for the crystals; possibly the free -lysyl and -hydroxylysyl amino acid group in the collagen molecule (Solomons and Irving, 1958) and a fat-soluble inductor substance (Urist and McLean, 1951; Urist, 1965) activate this reaction. The initial precipitate appears to be octacalcium phosphate (MacGregor and Brown, 1965).

As bone and tooth mature, the apatite crystals grow and solubilize and recrystallize, with gradual replacement of water until the space between crystals will no longer permit the passage of ions. Mature cortical bone is more dense than cancellous bone. As the bone growth is completed, haversian remodeling takes over and preserves the bone against complete calcification in order that it may continue to function as a reservoir and buffer. The parathyroids act to control the eroding action of the bone cells.

Though the hydroxyapatite structure of dentin and enamel is the same as that of bone, the positioning and spacing of the crystals differ (Neuman and Neuman, 1958). The density of dentin is higher because the crystals are deposited within the fiber bundles and the organic fraction is smaller. The crystals in enamel are larger than in bone and contain only half as much water.

The parathyroid glands are important to calcification, primarily because they control the level of ionized serum calcium. When the level falls, bone salt is mobilized by the parathyroid hormone and the level returns to normal. When parathyroid action stops, mobilization stops. This hormone action is necessary because hydroxyapatite is not soluble at the acidity (pH 7.4) of the extravascular fluid. It acts by the nonoxidative hydrolysis of carbohydrates in the bone cells with the production of lactic acid. Vitamin D assists in the conversion of carbohydrate to citrate, which increases the solubility of the bone salt. Johnston et al. (1962) have demonstrated the erosion of bone in vitro by transplanting grafts of parathyroid gland tissue.

Parathyroid also acts on kidney tubules to block the reabsorption of phosphates and increase reabsorption of calcium (Lavender et al., 1959) and assists in the active transport of calcium in the gut (Dowdle et al., 1960).

It is evident that calcium and phosphorus are required for the growth and maintenance of bones and teeth, but apparently all the nutrients required for animal growth are required for the growth of these hard tissues: ascorbic acid (Wolbach and Bessey, 1942), choline (Wolbach and Hegsted, 1953), mag-

nesium (Duckworth, 1939), manganese (Underwood, 1953), vitamin A (Wolbach, 1947), as well as vitamin D, protein and others.

CALCIUM ABSORPTION

The principal site of calcium absorption is the anterior portion of the small intestine. The absorbed calcium is rapidly transferred by the blood circulation throughout the body (Fig. 8-1). It is withdrawn into the bones and teeth, especially during periods of rapid growth, or into the bone trabeculae to replenish body stores of calcium. Marcus and Lengemann (1962) observed in rats that 88 per cent of Ca^{45} in a solid food diet was absorbed in the ileum, 8 per cent in the duodenum and 4 per cent in the jejunum. When the Ca^{45} was

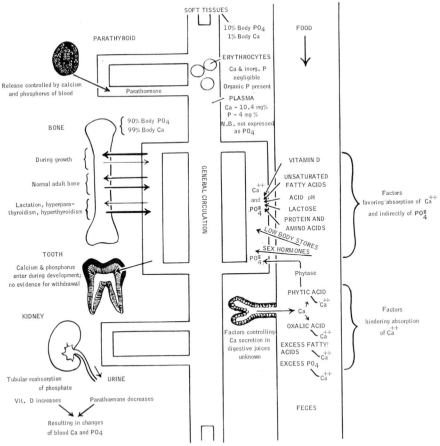

FIGURE 8-1. *Diagram summarizing the main factors in calcium and phosphorus metabolism. (From Jenkins, G. N.: The Physiology of the Mouth. 2nd. ed. Charles C Thomas, 1960.)*

given in liquid diet, the absorptions were ileum 62, jejunum 23 and duodenum 15 per cent. Thus, the liquid diet was being absorbed lower in the intestinal tract, presumably because of rapid passage through the gut.

Factors Affecting Calcium Absorption. The absorption of calcium from the intestine takes place against a concentration gradient. This was demonstrated by the use of inverted intestine sacs (Wilson and Wiseman, 1954; Schacter and Rosen, 1959). Present evidence indicates that this energy-requiring process involves vitamin D.

Human beings absorb calcium inefficiently. A growing rat may absorb nearly all ingested calcium, but in humans the absorption is usually only 20 to 30 per cent. Only a few of the many factors which influence the absorption of calcium will be discussed here.

Calcium Needs. The body absorbs calcium more efficiently if the stores in the bones are depleted. Children absorb calcium more efficiently than adults, but children on low calcium intakes are more efficient in the utilization of dietary calcium than those on adequate intakes (Nicholls and Nimalasuriya, 1939). Boys on a standard diet absorbed three times as much calcium as boys of similar age who had received high calcium diets for several months previous to testing (Macy, 1942).

Calcium Intake and Food Mass. The efficiency of absorption of calcium (as Ca^{45}) by fasted children was reduced from nearly 100 to about 40 per cent when the intake was increased from 1 mg. to 200 mg. (Bronner et al., 1956). The same research group had earlier reported that the absorption of minerals by children was reduced when the amount of food in the test meal was increased. For example, five times more iron was absorbed when iron was fed in a glass of water than when the same amount was fed in a breakfast meal. Thus, the absorption of calcium by human beings is affected by the quantity of calcium taken in a meal as well as by the mass of the meal.

Gastric Juice. Mineral elements are more soluble in acid solutions. The acids in gastric juice may hasten the solution of dietary calcium and thus increase the efficiency of calcium absorption. There is no proof that the absorption of calcium is impaired in patients with achlorhydria, however.

Carbohydrates. Sugar may alter the efficiency of calcium absorption. Studies with several species of animals indicate that lactose promotes the absorption of calcium and the calcification of bones. The pH of the intestines of rats receiving a high lactose diet was lower than that of the controls. In an experiment on boys, Duncan (1955) reported an improved retention of calcium with high lactose feeding and demonstrated that this was the result of improved retention rather than better absorption.

Fat. Moderate levels of fat favor the utilization of calcium (Jones, 1940). However, large amounts of fat in the diet (French, 1942), especially fats with higher melting points (Westerlund, 1934), interfere with calcium absorption. Only two studies of the effect of fats on calcium absorption have been conducted on human subjects, and both have been negative (Mallon et al., 1930; Steggerda and Mitchell, 1951).

Sprue, steatorrhea, biliary fistula and other disorders in which fat is poorly absorbed also interfere with calcium absorption (Snapper, 1950). This effect is due, at least in part, to the reaction to long-chain saturated fatty acids to form insoluble calcium soaps (Nicolaysen et al., 1953). In normal subjects the effect of fats on calcium absorption will be of minor importance if the intake of hard fats is not excessive.

Proteins and Amino Acids. Food proteins may improve the absorption of calcium as a result of the solubilizing actions of some of their constituent amino acids. Wasserman et al. (1956) noted that the absorption of calcium was improved in rats and chickens when amino acids, especially lysine and arginine, were added to their diets. The mechanism of this action is not known; it is not the result of increases in the solubility of calcium.

Oxalates. Foods from vegetable sources contain oxalates in varying amounts, depending primarily on the variety. Rhubarb, spinach, tomatoes and related plants are rich oxalates. The free oxalate in these foods may react with calcium during passage through the gut to form calcium oxalate, which is relatively insoluble, and thus reduce the absorption of the calcium.

Phytates. Between 60 and 90 per cent of the phosphorus in cereals, and lesser amounts of the phosphorus in foods from other vegetable sources, is present as phytates, hexaphosphate esters of inositol. Significant amounts of these phytates are converted to orthophosphates by the action of native phytases during food preparation, and by intestinal phosphatases during digestion. However, the phytates in certain foods (i.e., oatmeal) resist breakdown and react with food calcium as it passes through the gut. Bronner et al. (1954) demonstrated that about 50 per cent of the phytate in cooked oatmeal interfered with the absorption of calcium when it was fed in the breakfast of children. McCance and Widdowson (1942) demonstrated that calcium absorption by human subjects was significantly less when they were fed dark bread made with 92 per cent extraction wheat flour than when fed white bread made with 70 per cent extraction flour; the calcium absorption was significantly less when men were fed white bread fortified with sodium phytate than when fed unfortified white bread. Harris (1955) evaluated the significance of phytates in human nutrition and concluded that phytates are a potential nutritional hazard only in those areas of the world where people consume large amounts of whole grain cereals and small amounts of calcium. They do not cause a problem in the United States where the people customarily consume refined cereals and generous amounts of calcium-rich foods. As pointed out elsewhere, phytates have been shown to be effective in the reduction of caries development when fed in the diet of rodents.

Calcium/Phosphorus Ratio. The absorption of calcium is significantly impaired when excess phosphorus is present in a diet low in calcium content. Similarly, the absorption of phosphorus is impaired when an excess of calcium is present in a diet already low in phosphorus content. A normal diet for a rat contains about 0.5 per cent calcium. A diet containing 1.0 per cent calcium and 0.25 per cent phosphorus (Ca/P ratio = 4/1), or a diet containing 0.25 per cent calcium and 1.00 per cent phosphorus (Ca/P ratio = 1/4), will produce rickets in rats. On the other hand, rickets will not develop if rats are fed a diet containing 2.0 per cent calcium and 0.5 per cent phosphorus (Ca/P ratio = 4/1) or a diet containing 0.5 per cent calcium and 2.0 per cent phosphorus (Ca/P ratio = 1/4). In addition to a disturbed Ca/P ratio, it is necessary that the limiting element be present in deficient amounts. In fact, Harris (1959) reported normal bone ash and absence of rickets in rats fed a diet with a Ca/P ratio of 20/1; this result was obtained because the limiting element, P, was present in the diet in adequate (0.5 per cent) amounts. Even when the diet is deficient in calcium or phosphorus, rickets will not develop in rats if vitamin D is present in the diet. Therefore, the Ca/P ratio is of nutritional importance only if (1) the diet is deficient in one of these elements, (2) if the other element is present in excess and (3) if the diet is deficient in vitamin D content. These three conditions seldom occur together in human dietaries because diets are almost never deficient in phosphorus and are seldom sufficiently deficient in both calcium and vitamin D to cause metabolic disturbance. Even when human subjects were fed an abnormal experimental diet deficient in calcium and with a high phosphorus content (Ca, 202 mg., P, 972 mg./day; Ca/P ratio = 1/4.3), only minor changes in calcium balance were noted, and the simultaneous addition of calcium and phosphorus did not measurably affect the intestinal absorption of either ion (Spencer et al., 1965). After an extensive discussion of the subject, Harris and Nizel (1964) concluded that the amounts and ratios of calcium and phosphorus in normal diets are rarely sufficiently disturbed to cause nutritional problems.

Vitamin D. The mechanism of action of vitamin D is not yet clear. Studies with animals have revealed that one of its actions is to increase the absorption of calcium, especially when the dietary intake of this element is low. Apparently, vitamin D does not have a direct effect on phosphorus absorption (Nicolaysen et al., 1953). Very small amounts of vitamin D are required to obtain a maximum effect on calcium absorption, and it continues to effect the gut mucosa over several weeks (Mellanby, 1949).

Citric Acid. The beneficial action of citric acid and citrates in the treatment of experimental rickets has been used in the treatment of rickets and osteomalacia (Albright and Reifenstein, 1948). Although recent studies have failed to demonstrate an increase in calcium absorption, the citrate content of bones is low in rickets and is raised during vitamin D therapy.

CALCIUM EXCRETION

The kidney threshold for calcium is approximately 7 mg./100 ml. (Albright and Reifenstein, 1948), but it does not appear that the threshold is the primary factor that governs excretion. Children fed low levels of dietary calcium excreted very little in the urine, yet normal blood Ca levels were maintained (Holmes, 1945). The amount of calcium excreted in the urine is related to the calcium intake and the calcium status of an individual. A portion of the ionized plasma calcium as well as a part of the complexed plasma calcium are filtered at the glomerulus. Since a 70 kg. man filters about 23 gm. calcium and excretes only 150 to 200 mg. of calcium per day, approximately 99 per cent of the filtered calcium is reabsorbed. Certain diuretic and chelating compounds as well as stresses (i.e., immobilization and weightlessness) may increase urinary calcium excretion. Recently it was reported that urine contains variable amounts of nonexchangeable calcium (Pearson, 1965).

Most of the fecal calcium is dietary calcium that was not absorbed, but a small portion represents calcium which has been excreted endogenously and not reabsorbed. Harris et al. (1960) have observed species differences in the routes of excretion of endogenous calcium. The urine/feces ratios of calcium excretion by four species were: human beings, 2/1; monkeys, 2/1; dog, 1/9; and rat, 1/22. Thus, the route of endogenous excretion was the same in humans and monkeys and very different in dogs and rats. These results lead one to question whether rodents or canines are suitable substitutes for human beings in studies of calcium physiology. Possibly only primates are useful for this (Harris et al., 1960).

General Pathology

SYSTEMIC DISORDERS OF CALCIUM METABOLISM

Rickets. It results from poor mineralization of the organic cartilaginous matrix which precedes hard bone formation. Unlike normal bone formation, the epiphyseal cartilage cells do not degenerate and there is no continuation of the expansion of capillaries followed by mineralization along lines of longitudinal growth. New cartilage forms, and the epiphysis becomes irregularly widened. Retarded or inadequate mineralization in combination with gravitational and mechanical stresses may eventually lead to skeletal malformations.

Rickets may be due (1) to inadequate calcium intake (rare in the United States), (2) to impaired absorption of calcium because of a deficiency in vitamin D intake or ultraviolet light exposure or (3) to developmental defects in cell metabolism (called "vitamin D resistant" rickets). Vitamin D deficiency rickets can be treated by administration of 2000 to 6000 IU of vitamin D daily, and can be prevented by a dietary intake of about 400 IU daily. In cases of malabsorption, a water-miscible vitamin D should be prescribed; failing this, it should be given parenterally.

Osteomalacia. This is a generalized rarefaction and demineralization of the bone which occurs in adults and is due primarily to calcium and vitamin D deficiency.

Osteoporosis. Osteoporosis is the most common disorder of the skeletal system. It is an increased porosity of bone which occurs mostly in elderly persons. It is usually recognized as the end result of a progressive loss in bone mass, usually with no change in volume or shape until pressures bring about a collapse of structures that have been weakened from within. Diagnosis is usually based on radiographic evidence of collapsed vertebrae, expanded intervertebral disks and thinning of bone cortex. It has been attributed to endocrine, mineral and/or vitamin deficiencies. Since few patients have been cured by these approaches, however, there is a tendency to regard most cases as idiopathic. In one study, about 75 per cent of patients with hip fractures had collapsed vertebrae and osteoporosis (Urist, 1959).

Whedon (1959) reported that human beings about 60 years old exhibited nearly as much active bone formation as young people, but that they were excreting more calcium than they were retaining. Bone resorption outruns new bone formation, and the bones decalcify and become thin and fragile. Some older subjects require 800 mg. of calcium to achieve calcium balance. The theory that osteoporosis is the result of low intakes of calcium during many years is attractive but has not yet been proved.

Steatorrhea. In all diarrheas, and especially *steatorrhea*, the excretion of calcium is increased and the absorption of vitamin D is impaired. Hypocalcemia may develop in severe chronic cases.

Acidosis. This causes an increased excretion of calcium in the urine which may result in rarefaction of the skeleton, even though normocalcemia is maintained.

Hypercalcemia. This occurs when the homeostatic mechanism which controls the calcium level of the blood is overcome. This condition develops when excess calcium chloride is fed, in hyperparathyroidism, hypervitaminosis, disuse atrophy and idiopathic hypercalcemia in infants.

Oral Pathology

CALCIUM AND DENTAL CARIES

Although Mellanby (1936) claimed that hypoplastic enamel is caries susceptible, and Dobbs (1932) and also Hawkins (1934) reported correlations between hypoplastic enamel and the incidence of caries in children, Schour (1938) was firm in the belief that the quality of the enamel structure bears no significant relation to caries incidence. McKay (1929) reported that mottled enamel was not more susceptible to decay than normal enamel, even when imperfections reflecting disturbed calcium metabolism were present. This confusion was clarified by McCall and Krasnow (1938), who suggested that the hyperplasia in deciduous teeth is different from that in permanent teeth. The erupting hypoplastic deciduous tooth is smooth, soft and chalky and lacks density. The hyperplastic permanent tooth is pitted and grooved, its enamel is defective, but it has normal density. Its physical properties are not changed sufficiently to cause it to be prone to caries development. Susceptibility to decay is greatly dependent on the presence of pits and fissures on the enamel surface which serve to retain food until it has stagnated.

Attempts to prevent dental caries development by calcium and phosphorus therapy were in vogue about 30 years ago. Then, Malan and Ockerse (1941) conducted a three-year study in which schoolchildren were given dietary supplements of 0.5 gm. of phosphorus and 0.5 gm. of calcium daily. No significant differences in caries susceptibility in either the deciduous or permanent dentition were observed.

Increases in dental decay were noted in young rats when the calcium content of the diet was made progressively more deficient (Haldi et al., 1959). This is to be expected since structural imperfections in enamel and dentin develop in growing animals maintained on diets deficient in calcium and/or vitamin D.

EFFECT OF CALCIUM DEFICIENCY ON ALVEOLAR BONE

The alveolar bone, being cancellous and labile, is especially sensitive to mineral deficiencies, especially calcium deficiency. Becks and Weber (1931) reported that the marrow spaces of alveolar bone became hemorrhagic and filled with uncalcified fibro-osteoid tissue when rats were fed calcium-deficient diets. The teeth became loosened as a result of alveolar bone resorption and destruction of the periodontal ligament. An association seems to exist between periodontoclasia and faulty calcium absorption. Radusch (1947) observed an increase in alveolar resorption in patients with achlorhydria. Roentgenograms revealed that the lamina dura was characteristically absent in patients with disturbed calcium metabolism resulting from hormonal disturbances involving the parathyroid, thyroid and adrenal glands.

Requirements

The U.S. National Research Council (1964) has defined the Recommended Dietary Allowance of nutrients as the daily amount which will afford a "margin of sufficiency above average physiological requirements to cover variations among all individuals in a population. . . . They provide a buffer against the increased needs during common stresses and permit full realization of growth and reproductive potential; but they are not to be considered adequate to meet additional requirements of persons depleted by disease or traumatic stress."

The 1964 Recommended Dietary Allowance for calcium was:

Age Group (yr.)	Recommended Dietary Allowance (mg. per day)
0 to 1	700
1 to 9	800
9 to 12	1100
12 to 18 (boys)	1400
12 to 18 (girls)	1300
18 to 75	800
Pregnancy (late) and lactation	1300

An FAO/WHO Expert Group (1962) defined the Minimum Practical Allowance of calcium as "intake at which the needs of the great majority of persons in any defined population group are likely to be adequately met. . . . It can, therefore, be considered a safe allowance." However, the group noted that "the possibility of undesirable effects on the health of a population with an average calcium intake habitually below this level should be borne in mind." The FAO/WHO recommended a range of practical allowances for calcium as follows:

Age Group	Suggested Minimum Practical Allowance (mg. per day)
0 to 12 mos. (not breast fed)	500 to 600
1 to 9 years	400 to 500
10 to 15 years	600 to 700
16 to 19 years	500 to 600
Adults	400 to 500

The Recommended Dietary Allowances for calcium are generally higher than the Minimum Practical Allowances of the FAO/WHO because a larger factor of safety is used. A dietary allowance of 600 to 800 mg. of calcium for adults has been adopted by most of the countries in which dietary standards have been established.

Requirements for calcium are increased during growth to meet the needs for bone formation, and during late pregnancy and lactation. During the last seven lunar months, a pregnant woman is considered to require 25, 50, 84, 125, 175, 234 and 300 mg. of calcium in addition to her normal adult requirement (Mitchell and Curzon, 1939). Thus, she must ingest 750 mg. of food calcium (assuming 40 per cent utilization) to obtain 300 mg. of calcium for the fetus during the last month.

The requirement of calcium for lactation is determined by the volume of milk secreted. Breast milk contains 300 mg. of calcium per liter, and a liter approximates the average daily production by lactating women. Assuming 40 per cent absorption, the average woman must ingest an extra 750 mg. of calcium daily to meet lactation needs. The occurrence of osteomalacia in pregnant and lactating women in regions in which the diet is quite low in calcium content is proof that the calcium requirements are high during this stress period.

PHOSPHORUS

DISTRIBUTION OF PHOSPHORUS IN TISSUES

Of the approximate 600 gm. of phosphorus in the normal human body, 80 to 90 per cent is present in the bones and teeth and 10 to 20 per cent in the soft tissues. Most of the phosphorus in hard tissues is present as hydroxyapatite. Some of the soft tissue phosphorus is in the striated muscles and occurs as orthophosphates (hexose phosphates, adenylic acid, guanylic acid, thymidylic acid, cytidylic acid, uridylic acid, deoxyribonucleic acid, glycerophosphoric acid, phosphatidyl serine, pyridoxal-S-phosphate, urositol-6-phosphate) and as pyrophosphates (adenosine-s'-diphosphate, thiamine-pyrophosphate, coenzyme A, coenzyme I (DPN), coenzyme II (TPN), flavin adenine dinucleotide (FAD) and others). A small amount of soft tissue phosphorus is in brain tissue in the form of nucleoprotein, lecithin and cephalin.

The blood contains less than 2 gm. phosphorus, present as inorganic orthophosphate (mostly in serum), or as esters (mostly in cells) and phospholipids (two thirds in cells). In contrast with calcium, almost all inorganic serum phosphorus is filterable and ionizable, and therefore the level is not closely regulated. Unlike calcium, the ionizable phosphorus has no pharmacodynamic action of its own. Although a part of many important enzymes and complexes, phosphorus has no effect on nerve excitation or membrane permeability beyond its ability to interfere with the ionization of calcium.

The average biological life span of phosphorus in the animal body is only 30 days, in spite of the fact that it takes about 8 months to replace only 1 per cent of the phosphorus in a tooth. The turnover of phosphorus in hard tissues requires months, while that in soft tissues takes only minutes. This explains why it takes more dietary phosphorus day by day to maintain the phosphorus in soft tissues than in hard tissues, even though the soft tissue phosphorus is only 10 per cent of the total in the body.

FOOD SOURCES
OF PHOSPHORUS

Phosphorus is widely distributed in foods from plant and animal sources, the richest (mg./100 gm. fresh weight) being: wheat germ, 1084; cheeses, 95 to 875; Brazil nuts, 693; soybean flour, 558 to 655; walnuts, 570; enriched farina, 561; almonds, 504; beans, 406 to 457; cow peas, 65 to 426; rolled oats, 405; wheat, 354 to 400; peanuts, 393; peas, 388; lima beans, 385; lentils, 377; rye flour, 376; liver, 352; corn meal, 256; fish, 89 to 256; chicken, 188 to 220; eggs, 205; white bread, 87 to 102; and milk, 93 to 95 (Watt and Merrill, 1963). The above data indicate that nuts, legumes, meats, fish, cereals, milk and milk products are rich in phosphorus content. The phosphorus content of cereals decreases as the bran and germ are removed by milling. Most of the phosphorus in cereals and legumes is present as phytic acid compounds which are not efficiently utilized by human beings. Data on the phosphorus content of foods are of more practical value when expressed in terms of household units or the amount normally consumed per day. For example, a quart of

milk contains 908 mg. phosphorus. If a child consumes three glasses of milk daily he will ingest 680 mg. of phosphorus. Thus, milk is a much more important source of phosphorus than the above tabulation indicates.

Physiology

ABSORPTION OF
PHOSPHORUS

More than 70 per cent of the phosphorus ingested in foods is absorbed from the intestine into the blood. Thus, food phosphorus is utilized more efficiently than food calcium. Intestinal phosphatases release the food phosphorus and convert most of it to orthophosphates preliminary to absorption. Excesses of aluminum, calcium, iron, magnesium, etc., may interfere with the absorption of phosphorus because of the formation of insoluble phosphates. Conversely, high intakes of phosphorus may interfere with the absorption of these minerals, even causing conditioned malnutrition. Elsewhere in this chapter the effects of phytates upon mineral metabolism have been discussed. Food phytates as well as inorganic phosphates were shown to interfere with the absorption of iron (Sharpe et al., 1950) and calcium (Bronner et al., 1954) by children.

The earlier reports that vitamin D increases the absorption of dietary phosphorus, decreases the excretion of phosphorus, and is essential for the conversion of organic phosphorus in bone are now considered to be inaccurate. Apparently, vitamin D does not have a direct effect on phosphorus metabolism; however, the metabolic effects of vitamin D on calcium have an indirect effect on phosphorus metabolism.

EXCRETION

Fecal phosphorus represents that which was not absorbed, or was secreted endogenously into the intestinal tract. Urinary phosphorus is mostly inorganic phosphate, and the amount of urinary phosphorus varies according to the quantity of phosphorus absorbed and the amounts released during tissue breakdown (stresses, starvation, carbohydrate metabolism, etc).

General Pathology

SYSTEMIC DISTURB-ANCE IN PHOS-PHORUS METABOLISM

Since phosphorus plays a central role in the structure of cytoplasm and cell nuclei, all biological disorders involve phosphorus metabolism to some extent. Bone physiology involves three main processes: (1) matrix formation, (2) deposition of apatite and (3) bone destruction. Matrix formation involves the deposition of collagen and ground substance of the right composition; impaired matrix formation may lead to osteoporosis. Apatite formation involves deposition of crystals on the protein matrix; impaired apatite deposition may lead to osteomalacia and rickets. Bone destruction is a normal remodeling process; excess destruction may lead to osteitis fibrosa cystica. All these disorders except osteoporosis involve faults in phosphorus metabolism.

DEFICIENCY DISEASES

Phosphorus deficiency diseases are rather common in animals, especially ruminants, because forage crops tend to contain much more calcium than phosphorus. They become "unthrifty," their food intake is reduced and their bones become highly fragile. Phosphorus deficiency seldom occurs in human populations because the foods eaten by mankind in all parts of the world are adequate and even rich in phosphorus content.

Oral Pathology

PHOSPHORUS AND DENTAL CARIES

Investigators were late in finding a role for phosphorus deficiency in caries production, and for phosphorus abundance in caries control, primarily because too great importance was being placed on the Ca/P ratio of diets (Harris and Nizel, 1964). As a result, until recently, few experiments were conducted to evaluate the possible role of phosphorus in dental caries development. Without present-ing supporting evidence, Lennox (1931) and Hewat (1931) concluded that populations in South Africa and New Zealand, respectively, had a high incidence of caries because their diets were deficient in phosphorus.

The earliest experimental evidence of a correlation between low dietary phosphorus and dental caries was that presented by Klein and McCollum (1931), though they later suggested that the particle size of the corn in the diet was the cause (1933). Klein and Shelling (1931), Blackberg and Berke (1932), Agnew et al. (1933) and Rosebury and Karshan (1935) all observed an inverse relationship between low dietary phosphorus and high dental caries incidence in rodents; only Shelling and Asher (1933) dissented.

Ten years ago Harris et al. (1957) observed that the ash of food is caries-preventing when added to the diet of hamsters. On the basis of spectrographic and chemical analyses, they compounded a salt mixture which duplicated the mineral elements of the food ash. This synthetic ash was also cariostatic. They then observed that when phosphorus was omitted from this salt mixture the cariostatic activity was completely lost; in fact, the ash became cariogenic. By this experimental approach they demonstrated that potassium orthophosphate is an effective cariostatic agent, even when added to a diet already nutritionally adequate in phosphorus. During the past decade, more than 150 studies have been conducted by a score of investigators, and the results are nearly unanimous in indicating that various types of inorganic and organic phosphates are effective in preventing the development of caries in rodents (Nizel and Harris, 1964) (Fig. 8-2). The activities of the sodium salts of the different anions decrease in the following order: trimeta-, tripoly-, hexameta-, ortho- and pyro- (Nizel et al., 1962). The activities of salts of the same anion decrease as follows: hydrogen, sodium, potassium, calcium and magnesium (Nizel and Harris, 1964). Organic phosphates, such as sodium glycerophosphate and sodium phytate, are equally as active against caries development as sodium orthophosphate. The results have been the same whether the studies were conducted on hamsters, rats or cotton rats.

Present evidence indicates that the action of phosphates is primarily a local one, affect-

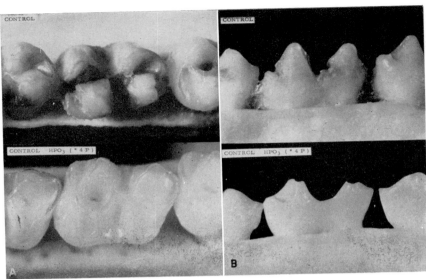

FIGURE 8-2. A, *Occlusal view of hamster molars—note morphologic as well as caries difference between group on control diet and experimental (control* + *HPO₃) diet.* B, *Lateral view of same teeth.*

ing the surface layers of the teeth. Navia and Harris (1966) observed little effect when sodium trimetaphosphate was fed to rats during gestation, a modest effect when fed before tooth eruption, and a major effect when fed immediately following eruption. The mechanism of action of phosphorus is different from that of fluorine; in fact, the caries controlling actions of these two elements are improved synergistically when they are administered together.

The anticaries action of phytates deserves comment. When wheat is milled in the manufacture of white flour, about 80 per cent of the phosphorus is removed, and about two-thirds of this is phytate phosphorus. When the phosphorus content of white flour was raised to whole wheat levels by the addition of either sodium phytate or sodium orthophosphate, the flour was no longer cariogenic when fed to rats. It is quite possible, therefore, that the increased cariogenicity of sophisticated diets is due, partly at least, to the removal of phosphorus during refinement of food products.

The above conclusion will not be valid until it has been proved that phosphates are effective in the control of dental caries in human populations. Stralfors (1964) added 2 per cent dibasic calcium phosphate to the bread and sugar in school lunches and fed these lunches to 1299 children on school days during two years. A control group of 963 children received the same lunch, but without phosphate supplement. Stralfors examined four proximal surfaces of the incisor teeth of these nine year old children by x-ray at the beginning and end of the first and second years. He noted a 50 per cent decrease in new caries development in the phosphate-fed group in the first year, and a 40 per cent decrease by the end of the second year.

Shipp and Mickelsen (1964) added dibasic calcium phosphate to the salt and bread consumed in the school lunches of two groups of children (390 control, 350 supplemented with phosphate). The whole dentition was examined for caries, using probe, mirror and bitewing x-rays of the posterior teeth. No differences in the development of new caries were noted.

It has been suggested that the Stralfors study was the more sensitive because children of a more homogeneous age group (8 to 10 years, mean 9 years) were tested and only one type of recently erupted tooth (maxillar incisor) was used to assay caries development. On the other hand, Shipp and Mickelsen used a less homogeneous group (7 to 14 years, mean 10.19 years), and the DMFT and DMFS increments of all teeth were scored. Both clinical studies were conducted using dicalcium orthophos-

TABLE 8-1.   EFFECTS OF LEVELS OF CALCIUM AND PHOSPHORUS UPON THE DEVELOPMENT OF CARIES IN RATS

Diet	No. Animals	Ca in Diet	P in Diet	C/P Ratio	Average No. Carious Lesions	Average Caries Score	% Caries Change vs. Control	Reference
1	40	0.49%	0.24%	2:1	22	44	+80%	Wynn et al., 1956
2	40	0.50	.52	1:1	18	25	—	
3	40	0.50	.98	1:2	13	16	—36%	
4	40	0.50	1.48	1:4	11	13	—48%	
1	20	0.29%	0.52%	1:2	23	34	26%	Wynn et al., 1959
2	20	0.57	.50	1:1	21	27	—	
3	20	1.01	.49	2:1	19	21	—22%	
4	20	1.57	.52	2:4	17	20	—26%	
1	15	0.25	0.25	1:1	24	47	+42%	Wynn et al., 1960
2	15	0.50	.50	1:1	24	33	—	
3	15	1.00	1.00	1:1	18	21	—36%	
4	15	1.50	1.50	1:1	17	20	—39%	

phate, yet animal studies have shown this to be only 20 per cent as active as sodium trimetaphosphate.

At the present time it is not known whether caries development in human beings can be controlled by supplementing foods with phosphates.

Gaunt et al. (1939) fed to rats during four generations a diet which duplicated the foods currently being eaten by working class people in Scotland. This diet did not support good rat growth. When it was supplemented with milk and green vegetables, the rat growth was much improved. When the diet was supplemented with calcium and phosphorus salts in amounts equivalent to the calcium and phosphorus in the milk and vegetables, the rat growth was greatly improved also. Thus, the poor growth observed in the rats fed the "working class diet" was mainly the result of deficiencies in calcium and phosphorus.

The rats fed the "working class diet" were unable to maintain a normal ash content in bones and incisors. At 100 days of age, the predentin width in these rats was 50 per cent greater than the predentin width in the teeth of rats fed calcium and phosphorus supplement and in rats whose teeth were normal.

Unfortunately, very few studies have been attempted in order to demonstrate the effects of calcium or phosphorus deficiency on human beings. Wynn et al. (1956; 1959; 1960) conducted three classical studies on rats which indicate that deficiencies or excesses of phosphorus and/or calcium influence dental caries

development. Rats normally require about 0.5 per cent calcium and 0.5 per cent phosphorus in the diet to support full growth and development.

The calcium and phosphorus content of diet No. 2 in each of the three series of experiments (Table 8-1) was 0.5 per cent and, therefore, diet No. 2 served as a normal control. The calcium and phosphorus contents of the remaining diets were varied either by holding the calcium content constant and varying the phosphorus, or by holding the phosphorus constant and varying the calcium, or by varying both calcium and phosphorus together. In each case, diet No. 1 was 50 per cent deficient in phosphorus or calcium, or both. It will be noted that dental caries was increased by all three deficient diets. The increased caries which resulted from feeding the phosphorus-deficient diet (+80 per cent) was greater than that resulting from the calcium-deficient diet (+26 per cent) and greater even than that which resulted from feeding the diet deficient in both phosphorus and calcium (+42 per cent). These results indicate that phosphorus-deficient diets are more cariogenic than calcium-deficient diets; in other words, phosphorus deficiency is more conducive to caries development than calcium deficiency.

In parallel studies, rats were fed diets which contained two (diet No. 3) and three (diet No. 4) times the normal requirement of phosphorus or calcium or both. The data obtained from diets Nos. 3 and 4 indicate that the addition of excess calcium to the diet reduced dental caries development significantly and that excess phosphorus reduced dental caries development even more.

These three studies indicate clearly that dietary deficiencies and excesses of calcium and phosphorus significantly affect dental caries development, and that phosphorus is the more important element in this regard. Many reports in the literature conducted earlier and later than these classical studies give strong support to this conclusion. Hartles (1962) reported that diets low in phosphorus cause a greater disturbance in bone than diets low in calcium. Ferguson and Hartles (1964) found that phosphorus deficiency alone caused greater disturbance to the formation of secondary dentin than did calcium.

Requirements

The requirements of human beings for phosphorus are of little concern to the nutritionist because the diet is generally more than adequate in this element, and phosphorus deficiency has not been found in human beings.

The requirements of phosphorus are not known with precision. The U.S. National Research Council (1964) recommends that the "phosphorus intake should be at least equal to that for calcium in the diets of children and of women during the latter part of pregnancy or in lactation periods." Dietary surveys have shown that the dietary phosphorus intake by adults is approximately 1.5 times that of calcium (Sherman, 1947). Since the NRC made no recommendations for other age groups, it may be assumed that intakes of 800 to 1200 mg. of phosphorus daily are adequate.

When populations obtain most of their calories from unrefined cereals, data on the total phosphorus intake are misleading since much of the phytin phosphorus will be poorly absorbed, even when the diet is adequate in vitamin D content. The NRC bulletin (1964) states: "In general, it is safe to assume that, if the calcium and protein needs are met through common foods, the phosphorus requirement also will be covered, because the common foods richest in calcium and protein are also best sources of phosphorus."

MAGNESIUM

Magnesium was first demonstrated to be present in living things about one century ago (Holmes, 1859). It is one of the major cations in animal and plant tissues and appears to be essential for all plant and animal life. The knowledge that magnesium is present in the chlorophyll molecule has stimulated many studies of its role in soil economy and in agriculture. The essentiality of magnesium as a nutrient for rats was reported by Leroy (1926). Since Erdtmann (1948) first observed that magnesium activates alkaline phosphatase, hundreds of other enzymes have been found to be activated by magnesium. In spite of this, the action of magnesium in living tissues is still little understood, and knowledge of its role in human nutrition is scant.

DISTRIBUTION OF MAGNESIUM IN TISSUES

The human body contains 20 to 30 gm. of magnesium (Widdowson et al., 1951), with the concentration increasing rapidly following birth (Greenberg et al., 1936). Half the magnesium in the body is present in bone and, as a result, it represents 0.5 to 0.7 per cent of the bone ash. The results of studies with radioactive Mg^{28} in dogs and human beings indicate that the metabolic half-life of magnesium is 12.8 hours. It is concentrated in the intracellular space of soft tissue and is present in largest quantities in the liver, striated muscle, kidney and brain. Red cells contain 6 mEq.Mg/liter.

FOOD SOURCES OF MAGNESIUM

The foods richest in magnesium (mg. per 100 gm. of edible portion) are: cocoa, 420; cashew nuts, 267; almonds, 252; Brazil nuts, 225; soya flour, 223; lima beans, 181; whole barley, 171; peanuts, 167; whole wheat, 165; pecans, 152; oatmeal, 145; hazelnuts, 140; walnuts, 134; corn, 121; and brown rice, 119. It is evident that nuts, cereals and legumes are important sources of magnesium.

Physiology

METABOLIC FUNC-TIONS OF MAGNESIUM

Magnesium functions chiefly as an activator of numerous important enzymes which split and transfer phosphate groups (i.e., phosphatases, enzymes involving ATP reactions). Since ATP is required in muscle contraction, in the syntheses of proteins, fats, nucleic acid and coenzymes, in glucose metabolism, in methyl transfer and in the activation of acetates, formates and sulfates, etc., as well as being a cofactor in decarboxylation, the importance of magnesium in human metabolism is obvious.

EXCRETION OF MAGNESIUM

Most of the magnesium excreted in the feces represents unabsorbed magnesium. McCance and Widdowson (1939) noted no fecal excretion when magnesium was given intravenously during 14 days. About 30 per cent of 60 to 120 mg. daily doses of magnesium were recovered in the urine (Leichsenring et al., 1951).

General Pathology

SYSTEMIC EFFECTS OF MAGNESIUM DEFICIENCY

Young mice fed a magnesium-deficient diet stopped growing after 9 to 13 days and died in 24 to 35 days (Leroy, 1926). Mice resumed normal growth when magnesium was restored to the diet, even late in the experiment. Kruse et al. (1932) induced deficiency in weanling rats fed diets containing 1.18 ppm magnesium. They showed vasodilation (erythema and hyperemia) within three to five days, pallor and cyanosis in 12 days, neuromuscular hyperirritability and seizures in 18 days and death after about 28 days. Chronic deficiency was manifested by alopecia, trophic skin lesions, hematomas and hyperemic gums.

Large amounts of calcium in the diet aggravate magnesium deficiency and increase the magnesium requirement (Tufts and Greenberg, 1937). High protein intake hastens the appearance of deficiency symptoms and increases the magnesium requirement (Colby and Frye, 1951). On the other hand, magnesium deficiency interferes with protein synthesis.

Many neuromuscular disorders in human beings have features resembling magnesium deficiency in animals (Wacker and Vallee, 1957). The magnesium deficiency syndrome in human beings is nearly identical to that observed in other species. The tetany in magnesium-deficient animals is much like the hypocalcemic tetany in humans.

Magnesium deficiency syndromes are known to result from severe renal disease, pregnancy toxemia, chronic alcoholism, losses of gastrointestinal secretions, severe diuresis induced by drugs and continued parenteral feeding of magnesium-free fluids. These conditions generally respond promptly to magnesium injections. All these are induced deficiencies; there has been no valid report of a true magnesium deficiency in a normal human being.

Oral Pathology

MAGNESIUM IN TEETH

Magnesium is present in calcified tissues and ranks after calcium and phosphorus in the abundance of mineral elements in teeth. The magnesium content of teeth is approximately 0.4 per cent (enamel) and 0.9 per cent (dentin), expressed on a fresh tooth basis. Since the moisture content of enamel is approximately 3 per cent and that of dentin is about 10 per cent, and since the tooth also contains significant amounts of "bound water," the above values are only approximate. The report of McCann and Bullock (1957) of increased concentrations of magnesium in the enamel of rats given water containing 100 ppm fluoride was recently discounted by Brudevold et al. (1965).

The form in which magnesium occurs in teeth is still being debated. Magnesium serves as cofactor in several enzyme systems which may be involved in the growth and development of immature teeth. It is unlikely, however, that the magnesium in mature teeth is serving a biological function. It is unlikely, also, that magnesium exchanges with calcium in tooth apatite, since in vitro experiments have shown that it dissolves out at a rate different from calcium. Apparently, magnesium is present in teeth as a complex ion $(MgOH)^+$ which is too large to fit into the lattice; it is adsorbed by the tooth instead.

EFFECT OF MAGNE-
SIUM DEFICIENCY ON
TOOTH GROWTH
AND DEVELOPMENT

Irving (1940) fed magnesium-deficient diets to young rats and noted during 12 days that the rate of incisor tooth growth was slightly reduced and that the absolute amount of magnesium was unchanged or slightly increased. Thus, magnesium was entering the actively growing incisor at approximately the same rate as it was being lost by attrition on the tooth surface. The incisor teeth behaved quite differently from the bones in the same animal; the bones lost one-third of their magnesium content within six days. When magnesium was restored to the diets of deficient rats, the rate of deposition of magnesium in the teeth and bone was high, and within four days the tooth concentration returned to a normal value.

The histological effects of magnesium deficiency are noteworthy. Depending on the duration of the deficiency, the predentin became two to three times wider 3 mm. above the basal portion of the tooth, causing the dentin-predentin junction to be in different levels of the tooth. Irving remarked that this "predentine step" is unique for magnesium deficiency. The calcified dentin beyond this area showed faint stratification lines running parallel to the dentin-predentin junction.

Apparently, the dentin was improperly calcified since it was less intensely stained with hematoxylin. The dentin in the labial portion of the tooth was more seriously affected as the magnesium deficiency became prolonged, and the shape of the pulp became distorted. The odontoblasts gradually became atrophied, so that at 23 days they were 33 μ long rather than a normal 50 μ. In certain areas these cells failed to recede from the predentin and became imbedded in it. After prolonged deficiency the dentin and the odontoblastic layer became folded. Magnesium deficiency provoked a degeneration of the enamel organ. The ameloblasts appeared as low cubical epithelium filled with calcareous granules, a condition which resembled other forms of enamel organ degeneration, such as that resulting from vitamin A deficiency hypophysectomy.

When magnesium deficiency was continued for as long as six months, the enamel organ of the incisal area became completely atrophied, and the cells were replaced by a noncellular structure. The enamel organ disappeared completely from the apical portion of the tooth, and the connective tissue came into direct contact with the outer dentin. The odontoblasts recovered rapidly if the magnesium deficiency was of short duration, but not when the deficiency was protracted.

MAGNESIUM AND
DENTAL CARIES

Magnesium is widely distributed in the foods consumed by mankind, and magnesium deficiency rarely, if ever, occurs in human populations. Even in the event it should occur, the tooth has priority over bone for new magnesium during nutritional crises and avidly retains it. Magnesium is of dental interest because it affects tooth solubility and thus dental caries development.

The evidence relating to the caries activity of magnesium is quite unclear. McClure (1948) observed cariogenic effects in rats when magnesium was added to their drinking water. Parma et al. (1957) reported that the cariostatic effect of fluoride is enhanced when magnesium is added to the drinking water. McClure and McCann (1960) noted no reduction in caries when $MgCO_3$ or $Mg_3(PO_4)_2$ were added to a caries-producing diet. Ritchie (1961) was able to increase the magnesium content of human teeth by feeding $Mg_3(PO_4)_3$, and reported that this increase was associated with improved caries resistance. Forbes (1963) pointed to the necessity of controlling the amounts of calcium and phosphorus in diets when the caries activity of magnesium is being studied.

These chronological citations indicate that magnesium is either inert or is cariogenic when added to the diets or water supplies of animals. The interesting report by Toth (1964) that the magnesium content of salivas from gypsies who had active caries was three times higher (1.59 mg. versus 0.5 mg./100 ml.) than salivas from gypsies whose teeth were caries resistant indicates that the relation of magnesium to dental caries is worthy of further study.

Requirements

The average daily intake of magnesium by adult human beings approximates 300 mg. Large amounts may be obtained from green vegetables (from their chlorophyll). Tibbetts and Aub (1937) demonstrated by balance studies that the daily requirement of magnesium by dental students was 222 mg. Duckworth and Warnock (1942) estimated that the requirement of infants is 150 mg./day, while that of women during pregnancy and lactation is 400 mg./day.

The soluble salts of magnesium are readily absorbed from the small intestine. Absorption is not affected by pH or the concurrent feeding of sodium carbonate or dilute hydrochloric acid (Duckworth, 1939).

REFERENCES

Agnew, M. C., Agnew, R. S., and Tisdall, F. F.: The production and prevention of dental caries. J. Amer. Dent. Assn., 20:193, 1933.

Albright, F., and Reifenstein, E. C., Jr.: The Parathyroid Glands and Metabolic Bone Disease. Baltimore, The Williams & Wilkins Co., 1948.

Anonymous: Calcium supplements. J. A. M. A., 178:870, 1961.

Becks, H., and Weber, M.: Influence of diet on bone system with special reference to alveolar process and labyrinthine capsule. J. Amer. Dent. Assn., 18:197, 1931.

Blackberg, S. N., and Berke, J. D.: Dental caries experimentally produced in the rat. J. Dent. Res., 12:609, 1932.

Bronner, F., and Harris, R. S.: Absorption and metabolism of calcium in human beings, studied with calcium⁴⁵. Ann. N. Y. Acad. Sci., 64:314, 1956.

Bronner, F., Harris, R. S., Maletskos, C. J., and Benda, C. E.: Effect of food phytates on calcium⁴⁵ uptake in children on low-calcium breakfasts. J. Nutr., 54:523, 1954.

Bronner, F., Moor, J. R., Harris, R. S., and Benda, C. E.: Differential absorbability of calcium salts by children. Fed. Proc., 14:428, 1955.

Brudevole, F., McCann, H. G., and Grøn, P.: Caries resistance as related to the chemistry of the enamel. In Wolstenholme, G. E. W., and O'Connor, M. (ed.): Caries-Resistant Teeth. Ciba Foundation Symposium. Boston, Little, Brown and Co., 1965, p. 121.

Colby, R. W., and Frye, C. M.: Effect of feeding high levels of protein and calcium in rat rations on magnesium deficiency syndrome. Amer. J. Physiol., 166:408, 1951.

Dixon, M., and Webb, E. C.: Enzymes. London, Longmans, Green & Co., 1958.

Dobbs, E. C.: Studies of blood calcium and phosphorus from a dental aspect. Dent. Cosmos, 74:867, 1932.

Dowdle, E. B., Schacter, D., and Schenker, H.: Requirement for vitamin D for the active transport of calcium by the intestine. Amer. J. Physiol., 198:269, 1960.

Drinker, N., Green, A. A., and Hastings, A. B.: Equilibria between calcium and purified globulins. J. Biol. Chem., 131:641, 1939.

Duckworth, J.: Magnesium in animal nutrition. Nutr Abstr. Rev., 8:841, 1939.

Duckworth, J., and Warnock, G. M.: Normal requirements of magnesium. Nutr. Abstr. Rev., 12:167, 1942.

Duncan, D. L.: Physiological effects of lactose. Nutr. Abstr. Rev., 25:309, 1955.

Erdtmann, H.: Glycerophosphatspaltung durch nierenphosphatase und ihre aktivierung. Z. Physiol. Chem., 172:182, 1948.

FAO Nutrition Meetings Report Series No. 30: Calcium Requirements. FAO/WHO Expert Group, Rome, 1962.

Ferguson, H. W., and Hartles, R. L.: The effects of diets deficient in calcium or phosphorus in the presence and absence of supplements of vitamin D on the secondary cementum and alveolar bone. Arch. Oral Biol., 9:647, 1964.

Fincke, M. L., and Garrison, F. A.: Utilization of calcium of spinach and kale. Food Res., 3:575, 1938.

Fleisch, H., and Bisaz, S.: Isolation from urine of pyrophosphate, a calcification inhibitor. Amer. J. Physiol., 203:671, 1962.

Forbes, R. M.: Mineral utilization in the rat. I. Effects of varying dietary ratios of calcium, magnesium and phosphorus. J. Nutr., 80:321, 1963.

French, C. E.: The interrelation of calcium and fat utilization in the growing albino rat. J. Nutr., 23:375, 1942.

Gaunt, W. E., Irving, J. T., and Thompson, W.: A long-term experiment with rats on a human dietary. II. Calcium and phosphorus depletion and replacement. J. Hyg. (Camb.), 39:91, 1939.

Greenberg, D. M., Anderson, C. E., and Tufts, E. V.: Pathologic changes in the tissues of rats reared on diets low in magnesium. J. Biol. Chem., 114:43, 1936.

Haldi, J., Wynn, W., Bentley, K. D., and Law, M. L.: Dental caries in the albino rat in relation to the chemical composition of the teeth and of the diet.

IV. Variations in the Ca/P ratio of the diet induced by changing the calcium content. J. Nutr., *67*:645, 1959.

Harris, R. S.: Phytic acid and its importance in human nutrition. Nutr. Rev., *13*:257, 1955.

Harris, R. S.: High calcium intakes on urine in animals. Fed. Proc., *18*:1100, 1959.

Harris, R. S., Moor, J. R., and Wanner, R. L.: Calcium metabolism in normal rhesus monkeys. 5th Internat. Congr. Nutr. (abstract), 1960, p. 4.

Harris, R. S., and Nizel, A. E.: Metabolic significance of the ratio of Ca to P in foods and diets. J. Dent. Res., *43*:1090, 1964.

Harris, R. S., Nizel, A. E., and Gardner, D. S.: Effects of food ash and trace minerals on dental caries in hamsters. 4th Internat. Congr. Nutr. Proc., 1957, p. 195.

Hartles, R. L.: The effects of dietary deficiencies of calcium or phosphorus on mineralisation. Brit. Dent. J., *113*:411, 1962.

Hawkins, H. F.: What is the cause of caries and systemic pyorrhea? J. Amer. Dent. Assn., *18*:943, 1934.

Hevesy, G. C.: Application of radioactive indicators in biology. Ann. Rev. Biochem., *9*:641, 1940.

Hewat, R. E. T.: Immunity to dental caries. New Zealand Dent. J., *27*:27, 1931.

Holmes, J. V.: The requirement for calcium during growth. Nutr. Abstr. Rev., *14*:597, 1945.

Holmes, O. W.: The Professor at the Breakfast Table. Cambridge. Houghton, Osgood and Co., 1859, p. 1880.

Irving, J. T.: Influences of diets low in magnesium on histological appearance of incisor teeth of rat. J. Physiol., *99*:8, 1940.

Johnston, C. C., Miner, E. R., Smith, D. M., and Deiss, W. P.: Influence of parathyroid activity on the chemical equilibrium of bone calcium *in vitro*. J. Lab. Clin. Med., *60*:689, 1962.

Jones, J. H.: The influences of fat on calcium and phosphorus metabolism. J. Nutr., *20*:367, 1940.

Kimizuku, H., and Kotetsu, K.: Binding of calcium ion to lecithin film. Nature, *196*:995, 1962.

Klein, H., and McCollum, E. V.: A preliminary note on the significance of the phosphorus intake in the diet and blood concentration, in the experimental production of caries immunity, and caries susceptibility in the rat. Science, *74*:662, 1931.

Klein, H., and McCollum, E. V.: The significance of food-particle size in the etiology of microscopic dental decay in rats. J. Dent. Res., *13*:69, 1933.

Klein, H., and Shelling, D. H.: The relation of diet to the toxicity of viosterol as a determining factor in the production of experimental caries in rats. J. Dent. Res., *11*:458, 1931.

Kruse, H. D., Orent, E. R., and McCollum, E. V.: Studies on magnesium-deficient animals. J. Biol. Chem., *96*:519, 1932.

Lavender, A. R., Aho, I., Rasmussen, H., and Pullman, T. N.: Evidence for a direct renal tubular action of parathyroid extract. J. Lab. Clin. Med., *54*:916, 1959.

Leichsenring, J. M., Norris, J., and Lamson, A.: Magnesium metabolism in college women. J. Nutr., *45*:477, 1951.

Lennox, J.: Observations on diet and its relation to dental disease: A further consideration of calcium and phosphorus metabolism in their relation to dental caries. S. Afr. Dent. J., *5*:156, 1931.

Leroy, J.: Nécessité du magnesium pour la croissance de la souris. Compt. Rend. Soc. Biol., *94*:431, 1926.

Lund, A. P., and Armstrong, W. D.: Effect of low calcium and vitamin D free diet on skeleton and teeth of adult rats. J. Dent. Res., *21*:513, 1942.

MacGregor, J., and Brown, W. E.: Blood bone equilibrium in calcium homeostasis. Nature, *205*:359, 1965.

Macy, I. G.: Nutrition and Chemical Growth in Childhood. Vol. I: Evaluation. Springfield, Ill., Charles C Thomas Co., 1942.

Malan, A. L., and Ockerse, T.: Effect of calcium and phosphorus intake of school children upon dental caries, body weights, and heights. S. Afr. Dent. J., *15*:153, 1941.

Mallon, M. G., Jordan, R., and Johnson, M.: A note on the calcium retention on a high and low fat diet. J. Biol. Chem., *88*:163, 1930.

Marcus, C. S., and Lengemann, F. W.: Absorption of Ca45 and Sr85 from solid and liquid food at various levels of the alimentary tract of the rat. J. Nutr., *77*:155, 1962.

McCall, J. O., and Krasnow, F.: The influence of metabolism on teeth. J. Ped., *13*:498, 1938.

McCance, R. A., and Widdowson, E. M.: The fate of calcium and magnesium after intravenous administration to normal persons. Biochem. J., *33*:523, 1939.

McCance, R. A., and Widdowson, E. M.: Mineral metabolism of healthy adults on white and brown bread dietaries. J. Physiol., *101*:44, 1942.

McCann, H. G., and Bullock, F. A.: The effect of fluoride ingestion on the composition and solubility of mineralized tissues of the rat. J. Dent. Res., *36*:391, 1957.

McClure, F. J.: Observations on induced caries in rats. VI. Summary results of various modifications of food and drinking water. J. Dent. Res., *27*:34, 1948.

McClure, F. J., and McCann, H. G.: Dental caries and composition of bones and teeth of white rats: Effect of dietary mineral supplements. Arch. Oral Biol., *2*:151, 1960.

McKay, F. S.: The establishment of a definite relation between enamel that is defective in its structure as mottled enamel, and the liability to decay. Dent. Cosmos, *71*:747, 1929.

Mellanby, E.: Rickets-producing and anti-calcifying action of phytate. J. Physiol., *109*:488, 1949.

Mellanby, M.: Influence of diet on caries in children's teeth. Med. Res. Council Special Report Series No. 221, London, 1936.

Mitchell, H. H., and Curzon, E. G.: "Actualities Scientifique et Industrielles," No. 771. Paris, Herman and Co., 1939.

Navia, J. M., and Harris, R. S.: Cariostatic effect of trimetaphosphate when fed to rats during different stages of tooth development. Proc. Internat. Assoc. Dent. Res., March, 1966.

Neuman, W. F., and Neuman, M. W.: The Chemical Dynamics of Bone Mineral. Chicago, University of Chicago Press, 1958.

Neuman, W. F., and Neuman, M. W.: Recent advances in bone growth and nutrition. Borden's Rev. Nutr. Res., 21:37, 1960.

Neuman, W. F., and Riley, R. F.: Uptake of radioactive phosphorus by calcified tissues of normal and choline-deficient rats. J. Biol. Chem., 168:545, 1947.

Nicholls, L., and Nimalasuriya, A.: Adaptation to low calcium intake in reference to calcium requirements of tropical population. J. Nutr., 18:563, 1939.

Nicolaysen, R., Eeg-Larsen, N., and Malin, O. J.: Physiology of calcium metabolism. Physiol. Rev., 33:424, 1953.

Nizel, A. E., Baker, N. B., and Harris, R. S.: The effect of phosphate structure upon caries development in rats. Proc. Internat. Assoc. Dental Res., Abst. 236, 1962, p. 63.

Nizel, A. E., and Harris, R. S.: The effects of phosphates on experimental dental caries. A literature review. J. Dent. Res., 43:1123, 1964.

Parma, C., Danek, J., and Hanusova, N.: Fluoridation of drinking water. Cesk. Stomat., 52:150, 1952.

Pearson, J. D.: Evidence of the existence of nonexchangeable calcium of human urine. Nature, 205:410, 1965.

Radusch, D. F.: Relationship of gastroacidity to periodontoclasia. J. Periodont., 18:110, 1947.

Reid, M. E.: Interrelations of calcium and ascorbic acid to cell surfaces and intercellular substances and to physiological action. Physiol. Rev., 23:76, 1943.

Ritchie, D. B.: Surface enamel magnesium and its possible relation to incidence of caries. Nature, 190:456, 1961.

Rosebury, T., and Karshan, M.: Susceptibility to dental caries in the rat. V. Influence of calcium, phosphorus, vitamin D, and corn oil. Arch. Path., 20:697, 1935.

Schachter, D., and Rosen, S. M.: Active transport of Ca^{45} by the small intestine and its dependence on vitamin D. Amer. J. Physiol., 196:357, 1959.

Schour, I.: Calcium metabolism and teeth. J. A. M. A., 110:870, 1938.

Sharpe, L. M., Peacock, W. C., Cooke, R. C., and Harris, R. S.: Effect of phytate and other food factors on iron absorption. J. Nutr., 41:433, 1950.

Shelling, D. H., and Asher, D. E.: Calcium and phosphorus studies. VIII. Some observations on the incidence of caries-like lesions in the rat. J. Dent. Res., 13:363, 1933.

Sherman, H. C.: Calcium and Phosphorus in Foods and Nutrition. New York, Columbia University Press, 1947.

Shipp, I. I., and Mickelsen, O.: The effects of calcium acid phosphate on dental caries in children: A controlled clinical trial. J. Dent. Res., 43:1144, 1964.

Snapper, I.: Calcium and phosphorus malnutrition. In Jolliffe, N., Tisdall, F., and Cannon, P. (ed.): Clinical Nutrition. New York, Paul B. Hoeber Inc., 1950.

Solomons, C. C., and Irving, J. T.: Studies in calcification; the reaction of some hard and soft-tissue collagens with 1-fluoro-2:4-dinitrobenzene. Biochem. J., 68:499, 1958.

Spencer, H., Menczel, J., Lewin, I., and Samachson, J.: Effect of high phosphorus intake on calcium and phosphorus metabolism in man. J. Nutr., 86:125, 1965.

Steggerda, F. R., and Mitchell, H. H.: The calcium balance of adult human subjects on high and low fat (butter) diets. J. Nutr., 45:201, 1951.

Stralfors, A.: The effect of calcium phosphate on dental caries in school children. J. Dent. Res., 43:1137, 1964.

Thomas, R. O., Litovitz, T. A., Rubin, M. I., and Geschickter, C. F.: Dynamics of calcium metabolism. Amer. J. Physiol., 169:568, 1952.

Tibbetts, D. M., and Aub, J. C.: Magnesium metabolism in health and disease. J. Clin. Invest., 16:491, 1937.

Toth, K.: Calcium, phosphorus and magnesium content of the saliva of caries active and caries resistant individuals. Acta med. (Acad. Sci., Hungary), 6:493, 1964.

Tufts, E. V., and Greenberg, D. M.: The biochemistry of magnesium deficiency. J. Bio. Chem., 122:715, 1937.

Underwood, E. J.: Trace elements. In Bourne, G. H., and Kidder, G. W. (ed.): Biochemistry and Physiology of Nutrition. Vol. II. New York, Academic Press, 1953.

United States National Research Council: Recommended Dietary Allowances. 6th ed. Food and Nutrition Board, National Academy of Sciences–National Research Council, Publ. No. 1146, 1964.

Urist, M. R.: The etiology of osteoporosis. J. A. M. A., 169:710, 1959.

Urist, M. R.: Bone: Formation by autoinduction. Science, 150:893, 1965.

Urist, M. R., and McLean, F. C.: Metabolic Interrelations. Trans. 3rd Conference Josiah Macy, Jr. Foundation, New York, 1951, p. 55. Osteogenic potency and osteogenetic inductor substances of periosteum, bone marrow, bone grafts, fracture callus, and hyaline cartilage transferred to the anterior chamber of the eye.

Wacker, W. E. C., and Vallee, B. L.: A study of magnesium metabolism in acute renal failure employing a multichannel flame spectrometer. New Eng. J. Med., 257:1254, 1957.

Wasserman, R. H., Comar, C. L., and Nold, M. M.: The influence of amino acids and other organic compounds on the gastrointestinal absorption of $calcium^{45}$ and $strontium^{90}$ in the rat. J. Nutr., 59:371, 1956.

Wasserman, R. H., Comar, C. L., Schooley, J. C., and Lengemann, F. W.: Interrelated effects of L-lysine and other dietary factors on the gastrointestinal absorption of $calcium^{45}$ in the rat and chick. J. Nutr., 62:367, 1957.

Watt, K., and Merrill, A. L.: Composition of Foods: Raw, Processed, Prepared. Handbook No. 8, Washington, D. C., U.S. Department of Agriculture, 1963.

Westerlund, A.: The effect of tripalmitin and triolein on the fecal output of grown rats. Lantbruks-Högskolans Ann., *1*:1, 1934.

Whedon, G. D.: Effects of high calcium intake on bone, blood and soft tissue; relationship of calcium intake to balance in osteoporosis. Fed. Proc., *18*: 1112, 1959.

Widdowson, E. M., McCance, R. A., and Spray, C. M.: The chemical composition of the human body. Clin. Sci., *10*:113, 1951.

Wilson, T. H., and Wiseman, G.: Use of sacs of everted small intestine for study of transference of substances from mucosal to serosal surface. J. Physiol., *123*:116, 1954.

Wolbach, S. B.: Vitamin-A deficiency and excess in relation to skeletal growth. J. Bone Joint Surg., *29*: 171, 1947.

Wolbach, S. B., and Bessey, O. A.: Tissue changes in vitamin deficiencies. Physiol. Rev., 22:233, 1942.

Wolbach, S. B., and Hegsted, D. M.: Perosis; epiphyseal cartilage in choline and manganese deficiencies in chick. A.M.A. Arch. Path., *56*:437, 1953.

Wynn, W., Haldi, J., Bentley, K. D., and Law, M. L.: Dental caries in the albino rat in relation to the chemical composition of the teeth and of the diet. II. Variation in the Ca/P ratio of the diet induced by changing the phosphorus content. J. Nutr., *58*: 325, 1956.

Wynn, W., Haldi, J., Bentley, K. D., and Law, M. L.: Dental caries in the albino rat in relation to the chemical composition of the teeth and to the diet. IV. Variations in the Ca/P ratio of the diet induced by changing the calcium content. J. Nutr., *57*: 645, 1959.

Wynn, W., Haldi, J., and Law, M. L.: Dental caries in albino rats fed diets containing different amounts of calcium and phosphorus with the same Ca/P ratio. J. Dent. Res., *39*:1248, 1960.

Chapter Nine

Iron and Micro Elements

ROBERT S. HARRIS, PH.D.

IRON

INTRODUCTION

The ancient Greeks believed that their god Mars had endowed iron with strength and that those who ate iron would become strong and masterful. The development of knowledge of the functions of iron as an essential element has been painfully slow during the intervening centuries. Important milestones in progress toward an understanding of the vital roles of iron have been: (1) the identification of iron as an integral part of the hemoglobin molecule, (2) the development of methods for the analysis of iron in foods and tissues, (3) the recognition that hypochromic microcytic anemia is the principal sign of iron deficiency, (4) proof of the function of iron in enzyme systems, and (5) the application of isotope techniques in studies of iron absorption and metabolism. Much is now known about the biology of iron in plant and animal systems; much more awaits discovery.

DISTRIBUTION OF IRON

The iron content of adult human beings varies between 3 and 5 gm. Moore and Dubach (1962) have calculated the distribution of iron in a 70 kg. man to be: 70.5 per cent as hemoglobin, 3.2 per cent as myoglobin, 26.0 per cent as storage iron (ferritin, hemosiderin), 0.1 per cent as transport-iron-transferrin-iron complex, 0.1 per cent as cytochrome, 0.1 per cent as catalase, and trace amounts in other cytochromes, flavoproteins and various iron-containing enzymes.

SERUM IRON

The level of iron in the serum ranges from 0.09 to 0.18 mg./100 ml. for men, and from 0.07 to 0.15 mg./100 ml. for women, yet the total amount of iron in the circulating plasma of an adult amounts to only 3 to 4 mg. The serum iron is high at birth, decreases by two days, increases by seven days, falls slowly to about 0.05 mg./100 ml. between 8 and 24 months of age, then slowly rises throughout childhood to adult levels. The turn-over of iron in the plasma is approximately 10 times each day. It is important to note that sufficient hemoglobin is broken down each day to release about 20 to 25 mg. of iron. Since the actual requirement of a normal man each day is only 1 mg., it is evident that essentially all the iron required for hemoglobin synthesis is recovered from discarded hemoglobin molecules.

SOURCES OF IRON

The iron content of vegetable foods varies considerably because of genetic and agricultural factors. Stiebeling (1932) classified the edible portions of plant foods as follows:

Poor in iron (0-4 ppm): Most fruits and fruit juices

Fair in iron (4-8 ppm): Some fruits, blanched leaves and stalks, roots and herbs

Good in iron (8-16 ppm): Potatoes, green stalks and leaves

Excellent in iron (over 16 ppm): Leguminous plants and leaves

When vegetables are boiled in liberal amounts of water, the iron content is reduced by 20 to 50 per cent through extraction.

When wheat is milled to produce white flour, the iron content is reduced by two thirds. Enriched white flour and enriched white bread contain more iron, and in a more available form, than whole wheat flour and bread.

Iron is partially released from its conjugates when foods are cooked or digested. Only small amounts of iron are absorbed from raw blood (Callender et al., 1957). The absorption is much improved by heat treatment to coagulate (Brading et al., 1957) or by cooking (Walsh et al., 1955). Preliminary heat treatment permits acids and pepsin to release considerable amounts of iron from hemoglobin (Kaldor, 1957) and from meat products.

The iron in foods from animal sources is approximately as available as that from vegetable tissues. Possibly 50 per cent of the iron in beef is absorbed by man (Johnston et al., 1948), but only 4 per cent of the iron in eggs (Moore and Dubach, 1951) and 13 per cent of the iron in spinach (McMillan and Johnston, 1941).

Iron in inorganic compounds is generally more available than iron present in complexes or in foods. Stitt et al. (1962) reported that rats absorbed iron from a ferric fructose chelate better than from ferrous sulfate. They observed that conditions in the duodenum favor the formation of a chelate which is stable and may be easily absorbed. Sugar chelates may be important vehicles for iron absorption.

The iron content of diets varies considerably. For instance, although the average United States diet was reported by Sherman (1935) to supply 14 to 20 mg. of iron per day, and diets in Australia supplied 20 to 22 mg. of iron (National Health Med. Res. Council, 1945), the diets of millions in India supplied only 9 mg. of iron daily (Health Bull., 1951), and diets consumed by the poor of Scotland contained only 11 mg. (Davidson et al., 1933), and these latter groups exhibited anemia which responded to iron therapy. Iron intakes tend to be inadequate when the people select their diets from high energy, low cost, low protein foods.

FUNCTIONS OF IRON COMPOUNDS

Iron in living organisms always appears to be bound to soluble organic molecules of high molecular weight, usually proteins. The activities of the five known heme proteins are mediated through one or several iron atoms in each molecule. The iron cannot be replaced by another element without loss of activity. Iron is in the center of the porphyrin ring and is bound to four pyrrole-nitrogens by ionic bonds.

Hemoglobin. This is the pigment of red cells and accounts for 75 per cent of the body iron; its molecular weight is 67,000, and it contains four iron atoms. The iron can be oxidized from the ferrous to the ferric state (as in methemoglobin) with loss of capacity to carry oxygen.

Myoglobin. Myoglobin is the heme-protein which occurs in skeletal and heart muscles. Its structure resembles hemoglobin, its molecular weight is 16,500, and it contains only one ferrous porphyrin group. Myoglobin has a far greater affinity for oxygen than hemoglobin; as a result, it can accept the oxygen released by hemoglobin in the tissues and serve as a reservoir of oxygen.

Catalases and Peroxidases. These are heme-containing enzymes which are able to liberate oxygen from peroxides; the iron in these compounds is in the ferric state. Most peroxidases are found in plant tissues. At least two animal peroxidases are known: verdoperoxidase in white cells, and lactoperoxidase in milk. Catalases (m.w., 250,000) contain four iron atoms per molecule while peroxidases have only one. Besides decomposing hydrogen peroxide, catalases may peroxidize large molecules, such as phenols.

Cytochromes. These are heme enzymes which occur in cell mitochondria and provide electron transfer through the ability of iron to undergo reversible oxidation. More than four components (a, b, c, a₃) are known. Cytochrome c (m. w., 12,000) is the only one which is easily extracted from tissues and purified.

Transferrin. Also called siderophilin, this nonheme plasma β_1-globulin (m. w., 90,000) selectively binds and transports iron.

Ferritin. Ferritin is an iron-protein complex found in significant amounts in the bone marrow, liver and spleen. More than 20 per cent of this molecule (m. w., 465,000) consists of iron as ferric hydroxide micelles attached to an apoferritin. Its molecule approximates $(FeOOH)_8 \cdot (FeOPO_3H_2)$. Radioactive iron-ammonium citrate was recovered in liver ferritin soon after injection into a dog (Hahn et al., 1943). The ferritin in intestinal mucosal cells increases following the ingestion of iron (Granick, 1946).

Hemosiderin. This is an iron-protein complex which contains 35 per cent iron and is made up of iron hydroxide units polymerized as in ferritin. In normal tissues there is more ferritin than hemosiderin; when iron is taken in excess, the proportion of hemosiderin increases progressively. Greenberg (1955) concluded that a significant portion of hemosiderin is ferritin. These compounds are equally available for hemoglobin synthesis.

Ferroflavoproteins. These contain iron and riboflavin as part of their prosthetic groups. Other forms of tissue iron compounds are xanthine oxidase, succinic dehydrogenase, and DPHN-cytochrome reductase.

ABSORPTION OF IRON

Iron is absorbed in the ionic state via the lymphatics, preferentially in the ferrous form because it is more efficiently absorbed than ferric iron. It is quite likely that ferric iron is reduced to the ferrous form in the stomach before absorption.

There is considerable evidence in support of the concept that the absorption of iron is related to physiological demand. Ferritin appears to serve as an intermediary binding site for iron after absorption into the intestinal wall and before release into the blood plasma (Granick, 1946). Although iron can be absorbed from any portion of the gastrointestinal tract, the uptake is greatest in the duodenum. The claim that partial gastrectomy reduces iron absorption has not been proved.

The presence of anemia (Choudhury and Williams, 1959) and the kinds and amounts of food ingested with the iron (Sharpe et al., 1950) reduce the absorption of food iron. Iron assimilation is reduced when the stomach empties quickly (Owren, 1953). The concurrent appearance of achlorhydria with anemia indicates that poor absorption of iron is a factor, but perhaps the achlorhydria is the result rather than the cause of the iron deficiency anemia (Hallén, 1938). The iron in most foods is present in complex forms. Previously, it was assumed that gastric juice is required for the release of iron from these protein complexes; however, since patients with achlorhydria can assimilate food iron efficiently this is not the explanation.

In rats, ferric iron was utilized only one fifth as well as ferrous iron (Venkatachalam et al., 1956). Moore et al. (1939) reported that in both humans and dogs the incorporation of iron into red cells was greater when ferrous iron was given rather than ferric. The ratio of effectiveness of ferrous/ferric is 5/1 in man (Moore et al., 1944). Possibly this ratio is a reflection of the time required for the reduction of ferric to ferrous iron in the gut wall.

Factors Which Increase Iron Absorption. The absorption of ferric iron is increased when ascorbic acid is given simultaneously; ferrous iron is not affected in this way (Moore and Dubach, 1956). Presumably, this effect is the result of the high reducing activity of ascorbic acid, but this has not been proved. The amount required for most efficient action is high—larger than the amount of ascorbic acid usually present in normal diets.

Iron absorption increases as the size of the test dose is increased, but the efficiency of absorption decreases (Bothwell et al., 1958). The absorption of iron appears to be improved by the concurrent ingestion of ascorbic acid, succinic acid, fructose, D-sorbitol and inosine, especially when iron is in the ferric form.

Efficiency of Iron Absorption. The efficiency of absorption of food iron varies with the types and amounts of food eaten, with physical activity, and especially with the adequacy of iron stores in the body tissues. The iron from hemoglobin, liver, muscle and iron-enriched bread is absorbed better than that from eggs and vegetables. Patients with an iron deficiency tend to absorb iron from foods more efficiently than normal people. The average absorption of food iron is only about 5.3 per cent in normal subjects, but may reach 21.6 per cent in iron-deficient patients.

Schulz and Smith (1958) observed 16 per cent absorption of iron when normal children were fed 30 mg. of ferrous sulfate, and 22 per cent when the same dose was given to anemic children. Chodos and Ross (1953) reported that normal subjects absorbed 10 to 34 per cent and anemic subjects 33 to 66 per cent of the iron administered. Schulz and Smith (1958) tagged foods with radioactive Fe^{59} and measured its absorption by children. The absorptions by normal children were: 9 per cent from milk, 12 per cent from mixed enriched cereals, 11 per cent from scrambled eggs, 7 per cent from raw eggs and 8 per cent from chicken livers. Absorption by iron-deficient children was nearly twice this.

It is evident that the availability of the iron in various foodstuffs is quite variable, that normal people absorb only 5 to 10 per cent of the food iron, and that deficient patients absorb perhaps 10 to 20 per cent. On this basis, a diet which contains 12 mg. of iron daily actually contributes only about 0.6 mg. of iron to normal subjects and possibly as much as 2 mg. to deficient subjects. The absorption of iron is regulated by some mechanism still not understood.

Factors Which Interfere with Iron Absorption. Calcium and phosphorus may interfere with iron absorption, since iron phosphates have low solubilities and calcium reduces the acidity of aqueous mixtures. Furthermore, iron and calcium compete for the same process in active transport during absorption (Mannis and Schachter, 1962). Anemia was produced in human beings when excess phosphorus was fed (Day and Stein, 1938); and the loss of iron through fecal excretion was increased (Brock, 1937). It should be emphasized that calcium, phosphorus and iron adversely affect one another only when one is in great excess and the other is present in limited amounts.

Phytic acid and its salts interfere with iron absorption in much the same way that they interfere with the absorption of calcium salts (see page 71). The solubility of iron phytate is low. The response of serum iron to large amounts of iron ingested in bread and jam was less when sodium phytate was fed concurrently (McCance et al., 1943). Both sodium phytate and the mixed phytate in oatmeal caused a reduction of absorption of iron when fed in the breakfasts of children (Sharpe et al. 1950). Hegsted et al. (1949) have reported that low phosphate diets increase iron utilization.

IRON UTILIZATION

Studies with radioiron have revealed that iron is utilized in hemoglobin synthesis within four hours following injection or ingestion (Hahn et al., 1939), and that 70 to 100 per cent of injected Fe^{59} is present as circulating hemoglobin within 12 days (Sharpe et al., 1950).

Isotope studies have proved also that most of the iron released from erythrocytes that are phagocytized by reticuloendothelial cells enters the plasma, and that the iron used in the synthesis of new hemoglobin comes mostly from the plasma. Although Bessis and Breton-Gorius (1957) have presented data which suggest that iron which has been released from destroyed cells may be used directly for new hemoglobin formation, it is nevertheless likely that most of the hemoglobin is formed using iron obtained from the plasma.

Isotope studies have demonstrated that iron is transferred from mother to fetus in the maternal plasma and in the maternal red cells (Pommerenke et al., 1942).

Iron is stored intracellularly in the bone marrow, liver, spleen and several other tissues, mostly as ferritin, and partly as hemosiderin. Both these compounds can be mobilized for hemoglobin synthesis, but the iron leaves at a slow rate. The iron stores in an adult man amount to about 1.2 to 1.5 gm., and nearly 20 per cent is in the liver (Haskins et al., 1952).

IRON EXCRETION

Since the body has only limited ability to excrete iron, insignificant amounts are excreted in the urine. Most of the iron in the feces represents unabsorbed iron. When large amounts of iron are released by hemolysis into the plasma, essentially none of it is excreted (McCance and Widdowson, 1938). Small amounts are nevertheless lost by endogenous excretion and from the skin. A maximum of 0.4 per cent of intravenously injected Fe[55] was recovered in the feces of rats (Copp and Greenberg, 1946). Similarly, Dubach et al. (1955) recovered 1 to 2 per cent of injected radioiron from the urine of normal human subjects, and less than 0.003 per cent of it was found in the sweat. Only 1.2 mg. of iron/year is lost in hair clippings and combings (Dubach et al., 1946).

It is evident that the body possesses no active regulatory mechanism for the excretion of iron, and yet a normal adult male excretes or loses 0.5 to 1.0 mg. of iron per day from sweat glands, in desquamated epidermal and mucosal cells, and in the urine and feces.

During menstrual periods, women lose 35 to 70 mg. of blood, equivalent to 16 to 32 mg. of iron/month or 0.5 to 1.0 mg./day/month. A normal fetus contains 273 mg. of iron (Widdowson and Spray, 1951). During delivery, a woman loses 80 to 120 mg. of iron in the placenta and 350 mg. in the blood (Fullerton, 1936).

SYSTEMIC EFFECTS OF IRON DEFICIENCY

Iron deficiency results in hypochromic microcytic anemia, a form of anemia in which the hemoglobin concentration is decreased more than the red cell count. This type of anemia is widespread among human populations, in some areas afflicting up to 50 per cent of the people. It is the most common form of anemia, and, in fact, is one of the most common deficiency diseases. This anemia results from inadequate iron intake or absorption, from blood loss, from frequent pregnancies and from poor food habits. It is seen most frequently among infants, children and women of child-bearing ages.

This anemia actually is a late sign of iron deficiency, since it does not become evident until the iron stores are exhausted. As the deficiency develops, the iron stores lower, transferrin levels rise, serum iron falls, hemoglobin decreases, and finally hypochromic anemia appears.

A common mild form of iron deficiency syndrome is characterized by chronic fatigue, depleted iron stores and slightly depressed blood hemoglobin. This syndrome is relieved shortly after iron therapy is begun (Coleman

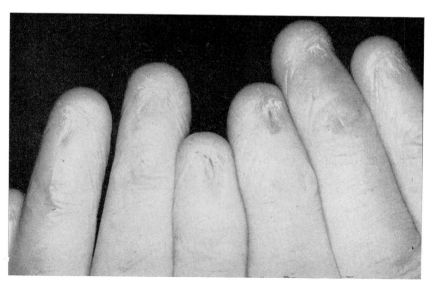

FIGURE 9-1. *Plummer-Vinson syndrome. Spoon-shaped appearance of fingernails (koilonychia) resulting from iron deficiency anemia.*

et al., 1955). Patients with iron deficiency anemia feel exhausted and are pallid; they may show glossitis, cheilosis, fingernail alterations (Fig. 9-1), paresthesias and stomach contraction. Many of these signs are so nonspecific that ailing patients are easily led into taking medicines that are of no therapeutic value.

IRON STORAGE DISEASES

High accumulations of iron have been observed in human beings as a result of (1) idiopathic hemochromatosis, (2) transfusion hemosiderosis, (3) prolonged iron therapy, or (4) excessive consumption of iron in diets. Finch and Finch (1955) reported the latter condition in Bantus who were consuming 200 mg. of iron per day in the diet, much of it from iron kettles used in cooking. An overload of iron has been produced in rats who received injections of inorganic iron over long periods of time (Brown et al., 1957). When subjects absorb iron in excess of 2 to 3 mg./day (which is more than is excreted), they will show a chronic buildup of iron amounting to over 1 gm. of iron in a year's time.

Idiopathic hemochromatosis is a rather rare disease, which occurs mainly in males. Possibly it is caused by genetic-related errors in metabolism. The liver becomes cirrhotic, the skin bronzes and diabetes mellitus commonly accompanies it (Finch and Finch, 1955). The iron is deposited mostly in the parenchyma of cells which are avid for it. Richter (1960) considers that the pathogenesis of this disease results from an abnormal pathway of ferritin metabolism.

ORAL PATHOLOGY IN IRON DEFICIENCY

Tongue and Mucous Membranes. Iron deficiency anemia or Plummer-Vinson disease is characterized by such oral signs as fissures in the labial commissures and superficial glossitis (Fig. 9-2). The papillae of the tongue are atrophied, thus giving the tongue a smooth, shiny and red appearance. All oral mucous membranes are atrophied and appear ashen gray. It is believed that oral tissues thus affected are more susceptible to carcinoma.

The clinical appearance of the tongue in iron deficiency resembles very closely that seen in niacin, riboflavin or other B complex deficiencies. The tongue has been variously described as a "blotching irregular denudation of the papillae," as having an "absence of epithelial tufts" of the filiform papillae, or as containing "dusky red irregular spots." Darby (1946) feels that the clinician should be aware that angular cheilosis and glossitis can be caused by an iron deficiency as well as by a vitamin B complex deficiency.

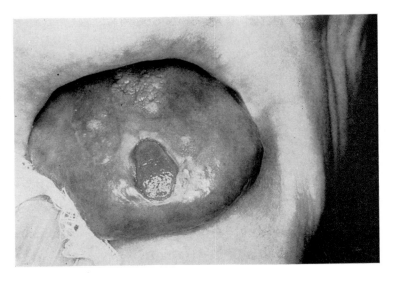

FIGURE 9-2. *Plummer-Vinson syndrome. Glossitis and ulceration resulting from iron deficiency anemia. (Courtesy Dr. P. McCarthy.)*

IRON REQUIREMENTS

An adult man and a postmenopausal woman in a temperate climate require approximately 1 mg. of iron daily, while a woman of menstrual age requires a total of about 2 mg. of iron daily. In order to meet growth requirements, children must have a net retention of about 0.6 mg. per day, which means that allowing for that lost by excretion, they need 1.6 mg./day (see Woodruff, 1958). Since the efficiency of iron absorption is only 5 to 10 per cent, a daily intake of 15 mg. of food iron is needed to supply the 0.7 to 1.5 mg. of absorbed iron.

MICRO OR TRACE ELEMENTS

The amounts of minerals in plant and animal tissues range from grams to micrograms per 100 grams of substance. Those that are present in abundance have been called macro elements, while those that are present in very small quantities have been referred to as micro elements. Years ago, chemical and spectrographic studies of tissues revealed the presence of certain elements, and since these minerals could not be quantitated by the methods then available, investigators referred to them as "trace elements." This term has remained in popular usage, even though most of the minerals now can be estimated with precision even when present in microgram or even 0.01 microgram amounts.

Many food chemists and nutritional biochemists have abandoned the term trace elements because it may be misleading. Mineral elements are equally essential whether needed in milligram or microgram quantities. Some elements occur in trace amounts in some tissues, but in rather high concentrations in other tissues. Also scientists cannot agree on the definition of a "trace element," primarily because some elements cannot easily be placed in such a category. For instance, in a recent publication Underwood (1962) placed iron and iodine in the "trace element" category, while Moore (1964) excluded them.

For present purposes, the term "trace element" will be used interchangeably with "micro element," all others being considered as macro elements.

ESSENTIALITY OF MICRO ELEMENTS

It has been difficult to establish the essentiality of micro elements in the nutrition of human beings because these elements are generally furnished in adequate amounts in mixed diets (Lowe et al., 1960), and the feeding of highly refined diets to human subjects is fraught with difficulties. However, the clear evidence from experimental animals furnishes a reliable indication that many of these micro elements are required by man also.

Many of the micro elements have been demonstrated to be required in enzyme systems in human subjects. Certain of these are required in low concentrations and may even act as antinutrients if present in excess. Unlike macro elements, the micro elements do not have important structural functions. Their role is primarily catalytic. Although they may occur as components of biological molecules (e.g., zinc in insulin), their main functions are as components of catalytic systems which control biochemical and physiological processes in cells. When absent, they may reduce enzyme activities and thereby produce pathologies and even death.

MICRO ELEMENTS AS ENZYME ACTIVATORS

Enzymes are organic catalysts synthesized by protoplasm. A cell contains thousands of different enzymes which regulate the chemical activities vital to its life. Enzymes are proteins. Very few enzymes (e.g., pepsin, trypsin) consist only of protein, yet the activities of even these may be accelerated by mineral ions such as manganese or calcium. Most of the enzymes that have been studied carefully have been shown to be conjugated proteins. The entire enzyme (holoenzyme) consists of protein (apoenzyme) linked to a nonprotein group (prosthetic group). Some prosthetic groups can be removed readily from the enzyme, and these are termed co-enzymes. Some co-enzymes are complex molecules (e.g., co-enzyme I, or DPN), and some are simple ions (e.g., copper in ascorbic acid oxidase). Because ions are chemically simple they are commonly referred to as "activators."

Not all micro elements are activators; they may be incorporated into the apoenzyme fractions or may be only parts of specific enzyme systems and play important roles in metabolism.

Several hundred enzyme systems containing micro elements are known, but only a few are understood. Schütte (1964) has listed a number of these enzymes and has cataloged them according to: (1), micro elements which constitute the prosthetic group (Ca, Cl, Cu, Mg, Zn); (2), micro elements which are an active part of the prosthetic group or are incorporated into the enzyme itself (Cd, Co, Cu, F, Fe++, Fe+++, K, Mg, Mn, Mo, Ni, NH_4, PO_4, Rb, V, Zn); (3), elements which have functions not yet understood (Al, B, Ba, Br, Ca, Cd, Cl, CNS, Cu, Fe++, Fe+++, I, K, Li, Mn, Mg, Mo, Na, Ni, NH_4, NO_3, PO_4, SO_4, Rb, Sr, V, W, Zn); (4), elements which serve as facultative activators (Al, As, AsO_4, B, BO_3, Ba, Br, Ca, Ce, Cl, Co, CN, F, Fe, Hg, I, K, La, Mg, Mn, Mo, Na, NH_4, NO, NO_2, NO_3, Nd, Ni, PO_4, Pb, Pr, Rb, S, SH, SO_2, SO_3, SO_4, S_2O_3, Si, Sm, Th, Ti, V, W, Y, Zn, Zr); and (5), elements which have been observed to inhibit the activity of these enzymes (Ag, AsO_3, B, Ba, Ca, Cd, CN, Co, Cs, Cu, Cu++, F, Fe++, Hg, Hg++, I, Cd++, Mg, Mn, Mo, Ni, Pb++, PO_3, PO_4, P_2O_7, S, SeO_2, SO_4, SH, Zn). This information is presented in detail here to emphasize the fact that many mineral elements are able to influence enzyme activities. It is likely that some of the micro elements now considered to be nonessential will eventually be proved to be essential to man's nutrition.

COPPER

The first demonstration that copper is essential in animal nutrition was reported by Hart et al. (1928), who observed that copper is required for the synthesis of hemoglobin. Since then, copper has been shown to be a component of several vital metalloenzymes and oxygen-carrying proteins, to play an essential role in bone formation, and to be a part of several respiratory enzymes.

Copper is essential to the life of plants, invertebrates and vertebrates. Currently, there is considerable research interest on the mineral interactions involving copper (i.e., meat anemia), and abnormalities of copper metabolism (i.e., hepatolenticular degeneration, or Wilson's disease).

DISTRIBUTION OF COPPER

Copper has been detected in all tissues of plant and animal origin. The data in Table 9-1 indicate that oysters, liver, legumes and cereals are the best sources. It is estimated that most diets supply between 2 and 5 mg. of copper per day and that any diet adequate in calorie content will provide the copper requirement of about 0.05 to 0.1 mg. per kg. body weight per day.

Chan (1935) analyzed human cadavers and reported values of 100 to 160 mg. of copper. Highest concentrations were noted in the internal organs (liver, kidney, heart) and the central nervous system, and lowest amounts in lungs and muscles (Gubler, 1956; Wintrobe et al., 1953). However, because of their large mass, bones and muscles account for 75 per cent of the total body copper.

The level of copper in the livers of human infants is high during the first two months of life, then decreases slowly over a period of many months (Ramage et al., 1933).

FUNCTIONS OF COPPER

Copper is an essential component of respiratory pigments in the blood and body fluids of many invertebrates. These compounds are equivalent to the hemoglobins in the blood

TABLE 9-1.   COPPER CONTENT OF REPRESENTATIVE FOODS*†

Oysters	3.62	Rye, whole	0.66
Liver	2.45	Corn	.45
Beans, dry	.96	Flour, whole wheat	.44
Lima	.92	Kale	.33
Peas, dry	.80	Prunes, dried	.29
Wheat	.79	Eggs	.25
Oats	.74	Mackerel	.23
Avocado	.69	Bread, white	.21

* Mg./100 gm. edible portion.
† From Sherman, 1952.

of man, functioning as carriers of oxygen to the tissues. Copper is also a constituent of respiratory enzymes, such as ascorbic acid oxidase, an enzyme commonly present in vegetable tissues and phenol oxidases. The only enzyme of this type in animal tissues is tyrosinase, which oxidizes a series of monohydric phenols to o-dihydric phenols, and to o-quinones. Tyrosinase can oxidize tyrosine and dihydroxyphenylalanine (dopa) and is, therefore, involved in the initial stages of melanin synthesis.

Copper plays a fundamental role in the utilization of iron in the early steps of hemopoiesis in young erythrocytes.

ABSORPTION, TRANSPORT AND EXCRETION OF COPPER

Copper is probably absorbed from the upper section of the small intestine, but little is known of the mechanism involved (Bearn and Kunkel, 1955). Absorption is decreased by the presence of excess calcium ions or phosphates in the gut and is enhanced by low pH (Tompsett, 1940). Organic complexes of copper in vegetable tissues may be absorbed even better than copper sulfate (Mills, 1958).

Cartwright et al. (1954) have estimated that human subjects consuming 2.6 mg. of copper per day in the diet retain approximately 0.01 to 0.03 mg., or 1 to 3 per cent.

Copper is transported via the plasma. When radioactive Cu^{64} was given orally or intravenously to normal human beings, the radioactivity of the serum first rose, then fell rapidly during the following four hours, and then showed a slower secondary rise (Bearn and Kunkel, 1955). Copper is loosely bound to serum albumin and transported to the liver and bone marrow where it is stored and incorporated into proteins. The slow secondary rise in serum radioactivity was associated with ceruloplasmin and probably represents the incorporation of copper into this protein by synthesis, or by exchange.

Copper is normally excreted in the bile (35 to 205 μg./liter) (Butler and Newman, 1956). In addition, small amounts are lost in the sweat, menstrual flow, milk and by desquamation.

COPPER DEFICIENCY

As mentioned above, Hart et al. (1928) demonstrated that young rats raised on a milk diet did not gain weight unless copper supplements were added. Copper deficiency in farm animals produces anemia, neurologic disorder, enzootic ataxia (swayback), myocardial fibrosis (falling disease), diarrhea (scours), depigmentation and loss of quality of hair, wool and feathers, and skeletal abnormalities. There have been no reports of similar symptoms in human beings. Presumably, this is because human subjects have not been maintained on diets sufficiently low in copper content for a sufficiently long period of time. Hypocupremia has been observed in patients with nephrosis, in whom large losses of ceruloplasmin in the urine have been noted. However, a diagnosis of copper deficiency has not been established in any of these patients (Gubler, 1956).

The kind and severity of copper deficiency syndrome appears to vary with animal species, age, environment and duration of copper deprivation. Neonatal ataxia has not yet been described as a consequence of copper deficiency in rats, rabbits and dogs, yet it has been noted in lambs born of copper-deficient ewes.

A copper deficiency anemia has been induced in rats, rabbits, chickens, pigs, dogs, sheep, goats and cattle. It is usually a hypochromic microcytic anemia (Smith and Medlicott, 1944). This type of anemia in rats, rabbits and pigs is indistinguishable from that induced by iron deficiency. This is to be expected if copper deficiency interferes with the early stages of hemoglobin formation, previous to the incorporation of iron into this molecule.

Baxter et al. (1953a, b) have studied the effects of copper deficiency on the bones of dogs. Deficiency bones were characterized by a decrease in thickness of the cortex, especially in the shaft of the femur. Histologically, the bones showed metaphyseal thinning and an increase in the width of the epiphyseal cartilage. A diffuse osteoporosis with alteration in osteoblastic activity was seen. There was also a cessation of destruction of calcified cartilaginous matrix, yet the chondroblastic activity was not impaired. Bone changes similar to these were not noted in dogs that had been made anemic by feeding iron-deficient diets.

FAULTY COPPER METABOLISM

Hepatolenticular degeneration (Wilson's disease) is a rare genetic fault in copper metabolism which develops in human beings maintained on diets of normal copper content. The absorption and retention of copper is abnormally high, the copper level in the serum is subnormal, and excretion in the feces is decreased. Copper accumulates in unusual amounts in various tissues, precipitating pathologic changes. The liver parenchyma degenerates and is replaced by fibrous tissue, producing a condition which resembles liver cirrhosis (Anderson and Popper, 1960). Progressive brain damage, particularly in the basic ganglia, is observed. Kidney function is impaired as a result of tubular and glomerular pathologies which develop during later stages of the disease. Eventually, the copper deposits in the cornea produce the characteristic Kayser-Fleischer rings. The tissues of untreated patients become progressively more abnormal, and death may finally ensue.

Patients with this disorder may be treated by prescribing (1) diets low in copper content, (2) agents such as ion exchange resins and potassium sulfide, which interfere with copper absorption, and (3) compounds such as dimercaprol (BAL) which chelate copper in the body fluids and increase its excretion in the urine.

TOXICITY OF COPPER

Copper salts are used in agriculture to control infections and infestations, and these may contaminate food supplies. Workers may accumulate toxic amounts of copper from industrial processes. Acute copper poisoning is usually characterized by a gastroenteritis accompanied by severe shock and rapid prostration. Copper poisoning results when farm animals or men are exposed continuously to soluble copper salts. For instance, sheep ex-

posed to 20 mg. of copper sulfate per kg. body weight were observed to develop symptoms of chronic poisoning.

Subacute doses of copper caused an increase in liver copper until a maximum was reached. The copper then liberated into the blood stream caused extensive hemolysis and jaundice. Death resulted in animals fed higher levels of copper.

INTERACTIONS OF COPPER WITH OTHER MINERALS

Molybdenum exerts a profound effect on copper metabolism. Farm animals raised on pastures containing low levels of molybdenum accumulated toxic quantities of copper in their livers. On the other hand, animals on pastures containing excess amounts of molybdenum developed copper deficiency disease and showed reduced levels of copper in the tissues.

This interrelationship is further complicated by the fact that increased amounts of other minerals in the diet (e.g., inorganic sulfates) aggravate the toxicity of molybdenum. These interrelations have been investigated by Dick (1956), Thaker and Beeson (1958) and Davis (1958), who concluded that molybdenum plus sulfate reduces the availability of copper, probably by interfering with its utilization by the tissues and/or by increasing the excretion of copper.

ZINC

Though zinc is a relatively abundant element in nature, and is present in all mammalian cells, its essentiality was not established until 1934 when two groups of investigators presented convincing evidence that it is essential for the growth and development of animals (Todd et al., 1934; Bertrand and Bhattacherjee, 1934).

The fundamental role of zinc appears to be as a cofactor in metalloenzymes, such as carbonic anhydrase (Kielin and Mann, 1940) and alcohol dehydrogenase (Vallee and Hoch, 1955). The finding of zinc in these and many other enzymes assisted greatly in explaining its role in metabolism.

Tucker and Salmon (1955) succeeded in inducing in pigs a nutritional deficiency disease which resembled the endemic disease parakeratosis which had long affected farm animals. They proved that it is a result of zinc deficiency. It was noted that a high calcium intake would aggravate this disease syndrome and that supplements of zinc carbonate would counteract it. About ten years later, Vallee et al. (1956, 1957) discovered that postalcoholic cirrhosis in human beings will respond to zinc therapy.

Thus, zinc is essential to the normal growth and development of the mouse, rat, pig, and possibly other animal species, and is effective in the treatment of several disorders in human beings.

DISTRIBUTION

Widdowson et al. (1951) studied the mineral composition of carcasses of human beings of different ages. They reported that the zinc content (fat-free basis) of 70 kg. adults ranged between 1.36 and 2.32 gm. Zinc was present in higher proportions in the liver, striated muscle and bone. Bergel et al. (1957) reported that the livers of newborn rats contained twice as much zinc as adult rat livers, and that the concentration decreased so rapidly that at two weeks of age it had dropped to the value (100 ppm) of adult rat livers.

The distribution of zinc in the tissues of dogs was studied by Montgomery et al. (1943) and Sheline et al. (1943) using radioactive Zn^{65}. When intravenously injected, Zn^{65} disappeared rapidly from the plasma. A rapid turnover was noted in the liver, pancreas, kidney and pituitary. The Zn^{65} did not accumulate in any organ during a 160-hour test period, except possibly in the red cells and bones. Radioactivity was detectable in the tissues, especially in the hair, even eight months after injection.

Radioactive zinc was observed by Feaster et al. (1955) to cross the placenta of dogs. Also, more than 50 per cent of an injected dose was observed to reach the young through the milk of dogs within 96 hours.

TABLE 9-2.   ZINC CONTENT OF REPRESENTATIVE FOODS*

Oysters	160	Egg yolk	3–4
Herring	70–120	Beets	3
Oatmeal	14	Corn	3
Wheat bran	14	Bread	2–4
Liver, pork	3– 15	Carrots	1–4
beef	3– 9	Clams	2
Wheat, whole	3– 9	Rice	2
Peas	3– 5	Milk	tr–3
Beef	2– 5	Cherries	1–2

* Mg./100 gm. edible portion.

Tissues of the male genital tract (testes, seminal vesicles, epididymis, prostate) are rich in zinc content, presumably in the form of carbonic anhydrase enzyme. But this does not explain the high level of zinc in semen (Mawson and Fischer, 1953). It is noteworthy that zinc deficiency in rats provokes degeneration of the testes and hypoplasia of the seminal vesicles and prostate (Millar et al., 1958). A possible competition between zinc and cadmium was noted by Pařízek (1957), who observed that the toxic action of cadmium on testicular tissue was counteracted by feeding zinc acetate.

Zinc has been implicated in the action of insulin and glucagon in the pancreas, but the mechanism is not yet understood. Weitzel et al. (1955) have isolated a zinc-cysteine-monohydrate from the choroid of the eye, which may explain why this tissue is unusually rich in zinc.

Zinc is widely distributed in foods, and normal dietaries are rarely deficient in this element. Oysters and herring are the best food sources of this nutrient (Table 9-2).

ABSORPTION AND EXCRETION OF ZINC

McCance and Widdowson (1942) estimated that the normal human being ingests 10 to 15 mg. of zinc daily and that 10 to 50 per cent of it is absorbed. A major portion of the zinc in food is excreted in the feces and is not absorbed. These investigators reported that 320 to 560 μg. of zinc are excreted daily in the urine. This report was confirmed by Vallee et al. (1957), who made serial estimations of zinc excreted in the urine of normal human beings day by day and reported an average of 550 μg. per day.

ZINC DEFICIENCY

Experimental zinc deficiency was induced in the rat by Stirn et al. (1935). Growth retardation, alopecia and graying of fur were the earliest deficiency effects observed when rats were fed a diet containing 1.5 ppm of zinc. They required about 50 per cent more food per gm. of body weight gained in comparison with control rats. Hyperkeratosis, thickening of the skin and loss of hair follicles were striking effects of zinc deficiency, yet the sebaceous glands were unaffected (Follis et al., 1941). An extreme parakeratosis of the esophagus which interfered with food consumption was observed. Zinc-deficient rats also develop a testicular degeneration which cannot be reversed by zinc therapy.

Zinc deficiency in man has been implicated in a syndrome which Prasād et al. (1963) have described in populations living in Iran and Egypt. Male subjects with this condition show dwarfism, hypogonadism, severe anemia, geophagia (eating of clay or earth) and hepatosplenomegaly. Their plasma zinc levels are low, they excrete less zinc in the urine and feces, and the zinc content of their hair is subnormal. It is thought that this syndrome results from a low content of zinc in the diet, from the practice of geophagia, from chronic blood loss, and from increased excretion of zinc in the sweat.

Postalcoholic cirrhosis appears to be associated with zinc deficiency. The concentration of zinc in the serum of patients with this disorder was decreased from a normal value of 121 μg./100 ml. to 66 μg./100 ml. (Vallee et al., 1957), and they excreted large quantities of zinc in the urine (1016 $\pm$ 196 μg./24 hours). This excretion pattern was counteracted by therapy with zinc sulfate.

The healing of wounds and burns was reported to be benefited by zinc (Strain et al., 1953, 1960). On the basis of studies with Zn^{65}, Savlov et al. (1962) concluded that zinc compounds have a powerful effect on the healing process of burns and surgically-produced wounds.

COBALT

DISTRIBUTION OF COBALT

Cobalt is widely distributed in plant and animal tissues, but usually in very small quantities. In many instances, the amounts are so small that accurate quantitation has not been possible by current methods of assay. Sylvester and Lampitt (1940) and Young (1948) estimated the cobalt content of several foods and reported values ranging from 0.01002 mg./100 gm. in coffee beans, to 12 mg./100 gm. in spinach and turnip greens (Table 9-3). The cobalt content of plant tissues is determined largely by the cobalt content of the soils in which they grow. For instance, the cobalt content of lettuce was raised from 0.03 to 0.51 mg./100 gm., dry basis, by the simple addition of cobalt sulfate to the soil (Smith, 1962).

ABSORPTION AND EXCRETION OF COBALT

Comar et al. (1946a, b) and Comar and Davis (1947) fed radioactive Co^{60} to rats and recovered 80 per cent in the feces, 10 per cent in the urine and most of the remainder in the liver. Brande et al. (1949) fed Co^{60} to pigs over a period of six weeks and studied the distribution of radioactivity in various organs. Radioactivity was detected in all tissues, with

TABLE 9-3.   COBALT CONTENT OF FOODS*†

Spinach	0.7–12.0	Beets	0.5 –1.0
Turnip greens	0.3–11.0	Chard, Swiss	1.0
Beet tops	4.0	Spinach,	
Buckwheat	3.5	New Zealand	1.0
Cocoa	0.3– 4.0	Tomatoes	1.0
Cabbage	0.7– 2.5	Milk	0.06–0.25
Pears	2.0	Corn	0.1 –0.2
Figs	2.0	Wheat	0.1
Cress, water	1.5	Rice	0.06
Onions	1.3	Flour, white	0.03

* Mg./100 gm. edible portion.
† From Sylvester and Lampitt, 1940; Young, 1948

particularly high levels in the livers, kidneys and adrenal glands.

The limited data indicate that cobalt is poorly absorbed from the gut and that most of the endogenous cobalt is excreted by way of the urine. Apparently, no studies on the absorption and excretion of cobalt have been published.

The average diet supplies about 5 to 8 μg. of cobalt daily to human beings. Since only 10 to 20 per cent of this is absorbed from the gut, possibly 1 to 3 μg. are utilized daily.

METABOLIC FUNCTIONS OF COBALT

Cobalt is essential to all animals and is especially important as a component of vitamin B_{12}. Ruminants do not require vitamin B_{12} because microorganisms in the rumen can use dietary cobalt in the synthesis of the vitamin. Since no measurable synthesis of vitamin B_{12} occurs in the gastrointestinal tract of nonruminants, a major portion of their cobalt requirement must be supplied as vitamin B_{12}.

Cobalt is required specifically as an activator of glycylglycine dipeptidase by forming a coordination compound with the substrate. Cobalt can replace other cations (e.g., iron, magnesium, manganese) in several enzymes, one of which is trimetaphosphatase (Dixon and Webb, 1958).

COBALT IN NUTRITION

Over many years there were frequent reports from Australia, New Zealand, Kenya, and the United States of a disease diagnosed as enzootic marasmus in sheep and cattle which was characterized by restricted growth, anorexia, muscular and visceral atrophy, and macrocytic anemia. Autopsy of animals dying with this syndrome revealed that their livers were low in cobalt content but high in iron. This disease is now known to result from cobalt deficiency (Underwood, 1962).

Cobalt is an effective therapeutic agent when given to ruminants orally but not when administered parenterally. However, parenteral vitamin B_{12} is effective (Smith et al.,

1951). This can be taken as proof that enzootic marasmus results from a vitamin B_{12} deficiency; cobalt is effective against this disease only because rumen microorganisms can use dietary cobalt in the synthesis of vitamin B_{12}.

Excess cobalt may cause abnormalities. A polycythemia developed when cobalt salts were fed at a level of 1 mg./kg./day to rats, mice, rabbits, pigs, dogs and human beings. Davis (1943) and especially Orten and Bucciero (1948) have suggested that the increase in red cell count is an attempt by the animal body to compensate for the anoxia which is produced by cobalt in the tissues as a result of binding of sulfhydryl groups. Polycythemia has been observed in subjects living under low atmospheric pressure, and also following violent exercise. This condition is relieved by oxygen inhalation or by the administration of lecithin or vasodilator drugs.

Though cobalt has been used to treat anemias, it should be emphasized that the effect is pharmacological rather than nutritional, since the level at which cobalt is effective is above the physiological range (Davis and Fields, 1955).

IODINE

Iodine is the heaviest element known to be required for animal growth and development. It was discovered by Courtois in 1811 while he was attempting to extract saltpeter from seaweed vats for Napoleon's army.

Several years later iodine was demonstrated to be present in animal tissues, and in 1820 Coindet discovered that iodine was effective in the treatment of simple goiter. David Marine has been given major credit for stimulating interest in iodine. Kimball and Marine (1918) conducted a therapeutic trial in the schools of Akron, Ohio, and found that small doses of iodine given during a 10-day period twice yearly would greatly reduce the incidence of simple goiter in children. Remarkable reductions in goiter incidence in other areas have been observed following prophylaxis with iodinated salt. The addition of iodine to salt has become an accepted public health practice.

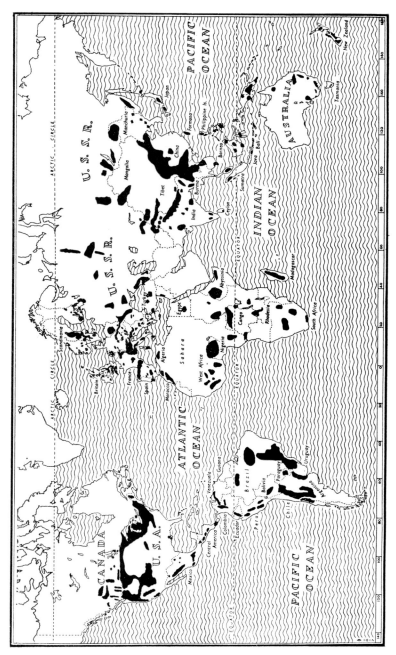

FIGURE 9-3.   Goitrous areas of the world. (Chilean Iodine Educational Bureau: Iodine Facts. World Goiter Survey. London, Facts, 271-280, 1946.

Endemic goiter occurs with varying intensity in most parts of the world (Figure 9-3), and close to 200 million people are afflicted with it (Kelly and Snedden, 1960). While the geographical distribution has not changed during the last century, the severity of this deficiency disease has declined considerably in those countries (United States, New Zealand, Switzerland) where iodized salt has been supplied as a public health measure. In neglected areas, simple goiter continues to be a very serious problem, and is often associated with feeblemindedness, physical degeneration and cretinism (Scrimshaw, 1960).

DISTRIBUTION OF IODINE

Iodine is widely distributed in plant and animal tissues. The amount in each species varies greatly, depending upon the iodine content of the soil or of ingested foods. Foods produced in areas adjacent to oceans with moderate to high rainfall or foods harvested from the sea are significantly higher in iodine content than foods grown in dry climates in the interior of land masses. Although drinking water does not contribute important amounts of iodine, data on the iodine content of water can be used as indices of the levels of iodine in the plants grown in a particular area.

The data in Table 9-4 indicate that the iodine content of ocean fish and shellfish is very high compared to other foods.

Most of the iodine in the human body is present in the thyroid gland, where it becomes a part of the thyroid hormone. It is found in lesser amounts in the kidneys, the salivary glands and the stomach.

ABSORPTION AND EXCRETION OF IODINE

Most of the iodine in the body is derived from ingested foodstuffs and some from water.

Iodine is absorbed chiefly from the small intestine. It circulates as both inorganic iodide and organic iodine but seems to enter the thyroid gland as the iodide ion. The thyroid tissues are iodine-seekers, concentrating the iodine from the blood to such an extent that the thyroid/blood ratio of iodine in human beings is about 25/1. Nearly 100 per cent of ingested iodide enters the thyroid gland directly or is excreted in the urine. The gland quickly oxidizes it, binds it to protein in the colloid in the form of mono-iodotyrosine, diiodotyrosine, thyroxine, diiodothyronine and triiodothyronine. These five compounds form a complex commonly called thyroglobulin. The metabolic circuit of iodine is shown in Figure 9-4.

The half-life of the thyroid hormone, once it has been secreted from the gland, is six to eight days (Riggs, 1952). About 40 per cent

TABLE 9-4. IODINE CONTENT OF FOODS*†

Fish	haddock	0.318	Eggs and	eggs	0.009
	cod	.416	dairy products	butter	0.006
	sea perch	.074		cheese	0.005
	halibut	.052		milk, whole	0.004
	mackerel	.037	Vegetables	spinach	0.020
	sole	.016		cabbage	0.005
Shellfish	shrimp	0.130		potatoes	0.005
	lobster	.102		asparagus	0.004
	clams	.078		beans	0.004
	oysters	.058		carrots	0.004
	crabmeat	.031		turnips	0.004
Meats	bacon	0.008		cucumbers	0.003
	pork	.005		lettuce	0.003
	beef	.003		peas	0.004

* Mg./100 gm. edible portion.
† From Chilean Iodine Educational Bureau, 1952.

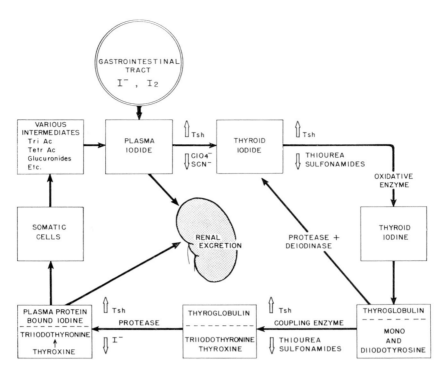

FIGURE 9-4.   *The metabolic circuit of iodine. Solid arrows indicate the direction of flow; open arrows point-ing upward indicate stimulation, and those pointing downward indicate inhibition. (Stanbury, J. B.: Symposium on Endemic Goiter. Fed. Proc., 17 (Supp. No. 2): 84, 1958.)*

is excreted in the urine as iodine, some 10 per cent is excreted in the feces in a protein-precipitative form, and a small amount is excreted in the sweat.

IODINE IN NUTRITION

The human need of iodine is estimated to be 75 to 200 gamma per day, the amount being determined by body size, age and several conditions which affect thyroid function. The need for iodine is increased during puberty and pregnancy. The requirement is also increased by diet and environment. Certain foods (e.g., the cabbage family) contain goitrogens which increase the iodine requirement. Milk may contain goitrogens which have come from cruciferous fodder consumed by cows (Clements and Wishart, 1956). The hardness of water may also interfere with iodine absorption.

Iodine deficiency reduces the supply of thyroid hormone; the gland enlarges because of hyperplasia of the cells lining the follicles and increase of colloid material. As a sequel, the energy metabolism is reduced, physical and mental development are retarded, the differentiation and maturation of tissues are delayed, the activities of endocrine glands (gonads, hypophysis, etc.) are altered, and the metabolism of carbohydrates, proteins and fats is disturbed.

Iodine deficiency, together with other conditioning factors, may produce simple goiter (Scrimshaw, 1957). This type of goiter responds rapidly when therapeutic doses of iodine in the form of supplements are given to population groups (Brush and Altland, 1952; Scrimshaw et al., 1953).

Iodine appears to have functions other than those concerning the thyroid gland. Ashing and Evans (1962) studied the skeletal response of young thyroidectomized female rats to iodine given as 5 μg. iodine/day as potassium iodide or 0.25 μg./day as *l*-thyroxine. The thyroidectomized controls grew only one third as rapidly as normal animals, but the thyroidectomized rats receiving either potassium oxide or thyroxine grew equally as well as the controls, and their skeletal maturation was normal. The biochemical mechanism of this stimulatory action on the skeleton is not clear.

MICRO ELEMENTS AND DENTAL CARIES

This subject has been reviewed by Hein (1955), Adler (1964) and Harris and Navia (1964). Investigations of the effects of trace elements other than fluorine upon the development of caries were stimulated by the report of Nizel and Bibby (1944), and others, which indicated that the geography of caries in human populations is not identical to the geography of fluorine in the soils of different regions. There is no doubt that approximately 1 ppm of fluorine in foods and water supplies reduces caries development significantly. However, some caries are found in people living in areas where fluorine is present in significant amounts, and caries may not be found in people living in areas where fluorine is essentially absent. Thus, there must be factors in addition to fluorine which influence caries development in animals and in mankind. For instance, on the basis of an epidemiological survey Ludwig et al. (1960) concluded that molybdenum reduced caries in children, while Tank and Storvick (1960) pointed out that selenium in foods and water supplies increased caries development in children.

The results of several laboratory investigations have a bearing on this point. Sognnaes and Shaw (1954) noted that an ash prepared from natural foods reduced caries development when added to a caries-preventive diet and fed to experimental animals. They explained this effect as being due to "the presence of certain trace elements in the ash mixture." A similar observation was made by Nizel and Harris in 1953. They had previously observed that milk and corn grown in Texas produced only 40 per cent as much caries as milk and corn grown in New England when fed in caries-producing diets to hamsters (Nizel and Harris, 1950, 1951). The fluorine content of these diets had been equalized. It may be

important that the New England corn contained more aluminum, copper, iron, manganese and molybdenum and less nickel than the Texas corn.

Pursuing the possibility that the cariogenic factor in these New England foods was attributable to a trace element, they ashed the two diet mixtures and added these ashes to equivalent amounts of the diets, thereby doubling the content of trace elements not volatile at the ashing temperature (550° C.). When the ash-supplemented diets were fed to hamsters to provide a comparison with the control caries-producing diets, a marked decrease in the incidence and severity of dental decay was noted in both ash-supplemented diets (Nizel and Harris, 1953). Apparently, the cariogenic factor in the New England corn and milk which they sought to identify was an organic substance that was destroyed in the ashing process, or possibly it was over-powered by one or more of the elements in the ash supplement.

Based on the quantitative analysis of these ashes by spectrographic and chemical methods, a "synthetic ash" was prepared using chemically pure oxides of the ten mineral elements that could be quantitated. Phosphorus was supplied as monopotassium phosphate. The hamsters fed this synthetic salt mixture showed a 95 per cent reduction in dental caries in comparison with the control hamsters fed the caries-producing diet. Thus, the synthetic salt mixture was fully as cariostatic as the "natural ash" which it duplicated.

In a concurrent experiment, it was noted that the cariostatic activity of this synthetic salt mixture was completely lost when the phosphate component was omitted. In fact, the mixture of elements in the phosphorus-free mixture was definitely cariogenic. Phosphorus demonstrated strong cariostatic properties.

IRON AND DENTAL CARIES

McClure (1948) added 250 ppm ferric chloride and ferric citrate to the drinking water of rats which were maintained on a caries-producing diet for 15 weeks, and the caries scores were compared with those of a control group offered unsupplemented water. A 50 per cent reduction in caries development was noted in the group fed 250 ppm iron as ferric chloride, and no reduction was noted in the group fed 250 ppm iron as ferric citrate; moreover, a 30 per cent increase in caries development was observed in the group fed 500 ppm iron as ferric citrate.

On the basis of in vitro studies Torell (1956a) concluded that iron salts affect enamel solubility. He claimed that ferric iron solutions can establish layers of hydrolyzed ferric precipitates on enamel and that phosphate ions on the enamel surface can be bound chemically by the ferric iron, increasing the rate of dissolution of apatite. Thus, iron increases enamel solubility and the capacity to retain sugar and promote smooth surface caries (Torell, 1956b).

Two carious teeth from a girl with hypoplastic anemia and transfusional hemosiderosis were analyzed for iron content. For seven years, beginning at nine months of age, she had received bimonthly blood transfusions (total, 21,600 ml. blood). The teeth contained 50.8 μg. of iron per gm., while similar control teeth contained 16.1 μg. per gm. (Landing et al., 1956). These investigators suggested that the iron was absorbed from the saliva, on the basis that salivary glands commonly show hemosiderosis in iron storage disease.

COPPER AND DENTAL CARIES

Brudevold and Steadman (1955) estimated the copper content of four successive layers of enamel from a pooled sample of fully erupted teeth and found a random distribution in the layers ranging between 15 to 30 ppm, with an over-all average of 20 ppm. No correlation was found between the copper content and the degree of tooth pigmentation or between the copper content and caries development.

In in vitro studies, Forbes and Smith (1952) observed that 0.25 mg. per cent of copper in sucrose-containing saliva partially inhibited acid formation, while 3 to 4 mg. per cent of copper prevented it completely. Dreizen et al. (1952) found no relation between the copper content of saliva and dental caries activity in human subjects. The level of copper normally observed in saliva did not prevent the growth of *Lactobacillus acidophilus*.

Shaw (1950) noted no effect when 0.5 per cent sodium-copper chlorophyllin was added to the drinking water of rats. On the other hand, Hein and Shafer (1951) observed a caries-reducing action when hamsters were fed sodium-copper chlorophyllin of 93.2 per cent purity.

Hein (1953) reported that copper-sulfate reduced dental caries development as the level of copper in drinking water was increased from 0, to 10, to 25 to 50 ppm. A straight-line relationship was observed when the caries scores were plotted against the logarithm of the copper concentration, indicating that tooth destruction was an inverse function of the copper sulfate concentration. Using a different approach, Kruger (1959) injected 0.005 to 0.02 mg. of copper nitrate intraperitoneally into rats and noted a predevelopmental effect which reduced dental caries development.

The evidence indicates that copper fed in the diet or in the drinking water, or injected intraperitoneally, may decrease caries development in experimental animals, but more data are needed. No studies of the effect of copper on caries development in man have yet been conducted.

ZINC AND DENTAL CARIES

The zinc content of the enamel and dentin of teeth and of bones is higher than in most tissues of the body. Cruickshank (1936, 1937, 1940) reported concentrations of 130 to 280 ppm of zinc in human teeth. Higher values were found in the teeth of tuberculosis patients than in normal subjects. Brudevold et al. (1963) suggested that these differences were the result of levels of intake of zinc in the diet.

Much more zinc is found in the surface layers of teeth than in the lower layers, a distribution pattern which resembles that of fluorine and lead. Zinc is readily taken up by synthetic hydroxyapatite and competes with calcium for positions on the surface of apatite crystals.

Knowledge of the effect of zinc on dental caries development is quite incomplete. McClure (1948) reported that 250 ppm of zinc sulfate in the drinking water of rats stimulated caries development. Mansell and Hendershot (1960) noted no effect on caries development when 680 ppm of zinc sulfate was fed to rats, They reported 190 ppm of zinc in the molars of these animals. Thus, it has not been established whether zinc affects tooth mineralization and dental caries development when it is fed in the diet or in the water supply of experimental animals. The effect of zinc on the teeth of human beings has not been studied.

COBALT AND DENTAL CARIES

Dreizen et al. (1952) noted that the salivas of caries-susceptible human subjects tended to be low on cobalt content, but they were not able to establish a correlation between low saliva cobalt values and dental caries in human subjects.

Goldenberg and Sobel (1952) suspended epiphyseal cartilage in a basal mineralizing solution containing sodium chloride, potassium chloride and sodium bicarbonate and observed that additions of cobalt accelerated the inactivation of the calcifying mechanism. Bird and Thomas (1963) reported that low concentrations of cobalt (0.10 mM or less) were unique among mineral elements in preventing the formation of apatite crystals.

Thus, there is preliminary evidence from in vitro studies that ionized cobalt may interfere with calcification mechanisms. However, no one has demonstrated that dietary cobalt affects dental caries development.

IODINE AND DENTAL CARIES

Although inorganic salts of iodine have not been found to affect dental caries development, McClure and Arnold (1941) reported that iodoacetic acid reduced caries when added to the drinking water of rats. Unfortunately, the LD$_{50}$ of iodoacetic acid is 116 ± 12 mg./kg. body weight, and the margin of safety of this compound is too narrow to permit its use as a prophylactic agent (Lundqvist, 1951).

Kruger (1959) observed no effect on dental caries development when weanling rats were fed a cariogenic diet, given injections of 0.002 to 0.005 mg. of iodine daily from the fifth to seventeenth day on this diet, and scored for dental caries at 20 weeks of age.

In testing the theory that the reduction in dental caries produced by fluorine may be mediated through the thyroid gland by increasing saliva flow, Muhler and Shafer (1954) fed desiccated thyroid in the caries-producing diet of rats in comparison with sodium fluoride. Sixty milligrams of the thyroid in the diet daily produced the same degree of caries reduction as 20 μg. of fluoride (as sodium fluoride) added to the drinking water. When the activity of the thyroid gland was markedly decreased by the administration of thiouracil, the incidence of caries in rats was decidedly increased. This effect of iodine was **hormonal** rather than nutritional or pharmacological.

The limited literature on the subject indicates that iodine has no important effect on the development of experimental caries.

MOLYBDENUM AND DENTAL CARIES

Adler (1964) concluded that molybdenum is a cariostatic agent and that selenium is cariogenic. The caries resistance of children in certain Hungarian towns was ascribed to the molybdenum present in their water supply (Adler and Straub, 1953). Similarly, the low caries incidence in children in Napier, New Zealand, was presumed to be the result of molybdenum content of the soils, since similar children in neighboring Hastings township had significantly more caries and the soils were much lower in molybdenum content (Ludwig et al., 1960). It has been suggested that molybdenum is effective because it synergistically enhances the reaction of fluoride with the tooth surface (Roberts, 1961; Buttner, 1961). Because Van Reen et al. (1962) observed no effect on caries development when molybdenum was fed in the diet of rats, and because few studies have been conducted, a conclusion as to the efficacy of molybdenum in the control of dental caries must await further experimentation.

*SELENIUM AND
DENTAL CARIES*

Hadjimarkos and Bonhorst (1958) presented evidence that selenium promotes the development of caries on the basis of a positive correlation between the selenium content of the urine and the incidence of caries in children in Oregon. Later, Hadjimarkos and Bonhorst (1961) reported a similar correlation between the selenium content of foods and caries incidence.

Selenium was fed in the drinking water of rats during the period of tooth formation and their teeth developed significantly more caries than the control group (Buttner, 1961).

Thus, from the evidence to date, it can be concluded that selenium is one trace element that has significant caries-producing potential if ingested in slightly excessive amounts.

*VANADIUM AND
CADMIUM AND
DENTAL CARIES*

In his review, Hein (1955) commented that several investigators found vanadium salts to be effective in reducing caries and that cadmium chloride was cariogenic. However, Muhler (1957) and Shaw (1962) reported that vanadium pentoxide had no effect on caries incidence when fed to rodents.

REFERENCES

Iron

Bessis, M., and Breton-Gorius, J.: Trois aspects du fer dans des coupes d'organes examinées au microscope electronique (ferritine et dérivé, dans les cellules intentinales, les erythroblastes et les cellulares reticulaires). Compt. Rend. Acad. Sci., *245*:1271, 1957.

Bothwell, T. H., Pirzio-Biroli, G., and Finch, C. A.: Iron absorption. I. Factors influencing absorption. J. Lab. Clin. Med., *51*:24, 1958.

Brading, I., Kaldor, I., and George, E. P.: The absorption of iron from Fe59-labelled hemoglobin by rats. Austral. Ann. Med., *6*:247, 1957.

Brock, J. F.: The effect of large doses of iron on the absorption of phosphorus. Clin. Sci., *3*:37, 1937.

Brown, E. B., Dubach, R., Smith, D. E., Reynafarje, C., and Moore, C. V.: Studies in iron transportation and metabolism. X. Long-term overload in dogs. J. Lab. Clin. Med., *50*:862, 1957.

Callender, S. T., Mallett, B. J., and Smith, M. D.: Absorption of hemoglobin iron. Brit. J. Hematol., *3*:186, 1957.

Chodos, R. B., and Ross, J. F.: Absorption of radioactive iron in normal, anemic and hemochromatotic subjects. Amer. J. Med., *14*:499, 1953.

Choudhury, M. R., and Williams, J.: Iron absorption and gastric operations. Clin. Sci., *18*:527, 1959.

Coleman, D. H., Stevens, A. R., and Finch, C. A.: The treatment of iron deficient anemia. Blood, *10*:567, 1955.

Copp, D. H., and Greenberg, D. M.: A tracer study of iron metabolism with radioactive iron. I. Methods: Absorption and excretion of iron. J. Biol. Chem., *164*:377, 1946.

Darby, W. J.: The oral manifestations of iron deficiency. J.A.M.A., *130*:830, 1946.

Davidson, L. S. P., Fullerton, H. W., Howie, J. W., Croll, J. M., Orr, J. B., and Godden, W.: Observations on nutrition in relation to anaemia, Brit. Med. J., *1*:685, 1933.

Day, H. G., and Stein, H. J.: The effect upon hematopoiesis of variations in the dietary levels of calcium, phosphorus, iron and vitamin D. J. Nutr., *16*:525, 1938.

Dubach, R., Moore, C. V., and Munnich, V.: Studies in iron transportation and metabolism. J. Lab. Clin. Med., *31*:1201, 1946.

Finch, S. C., and Finch, C. A.: Idiopathic hemochromatosis, an iron storage disease. Medicine, *34*:381, 1955.

Fullerton, H. W.: Hypochromic anaemias of pregnancy and the puerperium. Brit. Med. J., *ii*:577, 1936.

Granick, S.: Ferritin: its properties and significance for iron metabolism. Chem. Rev., *38*:379, 1946.

Granick, S.: Ferritin. 9. Increase of the protein apoferritin in the gastrointestinal mucosa as a direct response to iron feeding. The function of ferritin in the regulation of iron absorption. J. Biol. Chem. *164*:737, 1946.

Greenberg, D. M.: Intermediary metabolism and biological activities of ferritin. *In* Metabolism and Function of Iron, Report of 19th Ross Pediatric Research Conference, Portland, Oregon, 1955, pp. 33-35.

Hahn, P., Bale, W. F., Lawrence, E. O., and Whipple, G. H.: Radioactive iron and its metabolism in anemia; its absorption, transportation and utilization. J. Exp. Med. *69*:739, 1939.

Hahn, P. F., Bale, W. F., Ross, J. F., Balfour, W. M., and Whipple, G. H.: Radioactive iron absorption by gastro-intestinal tract. Influence of anemia, anoxia, and antecedent feeding distribution in growing dogs. J. Exp. Med., *78*:169, 1943.

Hallen, L.: Gastric secretion. Acta Med. Scand., *96* (Suppl 90):398, 1938.

Haskins, D., Stevens, A. R., Finch, S. C., and Finch, C. A.: Iron metabolism. Iron stores in man as measured by phlebotomy. J. Clin. Invest., *31*:543, 1952.

Health Bulletin No. 23, Gov't India Press, Simla, 1951.

Hegsted, D. M., Finch, C. A., and Kinney, T. D.: The influence of diet on iron absorption. J. Exp. Med., *40*:147, 1949.

Johnston, F. A., Frenchman, R., and Boroughs, E. D.: The absorption of iron from beef by women. J. Nutr., *35*:453, 1948.

Kaldor, I.: Hemoglobin as a source of iron in nutrition: some *in vitro* experiments. Austral. Ann. Med., *6*:244, 1957.

Mannis, J. G., and Schachter, D.: Active transport of iron by intestine: features of the two-step mechanism. Amer. J. Physiol., *203*:73, 1962.

McCance, R. A., Edgecombe, C. N., and Widdowson, E. M.: Phytic acid and iron absorption. Lancet, *ii*:126, 1943.

McCance, R. A., and Widdowson, E. M.: The absorption and excretion of iron following oral and intravenous administration. J. Physiol., *94*:148, 1938.

McMillan, T. J., and Johnston, F. A.: The absorption of iron from spinach by six young women, and the effect of beef upon the absorption. J. Nutr., *44*:383, 1941.

Moore, C. V., Arrowsmith, W. R., Welch, J., and Minnich, V.: Observations on the absorption of iron from the gastrointestinal tract. J. Clin. Invest., *18*:553, 1939.

Moore, C. V., Dubach, R., Minnich, V., and Roberts, H. K.: Absorption of ferrous and ferric radioactive iron by human subjects and by dogs. J. Clin. Invest., *23*:755, 1944.

Moore, C. V., and Dubach, R.: Observations on absorption of iron from foods tagged with radioiron. Trans. Assn. Amer. Physicians, *64*:245, 1951.

Moore, C. V., and Dubach, R.: Metabolism and requirements of iron in the human. J.A.M.A., *162*:197, 1956.

Moore, C. V., and Dubach, R.: *In* Comar, C. L., and Bronner, F., (ed.): Mineral Metabolism. Vol. 2, Part B. New York, Academic Press, Inc., 1962, Chapter 30.

National Health Med. Res. Council, 1945. (Austral) Spec. Rept. No. 1, 1946.

Owren, P. A.: The pathogenesis and treatment of iron deficiency anemia after partial gastrectomy. Acta Chir. Scand., *104*:206, 1953.

Pommerenke, W. T., Hahn, P. F., Bale, W. F., and Balfour, W. M.: Transmission of radioactive iron to human fetus. Amer. J. Physiol., *137*:164, 1942.

Recommended Dietary Allowances. 1963 Revision. Washington, D. C., Food and Nutrition Board, National Academy of National Sciences–Research Council, 1964.

Richter, G. W.: The nature of storage iron in idiopathic hemochromatosis and in hemosiderosis. J. Exp. Med., *112*:551, 1960.

Schulz, J., and Smith, N. J.: A quantitative study of the absorption of food iron in infants and children. Amer. J. Dis. Child., *95*:100, 1958.

Sharpe, L. M., Peacock, W. C., Cooke, R., and Harris, R. S.: The effect of phytate and other food factors on iron absorption. J. Nutr., *41*:433, 1950.

Sherman, H. C.: Chemistry of Food and Nutrition. New York, The Macmillan Company, 1935.

Stiebeling, H. K.: The iron content of vegetables and fruits. U. S. Department of Agriculture Circular No. 205. Washington, D. C., Gov't Printing Office, 1932.

Stitt, C., Charley, P. J., Butt, E. M., and Saltman, P.: Rapid induction of iron deposition in spleen and liver with an iron-fructose chelate. Proc. Soc. Exp. Biol. Med., *110*:70, 1962.

Venkatachalam, P. S., Brading, I., George, E. P., and Walsh, R. J.: An experiment in rats to determine whether iron is absorbed only in the ferrous state. Austral. J. Exp. Biol. Med. Sci., *34*:389, 1956.

Walsh, R. J., Kaldor, I., Brading, I., and George, E. P.: The availability of iron in meat: some experiments with radioactive iron. Austral. Ann. Med., *4*:272, 1955.

Widdowson, E. M., and Spray, C. M.: Chemical development *in utero*. A.M.A. Arch. Dis. Child., *26*:205, 1951.

Woodruff, C. W.: Multiple causes of iron deficiency in infants. J.A.M.A., *167*:715, 1958.

Micro Elements

Lowe, C. W.: Trace elements in infant nutrition. Pediatrics, *26*:715, 1960.

Moore, C. V.: Iron and essential trace elements. *In* Wohl, M. G., and Goodhart, R. S., (ed.): Modern Nutrition in Health and Disease. 3rd ed. Philadelphia, Lea & Febiger, 1964, p. 278.

Schütte, K. H.: The Biology of the Trace Elements: Their Role in Nutrition. (International Monographs: Aspects of Animal and Human Nutrition. Consultant ed.: André Voisin) London, Crosby Lockwood & Son, Ltd., 1964.

Underwood, E. J.: Trace Elements in Human and Animal Nutrition. 2nd ed. New York, Academic Press, Inc., 1962.

Copper

Anderson, P. J., and Popper, H.: Changes in hepatic structure in Wilson's disease. Amer. J. Path., *36*:483, 1960.

Baxter, J. H., and Van Wyk, J. J.: A bone disorder associated with copper deficiency. I. Gross morphological, roentgenological and chemical observations. Bull. Johns Hopkins Hosp., *93*:1, 1953a.

Baxter, J. H., Van Wyk, J. J., and Follis, R. H., Jr.: A bone disorder associated with copper deficiency. II. Histological and chemical studies on the bones. Bull. Johns Hopkins Hosp., *93*:25, 1953b.

Bearn, A. G., and Kunkel, H. G.: Metabolic studies in Wilson's disease using Cu^{64}. J. Lab. Clin. Med., *45*:623, 1955.

Butler, E. J., and Newman, G. E.: The urinary excretion of copper and its concentration in the blood of normal adults. J. Clin. Path., *9*:157, 1956.

Cartwright, G. E., Gubler, C. J., and Wintrobe, M. M.: Studies on copper metabolism. XI. Copper and iron metabolism in the nephrotic syndrome. J. Clin. Invest., *33*:685, 1954.

Chan, T. P., and Adolph, W. H.: Copper metabolism in man. Biochem. J., *29*:476, 1935.

Davis, G. K.: Mechanisms of trace element function. Soil Sci., *85*:59, 1958.

Dick, A. T.: Molybdenum in animal nutrition. Soil Sci., *81*:229, 1956.

Dreizen, S., Spies, H. A., and Spies, T. D.: The copper and cobalt levels of human saliva and dental caries activity. J. Dent. Res., *31*:137, 1952.

Gubler, C. J.: Copper metabolism in man. J.A.M.A., *161*:530, 1956.

Hart, E. B., Steenbock, H., Waddell, J., and Elvehjem, C. A.: Iron in nutrition. VII. Copper as a supplement to iron for hemoglobin building in the rat. J. Biol. Chem., 77:797, 1928.

Mills, C. F.: Comparative metabolic studies of inorganic and herbage-complex forms of copper in rats and sheep. Soil Sci., *85*:100, 1958.

Ramage, H., Sheldon, J. H., and Sheldon, W.: A spectrographic investigation of the metallic content of the liver of childhood. Proc. Roy. Soc. (B), *113*: 308, 1933.

Sherman, H. C.: Chemistry of Food and Nutrition. 8th ed. New York, The Macmillan Co., 1952.

Smith, S. E., and Medlicott, M.: The blood picture of iron and copper deficiency anemias in the rat. Amer. J. Physiol., *141*:354, 1944.

Smith, S. E., and Medlicott, M.: The blood picture of iron and copper deficiency anemias in the rabbit. Amer. J. Physiol., *142*:179, 1944.

Thaker, E. J., and Beeson, K. C.: Occurrence of mineral deficiency and toxicity in animals in the United States and problems of their detection. Soil Sci., *85*: 87, 1958.

Tompsett, S. L.: Factors influencing the absorption of iron and copper from the alimentary tract. Biochem. J., *34*:961, 1940.

Wintrobe, M. M., Cartwright, G. E., and Gubler, C. J.: Studies on the function and metabolism of copper. J. Nutr., *50*:395, 1953.

Zinc

Bergel, F., Everett, A. J. L., Martin, J. B., and Webb, J. S.: Cellular constituents: Major and minor metals in normal and abnormal tissues. Part I. Analysis of Wistar rat livers for copper, iron, magnesium, manganese, molybdenum and zinc. J. Pharm. Pharmacol., *9*:522, 1957.

Bertrand, G., and Bhattacherjee, R. C.: L'action combinée du zinc et des vitamines dans l'alimentation des animaux. Compt. Rend. Acad. Sci., *198*:1823, 1934.

Feaster, J. P., Hansard, S. L., McCall, J. T., and Davis, G. K.: Absorption, deposition and placental transfer of Zn^{65} in the rat. Amer. J. Physiol., *181*: 287, 1955.

Follis, R. H., Day, H. G., and McCollum, E. V.: Histologic studies of the tissues of rats fed a diet extremely low in zinc. J. Nutr., *22*:223, 1941.

Keilin, D., and Mann, T.: Carbonic anhydrase: Purification and nature of the enzyme. Biochem. J., *34*: 1163, 1940.

Mawson, C. A., and Fischer, M. I.: Zn and carbonic anhydrase in human semen. Biochem. J., *55*:696, 1953.

McCance, R. A., and Widdowson, E. W.: The absorption and excretion of zinc. Biochem. J., *36*:392, 1942.

Millar, M. J., Fischer, M. I., Elcoate, P. V., and Manson, C. A.: The effects of dietary zinc deficiency on the reproductive system of male rats. Canad. J. Biochem. Physiol., *36*:557, 1958.

Montgomery, M. L., Sheline, G. E., and Chaikoff, I. L.:

The elimination of administered zinc in pancreatic juice, duodenal juice and bile of the dog as measured by its radioactive isotope (Zn^{65}). J. Exp. Med., *78*:151, 1943.

Pařízek, J.: The destructive effect of cadmium ion on testicular tissue and its prevention by zinc. J. Endocrinol., *15*:56, 1957.

Prasad, A. S., Schulert, A. R., Miale, A., Farid, Z., and Sandstead, H. H.: Zinc and iron deficiencies in male subjects with dwarfism and hypogonadism but without ancylostomiasis, schistosomiasis or severe anemia. Amer. J. Clin. Nutr., *12*:437, 1963.

Savlov, E. D., Strain, W. H., and Huegin, F.: Radiozinc studies in experimental wound healing. J. Surg. Res., *2*:209, 1962.

Sheline, G. E., Chaikoff, I. L., Jones, H. B., and Montgomery, M. L.: Distribution of administered zinc in tissues of mice and dogs. J. Biol. Chem., *149*: 139, 1943.

Stirn, F. E., Elvehjem, C. A., and Hart, E. B.: The indispensability of zinc in the nutrition of the rat. J. Biol. Chem., *109*:347, 1935.

Strain, W. H., Dutton, A, M., Heyer, H. B., and Ramsey, G. H.: Experimental studies on the acceleration of burn and wound healing. Rochester, University of Rochester Rept., 1953.

Strain, W. H., Pories, W. J., and Hinshaw, J. R.: Zinc studies in skin repair. Surg. Forum, *11*:291, 1960.

Todd, W. R., Elvehjem, C. A., and Hart, E. B.: Zinc in the nutrition of the rat. Amer. J. Physiol., *107*: 146, 1934.

Tucker, H. F., and Salmon, W. D.: Parakeratosis or zinc deficiency disease in pig. Proc. Soc. Exp. Biol. Med., *88*:613, 1955.

Vallee, B. L., and Hoch, F. L.: Yeast alcohol dehydrogenase, a zinc metalloenzyme. J. Amer. Chem. Soc., *77*:821, 1955.

Vallee, B. L., Wacker, W. E. C., Bartholomay, A. F., and Hoch, F. L.: Zinc metabolism in hepatic dysfunction. II. Correlation of metabolic patterns with biochemical findings. New Eng. J. Med., *257*:1055, 1957.

Vallee, B. L., Wacker, W. E. C., Bartholomay, A. F., and Robin, E. D.: Zinc metabolism in hepatic dysfunction. I. Serum zinc concentrations in Laennec's cirrhosis and their validation by sequential analysis. New Eng. J. Med., *255*:403, 1956.

Weitzel, G., Buddecke, E., Fretzdorff, A. M., Strecker, F. J., and Roester, U.: Struktur der im Tapetum Lucidum von Hund und Fuchs enthaltenen Zinkverbindung. Z. Physiol. Chem., *299*:193, 1955.

Widdowson, E. M., McCance, R. A., and Spray, C. M.: The chemical composition of the human body. Clin. Sci., *10*:113, 1951.

Cobalt

Brande, R., Free, A. A., Page, J. E., and Smith, E. L.: The distribution of radioactive cobalt in pigs. Brit. J. Nutr., *3*:289, 1949.

Comar, C. L., and Davis, G. K.: Cobalt metabolism studies. IV. Tissue distribution of radioactive cobalt administered to rabbits, swine, and young calves. J. Biol. Chem., *170*:379, 1947.

Comar, C. L., Davis, G. K., and Taylor, R. F.: Cobalt

metabolic studies. I. Radioactive cobalt procedures with rats and cattle. Arch. Biochem. Biophys., 9: 149, 1946a.

Comar, C. L., Davis, G. K., Taylor, R. F., Huffman, C. D., and Ely, R. E.: Cobalt metabolism studies. II. Partition of radioactive cobalt by a rumen fistula cow. J. Nutr., 32:61, 1946b.

Davis, J. E.: Effect of oxygen, soybean lecithin carbamyl choline and furfuryl trimethyl ammonium iodide on experimental polycythemia. J. Pharmacol. Exp. Therap., 79:37, 1943.

Davis, J. E., and Fields, J. P.: Cobalt polycythemia in humans. Fed. Proc., 14:331, 1955.

Dixon, M., and Webb, E. C.: Enzymes. New York, Academic Press, Inc., 1958.

Orten, J. M., and Bucciero, M. C.: The effect of cysteine, histidine and methionine on the production of polycythemia by cobalt. J. Biol. Chem., 176:961, 1948.

Smith, E. L.: Cobalt. In Comar, C. L., and Bronner, F. (ed.): Mineral Metabolism, Vol. 2, Part B. New York, Academic Press, Inc., 1962, Chapter 31.

Smith, S. E., Koch, B. A., and Turk, K. L.: The response of cobalt-deficient lambs to liver extract and vitamin B_{12}. J. Nutr., 44:455, 1951.

Sylvester, N. D., and Lampitt, L. H.: The determination of cobalt in foods. J. Soc. Chem. Ind., 59:57, 1940.

Underwood, E. J.: Trace Elements in Human and Animal Nutrition. New York, Academic Press, Inc., 1962.

Young, R. S.: Cobalt. New York, Reinhold Publishing Co., 1948.

Iodine

Ashing, C. W., and Evans, E. S.: Maintenance of skeletal growth and maturation in thyroidectomized rats by KI infections. Anat. Rec., 142:211, 1962.

Brush, B. E., and Altland, J. K.: Goiter prevention with iodized salt: results of a thirty-year study. J. Clin. Endocrinol., 12:1380, 1952.

Chilean Iodine Education Bureau. Iodine Content of Foods. London, Annotated Bibliography, 1952.

Clements, F. W., and Wishart, J. N.: A thyroid-blocking agent in the etiology of endemic goiter. Metabolism, 5:623, 1956.

Coindet, J. R.: Découverte d'un nouveau remède contre le goître. Ann. Chim. Phys. (Paris), 15:49, 1820.

Ferguson, M. H., Naimark, A., and Hildes, J. A.: Parotid secretion of iodide. Canad. J. Biochem. Physiol., 34:633, 1956.

Hamilton, J. G.: The rates of absorption of the radioactive isotopes of sodium, potassium, chlorine, bromine and iodine in normal human subjects. Amer. J. Physiol., 124:667, 1948.

Kelly, F. C., and Snedden, W. W.: Prevalence and geographical distribution of endemic goiter. Monograph No. 44. World Health Organization, 1960.

Kimball, O. P., and Marine, D.: The prevention of simple goiter in man. Arch. Intern. Med., 22:41, 1918.

Riggs, D. S.: Quantitative aspects of iodine metabolism in man. Pharmacol. Rev., 4:284, 1952.

Scrimshaw, N. S.: Endemic goiter. Nutr. Rev., 15:161, 1957.

Scrimshaw, N. S.: Endemic goiter in Latin America. Pub. Health Reports, 75:731, 1960.

Scrimshaw, N. S., Cabezas, A., Castillo, F., and Mendez, J.: Effect of potassium iodate on endemic goiter and protein-bound iodine levels in school children. Lancet, 265:166, 1953.

Stanbury, J. B.: Symposium on endemic goiter. Fed. Proc., 17 (Supp. 2):84, 1958.

Micro Elements and Dental Caries

Adler, P.: Der Einfluss von Spurenelementen auf den Kariesbefall. Bibl. Nutr. Dieta, 5:54, 1964.

Adler, P., and Straub, J.: A water-borne caries-protective agent other than fluorine. Acta Med. Acad. Sci. Hung., 4:221, 1953.

Bird, E. D., and Thomas, W. C.: Effect of various metals on mineralization in vitro. Proc. Soc. Exp. Biol. Med., 112:640, 1963.

Brudevold, F., Steadman, L. T., Spinelli, M. A., Amdur, B. H., and Grøn, P.: A study of zinc in human teeth. Arch. Oral Biol., 8:135, 1963.

Brudevold, F., and Steadman, L. T.: A study of copper in human enamel. J. Dent. Res., 34:209, 1955.

Buttner, W.: Effects of some trace elements on fluoride retention and dental caries. Arch. Oral Biol., 6:40, 1961.

Cruickshank, D. B.: The natural occurrence of zinc in teeth. I. Preliminary experiments. Brit. Dent. J., 61:530, 1936.

Cruickshank, D. B.: The natural occurrence of zinc in teeth. II. Some general considerations. Brit. Dent. J., 63:395, 1937.

Cruickshank, D. B.: The natural occurrence of zinc in teeth. III. Variations in tuberculosis. Brit. Dent. J., 68:257, 1940.

Dreizen, S., Spies, H. H., and Spies, T. D.: The copper and cobalt levels of human saliva and dental caries activity. J. Dent. Res., 31:137, 1952.

Forbes, J. C., and Smith, J. D.: Studies on the effect of metallic salts on acid production in saliva. J. Dent. Res., 31:129, 1952.

Goldenberg, H., and Sobel, A. E.: Calcification. 9. Influence of alkaline earths on survival of the calcifying mechanism. Proc. Soc. Exp. Biol. Med., 81:695, 1952.

Hadjimarkos, D. M., and Bonhorst, C. W.: The trace element selenium and its influence on dental caries susceptibility. J. Pediat., 52:274, 1958.

Hadjimarkos, D. M., and Bonhorst, C. W.: The selenium content of eggs, milk, and water in relation to dental caries in children. J. Pediat., 59:256, 1961.

Harris, R. S., and Navia, J. M.: Foods, nutrition, trace metals and dental caries. Proc. Amer. Inst. Oral Biol., 21:32, 1964.

Hein, J. W.: Effect of copper sulfate on inhibition and progression of dental caries in the Syrian hamster. J. Dent. Res., 32:654, 1953.

Hein, J. W.: Effect of various agents on experimental caries. A résumé. In Sognnaes, R. F. (ed.): Advances in Experimental Caries Research. Washington, D. C., American Academy for the Advancement of Science, 1955, pp. 197-222.

Hein, J. W., and Shafer, W. G.: Further studies on the inhibition of experimental caries by sodium-copper chlorophyllin. J. Dent. Res., 30:510, 1951.

Kruger, B. J.: The effect of trace elements on experimental dental caries in the albino rat. Univ. Queensland Press, 1:1, 1959.

Landing, B. H., Lahey, M. E., Schubert, W. K., Spinanger, J.: Iron content of teeth in hemosiderosis. J. Dent. Res., 36:750, 1956.

Ludwig, T. G., Healy, N. B., and Losee, F. L.: An association between dental caries and certain soil conditions in New Zealand. Nature, 186:695, 1960.

Lundqvist, C.: The toxicity of iodoacetic acid and its quantitative relations to inhibition of glycolysis and dental caries. J. Dent. Res., 30:203, 1951.

Mansell, R. E., and Hendershot, L. C.: The spectrochemical analysis of metals in rat molar, enamel, femur and incisors. Arch. Oral Biol., 2:31, 1960.

McClure, F. J.: Observations on induced caries in rats. VI. Summary results of various modifications of food and drinking water. J. Dent. Res., 27:34, 1948.

McClure, F. J., and Arnold, F. A.: Observations on induced caries in rats—reductions by fluorides and iodoacetic acid. J. Dent. Res., 20:97, 1941.

Muhler, J. C.: The effect of vanadium pentoxide, fluorides, and tin compounds on dental caries experience in rats. J. Dent. Res., 36:87, 1957.

Muhler, J. C., and Shafer, W. G.: Experimental dental caries. IV. The effect of feeding desiccated thyroid and thiouracil on dental caries in rats. Science, 119:687, 1954.

Nizel, A. E., and Bibby, B. G.: Geographic variations in caries prevalence in soldiers. J. Amer. Dent. Assn., 31:1619, 1944.

Nizel, A. E., and Harris, R. S.: Effect of foods grown in different areas on prevalence of dental caries in hamsters. Arch. Biochem., 26:155, 1950.

Nizel, A. E., and Harris, R. S.: The caries producing effect of similar foods grown in different soil areas. New Eng. J. Med., 244:361, 1951.

Nizel, A. E., and Harris, R. S.: Cariostatic effects of ashed foodstuffs fed in the diet of hamsters. J. Dent. Res., 32:672, 1953. (Abstract.)

Nizel, A. E., and Harris, R. S.: Effects of ashed foodstuffs on dental decay in hamsters. J. Dent. Res., 34:513, 1955.

Roberts, R. A.: The synergistic effect of molybdenum and fluorine upon caries in the rat. J. Dent. Res., 40:724, 1961.

Shaw, J. H.: Ineffectiveness of Na-Cu chlorophyllin in prevention of experimental dental caries. New York Dent. J., 16:503, 1950.

Shaw, J.: Chemistry of caries prevention. In Sognnaes, R. F. (ed.): Chemistry and Prevention of Dental Caries. Springfield, Ill., Charles C Thomas, 1962.

Sognnaes, R. F., and Shaw, J. H.: Experimental rat caries. IV. Effect of a natural salt mixture on the caries conduciveness of an otherwise purified diet. J. Nutr., 53:195, 1954.

Tank, G., and Storvick, C. A.: Effect of naturally occurring selenium and vanadium on dental caries. J. Dent. Res., 39:473, 1960.

Torell, P.: The possible caries activating effect of chemical reactions between sucrose and ferric iron. Odont. Tidsk., 64:165, 1956a.

Torell, P.: The possibility of using iron fluorides in topical applications. Odont. Tidsk., 64:189, 1956b.

Van Reen, R., Ostrom, C. A., and Barzinskas, V. J.: Studies of the possible cariostatic effect of natrium molybdate. Arch. Oral Biol., 7:35, 1962.

Chapter Ten

The Metabolism and Safety of Fluorides[*]

Isadore Zipkin, Ph.D.

FLUORINE

INTRODUCTION

Fluorine is a ubiquitous element found in fluoride form in plants, soil and water. In the earth's crust it occurs as fluorspar or fluorite (calcium fluoride), cryolite (sodium aluminium fluoride) and fluorapatite. Its presence was detected in animal body tissues at the beginning of the nineteenth century by Morichini and Gay-Lussac, both working independently. This element is found particularly in bones, and in minute amounts appears in all soft tissues and body fluids.

In excessive amounts it is an enzyme inhibitor in experimental situations and, therefore, an inhibitor of cellular metabolism. In fact, in large doses it is a poison. But in the small concentrations in which it occurs in human food and water, it is not deleterious; on the contrary, it is beneficial—so much so that it is considered a necessary nutrient in the sense that it is essential for optimal dental health. Fluoride, like so many other dietary constituents, is potentially harmful only if it is ingested in excessive and unusually large amounts.

Since fluoride is regarded, on the basis of our present knowledge, as the most important micro-mineral nutrient for promoting dental

health, it is extremely important to have a complete understanding of the deposition, metabolism and safety of this element from a human nutritional standpoint. The purpose of this chapter, therefore, is to discuss the effect of fluoride on the dental, skeletal and other tissues when it is present in a concentration of 8 ppm (which had been found naturally in some communal water supplies) as well as the effect of a concentration of 1 ppm fluoride (as it is presently being added to the drinking water supplies of 58 million people in the United States). Data on the excretion of fluoride in the urine, feces and perspiration will also be presented. In addition to discussing the deposition of fluoride in the bones and teeth, there will be discussion of the evidence concerning its built-in safety feature, namely, rapid elimination in the urine.

DEPOSITION OF FLUORIDE IN THE TEETH

Mottling. Historically, the importance of the effect of fluoride on dental health was not really appreciated until research into the cause of "Colorado brown stain" in this country and "denti neri" or "denti di Chiaie" (so called after Professor Stefano Chiaie, who first described the defect) in Italy was undertaken (McClure, 1962). This mottling is characterized by a defect in the enamel which consists of a discoloration which varies from white spots to brown or even black stains

* This article was writtend by Dr. Zipkin in his private capacity. No official support or endorsement by the National Institute of Dental Research, National Institutes of Health, Public Health Service, is intended or should be inferred.

111

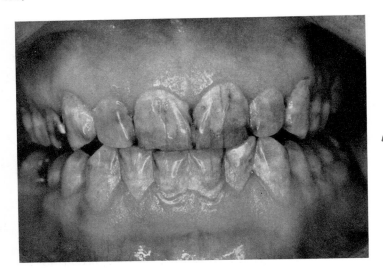

FIGURE 10-1. *Mottled enamel due to fluorosis.*

sometimes accompanied by a pitting of the surface.

Black and McKay (1916) in their several surveys came to the conclusion that these stains and defects were found in individuals, particularly youngsters, who drank water from deep artesian wells. They also noted a "curious absence of decay" in the teeth of children with mottled enamel. It was in 1931 when Churchill, a chemist for the Aluminum Company of America, demonstrated a more than usual amount of fluoride in the water supplies of Bauxite, Arkansas, that this element was suspected as the causative agent. Further findings of excessive fluoride in water supplies of other areas where mottled teeth were endemic confirmed the fluoride–mottled tooth relationship. There were other bits of evidence that corroborated this hypothesis. For example, at the same time two other laboratories found fluoride to be the etiological factor of mottled enamel (Smith et al., 1931; Velu, 1931). Smith and his associates conducted a feeding experiment in which the drinking water of St. David, Arizona, was fed to one group of rats while littermates were given drinking water with a known amount of sodium fluoride. Both groups of rats developed the same type of mottled effect on the incisor teeth. It was, therefore, concluded that fluoride was the cause of mottled enamel. A second bit of evidence was the lack of appearance of mottled teeth in children of Oakley,

Idaho, and Bauxite, Arkansas, who drank water from which the fluoride had been removed eight years previously. Their older brothers and sisters had this defect because they had drunk water supplies that had been high in fluoride content.

McKay (1932) stated, "It has become established through this investigation of mottled enamel that some ingredient of water is capable of profoundly altering the intrinsic structure of enamel during the calcification period, and that this ingredient invariably so acts. To a reasonable degree these researches that I am describing have satisfied us that this responsible ingredient is fluorine."

In 1933 Dean began an extensive study of the relation of fluorides in the community drinking water and tooth mottling or, as he

TABLE 10-1. CALCULATION OF INDEX OF DENTAL FLUOROSIS[*]

Classification	Weight (w)	Frequency (f)	Frequency × Weight (fw)
Normal	0	9	0
Questionable	0.5	19	9.5
Very mild	1	44	44
Mild	2	81	162
Moderate	3	98	294
Severe	4	38	152
		$\Sigma (f) = N = 289$	$\Sigma (fw) = 661.5$

Index of Dental Fluorosis $\dfrac{(fw)}{N} = \dfrac{661.5}{289} = 2.3$

[*] Taken from Dean, 1942.

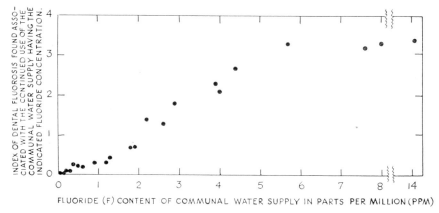

FIGURE 10-2. *Variation of index of dental fluorosis with the fluoride concentration of the communal water supply. (Observations on 5824 white children in 22 cities in 10 states.) After Dean (1942).*

termed it, "endemic dental fluorosis." He later developed a quantitative score called the "Index of Dental Fluorosis (Dean, 1942). An example of this scoring method taken from data obtained in Amarillo, Texas, whose drinking water averaged 3.9 ppm fluoride is presented in Table 10-1. It can be seen from this that the weighted mean Index of Dental Fluorosis for Amarillo is between mild and moderate. The relation between the fluoride content of the drinking water and the Index of Dental Fluorosis in 22 cities is shown in Figure 10-2.

Fluorosis more severe than "questionable" (an Index of 0.5) is seen only at fluoride concentrations exceeding about 1.5 ppm. When the drinking water contains 1 ppm fluoride or less, the Index of Dental Fluorosis is between "normal" and "questionable." "Idiopathic opacities" or nonfluoride dental opacities may be confused with dental fluorosis, and criteria have been formulated for their differential diagnosis (Russell, 1961). It has also been reported that nonspecific dental opacities are less frequent in areas in which the water contains optimum amounts of fluoride than in areas where it does not.

Dental fluorosis develops pre-eruptively since calcification of the teeth is presumed to be complete prior to eruption. The maturation schedule of the deciduous teeth is rapid and short compared with that of the permanent teeth and, in addition, little fluoride passes the human placenta (Zipkin and Babeaux, 1965) so that dental fluorosis in the deciduous teeth is rarely seen at levels of below 4 or 5 ppm fluoride in the drinking water.

The mechanism by which fluoride produces fluorosis is not known except that it is presumed that normal amelogenesis is interrupted. Microscopically, the enamel is characterized by the presence of poorly developed interprismatic cementing substance and incompletely calcified roots; and, in some cases, the interprismatic cementing substance may be absent.

Effect of Age on Deposition of Fluoride in the Teeth. The fluoride content of the dry, fat-free enamel and of the dentin of deciduous or permanent teeth in a low fluoride area is about 0.01 and 0.02 per cent, respectively, and increases with the fluoride content of the drinking water. Concentrations of fluoride as high as 0.11 per cent in the enamel and 0.21 per cent in the dentin have been observed where the drinking water contained 8 ppm fluoride. The level of fluoride in teeth under continuous exposure to a given level of fluoride in the drinking water increases with age until a plateau is reached, as shown in Figures 10-3 and 10-4.

No data appear to be available, however, on deposition of fluoride in teeth of individuals at various ages who have had no prior exposure to fluoride. It would be expected, however, on the basis of animal work, that less fluoride would be deposited in the teeth of older individuals without prior exposure to fluoride than in younger persons if intakes were equal over the same interval of time.

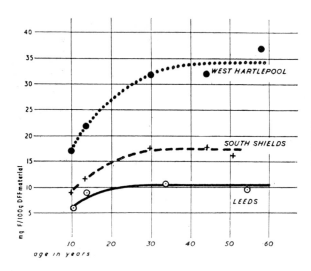

FIGURE 10-3. *The relation between age and fluoride content of human premolar enamel. Fluoride in drinking water: West Hartlepool, 1.9 ppm, South Shields, 0.8 ppm; Leeds, < 0.5 ppm. After Jackson and Weidmann (1959).*

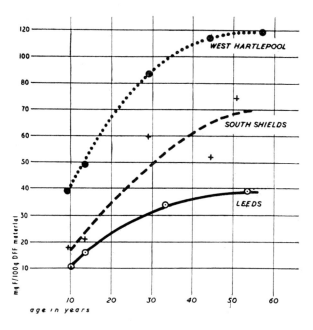

FIGURE 10-4. *The relation between age and the fluoride content of human premolar dentin. Fluoride in the drinking water: West Hartlepool, 1.9 ppm; South Shields, 0.8 ppm; Leeds, < 0.5 ppm. After Jackson and Weidmann (1959).*

It should be pointed out that fluoride is concentrated in the outer layers of the enamel and gradually decreases to the dentinoenamel junction and then increases again in concentration toward the pulp. The importance of these findings to caries will be discussed in Chapter Twenty-seven.

EFFECT OF FLUORIDE ON THE CHEMISTRY OF THE TEETH

It is generally agreed that fluoride reacts with the hydroxyapatite of teeth to form fluorapatite and not calcium fluoride. Fluorapatite is less soluble than hydroxyapatite and its formation may partially explain the cariostatic effect of fluoride.

Whereas fluoride has been shown to decrease the carbonate and citrate content of bone and elevate slightly the magnesium content (Zipkin, McClure and Lee, 1960), its effect on the chemical composition of teeth is equivocal. For example, Nikiforuk and Grainger (1965) reported that fluoride reduced the concentration of carbonate and citrate in enamel but Brudevold, McCann and Grøn (1965) did not agree with these findings. Brudevold, McCann and Grøn (1965) also corroborated the findings of others that fluoride did not affect the magnesium content of enamel. It is interesting to note that fluoride is highest in the outermost layer of enamel and decreases to the dentinoenamel junction. Carbonate, however, shows an inverse trend and increases in concentration from the outermost layer to the dentinoenamel junction. No such relation is apparent for either calcium or phosphorus.

SAFETY OF FLUORIDE WITH RESPECT TO THE TEETH

It has been demonstrated in a number of carefully controlled studies that fluoride at a level of 1 ppm in the drinking water will reduce the incidence of dental decay by about 60 per cent in individuals exposed since birth. The epidemiological studies of Dean (1942) and others have shown that at this level of fluoride intake the degree of mottling is so slight on tips of the molar cusps as to be seen only by the trained eye.

DEPOSITION OF FLUORIDE IN BONE

RELATION TO CONCENTRATIONS IN THE DRINKING WATER

A linear relationship was observed between the concentration of fluoride in the drinking water up to at least 4 ppm and its concentration in a number of human bones, as shown in Figure 10-5 (Zipkin et al., 1958).

Specimens of iliac crest, vertebra, rib, and sternum were obtained at autopsy from individuals who had been drinking water containing < 1.0, 1.0, 2.6 and 4.0 ppm fluoride for at least 10 years prior to death at an average age of 56, 76, 66 and 56 years, respectively, in each of the above fluoride areas. On a dry, fat-free basis, the bones averaged about 0.05, 0.15, 0.25 and 0.40 per cent fluoride, respectively. Departure from this linearity was seen in bones from one aged Bartlett, Texas, indi-

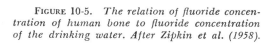

FIGURE 10-5. *The relation of fluoride concentration of human bone to fluoride concentration of the drinking water. After Zipkin et al. (1958).*

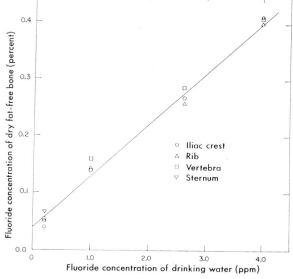

vidual whose drinking water contained about 8 ppm fluoride; that is, less fluoride had been deposited than would have been predicted by the straight line function seen in Figure 10-5.

EFFECT OF AGE ON DEPOSITION OF FLUORIDE IN BONE

The deposition of fluoride in bone is age dependent, as shown in Figure 10-6. A plateau at about 0.2, 0.25 and 0.4 per cent fluoride was observed in the human rib at levels of intake of < 0.5, 0.8 and 1.9 ppm fluoride, respectively, in individuals under continuous exposure to fluoride from birth. The values for bone fluoride obtained in these English subjects (Jackson and Weidmann, 1958) are somewhat higher than those obtained by Zipkin et al. (1958) and may well have resulted from the consumption of tea, a high source of fluoride.

No data appear to be available on the deposition of fluoride in the bones of individuals of different ages without previous exposure to fluoride—a situation that would arise when fluoride is first introduced into a community.

EFFECT OF FLUORIDE ON THE CHEMICAL COMPOSITION AND CRYSTALLINITY OF BONE

The bones obtained at demise from individuals previously mentioned who lived in areas containing < 1.0, 1.0, 2.6 and 4.0 ppm fluoride were analyzed for sodium, potassium, calcium, phosphorus, magnesium, carbonate and citrate, as shown in Table 10-2. Over an eightfold range in bone fluoride (Figure 10-5), no change was seen in the concentration of calcium, phosphorus, or potassium. A slight decrease was seen in the sodium content, about a 10 per cent over-all decrease in carbonate, and a striking 30 per cent decrease in citrate. The magnesium showed a small but consistent increase as the fluoride content of the bones increased. No data appear to be available on the effect of fluoride on the organic matrix of human bone.

Calcium and phosphorus are the chief components of the inorganic structure of bones and teeth and are present as very small crystals of apatite, $Ca_6(PO_4)_{10}(OH)_2$, in which the hydroxyls of random crystals can be partially or totally substituted for statistically by fluoride. A given weight of such small crystals will pre-

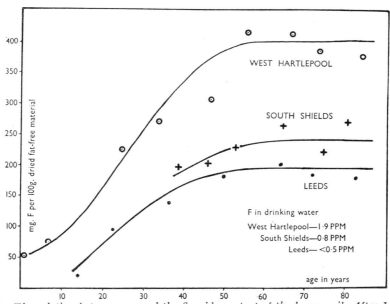

FIGURE 10-6. *The relation between age and the fluoride content of the human rib. After Jackson and Weid-mann (1958).*

TABLE 10-2. ASH, FLUORIDE, CALCIUM, PHOSPHORUS, MAGNESIUM, SODIUM, POTASSIUM, CARBONATE AND CITRATE CONTENT OF SELECTED HUMAN BONES AS RELATED TO FLUORIDE CONCENTRATION OF THE DRINKING WATER*

Bone Fluoride in H_2O (ppm)	Iliac Crest		Rib		Vertebra	
	< 1.0	4.0	< 1.0	4.0	< 1.0	4.0
Constituent %						
Ash†	53.5	58.6	56.0	57.4	49.8	52.3
Fluorine	0.08	0.69	0.08	0.70	0.10	0.80
Calcium	38.8	38.4	38.8	38.4	37.8	37.8
Phosphorus	17.5	17.7	17.5	17.7	17.5	17.5
Magnesium	0.50	0.57	0.50	0.60	0.50	0.61
Sodium	0.73	0.70	0.79	0.68	0.73	0.68
Potassium	0.09	0.08	0.09	0.10	0.18	0.12
Carbonate	5.83	5.20	5.41	5.19	5.31	4.55
Citrate	2.23	1.56	1.92	1.39	1.95	1.30

* Taken from Zipkin et al., 1960.
† Ash is expressed on a dry, fat-free basis. All other values are expressed on an ash basis.

sent a much larger surface area than an equal weight of fewer crystals of larger size. Since citrate will not fit into the crystal lattice and therefore is probably oriented on the surface, a smaller surface per unit mass occasioned by larger crystals would lead to a decrease in citrate. By x-ray diffraction techniques, it was shown that fluoride produced an increased resolution of the x-ray diffraction pattern, which was interpreted to indicate larger crystal size (Zipkin et al., 1962). The reduction in total surface area per unit weight of bone mineral could thus account for the decrease of bone citrate with an increase in fluoride. In general, larger crystals are presumed to be more stable and less soluble in dilute acids. If a similar relation between fluoride and crystal size can be demonstrated in enamel, then perhaps another parameter can be introduced to explain the cariostatic property of fluoride.

MOBILIZATION (RELEASE) OF FLUORIDE FROM THE SKELETON

Information is lacking on the release of fluoride from human bone when the level of fluoride in the drinking water is reduced. Any loss of fluoride from the skeleton would be reflected in the urinary excretion of fluoride and such evidence will be presented later, indicating that human bone may continue to lose fluoride gradually (Likins et al., 1956).

The deposition and mobilization of fluoride in bone appear to be related to its rate of growth. Young growing bone incorporates fluoride at a greater rate than older bone. This phenomenon is probably related to the relative rates of bone accretion (osteoblastic activity) and bone resorption (osteoclastic activity). Hence, as bone turnover or bone formation decreases, less fluoride is deposited since less apatite would be exposed for fluoride incorporation. In a sense, then, the inorganic phase may become more "buried" and less accessible for fluoride incorporation. Some of the fluoride which is deposited in bone may be incorporated into the crystal lattice of fluoride in place of hydroxyl and some may be adsorbed onto the crystal surfaces. The relative amounts of fluoride in these modifications are not known.

The slow release of fluoride from bone after discontinuance of fluoride administration may reflect the decreased turnover of old bone; that is, as bone metabolism and incorporation of calcium and phosphorus slow down as indicated by isotope studies, the bone crystals become more stable and more resistant to breakdown and recrystallization. For these reasons, then, fluoride tends to be held by the bone crystals and very little is lost.

It is difficult to invoke the same rationale to explain the decrease of fluoride deposition in enamel with age since enamel maturation is not similar to bone maturation. Enamel is avascular and shows no regenerative capacity. Although the inorganic phases are similar, the organic matrix differs markedly. In addition, enamel is not considered a "vital" tissue from a metabolic standpoint, so that deposition of fluoride in enamel may follow physicochemical principles rather than physiological processes.

SAFETY OF FLUORIDE
WITH RESPECT TO BONE

Since about 96 per cent of the fluoride in the animal body is found in the skeleton it is important to determine if any safety hazard is associated with elevated concentrations of fluoride in bone. It will be recalled that the bones obtained from individuals exposed to drinking water containing up to 4 ppm fluoride showed an approximately eightfold increase in fluoride concentration. Microscopic examination of these bones and intervertebral cartilage was made by Geever et al. (1958b) for focal calcification and cartilaginous changes of the periosteum and the adjacent tendon or fascia as well as for osteoclasia and osteophytosis. In addition, marrow sections were examined for hemopoiesis, and estimations were made for the degree of trabeculation and the thickness of the spongy bone of the medulla. No significant differences that could be related to fluoride intake were observed between the group exposed to fluoride and the control group.

No correlation has been observed between bone fracture experience of young military service inductees and the concentration of fluoride in the drinking water of their communities (McClure, 1946).

In another study, radiographs taken of the hands and wrists of children aged 7 to 14 years living in communities containing 3.3 to 6.2 ppm fluoride in the drinking water were compared with those of children of similar age drinking water containing 0.1 ppm fluoride. No differences in skeletal age and in the index of ossification of the carpal bones were evident (McCauley and McClure, 1954).

In an extensive pediatric study children drinking water containing 1.2 ppm fluoride in Newburgh, New York, were periodically compared with a similar group in Kingston, New York, whose drinking water was essentially fluoride-free (Schlesinger et al., 1956).

No differences were seen between the two groups in height, weight, x-rays of hands, knees and lumbar spines, and in bone age approximations. In addition, no differences were found in hemoglobin levels, red and white blood cell counts, routine urinalyses, ophthalmological and otological examinations, visual acuity or in audiometric testing. No differences were seen in the time of onset of menstruation, in the number of stillborns and in maternal and infant mortality rates.

A comprehensive examination was made of individuals aged 15 to 68 years who had had 15 years of continuous residence in either Bartlett, Texas (8 ppm fluoride), or Cameron, Texas (0.4 ppm fluoride), prior to initiation of a study (Leone et al., 1954). Ten years later the examinations were repeated on as many individuals as could be located. Except for dental fluorosis, no deleterious changes were present in individuals living in Bartlett. There was no unusual occurrence of bone fractures, arthritis, hypertrophic bone changes, exostoses or interference with fracture healing. About 10 to 15 per cent of the population examined in the high fluoride area showed some coarsened trabeculation and a slight increase in bone density, but without any functional impairment. Thus, "skeletal fluorosis" may be said to begin to appear in people exposed to 8 ppm fluoride in the drinking water, and the recommended fluoridation level of 1 ppm represents an eightfold safety factor.

The observation of increased bone density as a result of high fluoride intake has encouraged the use of high levels of fluoride in such bone resorptive conditions as osteoporosis, Paget's disease and multiple myeloma. As much as 100 mg. of fluoride per day in gelatin capsules as sodium fluoride have been given to such patients for extended periods of time. In addition to the seven studies enumerated by Zipkin, Bernick and Menczel (1965) and Zipkin (1965a), three others have been reported (Cohen and Rubini, 1965; Rich and Ivanovich, 1965; and Rose, 1965). While most of these studies report that fluoride partially prevented the loss of calcium from bone, some indicated little, if any, effect of fluoride, so that the use of fluoride in the bone diseases mentioned must be approached with some conservatism until the equivocation is resolved.

EXCRETION OF FLUORIDE

EXCRETION IN URINE, FECES AND PERSPIRATION

Two efficient mechanisms operate to maintain a low level of fluoride in the soft tissues and the circulating body fluids. The high affinity of the calcified structures for fluoride and their reluctance to mobilize any substantial amounts to the systemic circulation have been discussed and represent one of the mechanisms responsible for fluoride homeostasis.

The efficiency of the human kidney in excreting fluoride is the second mechanism available for maintaining continued low concentrations of fluoride in the body fluids (Armstrong, 1961). The clearance of fluoride by the kidneys exceeds that of chloride but is less than that of creatinine, indicating that some tubular resorption occurs (Carlson et al., 1960). An increased urinary flow enhances the excretion of fluoride, indicating that diuresis decreases tubular resorption of fluoride.

The urinary excretion of fluoride reflects its deposition in the skeleton, since less fluoride is excreted during periods of active bone growth coincident with rapid deposition of fluoride. Children 5 to 14 years of age who were drinking water containing 1 ppm fluoride required almost 3 years for their urine to reach excretion levels of 1 ppm fluoride (Zipkin et al., 1956), indicating that fluoride was being deposited in the growing bone. On the other hand, the urine of adults 30 to 34 years of age in the same communities reached 1 ppm fluoride within the first week after fluo-

ridation of the water supply to 1 ppm. These observations indicate that the age of the individual is an important factor in the urinary excretion of fluoride, and that the bones of adults, even without prior exposure to fluoride, do not incorporate appreciable amounts of fluoride.

The urinary excretion of fluoride can also serve as an index of release of fluoride from the bone, as already mentioned, since the fluoride lost would appear in the urine. The water supply of Bartlett, Texas, was defluoridated from 8 ppm fluoride to about 1 ppm, and urine from children aged 7 to 16 years and from adults over 20 years of age was examined for fluoride. Results of the examination are shown in Table 10-3. Even after 113 weeks of defluoridation of the water supply, the urinary concentration of fluoride had not declined to 1 ppm fluoride, as would be expected in individuals drinking water containing 1 ppm (Likins et al., 1956). These data suggest that fluoride continues to be lost slowly from the human skeleton.

The effect of elevated "challenge" doses of fluoride on its urinary excretion has been studied, with the fluoride given over a 24-hour period or given in a single dose. When up to 5 mg. of fluoride as NaF was ingested in water and food over a 24-hour period by young adults, approximately 65 per cent was found in the urine, about 20 per cent in the feces, and the remainder in the perspiration, so that about 80 per cent of the intake was eliminated in the urine and perspiration. In a "hot-moist" atmosphere, the excretion of fluoride in the perspiration was increased to about 30 per cent, with a corresponding decrease in the urinary concentration of fluoride. As a result of these studies, it was stated that "there was no significant retention of fluoride in the

TABLE 10-3. RELATION OF FLUORIDE CONCENTRATION OF HUMAN URINE TO FLUORIDE CONCENTRATION OF DRINKING WATER FOLLOWING DEFLUORIDATION*

Weeks after defluoridation	0	1	5	9	20	39	47	113
Fluoride in water (ppm)	8.0	0.8	1.3	1.0	0.9	1.0	1.2	0.9
Age Group (yrs.)				Fluoride in Urine (ppm)				
7-16	6.5	4.9	4.9	5.3	5.0	3.5	4.2	2.2
20-70	7.7	5.1	3.9	4.3	4.1	3.4	3.8	2.5

* Taken from Likins et al., 1956.

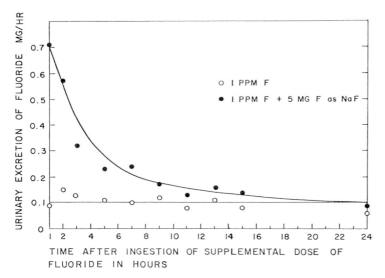

FIGURE 10-7.   *Urinary excretion of fluoride in eight normal individuals drinking water containing 1 ppm fluoride and receiving a supplement of 5 mg. fluoride as sodium fluoride. After Zipkin and Leone (1957).*

bodies of these young adult men, when total daily fluorine ingested did not exceed 4.0-5.0 mg. daily" (McClure et al., 1945).

An acute "challenge" dose of 5 mg. fluoride as NaF ingested in 200 ml. of water at one sitting was readily excreted in the urine. During the first 3 hours, about 20 per cent of the ingested fluoride was found in the urine. Ten hours after ingestion fluoride was being excreted at the same rate as the controls, namely about 0.1 mg. fluoride per hour during the 24-hour period, as shown in Figure 10-7.

SAFETY OF FLUORIDE WITH RESPECT TO KIDNEYS

It is apparent that the kidneys efficiently accommodate three to four times the daily amount of fluoride normally consumed in an area where the drinking water contains 1 ppm fluoride. The greatest proportion of fluoride deposited in bone is held in a stable configuration and is not readily mobilized and, hence, unusual amounts of fluoride are not excreted through the kidneys. Increased intake of a fluoridated water supply enhances flouride excretion in the urine by decreasing tubular resorption of this ion, as already mentioned. No increase in urinary calculi or damage to the kidneys has ever been observed as a result of intakes of fluoride as high as 8 ppm in the drinking water. Even individuals

who have a reduced urinary output as a result of kidney disease, which might lead to increased deposition of fluoride in bone, would be expected to tolerate an eightfold increase in bone fluoride without chemical or histopathological change or functional impairment.

CONCENTRATION OF FLUORIDE IN TISSUES OTHER THAN BONES AND TEETH

BLOOD, MILK, SALIVA AND BILE

These body fluids contain 0.1 to 0.2 ppm fluoride, which is not appreciably altered by increased intakes of fluoride, probably as a result of homeostatic mechanisms (Armstrong, 1961).

SOFT TISSUES

The heart, liver, lung and spleen of individuals drinking water containing < 1.0, 1.0, 2.6 and 4.0 ppm fluoride had a concentration of about 1 ppm fluoride in the wet tissue, which was not related to the concentration of fluoride in the drinking water (Smith et al., 1960). The high concentration of fluoride in the kidneys (3 ppm fluoride) may be due to adventitious retention of fluoride in the uriniferous tubules. Reported levels of fluoride in the aorta as high as 15 ppm are probably the result of the presence of calcified areas, which increase with age.

CARTILAGE

Dry, fat-free intervertebral cartilage from individuals in the four fluoride areas already mentioned contained 30, 40, 40 and 70 ppm fluoride, respectively. Intervertebral cartilage from one individual of Bartlett, Texas (8 ppm fluoride in the water at time of autopsy), contained 110 ppm fluoride, so that the concentration of fluoride in this tissue appears to increase with fluoride exposure. The very limited deposition of fluoride in intervertebral cartilage as compared to bone (1500 ppm in individuals drinking 1 ppm fluoride) is probably related to its low calcium, phosphorus and ash content (1, 0.5 and 4.0 per cent, respectively) compared to dry, fat-free bone (22, 11 and 60 per cent, respectively).

CALCULI

Urinary calculi from humans residing in an essentially fluoride-free area contained about 2500 ppm fluoride on a dry basis (ash content variable and as high as 85 per cent) compared to bone from individuals with the same fluoride exposure. It appears, therefore, that the mechanism for deposition of fluoride in urinary calculi is quite different from that for bone.

Biliary tract calculi averaged 20 ppm fluoride in individuals exposed to drinking water containing 2.6 ppm fluoride. This low value is probably due to the low calcium, phosphorus and ash content of these calculi and is in keeping with the low fluoride value (0.1 to 0.2 ppm) found in bile.

PLAQUE

Dental plaque has been reported to contain 6 to 180 ppm fluoride in the wet samples compared with 0.1 to 0.2 ppm in the saliva. It will be important to determine how much of

the fluoride is ionic since it may be sufficient to depress carbohydrate fermentation and thus exert a cariostatic effect on the underlying enamel.

MORTALITY STUDIES

Data presented by Hagan et al. (1954) have shown no statistically significant difference in mortality rates for five major causes of death between cities with fluoride in the water supply and those without (Table 10-4).

The "fluoride" cities, which had at least 0.7 ppm fluoride in the drinking water from natural sources, were paired with a neighboring city whose water supply contained less than 0.25 ppm fluoride. The total population of all the cities was in excess of 2,000,000. It can be seen that the death rates, adjusted for age, race and sex show no differences related to the fluoride concentration of the drinking water.

In another study, illustrated in Table 10-5, there was no significant change in the ratio of expected to recorded deaths in a number of communities before and after fluoridation of the drinking water (Hagan, 1955). The expected deaths were calculated using the United States age-race-sex mortality experience as a standard. Deaths resulting from accidents, suicides and homicides were excluded.

It has also been demonstrated that long-term exposure of individuals to 2.5 ppm fluoride in Colorado Springs, Colorado, did not result in an increased number of deaths from a variety of disease entities (Geever et al., 1958a); that is, the percentage of people dying from various major causes was no greater in individuals consuming 2.5 ppm fluoride in

TABLE 10-4. MORTALITY RATES PER 100,000 POPULATION (ADJUSTED FOR AGE, RACE AND SEX*

Death	"Fluoride" Cities	Control Cities
Heart disease	354.8	357.4
Cancer	135.4	139.1
Intracranial lesions	111.5	104.8
Nephritis	21.9	26.9
Cirrhosis of the liver	6.6	8.2
All causes	1010.6	1005.0

* Taken from Hagan et al., 1954.

TABLE 10-5.   RATIO OF RECORDED TO EXPECTED DEATHS IN FIVE CITIES INGESTING A
FLUORIDATED DRINKING WATER*

City and Year of Fluoridation	Year	Number of Deaths		Ratio of Recorded to Expected Deaths
		Recorded	Expected	
Grand Rapids, Ill., 1945	1940	1643	1674	0.98
	1950	1803	1764	1.02
Sheboygan, Wis., 1946	1940	394	380	1.03
	1950	383	416	0.92
Evanston, Ill., 1947	1940	611	660	0.93
	1950	672	765	0.88
Madison, Wis., 1948	1940	546	612	0.89
	1950	652	762	0.86
Charlotte, N. C., 1949	1940	806	782	1.03
	1950	964	933	1.03

* Taken from Hagan, 1955.

the drinking water of Colorado Springs for over 20 years than in those living there for only 5 years.

These three studies demonstrate that no relationship exists between mortality experience from a variety of causes and the presence of fluoride in drinking waters, whether naturally present or mechanically added. In addition, long-term exposures to levels as high as 2.5 ppm fluoride in the drinking water for over 20 years had no effect on mortality statistics when these were compared to figures for individuals with short-term exposure and drinking the same water.

SUMMARY

Fluorosis is limited almost exclusively to the permanent teeth and its presence is nearly imperceptible if the individual has been drinking water with concentrations of less than about 1.8 ppm fluoride. Fluorosis is rarely seen in the deciduous teeth unless there are concentrations of fluoride exceeding 4 to 5 ppm in the communal water supply. Idiopathic dental opacities are seen less frequently in areas in which the water contains an optimum amount of fluoride.

Enamel and dentin of teeth of people living in an area in which the water is essentially fluoride-free contain about 0.01 and 0.02 per cent fluoride, respectively, and the concentration of fluoride increases with elevated fluoride exposure. The iliac crest, rib and vertebra contain about 0.05 per cent fluoride in communities whose drinking water contains less

than 1 ppm fluoride, but the fluoride deposition increases proportionally to the fluoride content of the drinking water up to 4.0 ppm, when the bones contain about 0.4 per cent fluoride.

Under a constant regimen of exposure to fluoride from birth, the bones and teeth lose their capacity to incorporate fluoride as they age, so that the bones and teeth of the adult reach a plateau in fluoride deposition.

Fluoride in doses up to 100 ppm has been used in the management of resorptive bone disease to reduce the loss of calcium from bone. The data on some 10 studies, however, are still equivocal.

Urinary excretion data indicate that adults not previously exposed to fluoride incorporate less fluoride in the calcified structures than children also without previous exposure to fluoride; that is, the fluoride concentration of the urine of adults reaches 1 ppm within one week after fluoridation of a communal water supply to 1 ppm fluoride. Growing children, on the other hand, require about three years for the urine to attain a concentration of 1 ppm fluoride.

Urinary excretion data also indicate that the calcified structures lose fluoride very slowly and retain the major proportion of the incorporated fluoride when fluoride supplementation is discontinued.

The kidneys easily adjust to acute or chronic daily doses of fluoride as large as 5 mg. by rapid excretion of the ion. It is cleared from the blood at a much more rapid rate than is chloride. Increased intakes of fluoridated water enhance the excretion of fluoride by reducing resorption of fluoride by the kidney tubules.

Body fluids such as blood, milk, saliva and bile contain 0.1 to 0.2 ppm fluoride, and these concentrations are little influenced by moderate dietary supplementations of fluoride.

Soft tissues, with the exception of the kidney and aorta, contain about 1 ppm fluoride in the wet tissue, and this concentration is not increased by drinking water containing up to 4 ppm fluoride.

Dry, fat-free intervertebral cartilage normally contains about 0.005 per cent fluoride, but this becomes elevated as the fluoride concentration of the drinking water is increased.

Urinary calculi contain about 0.25 per cent fluoride on an oven-dry basis, whereas biliary tract calculi contain only about 0.002 per cent.

Wet samples of "dental plaque" contain 6 to 180 ppm fluoride, with a mean concentration of 67 ppm. The relative proportion of ionized fluoride in the plaque may have some relevance to the carious process in the underlying enamel.

The efficiency of the kidney in eliminating fluoride coupled with the affinity of the calcified structures for fluoride provides a homeostatic mechanism for the maintenance of low levels of fluoride in the circulating body fluids.

No differences between cities with fluoridated water supplies and those without have been found in the mortality rates for at least five major causes of death.

The data reviewed here reiterate the finding of many authors that fluoride in the drinking water at a level of 1 ppm presents no human health hazard.

REFERENCES

Armstrong, W. D.: Mechanism of fluoride homeostasis. Arch. Oral Biol. (Spec. Suppl.), *4*:156, 1961.

Black, G. V., and McKay, F. S.: Mottled teeth: an endemic developmental imperfection of the enamel of the teeth heretofore unknown in the literature of dentistry. D. Cosmos, *58*:129, 477, 627, 781, 894, 1916.

Brudevold, F., McCann, H. G., and Grøn, P.: Caries resistance as related to the chemistry of the enamel. *In* Wolstenholme, G. E. W., and O'Connor, M. (ed.): Caries-Resistant Teeth. Ciba Foundation Symposium. Boston, Little, Brown and Co., 1965, p. 121.

Carlson, C. H., Armstrong, W. D., and Singer, L.: Distribution and excretion of radiofluoride in the human. Proc. Soc. Exp. Biol. Med., *104*:235, 1960.

Cohen, M. B., and Rubini, M. E.: The treatment of osteoporosis with sodium fluoride. Clin. Orthopaed., *40*:147, 1965.

Dean, H. T.: The investigation of physiological effects by the epidemiological method. *In* Moulton, F. R. (ed.): Fluorine and Dental Health. Washington, D. C., Amer. Assn. Advance. Sci. Pub. No. 19, 1942, pp. 23-31.

Geever, E. F., Leone, N. C., Geiser, P., and Lieberman, J. E.: Pathological studies in man after prolonged ingestion of fluoride in drinking water. 1. Necropsy findings in a community with a water level of 2.5 ppm. J. Amer. Dent. Assn., *56*:499, 1958a.

Geever, E. F., Leone, N. C., Geiser, P., and Lieberman, J. E.: Pathological studies in man after prolonged ingestion of fluoride in the drinking water. Pub. Health Rep., *73*:721, 1958b.

Hagan, T. L.: Effects of fluoridation on general health as reflected in mortality data. *In* Muhler, J. C., and Hine, M. K. (ed.): Fluoride and Dental Health. Bloomington, Ind., Indiana University Press, 1955, pp. 157-165.

Hagan, T. L., Pasternack, M., and Scholtz, G. C.: Waterborne fluorides and mortality. Pub. Health Rep., *69*:450, 1954.

Jackson, D., and Weidmann, S. M.: The relationship of human bone as related to age and the water supply of different regions. J. Path. Bact., *76*:451, 1958.

Jackson, D., and Weidmann, S. M.: The relationship between age and fluorine content of human dentine and enamel: A regional survey. Brit. Dent. J., *107*: 303, 1959.

Leone, N. C., Shimkin, M. B., Arnold, F. A., Stevenson, C. A., Zimmerman, E. R., Geiser, P. R., and Lieberman, J. E.: Medical aspects of excessive fluoride in water supply. Pub. Health Rep., *69*:925, 1954.

Likins, R. C., McClure, F. J., and Steere, A. C.: Urinary excretion of fluoride following defluoridation of a water supply. Pub. Health Rep., *71*:217, 1956.

McCauley, A. B., and McClure, F. J.: Effect of fluoride in drinking water on the osseous development of the hand and wrist in children. Pub. Health Rep., *69*: 671, 1954.

McClure, F. J.: Non-dental physiological effects of trace quantities of fluorine. *In* Moulton, F. R. (ed.): Dental Caries and Fluorine. Washington, D. C., Amer. Assn. Advance. Sci., 1946, pp. 74-92.

McClure, F. J., Mitchell, H. H., Hamilton, T. S., and Kinser, C. A.: Balance of fluorine ingested from various sources in food and water by five young men. J. Ind. Hyg. Toxicol., *27*:159, 1945.

McKay, F. J.: Fluorine content of certain waters in relation to the production of mottled enamel. J. Amer. Dent. Assn., *19*:1715, 1932.

Nikiforuk, G., and Grainger, R. M.: Fluoride-carbonate citrate interrelations in enamel. *In* Stack, M. V., and Fearnhead, R. W. (ed.): Tooth Enamel. Bristol, England, John Wright and Sons, Ltd., 1965, p. 26.

Rich, C., and Ivanovich, P.: Response to sodium

fluoride in severe primary osteoporosis. Ann. Int. Med., *63*:1069, 1965.

Rose, G. A.: A study of the treatment of osteoporosis with fluoride therapy and high calcium intake. Proc. Roy. Soc. Med., *58*:436, 1965.

Russell, A. L.: The differential diagnosis of fluoride and non-fluoride opacities. Pub. Health Dent., *21*: 143, 1961.

Schlesinger, E. R., Overton, D. E., Chase, H. C., and Cantwell, K. T.: Newburgh Kingston caries-fluorine study. XIII. Pediatric findings after ten years. J. Amer. Dent. Assn., *52*:296, 1956.

Smith, F. A., Gardner, D. E., Leone, N. C., and Hodge, H. C.: The effects of the absorption of fluoride. V. The chemical determination of fluoride in human soft tissues following prolonged ingestion of fluoride at various levels. A.M.A. Arch. Ind. Health, *21*:330, 1960.

Smith, M. C., Lantz, D. M., and Smith, H. V.: The cause of mottled enamel; a defect of human teeth. Tucson, University of Arizona Press, Agr. Exp. Station Tech. Bull., *32*:253, 1931.

Zipkin, I.: Effects and metabolism of water-borne fluoride in man. *In* Blix, G. (ed.): Symposia of the Swedish Nutrition Foundation. III. Nutrition and Caries Prevention. Uppsala, Sweden, Almqvist and Wiksells, 1965a, pp. 96-11.

Zipkin, I.: Physiological effects of water-borne fluoride on the skeleton of man and mechanisms relating to its deposition and mobilization. WHO monograph on Fluoride and Human Health. 1965b, in press.

Zipkin, I., and Babeaux, W. L.: Maternal transfer of fluoride. J. Oral Therap. Pharmacol., *1*:652, 1965.

Zipkin, I., Bernick, S., and Menczer, L. F.: A morpho-logical study of the effect of fluoride on the perio-dontium of the hydrocortisone-treated rat. Perio-dontics, *3*:111, 1965.

Zipkin, I., and Leone, N. C.: Role of urinary fluoride output in normal adults. Amer. J. Pub. Health, *47*:848, 1957.

Zipkin, I., Likins, R. C., McClure, F. J., and Steere, A. C.: Urinary fluoride levels associated with the use of fluoridated water. Pub. Health Rep., *71*: 767, 1956.

Zipkin, I., McClure, F. J., and Lee, W. A.: Relation of the fluoride content of human bone to its chemi-cal composition. Arch. Oral Biol., *2*:190, 1960.

Zipkin, I., McClure, F. J., Leone, N. C., and Lee, W. A.: Fluoride deposition in human bones after pro-longed ingestion of fluoride in drinking water. Pub. Health Rep., *73*:732, 1958.

Zipkin, I., Posner, A. S., and Eanes, E. D.: The effect of fluoride on the X-ray diffraction pattern of the apatite of human bone. Biochim. Biophys. Acta, *59*:255, 1962.

Reviews

Hodge, H. C., and Smith, F. A.: Fluorine Chemistry. Vol. IV. Simons, J. H. (ed.). New York, Academic Press, Inc., 1965.

Jenkins, G. N.: The physiology of fluoride. Int. Dent. J., *12*:208, 1962.

McClure, F. J.: Fluoride Drinking Waters. Public Health Service Pub. No. 825. Washington, D. C., U. S. Government Printing Office, 1962.

Roholm, K.: Fluorine Intoxication: A Clinical Hygienic Study with a Review of the Literature and Some Experimental Investigations. London, England, H. K. Lewis & Co., Ltd., 1937.

Chapter Eleven

Fat Soluble Vitamins

STANLEY N. GERSHOFF, PH.D.

INTRODUCTION

During the second decade of this century, McCollum and Davis and Osbourne and Mendel observed that some of their rats fed purified diets grew poorly and developed inflamed eyes. When fats, such as butter or cod liver oil, or ether extracts of egg yolk were added to their rations, the animals grew normally. Thus, the concept of fat soluble vitamins was developed, and nutritionists wrote of "fat soluble vitamin A" and "water soluble vitamin B." During the succeeding 35 years, the multiplicity of the water and fat soluble vitamin complexes was slowly unraveled in a series of fascinating studies, whose history has been described by McCollum (1957).

There are four known fat soluble vitamins —A, D, E and K. In contrast to our extensive knowledge of the metabolic roles of the B vitamins, understanding of the metabolic functions of the fat soluble vitamins is fragmentary. Vitamin A deficiency remains a major health problem in many parts of the world. Large numbers of children lose their sight permanently every year in Southeast Asia, the Middle East and Africa because of a deficiency of vitamin A in their diets. Although vitamin D deficiency is relatively rare, there are still some areas of the world where rickets remains a medical problem. Vitamin K deficiency is a potential problem of the newborn child and may also be observed in individuals with medical problems affecting their ability to absorb fat. In recent years, the problem of vitamin E deficiency in human beings has become of interest to several groups of investigators.

Fortunately, inexpensive natural and synthetic sources of all the fat soluble vitamins are almost universally available. However, the formidable problem remains of educating the public, particularly in impoverished areas of the world, in nutritional habits which, within their economic circumstances, will provide them with adequate dietary sources of the fat soluble vitamins as well as other nutrients. Concomitantly, the problem remains of making workers in the various medical disciplines aware of the problems of deficiencies of the fat soluble vitamins and how to diagnose and treat them.

VITAMIN A

CHEMISTRY

Dietary vitamin A activity is found in foods of animal origin which contain vitamin A or in plant foods containing those carotenoid pigments which can be converted to vitamin A after ingestion.

There are several forms of vitamin A, by far the most common being vitamin A_1, a slightly yellow crystalline (m.p. 62–64°C) unsaturated cyclic alcohol, whose structure is shown below.

Vitamin A_1

(From White, A., Handler, P., and Smith, E.: Principles of Biochemistry. 3rd ed. Blakiston Division, McGraw-Hill Book Company. Used by permission.)

125

Other forms of the vitamin, A_2 and neo-vitamins A-a and A-b, vary slightly in their structure or in their steric configuration. Vitamin A is found in nature usually esterified with fatty acids, and occasionally bound to a protein molecule. Because of its high degree of unsaturation, it is easily oxidized in the presence of atmospheric oxygen. Its stability may be protected by the use of antioxidants such as hydroquinone or vitamin E. Vitamin A reacts with antimony trichloride to produce a transient blue color, the intensity of which can be easily measured in a colorimeter or a spectrophotometer. This reaction forms the most common basis for the estimation of vitamin A in foods and other biologic materials.

The carotenoid pigments or carotenes are found in colored plant materials. These compounds are structurally similar to vitamin A and at least ten of them have provitamin A activity. Their ability to replace vitamin A in the diet varies with their structure and with the animals being studied. The cat, for instance, is unable to significantly utilize carotenoids. For humans, β-carotene is the most important of the carotenes, but α-carotene and cryptoxanthin are also of importance as precursors of vitamin A. The structure of β-carotene is such that it would seem that each molecule of β-carotene should provide two molecules of vitamin A.

photometric determination. As in the case of vitamin A, the carotenes are not stable in the presence of atmospheric oxygen. This is of particular practical importance to food processors and to farmers who have the problem of preservation of the provitamin A activity in stored feeds.

ABSORPTION AND UTILIZATION

Since both Vitamin A and the carotenes are fat soluble, their utilization may be decreased by any physiological state which interferes with fat absorption. Thus, such diseases as sprue or liver diseases which interfere with bile production or flow may induce a vitamin A deficiency. Diarrhea and excessive intakes of mineral oil may also interfere with the absorption of vitamin A and its precursors. Within reasonable limits, the dietary level of fat does not have much affect on the absorption of vitamin A. On the other hand, studies done on several species of animals, including man, indicate that the utilization of carotenoids may be affected by dietary fat levels, being greater with increased fat intakes. Vitamin A alcohol is absorbed through the intestinal wall, during which process it is esterified with long chain fatty acids. The vitamin A is transported via the lymph to the blood

β-Carotene

γ-Carotene

(From White, A., Handler, P., and Smith, E.: Principles of Biochemistry. 3rd ed. Blakiston Division, McGraw-Hill Book Company. Used by permission.)

Unfortunately, as will be discussed later, the ability of humans to convert β-carotene to Vitamin A does not approach this degree of efficiency. The carotenes are deeply colored and are usually assayed by methods which involve their isolation followed by a spectro-

and then distributed to the tissues, the liver and kidneys being primary sites of deposition. At one time it was thought that the conversion of carotenes to vitamin A took place in the liver. It is now generally accepted that the main site of this conversion is in the walls of the intestines. All the ingested carotenoids with provitamin A activity may not be cleaved. Depending on the physiological state of the individual and the levels being fed, varying amounts of carotene are absorbed intact and appear in the liver, serum and sometimes in body fat. Both vitamin A and β-carotene are readily stored for long periods of time and, following a period of liberal ingestion of the vitamin or provitamin, these stores may remain available for periods of months to meet nutritional needs for vitamin A.

SOURCES

The major dietary sources of vitamin A are the fats of dairy products and eggs. For individuals who find them tasty, liver and kidneys may provide considerable quantities of the vitamin. Marine liver oils are particularly potent sources of vitamin A, and cod and halibut liver oil are often used as dietary supplements. Vitamin A_2, which occurs in the livers of many fresh water fishes, is of little importance in human nutrition.

Green leafy vegetables and yellow vegetables such as carrots, corn or sweet potatoes, are excellent sources of carotenoids. Red palm oil, which is found in abundance in many tropical parts of the world, is a particularly potent source of carotenoids. Nevertheless, because of a combination of custom and ignorance, vitamin A deficiency is a major health problem in many areas where red palm oil is available.

It should be pointed out that the vitamin A and carotene content of foods varies considerably, depending on the source of the food samples being analyzed. Food tables which are the most used source of food composition information may in some circumstances provide misleading figures.

REQUIREMENTS

By definition, 1 International Unit (IU) of vitamin A is equal to 0.3 micrograms of vitamin A alcohol, or 0.6 microgram of β-carotene. This would indicate that β-carotene is half as active as vitamin A alcohol. Unfortunately, these figures are confusing because they were obtained from data based on

a specific rat assay. Studies done on a wide variety of animals, including man, suggest that 20 IU of vitamin A alcohol or 40 IU of β-carotene per kilogram of body weight are the minimal daily requirements to prevent gross signs of deficiency. Thus, on a weight basis, β-carotene appears to be approximately one-fourth as active as vitamin A.

There are many factors which influence the conversion of β-carotene to vitamin A, including the state of nutriture with regard to the vitamin. Deficient animals are much more efficient in converting β-carotene to vitamin A than well fed ones. The form in which the carotene is fed is also of importance. For instance, the carotene in carrots is more available from cooked than from raw ones. As is generally true for all nutrients, the requirements of man and animals for vitamin A increase with growth, pregnancy and lactation. In the 1964 revised edition of Recommended Dietary Allowances, it was presumed that the average American diet contains two thirds of its vitamin A activity as carotene, and one third as vitamin A. On this basis, the recommended allowance for adults of 5000 IU per day is considered to be provided by 4000 IU of β-carotene (equivalent to 2000 IU of vitamin A) and 1000 IU of preformed vitamin A. This recommendation is about twice the minimum requirement of 1500 IU of preformed vitamin A. Allowances provided for children of different ages and pregnant and lactating women take into account their increased needs for the vitamin.

METABOLIC FUNCTIONS

Information concerning the metabolic roles of vitamin A are incomplete, but its function in the retina has been described in detail in a series of classic studies, many of them by Wald's laboratory. The information concerning the role of vitamin A in vision is one of the rare instances in which it has been possible to associate a well-defined biochemical abnormality caused by a vitamin deficiency with functional changes in an organ. Chronic vitamin A deficiency is associated with night blindness. The human retina contains two types of light sensitive cells: the cones and the rods. These cells contain several proteins conjugated with vitamin A aldehyde. The one which has received the most attention is rhodopsin, or visual purple, which is found in the cones and which is the main photosensitive material in the rods, the latter cells being responsible for vision in dim light.

In the presence of light, retinal rhodopsin is cleaved through a series of reactions to a protein-opsin, and transvitamin A$_1$ aldehyde-retinene. Opsin cannot recombine with trans-retinene but can with $\triangle''$ -cis-retinene to form rhodopsin. A small amount of the trans-retinene may be converted directly to the cis form. Most of the cis-retinene, however, is formed as a result of the following reactions:

$$\text{trans-retinene} + \text{DPNH} + \text{H}^+ \longrightarrow$$
$$\text{trans-vitamin A}_1 + \text{DPN}^+$$
$$\longrightarrow \text{cis-vitamin A}_1$$
$$\text{cis-vitamin A}_1 + \text{DPN}^+ \longrightarrow$$
$$\text{cis-retinene} + \text{DPNH} + \text{H}^+$$

It is thought that much of the formation of cis-retinene occurs in the liver. In vitamin A deficient animals, the regeneration of rhodopsin is slowed down. Since this primarily affects the rod cells, night blindness occurs.

With the exception of its functional characteristics in the retina, the active form in which vitamin A acts at the cellular level is obscure. If animals are fed diets containing vitamin A acid instead of vitamin A, they grow normally but become blind and sterile. Presumably, this means that vitamin A acid cannot be converted to vitamin A or retinene and that the alcohol or aldehyde forms of the vitamin are necessary for reproduction as well as vision.

Vitamin A is of importance in the biosynthesis of mucopolysaccharides. Some of the pathologic changes observed in hypo- and hypervitaminosis A may be the result of abnormalities in mucopolysaccharide metabolism.

PATHOLOGY OF VITAMIN A DEFICIENCY

Vitamin A deficiency is characterized in most species by growth failure, visual defects, an inability to maintain the integrity of epithelial structures in a variety of tissues and reproductive failure. Because vitamin A deficient people and animals are prone to infection, an anti-infection role has been improperly ascribed by some to the vitamin. Vitamin A does not affect pathogens and there is no evidence that vitamin A deficiency affects the immune response. However, to the extent that it is necessary for the maintenance of the integrity of mucous membranes, it plays a role in protecting tissues against invasion by pathogens and, therefore, protects against infection. This does not imply that the taking of vitamin A in amounts greater than the accepted requirements will provide added protection against disease.

Epithelial Tissues. Vitamin A deficiency is manifested in the skin as dryness or xerosis, which gives it a rough appearance. There is hyperplasia and hyperkeratinization of the epidermis and a plugging of the hair follicles, producing a papular eruption. This is called follicular hyperkeratosis and is looked for in surveys of the nutritional status of people.

Changes are seen in epithelial cells that have a secretory function. The mucosa of the digestive system, the respiratory tract and the genitourinary system and cells of the sensory organs and the endocrine system may undergo hyperkeratinization. Microscopically, there is an atrophy of the columnar or cuboidal cells, with gradual replacement by stratified squamous keratinized epithelium.

Eyes. Besides night blindness, vitamin A deficiency is associated with other eye disorders. Keratinizing metaplasia of the sclera and cornea occur, accompanied by diminution of tear secretion and increased vascularization of the substantia propria. This condition, called xerophthalmia, is extremely serious and if untreated may lead to the breakdown of the outer layers of the eye, resulting in permanent blindness as a result of loss of the vitreous body, displacement of the lens and infection.

Prolonged vitamin A deficiency may also result in permanent degeneration of the retina. Many investigators have associated vitamin A deficiency with Bitot's spots, white elevated, sharply outlined patches not wetted by tears and covered with a material resembling dried foam. The interpretation of the occurrence of Bitot's spots is controversial, but they are usually found in poorly nourished populations who often show other evidences of vitamin A deficiency.

Bone. In young experimental animals, vitamin A deficiency is accompanied by cessation of bone growth. There is normal intramembranous bone formation, but remodeling sequences become abnormal and stop. The bones of young vitamin A deficient animals may be short and thick. If vitamin A deficiency is induced in adult animals, bone disorders are usually not observed.

Nervous System. It is generally agreed that nerve lesions observed in experimental vitamin A deficiency are the result of disproportionate growth between bone and nervous tissue. Because bony growth is limited, the brain and spinal cord grow in an increasingly confined area which produces pressure changes in the central nervous system.

HYPERVITAMINOSIS A

Although there is considerable literature on the toxic effects of excessive injection of vitamin A in children, adults and animals, physicians continue to prescribe large doses of vitamin A, particularly for dermatologic problems not related to vitamin A, and lay people continue to misuse vitamin supplements.

The feeding of excessive amounts of vitamin A to experimental animals has been associated with anorexia, bone fragility and tenderness related to a speeding up of growth processes in bone, loss of hair, excessively dry scaly skin and hypoprothrombinemia. Similar effects have also been described in reports of infants, children and adults taking daily supplements of vitamin A varying from twenty-five to several hundred times their requirements. Fortunately, discontinuation of massive doses of vitamin A in these individuals has usually been followed by a dramatic disappearance of symptoms in a few days.

There have also been reports of discoloration of the skin without other symptoms in individuals consuming large amounts of foods containing carotenoid pigments for prolonged periods.

Oral Relevance of Vitamin A

ORAL MUCOUS MEMBRANE

The primary effect of vitamin A deficiency is on tissues of ectodermal origin, particularly the epithelial cells lining the mucosa of the mouth and salivary glands. The cells atrophy and are replaced by keratinized epithelium.

There has been some preliminary evidence that oral leukoplakia can be improved in some patients by local use of troches containing 75,000 units of vitamin A acetate, given in dosages of 10 troches a day for 2 to 5 weeks. This therapy did not cause any harm to any of the patients in spite of the fact that none actually showed clinical or laboratory evidence of a vitamin A deficiency (Silverman et al., 1963). Others have also reported an anti- or dekeratinizing effect upon normal or hyperkeratotic oral mucosa with administration of 400,000 units of vitamin A daily for 21 days (Johnson et al., 1963).

TEETH

Teeth of Experimental Animals. In the enamel organ of rats there is an atrophy and degeneration of ameloblasts, which ultimately change to keratin if the animals are fed a vitamin A deficient diet. This dietary A deficiency can slow down and even completely stop the growth of incisor teeth in rats (Wolbach and Howe, 1933; Mellanby, 1939; Schour et al., 1941). A characteristic of this deficiency is a disturbance in enamel formation which produces a hypoplastic and chalky white incisor as a result of loss of orange pigment (Irving and Richards, 1939).

The odontoblasts are also exceedingly sensitive to vitamin A deficiency and they also atrophy. Irregular activities in the labial and lingual odontoblasts take place, producing a thick, regular labial dentin with interglobular spaces and a thin atubular lingual dentin (Mohammed and Mardfin, 1961).

Dental caries is three times more prevalent in hamsters on a vitamin A deficient diet compared to those on a control diet, which can be attributed to the reduced salivary flow from atrophy of the salivary glands in the vitamin A deficient group (Salley et al., 1959).

The growth and development of jaws and teeth are affected by vitamin A deficiency, which is manifest by crowding of the teeth and by roots that are stunted and thickened (Mellanby, 1939).

Human Teeth. Dinnerman (1951) found that vitamin A deficiency in children pro-

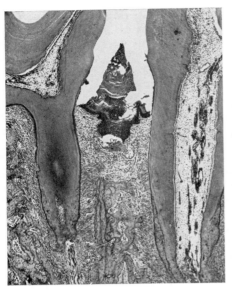

FIGURE 11-1. *Epithelial proliferation, pocket formation and abscess formation associated with irritation from food debris in vitamin A-deficient albino rat. Note pronounced hyperkeratosis. (Glickman: Clinical Periodontology. 1st ed.)*

duced atrophy of the enamel organ and metaplasia of the ameloblasts to nonspecialized stratified epithelium in the developing tooth germ which resulted in hypoplasia. The changes in the dentin consisted of poor calcification and wider than normal predentin areas.

Boyle (1933) described the following histologic changes in the tooth germs of a 3½ month old infant with vitamin A deficiency: atrophy of enamel and cessation of enamel formation, replacement of ameloblasts and stellate reticulum by squamous epithelium, defectively calcified dentin, and wide predentin with capillary cell inclusions.

There were no dental changes in a number of children who had the characteristic clinical signs of vitamin A deficiency (Block, 1930), nor does this avitaminosis have an effect on dental caries incidence (Marshall-Day, 1944).

PERIODONTIUM

In Animals. In dogs on a vitamin A deficient diet the gingival epithelium becomes hyperplastic (Mellanby, 1941). In rats there are also hyperkeratosis and hyperplasia of gingival tissue which is accompanied by atrophy

of the alveolar bone, calcification and irregular cementum resorption and ankylosis of the teeth (Miglani, 1959).

Glickman and Stoller (1948) reported that vitamin A deficient albino rats had deeper periodontal pockets than control animals, but only in the presence of a local irritating factor (Fig. 11-1) These deeper pockets were the result of a proliferation of the basal cells of the epithelium of the gingival sulcus. However, they did not see any evidence of periodontal membrane or alveolar bone involvement as a result of vitamin A deficiency without local irritants.

In Humans. A survey of population groups indicates that there may be an association of periodontal disease with lowered serum vitamin A levels. However, it was not at all certain that vitamin A or any other nutrient studied in this survey was truly independent of age or hygiene (Russell, 1963). Others have also reported that there seems to be a trend toward a positive correlation with vitamin A deficiency (Radusch, 1940; Marshall-Day, 1944).

HYPERVITAMINOSIS A

Irving (1949) suggests that tissues of mesenchymal origin are sensitive to hypervitaminosis A. The rate of formation of the alveolar bone is greatly reduced and active osteoblasts become less prominent, with the result that the bone becomes abnormally thin.

VITAMIN D

Rickets is a disease of which man has been aware for thousands of years. Until recently, a large percentage of children raised in the northern hemispheres suffered from rickets, although the curative value of liver and fish liver oils has been known for centuries. In 1922 McCollum and his associates found that heated and aerated cod liver oil would not cure ophthalmia in their experimental animals (since the vitamin A had been destroyed) but would cure rickets in rats fed diets with abnormal calcium:phosphorus ratios. These workers called their factor vitamin D. At about the same time, the relationship of ultraviolet light to the formation of vitamin D in animals and foods was being studied inten-

$$CH_3CH-CH=CH-CH-CH\begin{smallmatrix}CH_3\\\\CH_3\end{smallmatrix}$$

Vitamin D₂

$$CH_3CHCH_2CH_2CH_2CH\begin{smallmatrix}CH_3\\\\CH_3\end{smallmatrix}$$

Vitamin D₃

sively. During the next several years, the chemistry of the vitamin was elucidated.

CHEMISTRY

There are a number of steroid alcohols with vitamin D activity. The most important of these nutritionally are vitamin D_3, or cholecalciferol, and vitamin D_2, or ergocalciferol.

Compounds with vitamin D activity are white, odorless, crystalline substances, soluble in organic solvents but not water. They are stable to oxidation, heat, acid and alkali.

Biological methods are usually employed in analyzing materials for vitamin D content because of a lack of suitable chemical and physical methods. Since microorganisms do not require the vitamin, vitamin D assays are usually based on the cure or prevention of rickets in rats or chicks. The bases for these analyses are comparative measurements of bone or beak ash or, most frequently, the mineralization of cartilage in the tibial metaphyses (line test) of rachitic animals fed test materials containing vitamin D and others fed standard amounts of vitamin D.

ABSORPTION AND UTILIZATION

As in the case of vitamin A, conditions such as disease of the gallbladder, pancreas or liver which interfere with fat absorption also interfere with the absorption of vitamin D. Vitamin D is stored primarily in the liver. Although storage of this vitamin is relatively less than vitamin A, it is sufficient so that when adequate daily sources of vitamin D cannot be assured, children may be given massive single doses of the vitamin (approximately 600,000 IU) every 6 months to protect them against rickets.

SOURCES

There are few good food sources of vitamin D. Fish liver oils contain large variable amounts of the vitamin and have often been used as both vitamin A and D supplements. In many countries, including the United States, milk is routinely enriched with 400 IU of vitamin D per quart. Ergosterol, which on ultraviolet irradiation is converted to vitamin D_2, is found widely in the plant kingdom, being particularly concentrated in yeasts and fungi. Nevertheless, vitamin D is rarely found in vegetables. Most animals acquire vitamin D by sunlight irradiation of 7-dehydrocholesterol, a precursor of vitamin D_3 found in their skins. 7-Dehydrocholesterol can be synthesized by animals from cholesterol, so a dietary source of the provitamin is not necessary.

REQUIREMENTS

An International Unit of vitamin D is 0.025 microgram of vitamin D_3. It is apparent that there is an inverse relationship between the amount of dietary vitamin D required and exposure to sunshine. The bases for establishing vitamin D requirements for human beings are not very secure. It appears certain that dietary vitamin D is required by infants and growing children. The occasional appearance of a deficiency state in adults suggests that they also require vitamin D, although ordinarily their needs are probably met by exposure to sunlight. The amounts of vitamin D needed during pregnancy and lactation are not known. The current edition of Recommended Dietary Allowances suggests that 400 IU per day be given to infants, growing boys and girls, lactating women and women in their second and third trimesters of pregnancy.

METABOLIC FUNCTIONS

It is not as yet possible to explain the mode of action of vitamin D. The pathology of vitamin D deficiency is entirely related to its effect on bone formation. It is clear that vitamin D has a significant effect on calcium absorption from the gastrointestinal tract and, because of this, on phosphorus absorption. Rachitic animals show higher levels of fecal calcium and phosphorus and lower levels of serum calcium and phosphorus than normal animals.

There have been several theories as to how vitamin D affects calcium absorption, but they have not received general support. The feeding of vitamin D results in a lowering of the pH of the intestinal contents, and some calcium salts such as the phosphates are more soluble and more easily absorbed at a low pH. However, using other methods to lower the pH of the intestinal contents has not resulted in increased calcification in rachitic animals.

Vitamin D deficiency is accompanied by a marked lowering of serum and bone citrate. This may be related to abnormalities which have been reported in the mitochondria of vitamin D deficient rats. It has been claimed that the feeding of citrates increases calcium absorption and has an antirachitic effect in animals and infants. However, the injection of large amounts of citrate in rachitic animals has not been effective in curing rickets. Whether the effect of vitamin D on citrate metabolism is of importance in the etiology of rickets remains controversial.

Rickets is usually accompanied by a marked rise in serum alkaline phosphatase activity, thus providing a useful laboratory diagnostic test for rickets, although it must be remembered that other disease conditions may also result in increased serum alkaline phosphatase values. The relationships of alkaline phosphatase to rickets are not understood.

There is disagreement as to whether vitamin D has a primary effect on the calcification of bone. Histologic changes in the metaphyses of rachitic rat bones unrelated to serum calcium and phosphorus levels have been reported by some workers but have been denied by others. Calcification has been initiated in rachitic bone placed in serum from normal animals but not in that from rachitic animals, even though the calcium and phosphorus contents of the sera used have been the same. It has also been reported that the administration of vitamin D to rachitic rats initiates healing of rickets even when animals are not provided with dietary calcium or phosphorus. On the other hand, studies indicate that there is no difference in the amount of deposition of injected radioactive calcium in the bones of rachitic and normal chicks. Thus, it can be seen that the question of whether vitamin D has a direct effect on bone remains controversial.

There probably is little direct action of vitamin D on the parathyroids. In rickets, there is increased parathyroid activity which is effective in maintaining serum calcium near normal levels. This is accompanied by a decrease in renal tubular reabsorption of phosphorus and accounts in part for the low serum phosphorus values seen in rickets, which lower the solubility product of serum calcium and phosphate ions.

PATHOLOGY OF VITAMIN D DEFICIENCY

Vitamin D deficiency in children and growing animals is called rickets. It has been produced in a large number of animal species, although in some (such as the rat), it is necessary to feed vitamin D deficient diets containing very high or low ratios of calcium to phosphorus to produce the disease. In adults, vitamin D deficiency is known as osteomalacia. Rachitic animals grow poorly. They are deformed, walk unnaturally with the appear-

ance of being in pain, and are subject to bone fractures. The ash content of dry, fat-free normal bone of most species is over 50 per cent; in rachitic bone, it may be as little as 25 per cent. Rickets is seldom fatal but may predispose individuals to other diseases, particularly bronchopneumonia because of deformities of the rib cage.

To understand the microscopic changes that characterize rickets, it is necessary to understand the normal growth and structural pattern of long bones. In endochondral bone formation, the epiphyseal cartilage grows continuously both toward the diaphysis, which is the shaft, and toward the distal end of the bone. At the epiphyseal plate (metaphysis), the cartilage cells line up in a palisade-like arrangement to form vertical columns (Fig. 11-2). The provisional zone of calcification is made up of these cartilaginous spicules with capillaries and osteoblasts growing between them. Osteoid tissue, formed by the osteoblasts, replaces the cartilage, and the deposi-

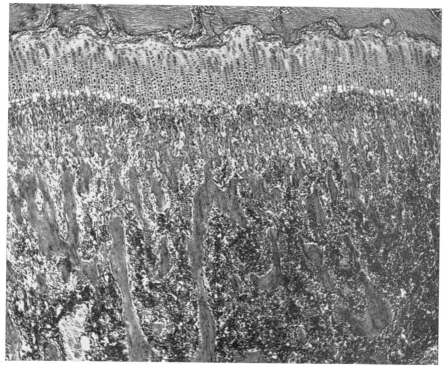

FIGURE 11-2. *Tibia, guinea pig. This shows the upper tibial epiphyseal cartilage and underlying bone. Note straight rows of cartilage cells and trabeculae of bone beneath. The marrow cells are abundant and are found up close to the cartilage. H. and E. (×60). (Follis, R. H., Jr.: Deficiency Disease. Charles C Thomas, 1958.)*

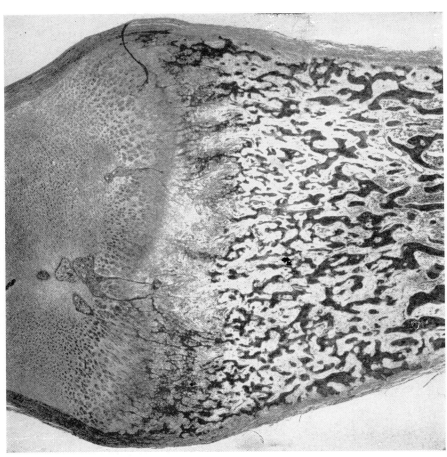

FIGURE 11-3. *Rib. Moderate rickets. Costochondral junction from a seven month old colored male dying acutely in three days of lobular pneumonia. Note increase in width of cartilage shaft junction. Especially prominent in the increase is width of zone of proliferative cartilage cells and irregularities in the calcification of the region. Note also tongues of cartilage projecting down toward the shaft surrounded on either side by invading vessels. The trabeculae beneath the cartilage are more numerous than usual. There is a great deal of osteoid on such trabeculae. (Follis, R. H., Jr.: Pathology of Nutritional Disease. Charles C Thomas, 1948.)*

tion of calcium and phosphate in the osteoid forms bone.

In rickets, the palisade-like arrangement of cartilage cells is replaced by irregular haphazard rows. There is a lack of orderly change from cartilaginous material to calcified bone which is the basic alteration that characterizes rickets (Fig. 11-3). Mineralization of the osteoid matrix does not occur, and the matrix is seen microscopically as a broad zone of pink-staining tissue surrounding bone trabeculae. As a result, in the metaphyseal area at the osteochondral junction, there is an overgrown and disorganized zone of cartilage, capillaries and fibroblasts.

Owing to the weight of the body and the softness of the osteoid and cartilaginous material, skeletal deformities result. The vitamin D deficient infant who places great stress upon his head as he lies in the crib develops soft spots in the skull in a condition called craniotabes. The frontal and parietal bones undergo a thickening or bossing, giving the head a squared appearance. Closing of sutures and anterior fontanelles may be delayed. There is an overgrowth of cartilage at the costochondral junction of the chest, giving rise to the rachitic rosary. The sternum protrudes and the sides of the rib cage are depressed, producing a "pigeon breast" deformity. At the lower margin of the rib cage, the diaphragm causes a sharp depression known as "Harrison's groove." When the child begins to walk, there may be changes in the vertebral column and long bones. There may be a lumbar lordosis, which is a forward curvature of the spine, as well as the typical bowing of the legs or even knock-knees.

This disturbed calcification of the epiphyseal cartilage is visualized on x-ray as irregular and broad epiphyseal disks. Decalcification is seen as a thinning laminating of the cortical bone.

Osteomalacia or adult rickets is characterized by an accumulation of uncalcified osteoid tissue in the costochondral joints. It is prevalent in women of the Orient, whose bodies have been drained of calcium by numerous pregnancies and prolonged nursing and who receive diets low in both calcium and vitamin D.

HYPERVITAMINOSIS D

As in the case of vitamin A, the use of excessive vitamin D therapy in humans for diseases unrelated to rickets has produced considerable information concerning vitamin D toxicity.

Undesirable side effects have been observed in adults receiving 100,000 IU or more of vitamin D daily for several months, and in children receiving 40,000 IU daily. A few years ago it was found that large amounts of vitamin D gave relief to people suffering from arthritis, and many took one million units a day over a period of months. Later it was found that these people developed side effects, not unlike those found in atherosclerosis, which were more serious than the arthritic condition that was being treated.

Huge amounts of vitamin D will induce a very intense calcification of the bone by assisting in the absorption and deposition of calcium in excessive amounts. Hypercalcemia, hyperphosphatemia, deposition of calcium in many of the organs, especially in the arteries, and dense calcification of the bone along the metaphysis are characteristic of overcalcification resulting from excess vitamin D intake. In addition, the formation of renal calculi and adrenal dysfunction are often seen. Vita-

min D is undoubtedly the most toxic of all the vitamins when ingested in excess of one's needs.

There is evidence that the difference between a toxic dose and the daily requirement of vitamin D may be much smaller than is commonly suspected. It is believed that idiopathic hypercalcemia reported in British infants may have been due in part to daily intakes of 2500 to 4000 IU, since they responded favorably to a reduction in their dietary calcium and elimination of their vitamin D supplements. Although some infants with hypercalcemia may be hypersensitive to vitamin D, it may be that the British experience was one of true intoxication.

Although the data may not be too strongly verified, there have been reports in the United States that 1800 IU per day may retard linear growth in children.

Oral Relevance of Vitamin D

The main effect of vitamin D deficiency is delayed and impaired calcification of tooth and bone.

TEETH

In Animals. In vitamin D deficient dogs Mellanby (1929) found that the tooth enamel was not only poorly calcified but that there was also faulty matrix formation. Becks and Ryder (1931) observed that ameloblasts underwent hydropic degeneration and there was a resultant irregular and poorly mineralized enamel surface. On the other hand, Weinmann and Schour (1945a) reported that enamel formation and mineralization are unaffected by vitamin D deficiency but that cysts do occur after the enamel is completely formed.

The dentin is a very sensitive indicator of vitamin D deficiency. Even in a mild deficiency there is a widening of the predentin, an irregular dentin-predentin border and formation of interglobular spaces in the dentin which represent uncalcified dentin matrix (Weinmann and Schour, 1945a; Mellanby, 1929). In a severe deficiency the widening of the dentin was found to be as much as 90

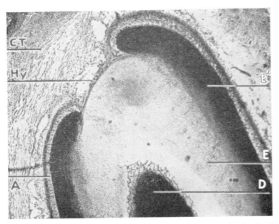

FIGURE 11-4. *Enamel hypoplasia near the incisal edge of a permanent tooth germ of a rachitic child. D, dentin; E', inner zone of enamel; E'', outer zone of enamel; A, ameloblasts, Hy, enamel hypoplasia; CT, connective tissue surrounding the germ. (Courtesy of J. R. Blayney from Kronfeld and Boyle: Histopathology of the Teeth. 4th ed. Lea & Febiger, 1959.)*

microns, compared to the normal value of 10 to 20 microns (Karshan, 1933; Weinmann and Schour, 1945a). The appearance of a calciotraumatic line in the dentin is the earliest sign of an acute deficiency of vitamin D.

In Humans

Eruption of Teeth. In children, a vitamin D deficiency retards the eruption of teeth (Speidel and Stearns, 1940).

Enamel Hypoplasia. The developing enamel in a child's tooth may become hypoplastic as a result of rickets. The ameloblasts are unable to function as a result of a lack of vitamin D, and the enamel calcifies poorly and may in some areas fail to form (Figure 11-4). The mechanism for the formation of enamel hypoplasia has been explained by two theories: the folding and collapse of uncalcified enamel matrix, and ameloblastic degeneration. It should be pointed out that not all rachitic children develop hypoplastic teeth. In fact, Eliot et al. (1934) showed that only about one-third of all the children with rickets they studied had developed enamel hypoplasia. Sarnat and Schour (1942) found in their study that only one of six children with rickets had

enamel hypoplasia. On the other hand, Marshall-Day (1944) found that 67 per cent of 200 Indian boys who had no rickets had hypoplasia of the teeth, and in those boys who showed clinical signs of rickets 62 per cent had enamel hypoplasia, indicating that there was no correlation found between enamel hypoplasia and rickets.

Hypoplasia–Dental Caries Relationship. Mellanby (1929) demonstrated hypoplastic tooth changes in rachitic rats, but not caries. In spite of this inability to produce caries in rats on a vitamin D deficient diet, she still suggested that, in humans, abnormalities of the surface enamel, which were called m-hypoplasia, predisposed the tooth to increased caries susceptibility. Some have confirmed this hypothesis, (Anderson et al., 1934; Bibby, 1943), and others have found no correlation between rachitic hypoplasia and caries susceptibility (Shelling and Anderson, 1936).

Most investigators agree that the initiation of dental caries is no more frequent in a hypoplastic tooth than in a normal tooth. However, once the caries has started in the hypoplastic tooth, it grows and spreads more rapidly. It is possible that the pitted areas allow for the retention of bacteria and dental plaque and thus can be considered areas of lowered caries resistance. In addition, the enamel at the depth of these pits is relatively thin and easily penetrable. Mellanby and Coumoulos (1944) as well as McBeath and Verlin (1942) observed that vitamin D supplements fed to children reduced the incidence of dental caries. However, neither Malan and Ockerse (1941) nor Marshall-Day and Sedwick (1934) found vitamin D supplements in children's diets an effective means of caries prevention. Theoretically, if adequate vitamin D is given before eruption of teeth it should improve the structure of the tooth and make it less caries prone. But since there is more than one factor involved in dental caries production or resistance, it is difficult to assess accurately the relative importance of vitamin D in this whole process. No doubt more good than harm will ensue if optimal amounts of vitamin D are included in the diet, but this is certainly only a partial answer to the problem of dental caries.

PERIODONTIUM

In Animals. Since the prime characteristic of vitamin D deficiency is a failure in calcification, animals that are on vitamin D deficient diets will show uncalcified osteoid tissue in alveolar bone with wide demineralized osteoid borders which resist resorption (Weinmann and Schour, 1945b).

In dogs, Becks and Weber (1931) found that a vitamin D deficiency produced osteoporosis of the alveolar bone, a replacement of the bone by fibro-osteoid tissue, and an obliteration of the lamina dura and the periodontal membrane around the tooth itself. The reason the width of the periodontal space becomes narrow and obliterated is that there is a continued growth of the osteoid border of the alveolar bone.

EFFECTS OF HYPER-VITAMINOSIS D ON TEETH AND PERIODONTIUM

Excessive intake of vitamin D will produce irregular dentin formation and pulp stones in the teeth (Harris and Innes, 1931). The alveolar bone, periodontal membrane and gingivae become hypercalcified (Becks, 1942). The cementum increases so that ankylosis between tooth and bone may result. Extensive amounts of calculus may also be present. Similar findings of calcified deposit in the periodontal membrane, cemental spurs, calcification of the principal periodontal fibers and resorption of alveolar bone were noted by Fahmy et al. (1961).

In humans, excessive vitamin D intake can produce the following oral findings: rampant caries; malocclusion; thinning of the enamel,

dentin and the alveolar process; rarefaction of molar roots (Dawson, 1937); atubular or poorly tubular root dentin; areas of resorption repaired by irregular dentin and pulpal calcification (Berkenhagen and Elfenbaum, 1955).

VITAMIN E

In 1922 Evans and Bishop reported a hitherto unrecognized dietary factor essential for reproduction in rats. The factor which was found in many foods was given the name vitamin E. In 1936 Evans and his associates isolated an alcohol from wheat germ oil having vitamin E activity. They called it alpha-tocopherol from the Greek words *tokos*, meaning "childbirth," and *pherin*, meaning "to carry," i.e., an alcohol which helps the bearing of young. In many ways, it has been unfortunate that the original work on vitamin E was done on rats, a species in which vitamin E deficiency causes sterility. To this day, food faddists and the unscrupulous individuals who thrive on the ignorance of the public and the prejudices of the faddist promote vitamin E for the treatment and prevention of sterility, not to mention heart disease and muscular dystrophy, although there is no evidence that a vitamin E deficiency is of any significance in the etiology of these diseases.

CHEMISTRY

Eight tocopherols have been isolated from seed oils. Of these, α-tocopherol has the most vitamin E activity. The others, which vary considerably in their biological activity, differ structurally in the methylation of their

α-Tocopherol

benzene rings and in the saturation of their isoprenoid side chains. Many compounds with structures similar to those of the tocopherols have been synthesized and have been found to have vitamin E activity. In addition, many antioxidants unrelated chemically to the tocopherols have vitamin E activity.

The tocopherols are fat soluble oily liquids at room temperature and stable to heat, acids and alkalies but are easily destroyed by oxidation, particularly in the presence of ultraviolet light. They are rapidly destroyed in the presence of rancid fats and iron salts.

There are several colorimetric chemical methods for the determination of tocopherols in tissues and foods which make use of their reducing properties. Until relatively recently, these methods did not differentiate between the tocopherols. There are now a variety of chromatographic techniques for separating the tocopherols and, by using these more sophisticated techniques in conjunction with the older methods, it is possible to conveniently measure the individual tocopherols.

Biological assays for vitamin E have most often been based on the ability of the vitamin to protect against sterility in rats. This type of assay is long and costly. In recent years, ability of dietary vitamin E to protect against hemolysis of red cells in the presence of peroxide or dialuric acid has been used as a bioassay for the vitamin. Although bioassays are expensive, time consuming and not very sensitive, they are often of value because they measure biological activity directly.

ABSORPTION AND UTILIZATION

The disease conditions which interfere with the absorption of vitamin A and D affect the absorption of vitamin E in the same way. Because of its role as an antioxidant, the ingestion of diets containing easily oxidized substances such as polyunsaturated fat may result in vitamin E destruction within the gastrointestinal tract and in the tissues. Conversely, the presence of vitamin E in the diet may protect against the oxidative destruction of vitamin A and the polyunsaturated fatty acids. Vitamin E is absorbed without being

esterified and is deposited primarily in adipose tissue. Total body tocopherol in adults amounts to several grams.

SOURCES

Vitamin E is widespread in nature. The richest sources of dietary tocopherols are the vegetable or seed oils. The total amount of tocopherol and the kind of tocopherol vary considerably from oil to oil. Wheat germ oil is a particularly potent source of alpha-tocopherol, containing about 260 mg. of total tocopherol per 100 gm., 60 per cent of which is alpha-tocopherol. Although vegetables, fruits, butter and eggs contain considerably less vitamin E than the seed oils, they are eaten in such larger quantities that they supply the major part of the human intake of the vitamin.

REQUIREMENTS

An International Unit of vitamin E is equivalent to 1 mg. of dl-α-tocopherol acetate. One milligram of d-α-tocopherol acetate is equal to 1.36 IU because l-α-tocopherol has less biopotency than the d form.

Because of a lack of sufficient information and because vitamin E requirements vary with the amount of polyunsaturated fat eaten, standards for vitamin E intake by humans have not been established. It has been estimated that the average daily consumption of d-α-tocopherol by American adults is approximately 14 mg. Until recently, there has been little concern about the possibility of vitamin E deficiency in man. The current, almost unrestrained, promotion of polyunsaturated fats may make dietary levels of vitamin E for Americans a future problem.

METABOLIC FUNCTIONS

It is clear that vitamin E functions at the tissue level as an antioxidant since it can be replaced in the diet to a large extent by many other antioxidants. Currently, there is considerable disagreement over whether vitamin

E has a specific metabolic function or whether its activity is simply related to its nonspecific antioxidant properties. Those who argue for a specific physiological role for vitamin E point out that, even in depleted animals, there are still detectable levels of vitamin E. They also point out that vitamin E administration is effective in protecting all species studied against all the often very different manifestations of the deficiency state; whereas the different antioxidant substitutes for vitamin E may be effective in one species and not another and may provide only partial protection in a given species. There is also some evidence that vitamin E may have a specific effect in the biosynthesis of coenzyme Q.

PATHOLOGY OF VITAMIN E DEFICIENCY

Vitamin E deficiency has been observed in a wide variety of animals, including man. The symptoms of vitamin E deficiency vary much more widely from species to species than do other vitamin deficiency states.

In Animals

Reproduction. As has already been pointed out, vitamin E deficiency results in sterility in rats. The female rat conceives normally, but the fetuses die during gestation and are reabsorbed. The injury to the female rats is not permanent and, following vitamin E therapy, they are able to reproduce normally. Vitamin E deficient male rats show permanent degeneration of testicular germinal epithelium. In other species, vitamin E deficiency may not be associated with reproductive failure. The male mouse, for instance, shows no sign of testicular damage when made vitamin E deficient.

Muscle. A common finding in vitamin E deficient animals of many species is a muscular dystrophy. In these animals, there are degenerative changes in muscle fibers followed by atrophy and replacement by connective tissue and fat. These changes are accompanied by a large increase in urinary creatine, a decrease in muscle creatine and an increase in

the oxygen consumption of muscle. Cardiac muscle and EKG changes have been reported in some species and death may be caused by cardiac failure.

Encephalomalacia and Exudative Diathesis in Poultry. In vitamin E deficient birds fed diets containing linoleic or arachidonic acids, lesions of the brain occur, caused in part by vascular changes. Encephalomalacia in poultry is characterized by convulsions, paralysis, ataxia, head retraction and sudden death.

Vitamin E deficient chickens also develop exudative diathesis characterized by hemorrhage and leakage of plasma from the capillaries, particularly those just under the skin.

Liver Necrosis. When rats are fed diets deficient in vitamin E and the sulfur amino acids, liver necrosis occurs. The protective effect of the sulfur amino acids is, in great part, due to their contamination with selenium. Selenium can also prevent exudative diathesis in chicks and partially protect against the muscular dystrophy of vitamin E deficiency.

Steatitis. In several species of animals, the feeding of vitamin E deficient diets containing large amounts of polyunsaturated fatty acids results in steatitis or yellow fat disease. Peroxidation and polymerization of the polyunsaturated fatty acids results in the production of acid-fast ceroid pigments which are deposited in the fat, giving it a brown or, in some cases, a yellow-orange appearance.

Red Blood Cells. The red blood cells of vitamin E deficient animals are more easily hemolyzed than those of normal animals in the presence of peroxide or dialuric acid. Hemolysis tests have been developed which are used in the assessment of vitamin E nutriture.

A macrocytic anemia has been reported in vitamin E deficient monkeys.

In Humans. There has been little evidence that vitamin E deficiency is an important problem in human nutrition. Studies of newborn infants and children with steatorrhea have related increased red blood cell hemolysis and low serum tocopherol levels. Creatinuria, ceroid, pigment deposition, increased hemolysis rates and muscle lesions have been reported in individuals with diseases, such as cystic fibrosis of the pancreas, which reduce vitamin E absorption. Recently, a macrocytic anemia which responded to α-tocopherol has been reported in malnourished Jordanian children.

TOXICITY

Hypervitaminosis E in animals and man has not been described. Adult humans have taken a gram a day for months without developing signs of toxicity.

Oral Relevance of Vitamin E

Prolonged deficiency of vitamin E causes a loss of pigmentation in the enamel of the rodent incisor because there is a degeneration

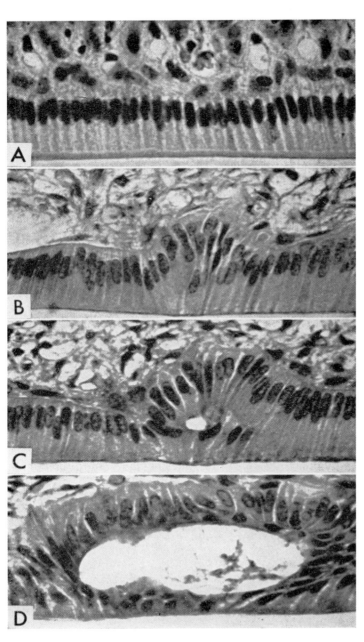

FIGURE 11-5. *Photomicrographs of the middle third of the enamel organ of upper incisors. Orig. mag. ×600. A. Control rat receiving stock diet. B. Rat (No. 437) receiving vitamin E-deficient diet (Group A) for 213 days. Note piling up of ameloblasts into edematous tissue. C. Rat (No. 437) receiving vitamin E-deficient diet (Group A) for 213 days. Note that folding of ameloblasts has progressed. D. Rat (No. 437) receiving vitamin E-deficient diet (Group A) for 213 days. Note cavity formed by folding ameloblasts. (Pindborg, J. J.: J. Dent. Res.: 31, No. 6, 1952.)*

of the enamel organ and consequent disappearance of the iron-containing granules of the ameloblasts. However, according to Irving (1958), "the loss of tooth color is merely incidental to the degeneration of the enamel organ, the same being seen in vitamin A deficiency, and vitamin E plays no specific role in the metabolism of the pigment."

Granados and Dam (1945) suggest that the depigmentation that occurs with hypovitaminosis depends on the presence of dietary fats and not iron and that the degree of depigmentation varies with different fats.

Pindborg (1952) described the histologic changes in the enamel organ of animals that had been subjected to a vitamin E deficiency as consisting of damage to the capillaries, edema and disorganization of both the stratum intermedium and outer enamel epithelium (the papillary layer). The ameloblasts were disarranged and, because they piled up and folded, they formed cystic cavities (Fig. 11-5). Grossly, a chalky white depigmentation appeared on the continuously growing incisor

of the rat (Granados and Dam, 1945).

The significance of these chemical and histopathologic manifestations of vitamin E deficiency in tooth structure is not clearly associated with its resistance or susceptibility to dental caries.

VITAMIN K

In 1929 Dam reported that chickens fed certain low fat diets developed a hemorrhagic condition. In succeeding years evidence was obtained in several laboratories that a fat soluble, nonsaponifiable, nonsterol vitamin protected chickens against this condition. Dam called it vitamin K and, in 1939, it was isolated by several groups.

CHEMISTRY

Vitamin K occurs in nature in two forms— K_1 and K_2—whose structures are as shown.

Vitamin K_1

Vitamin K_2

Menadione

(From White, A., Handler, P., and Smith, E.: Principles of Biochemistry. 3rd ed. Blakiston Division, McGraw-Hill Book Company. Used by permission.)

There are several forms of vitamin K_2 which vary in the number of isoprenoid units in their side chains. In addition, menadione, a synthetic vitamin K_3, is often used in medicine.

Vitamins K_1 and K_2 and menadione have similar activity on a molar basis. There have been a large number of synthetic derivatives of vitamin K prepared which have varying amounts of biological activity. Some of these are water soluble.

Vitamins K_1 and K_2 are yellowish oils, unstable in ultraviolet light and easily destroyed by oxidation.

Physicochemical methods for the determination of vitamin K involve the use of various types of extractions and chromatographic separations followed by colorimetric determinations making use of the quinone structure of the vitamin.

A biological assay based on the coagulation of chicken blood has often been used for the determination of vitamin K activity. The assay utilizes the ability of vitamin K and test substances to prevent or cure hemorrhagic disease in chickens.

ABSORPTION
AND UTILIZATION

As in the case of the other fat soluble vitamins, factors which interfere with the absorption of fat interfere with the absorption of vitamin K. In the absence of bile, menadione or the water soluble derivatives of vitamin K are more easily absorbed than vitamins K_1 or K_2.

Deposition of absorbed vitamin K is greatest in the liver and spleen and is associated primarily with the mitochondria of the cells.

SOURCES

Vitamin K is widespread in nature. Green vegetables are very rich sources of vitamin K_1. Although animal products are low in the vitamin, they provide a significant per cent of human intake. Some microorganisms synthesize large amounts of vitamin K_2. This is of importance because it is generally thought that the intestinal flora provides vitamin K

to its host animal. This is particularly true in coprophagous animals.

REQUIREMENTS

Because of variability in the dietary content of vitamin K and lack of knowledge concerning the amount of vitamin K made available by the intestinal flora, it has been impossible to set up nutritional standards for the vitamin.

METABOLIC FUNCTIONS

The way in which vitamin K is involved in the production of blood coagulation factors is not understood. There is considerable evidence suggesting that vitamin K is associated with electron transport and oxidative phosphorylation. Vitamin K is similar in structure to coenzyme Q and is found in the mitochondria. Dicumarol, a competitive inhibitor of vitamin K, is effective in uncoupling oxidative phosphorylation in mammalian mitochondria. Attempts to demonstrate vitamin K in coagulation factors have not been successful. It is therefore thought that the vitamin functions in one or more of the enzyme systems which are necessary for the production of several of the blood coagulation factors.

There is evidence that vitamin $K_{2(20)}$ with four isoprenoid units in its side chain is the most active form of the vitamin in animal tissues and that the various forms of vitamin K are converted to vitamin $K_{2(20)}$ in animal tissues.

PATHOLOGY OF
VITAMIN K DEFICIENCY

The only reported manifestation of vitamin K deficiency of importance is a defect in blood coagulation. The mechanism by which blood coagulates is extremely complicated and involves a large number of clotting factors. Simplified, it is a two-stage process involving the formation of thrombin from prothrombin followed by the formation of fibrin from fibrinogen in the presence of thrombin. The fibrin then forms the matrix of the clot. In the plasma of vitamin K deficient animals there are deficiencies in the amount of prothrombin and several factors needed for its conversion to thrombin.

Ordinarily, vitamin K deficiency is not a human health problem, but because it may occur when a disease condition affecting fat absorption exists, it is of importance that blood clotting capacity be determined before surgery.

The blood of the newborn child is deficient in several coagulation factors. In addition, the infant does not have a well developed gastro-intestinal flora and mother's milk is not a rich source of vitamin K. It is not an uncommon practice to administer vitamin K to expectant mothers shortly before delivery or to the new-born, particularly the premature, as a preventive measure against hemorrhagic disease of the newborn.

TOXICITY

There is evidence that excessive use of menadione in the newborn will produce hemolytic anemia, hyperbilirubinemia and kernicterus. Hemolytic anemia has also been produced in vitamin E deficient rats by excessive doses of menadione, but vitamin K_1 is not toxic.

Oral Relevance of Vitamin K

Since vitamin K is intimately associated with the clotting mechanism, it is not surprising to find that it has been used as a prophylactic measure against postoperative bleeding in patients undergoing oral surgical procedures. It has been used intravenously (50 mg. vitamin K dissolved in 25 ml. of distilled water). It has also been prescribed for oral use in 10 mg. tablets four times a day (Morgan and Christensen, 1963). In the opinion of the writer, the evidence is neither convincing nor is the procedure of routine prophylactic administration of vitamin K warranted in patients who demonstrate a normal clotting time.

It was speculated that vitamin K might have some anticaries properties since it can reduce the rate of acid formation in the mouth (Fosdick et al., 1942), but this has not been confirmed.

Vitamin K compounds have been found to be required for the growth of *B. melanino-genicus*, an organism closely associated with periodontal disease. It is speculated that a suitable antimetabolite of vitamin K might interfere with growth of this organism and consequently prevent the occurrence of perio-dontal disease (Macdonald and Gibbons, 1962).

REFERENCES

General

McCollum, E. V.: A History of Nutrition. Boston, Houghton Mifflin Co., 1957.
McLean, F. C., and Budy, A. M.: Vitamin A, vitamin D, cartilage, bones and teeth. Vitamins Hormones, *21*:51, 1963.
Moore, T.: Vitamin A. New York, Elsevier Publishing Co., 1957.
National Academy of Sciences–National Research Council: *Recommended Dietary Allowances,* 6th Edition, Pub. No. 1146, 1964.
Sebrell, W. H., Jr., and Harris, R. S. (ed.): The Vita-mins. New York, Academic Press, Inc., 1954.

Vitamin A

Block, C. E.: Vitamin A deficiency and dental anoma-lies in man. Acta Paediat., *11*:536, 1930.
Boyle, P. E.: Manifestations of vitamin A deficiency in a human tooth germ. Case report. J. Dent. Res., *13*:39, 1933.
Dinnerman, M.: Vitamin A deficiency in unerupted teeth of infants. Oral Surg., *4*:1024, 1951.
Glickman, I., and Stoller, M.: The periodontal tis-sues of the albino rat in vitamin A deficiency. J. Dent. Res., *27*:758, 1948.
Irving, J. T.: The effect of avitaminosis and hyper-vitaminosis A upon the incisor teeth and incisal alveolar bone of rats. J. Physiol., *108*:92, 1949.
Irving, J. T., and Richards, M. B.: Influence of age upon requirements of vitamin A. Nature, *144*:908, 1959.
Johnson, J. E., Ringsdorf, W. M., Jr., and Eheraskin, E.: Relationship of vitamin A and oral leukoplakia. Arch. Dermat., *88*:607, 1963.
Marshall-Day, C. D.: Nutritional deficiencies and dental caries in Northern India. Brit. Dent. J., *76*: 115, 1944.
Mellanby, H.: Preliminary note on defective tooth structure in young albino rats as a result of vita-min A deficiency in maternal diet. Brit. Dent. J., *67*:187, 1939.
Mellanby, H.: Effect of maternal dietary deficiency of vitamin A on dental tissues in rats. J. Dent. Res., *20*:489, 1941.
Miglani, D. C.: Effects of vitamin A deficiency on the

periodontal structures of rat molars, with emphasis on cementum resorption. Oral Surg., 12:1372, 1959.

Mohammed, C. I., and Mardfin, D. F.: Pulpal response of vitamin A-deficient rats to operative procedures. J. Dent. Res., 40:757, 1961. (Abstract.)

Radusch, D. F.: Nutritional aspect of periodontal disease. Ann. Dent., 7:169, 1940.

Russell, A. L.: International nutrition surveys: A summary of preliminary dental findings. J. Dent. Res., 42:233, 1963.

Salley, J. J., Bryson, W. F., and Eshleman, J. R.: Effect of chronic vitamin A deficiency on dental caries in the Syrian hamster. J. Dent. Res., 38:1038, 1959.

Schour, I., Hoffman, M. M., and Smith, M. C.: Changes in incisor teeth of albino rats with vitamin A deficiency and effects of replacement therapy. Amer. J. Path., 17:529, 1941.

Silverman, S., Jr., Renstrup, G., and Pindborg, J.: Studies in oral leukoplakia. III. Effects of vitamin A. Acta Odont. Scand., 21:271, 1963.

Silverman, S., Jr., Renstrup, G., and Pindborg, J.: Studies in oral leukoplakias. VII. Further investigations on the effects of vitamin A on keratinization. Acta Odont. Scand., 21:553, 1963.

Wolbach, S. B., and Howe, P. R.: The incisor teeth of albino rats and guinea pigs in vitamin A deficiency and repair. Amer. J. Path., 9:275, 1933.

Vitamin D

Anderson, P. B., Williams, C. H. M., Halderson, H., Summerfeldt, C., and Agnew, R. S.: The influence of vitamin D in the prevention of dental caries. J. Amer. Dent. Assn., 21:1349, 1934.

Becks, H.: Dangerous effects of vitamin D overdosage on dental and paradental structures. J. Amer. Dent. Assn., 29:1947, 1942.

Becks, H., and Ryder, W. B.: Experimental rickets and calcification of dentine. Arch. Path., 12:358, 1931.

Becks, H., and Weber, M.: Influences of diet on bone system with special reference to alveolar process and labyrinthine capsule. J. Amer. Dent. Assn., 18:197, 1931.

Berkenhagen, R., and Elfenbaum, A.: Dentin dysplasia associated with rheumatoid arthritis and hypervitaminosis. Oral Surg., 8:76, 1955.

Bibby, B. G.: The relationship between microscopic hypoplasia (Mellanby) and dental caries. J. Dent. Res., 22:218, 1943. (Abstract.)

Dawson, A. R.: Dangers of overdosage of synthetic vitamin D in relation to teeth. J. Calif. Dent. A., 13:188, 1937.

Fahmy, H., Rogers, W. E., Mitchell, D. F., and Brewer, H. E.: The effect of hypervitaminosis D on the periodontium of the hamster. J. Dent. Res., 40:870, 1961.

Eliot, M. M., Souther, S. P., Anderson, B. G., and Arnim, S. S.: A study of the teeth of a group of school children previously examined for rickets. Am. J. Dis. Child., 48:713, 1934.

Harris, L. J., and Innes, J. R. M.: Mode of action of vitamin D; studies on hypervitaminosis D; influence of calcium-phosphate intake. Biochem. J., 25:367, 1931.

Karshan, M.: Calcification of teeth and bones on rachitic and non-rachitic diets. J. Dent. Res., 13:301, 1933.

Kiguel, E.: Effect of vitamin D deficiency on tooth development. J. Dent. Res., 39:672, 1960.

Malan, A. E., and Ockerse, T.: Effect of calcium and phosphorus intake of school children upon dental caries, body weight, and heights. S. Afr. Dent. J., 15:153, 1941.

Marshall-Day, C. D.: Nutritional deficiencies and dental caries in Northern India. Brit. Dent. J., 76:115, 143, 1944.

Marshall-Day, C. D., and Sedwick, H. J.: Fat-soluble vitamins and dental caries in children. J. Nutr., 8:309, 1934.

McBeath, E. C., and Verlin, W. A.: Further studies on role of vitamin D in nutritional control of dental caries in children. J. Amer. Dent. Assn., 29:1393, 1942.

Mellanby, M.: Diet and teeth; an experimental study. London, Med. Res. Council, Special Reports Series, No. 140, 1929.

Mellanby, M., and Coumoulos, H.: Improved dentition of 5 year old London school children: comparison between 1943 and 1929. Brit. M. J., 1:837, 1944.

Sarnat, B. G., and Schour, I.: Enamel hypoplasia (chronologic enamel aplasia) in relation to systemic disease: chronologic, morphologic and etiologic classification. J. Amer. Dent. Assn., I, 28:1989, 1941; II, 29:67, 1942.

Shelling, D. H., and Anderson, G. M.: Relation of rickets and vitamin D to the incidence of dental caries, enamel hypoplasia and malocclusion in children. J. Amer. Dent. Assn., 23:840, 1936.

Spiedel, T. D., and Stearns, G.: Relation of vitamin D intake to age of infant at time of eruption of first deciduous incisor. J. Pediat., 17:506, 1940.

Weinmann, J. P., and Schour, I.: Experimental studies in calcification. I. Effect of rachitogenic diet on dental tissues of white rat. Amer. J. Path., 21:821, 1945a.

Weinmann, J. P., and Schour, I.: Experimental studies in calcification. II. Effects of rachitogenic diet on alveolar bones of white rat. Amer. J. Path., 21:833, 1945b.

Vitamin E

Granados, H., and Dam, H.: Inhibition of pigment deposition in incisor teeth of rats deficient in vitamin E from birth. Proc. Soc. Exp. Biol. Med., 59:295, 1945.

Granados, H., and Dam, H.: A method of evaluating the degree of incisor depigmentation in vitamin E deficient albino rats. Odont. Trdskr., 56:6, 1948.

Irving, J. T.: Effect of vitamin E deficiency on the enamel organ. J. Dent. Res., 37:972, 1958. (Abstract.)

Pindborg, J. J.: The effect of vitamin E deficiency on the rat incisor. J. Dent. Res., 31:805, 1952.

Symposium on Vitamin E and Metabolism. Vitamins Hormones, 20:373, 1962.

Vitamin K

Fosdick, L. S., Farder, O. E., and Calandra, J. C.: The effect of synthetic vitamin K on the rate of acid formation in the mouth. Science, *96*:45, 1942.

Isler, O., and Wiss, O.: Chemistry and biochemistry of the K vitamins. Vitamins Hormones, *17*:54, 1959.

Macdonald, J. B., and Gibbons, K. J.: The relationship of indigenous bacteria to periodontal disease. J. Dent. Res., *41*:320, 1962.

Morgan, D. H., and Christensen, R. W.: A clinical re-evaluation of the effectiveness of vitamin K_1 in oral surgery. J. South. Calif. Dent. A., *31*:333, 1963.

Chapter Twelve

The B Vitamins

SAMUEL DREIZEN, D.D.S., M.D.

INTRODUCTION

The modern history of the B vitamins begins in 1884 when, by the simple expedient of substituting a moderate amount of meat and legumes for part of the rice in the daily diet, Takaki (1887) succeeded in almost completely eradicating beriberi from the Japanese navy. In 1897 Eijkman induced the avian counterpart of human beriberi by restricting hens to a regimen of polished rice and then cured the characteristic leg paralysis by incorporating rice polishings in the experimental diet. Subsequently, it was shown that an extract of rice polishings was effective in the treatment of both paralyzed birds and human beriberi. The beriberi factor was named "water soluble vitamin B" by McCollum and Davis in 1915 to distinguish it from fat soluble vitamin A. Five years later Emmett and Luros (1920) differentiated between the heat labile antineuritic factor and the heat stable substance which they assumed to be the pellagra-preventive factor. The respective fractions were designated vitamin B_1 and vitamin B_2 by the British and vitamin B_1 and vitamin G by the Americans. Attempts to isolate and identify the supposedly single thermostable component disclosed a diversity of properties depending on the raw materials used, the methods of fractionation and the animals employed for test purposes. Today it is recognized that the heat stable factor is a mixture rather than a single entity which includes most of the 11 generally accepted members of the vitamin B complex.

Although the B vitamins differ substantially in their chemical composition ranging from choline ($C_5H_{15}NO_2$) with a molecular weight of 121.18 to vitamin B_{12} ($C_{63}H_{90}$

CoN$_{14}$O$_{14}$P) with a molecular weight of 1357.44, they all have certain features in common. They are universally distributed in all living cells whether of plant, animal or bacterial origin. Each is known or presumed to be an integral part of a biological catalytic system serving an essential role in metabolic processes. With rare exceptions their physiologic activity is associated with a single chemical structure. Accordingly, Williams and associates (1950) define the B vitamins as "those organic substances which act catalytically in all living cells and which function nutritionally for at least some of the higher animals."

THIAMINE (VITAMIN B_1)

The antiberiberi factor was first isolated by Jansen and Donath (1927), who extracted rice polish in acid solution, adsorbed the active principal onto activated fuller's earth, eluted the adsorbate in the presence of dilute acid and obtained approximately 200 mg. of crystals. The isolate proved prophylactic and therapeutic in tests against avian polyneuritis. Synthesis and elucidation of the exact chemical composition of vitamin B_1 was accomplished by Williams and Cline in 1936. Later that year the synthetic material was found to be effective in the treatment of human beriberi. The compound was named thiamine in the United States and aneurin in Europe.

STRUCTURE

The structural formula of thiamine (C_{12} $H_{17}N_4OSCl \cdot HCl$) is shown in Figure 12-1. It

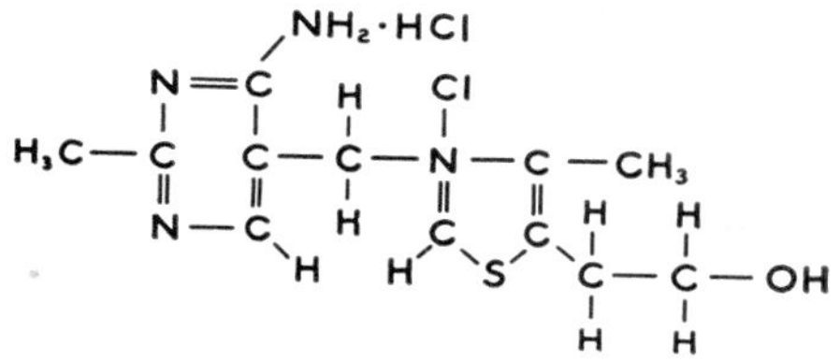

THIAMINE CHLORIDE HYDROCHLORIDE

FIGURE 12-1.

is chemically unique as it contains a thiazole nucleus which is otherwise unknown in nature, and it is the only one of the vitamins other than biotin to contain sulfur. The thiazole group is connected to a pyrimidine ring by a methylene bridge. The methylene bridge can be split by the enzyme thiaminase present in some fish and molluscs or by treatment with sulfite. Thiamine is highly soluble in water and forms a hydrate on exposure to air of average humidity. In the dry form it is very stable and can withstand heating without any diminution in potency. It is readily destroyed by alkalies or by boiling in aqueous solutions at any pH above 5.5.

FOOD SOURCES

Thiamine occurs in animal and plant tissues in amounts ranging from 0.1 to 2.0 μg. per gm. The richest food sources are pork, organ meats, lean meats, yeast, eggs, green leafy vegetables, whole cereals, nuts and legumes. A variable amount is synthesized by the bacteria in the intestinal tract.

The method of preparation may adversely affect the thiamine content of a food. Large losses are incurred by the milling of grains, which removes the vitamin-containing components. Thiamine is also lost by excessive cooking at an alkaline pH and by leaching into the cooking water, which is often discarded.

ABSORPTION, STORAGE AND EXCRETION

Thiamine is absorbed from the upper part of the small intestine either in the free form supplied by vegetable products or in the phosphated form prevalent in foods of animal origin. Individuals vary considerably in their efficiency of thiamine absorption. The process is probably one of simple diffusion as the amount absorbed is roughly proportional to the amount ingested at intake levels up to 5 mg. per day. On reaching the liver and kidneys, free thiamine is phosphorylated to the coenzyme cocarboxylase. The same organs can dephosphorylate cocarboxylase and make free thiamine available to the blood for excretion in the urine or for transport to other tissues for rephosphorylation.

The distribution of thiamine in the body extends from traces in the blood, spleen and lungs to comparatively large amounts in the muscles, liver, heart, kidneys and brain. None of the tissues can store this vitamin for any length of time. In periods of deprivation, depletion occurs most rapidly from the muscles and most slowly from the brain, heart and liver.

The primary excretory pathway of thiamine is through the kidneys. In the urine the bulk is in the free form, the remainder as phosphorylated thiamine and as thiazole and pyrimidine derivatives. Fecal excretion is slight and essentially constant over the usual range of intake. It rises appreciably when the intake is excessive. Minute amounts of thiamine are also excreted in the sweat and in expired air.

METABOLIC FUNCTIONS

Thiamine is intimately involved in intermediary carbohydrate metabolism and in biologic oxidative reactions. As the coenzyme cocarboxylase (thiamine pyrophosphate or diphosphothiamine) it combines with a substrate specific protein to act as a catalyst in the pure decarboxylation of pyruvic acid or in the simultaneous dehydrogenation and decarboxylation of pyruvic acid to acetic acid and carbon dioxide. Apparently all nucleated cells can form the necessary carboxylase systems provided they are supplied with thiamine. Thiamine has been shown to be part of a coenzyme of transketolase which is essential for carbohydrate utilization in the nonnucleated red blood cell (Racker, de La Haba and Leder, 1953). In the presence of a thiamine deficiency, impairment of carbohydrate metabolism is reflected by an elevated blood pyruvic acid level and by a diminution in erythrocyte transketolase content.

METABOLIC INTERRELATIONSHIPS

There is a metabolic interdependence of thiamine with manganese which acts as an oxidative catalyst in the utilization of thiamine by the tissues. There also appears to be a relationship between thiamine and zinc which, like manganese, can replace magnesium in the carboxylase complex.

Thiamine deficient individuals develop anorexia, gastric atony, spastic colon and achlorhydria, which may interfere with the ingestion and absorption of other nutrients. An apparent physiologic connection between thiamine and ascorbic acid has been demonstrated in dogs and in man. Erosive necrotic periodontal lesions have been produced experimentally in thiamine deficient dogs. The lesions healed promptly following the administration of ascorbic acid, indicating that thiamine is required for the synthesis of vitamin C in this animal. In man, the onset of scurvy in persons receiving minute amounts of vitamin C is delayed by ingestion of small quantities of thiamine. Conversely, the antineuritic potential of thiamine may be intensified by ascorbic acid.

THIAMINE DEFICIENCY

Systemic Manifestations. The most prominent human manifestations of thiamine deficiency are those encountered in beriberi. In the adult this disease is characterized by the clinical triad of neuritis, edema and cardiovascular disturbances. It is classified as dry, wet or cardiac beriberi, depending on which symptoms predominate. A combination of two or more types is called mixed beriberi.

The major clinical expression is an ascending symmetrical peripheral neuritis beginning in the lower extremities and marked by weakness, cramping, paresthesias, calf tenderness, altered reflexes, toe and foot drop and muscle atrophy. The upper extremities are not in-volved to any great degree until the leg symptoms become pronounced. In the most severe cases the process extends to include the muscles of the trunk and diaphragm. Beriberi heart failure is denoted by a paroxysmal onset of precordial pain, dyspnea, cyanosis, tachycardia and a small thready pulse. The heart is strikingly enlarged, usually to the right. Edema begins in the feet and legs and gradually ascends up the body into the abdomen (ascites) and thorax (hydrothorax and hydropericardium).

The beriberi syndrome usually develops after a long prodromal period of ill health. In the interim there is impairment of intellectual and cognitive faculties, weight loss, anorexia, fatigue and numerous other neurasthenic complaints which are typical of the subclinical phase of developing B vitamin deficiency states.

Thiamine deficiency has also been implicated in the etiology of Wernicke's disease, which is seen most frequently in the chronic alcoholic. Clinically, the condition is represented by vertigo, headache, nystagmus, photophobia, optic neuropathy, clouding of consciousness, oculomotor dysfunction, ataxia and bulbar paralysis. The pathologic findings are those of an acute hemorrhagic polioencephalopathy, and the lesions are similar to if not identical with those described in experimental thiamine deficiency.

Oral Manifestations. There are no specific oral lesions in a pure thiamine deficiency such as occur in some of the other avitaminoses. Cranial nerve involvement in human thiamine deficiency is usually confined to the vagus, auditory, abducens and oculomotor nerves. Patients with a generalized deficiency of the B group of vitamins may have a hypersensitivity of the oral mucosa which is relieved by thiamine. Oral and facial neuralgias and pain, herpes simplex and aphthous stomatitis have been treated with thiamine with variable success. Edematous swelling of the gingivae and lingual papillae and impairment of the sense of taste have been described in some cases of beriberi.

Occasionally, the clinical appearance of athiaminosis may mask the symptomatology of coexisting deficiencies of one or more of the other B vitamins. Administration of thiamine to such cases may precipitate oral lesions attributable to other members of the vitamin B complex.

REQUIREMENTS

The thiamine requirement of man is influenced by the carbohydrate, fat, protein and alcohol content of the diet. For practical purposes, thiamine requirements are best considered in terms of the total caloric intake. In the 1963 revision of the recommended dietary allowances (National Academy of Sciences–National Research Council), the thiamine requirement was reduced to 0.4 mg. per 1000 calories for all ages. When caloric consumption is less than 2000 per day the suggested thiamine intake is 0.8 mg. Additional allowances of 0.2 and 0.4 mg. per day above that of the nonpregnant woman are recommended for the last two trimesters of pregnancy and during lactation, respectively.

TOXICITY

Thiamine is nontoxic when taken by mouth, even in relatively large doses. Parenteral administration has caused reactions resembling anaphylactic shock in hypersensitive individuals. Only in rare instances has this reaction followed the first injection of thiamine. Contact dermatitis has also been reported in persons allergic to thiamine upon exposure of the skin to thiamine preparations.

RIBOFLAVIN (VITAMIN B₂)

An impure isolate of riboflavin was first obtained in 1879 but it was not until the discovery of the "yellow enzyme" by Warburg and Christian in 1932 that its biologic importance was recognized. The yellow enzyme proved to be a flavoprotein comprised of a flavin pigment linked to a protein carrier. Subsequent demonstration that ribose was an integral part of the flavin component led to the adoption of the name riboflavin.

STRUCTURE

Pure riboflavin was first isolated by Kuhn,

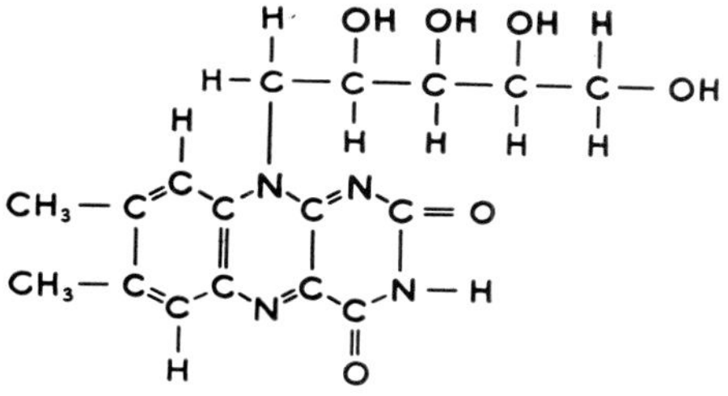

RIBOFLAVIN

FIGURE 12-2.

György and Wagner-Jauregg in 1933 and synthesized by Kuhn and Karrer and their associates in 1935. The riboflavin molecule ($C_{17}H_{20}N_4O_6$) consists of an isoalloxazine nucleus with a ribityl side chain attached to the middle ring as shown in Figure 12-2. The vitamin is yellow-orange in color, heat stable and only slightly soluble in water. The equally active phosphate salt is much more soluble in water. Both riboflavin and riboflavin phosphate decompose on exposure to light or to strongly alkaline solutions. Aqueous solutions are yellow and show a green fluorescence.

FOOD SOURCES

Riboflavin is widely distributed in plant and animal tissues. Among the best food sources are milk, eggs, liver, kidney, heart and green leafy vegetables. The leaves and growing parts of green vegetables contain the most riboflavin. As the leaves get older and dry out, the riboflavin content diminishes.

Riboflavin is lost in appreciable quantities in food preparation if the food is cooked while exposed to light, if the cooking water is not consumed and if vegetables are dehydrated before use. Considerable quantities of riboflavin may be destroyed in milk if it is bottled and exposed to sun or bright daylight for any length of time.

ABSORPTION, STORAGE AND EXCRETION

Riboflavin is readily absorbed from the upper intestinal tract. Phosphorylation occurs in the intestine, liver, blood and other tissues. The bulk of riboflavin in the body is concentrated in the liver, heart and kidneys. The quantities in the body organs cannot be increased above certain levels even by forced feeding. Blood contains between 0.28 and 0.55 micrograms per milliliter. Most is in combination with plasma globulin.

In man, riboflavin is excreted predominantly in the feces. Fecal riboflavin is derived from both the intestinal wall and from bacterial synthesis. It is passed out in the urine as riboflavin, riboflavin phosphate and uroflavin in amounts varying with the dietary intake and tissue storage. An inconsequential amount is lost in the sweat. A lactating woman excretes approximately 10 per cent of ingested riboflavin in the milk.

METABOLIC FUNCTIONS

Riboflavin is a component of flavoprotein enzyme systems composed of a vitamin-containing prosthetic group acting as a coenzyme and a substrate specific protein serving as an apoenzyme. The riboflavin coenzymes contain either riboflavin mononucleotide (riboflavin 5-phosphate) or riboflavin adenine dinucleotide. Many different riboflavin containing enzymes have been identified, all of which participate as catalysts in cellular respiration through the transfer of hydrogen by alternate oxidation and reduction.

RIBOFLAVIN DEFICIENCY

Systemic Manifestations. The principle target areas in ariboflavinosis are the eyes, skin, lips and tongue. The most frequently encountered eye manifestation is vascularization of the cornea. Proliferation of capillaries in the limbic plexus extends into the superficial layers of the cornea and anastamoses to form tiers of loops. In advanced cases the capillaries may invade the entire cornea and superficial corneal ulcerations and punctate opacities may develop. Other ocular changes include conjunctivitis, blepharitis, photophobia, lacrimation, burning and itching of the eyes, changes in pigmentation of the iris and visual disturbances.

The dermatitis of riboflavin deficiency begins most often in the nasolabial folds and spreads to the wings and vestibule of the nose. Other areas of involvement are the canthi of the eyes, eyelids, ears, bridge of the nose, chin, malar eminences and forehead. The facial dermatitis is scaly and greasy in type on a reddened base. An eczematous dermatitis of the scrotum and vulva has been reported.

Oral Manifestations. The characteristic lip lesions in ariboflavinosis are cheilosis or angular stomatitis. The word cheilosis was originated by Sebrell and Butler (1939) to describe a "morbid condition of the lips" in experimentally induced human riboflavin deficiency. In ariboflavinosis, the lip lesions begin as a pearly gray pallor of the epithelium at the angles of the mouth. After several days the pallor is succeeded in turn by erythema, maceration and radiating superficial fissures on a red base (Figure 12-3). Chronic lesions develop a yellowish crust. Both angles are usually involved although they may differ in degree. Infection with hemolytic streptococci, staphylococci, monilia and herpes virus is common in the acute stage. The infected lesions are frequently aggravated and extended onto the surrounding skin by licking and picking. The early subjective symptoms are mild and consist of a feeling of dryness or a slight burning sensation. Deeply infected lesions are painful and leave scars on clearing

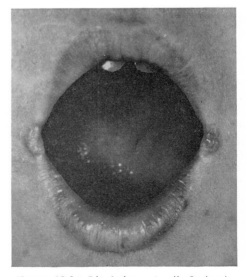

FIGURE 12-3. *Lip lesions of ariboflavinosis.*

following therapy. In advanced cases the exposed labial mucosa to the line of closure may be denuded, swollen and reddened, with increased vertical markings.

In riboflavin deficiency, the lingual papillae are initially swollen, flattened and mushroom-shaped, imparting a pebbly or granular appearance to the dorsum. Later, irregular patchy denudation may develop attributable to papillary atrophy. In advanced cases the tongue takes on a purplish red or magenta color due to dilation and proliferation of the capillaries with a concomitant slowing of the circulation. The tongue may be painful and burn when food is taken, but the relative lack of desquamation keeps the glossodynia from approaching the intensity associated with glossitides of some of the other B vitamin deficiencies. Loss or diminution of the sense of taste is rare in riboflavin deficiency.

REQUIREMENTS

In the 1963 revision of the recommended daily allowances (National Academy of Sciences–National Research Council), the recommendation for riboflavin is 0.6 mg. per 1000 calories regardless of age or sex. The caloric relationship supplants the protein relationship used in the previous computations of the recommended allowances for riboflavin. An additional 0.3 mg. per day of riboflavin is suggested for the last two trimesters of pregnancy and 0.6 mg. per day during lactation.

TOXICITY

In the human, riboflavin is nontoxic in amounts far in excess of the therapeutic dosage. The low solubility of riboflavin may be responsible for its relative innocuousness.

NICOTINIC ACID (NIACIN)

Nicotinic acid, an organic compound of relatively simple structure, was synthesized long before it was found in natural foodstuffs and discovered to be an important factor in human nutrition. It was not until it was prepared from nicotine by Huber in 1867 that

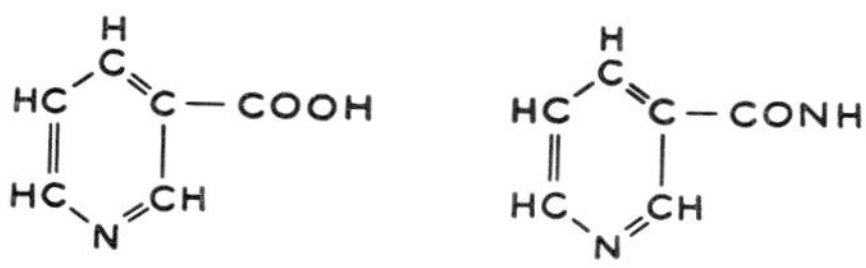

NICOTINIC ACID   NICOTINAMIDE

FIGURE 12-4.

its formula and chemical properties were established. The nutritional significance of nicotinic acid was not suspected until it was recovered from rice polishings in 1912 by Suzuki and associates and found to be part of triphosphopyridine nucleotide (codehydrogenase II) by Warburg and Christian in 1934. In 1937 Elvehjem and associates isolated nicotinic acid from liver extracts and showed it to be curative against canine blacktongue, a condition analogous to human pellagra. Within a year, nicotinic acid and its amide proved to be strikingly effective in the treatment of pellagra (Spies, et al., 1938). Pellagra had first been described by Casal (1762), who in 1735 had astutely related it to poverty and inadequate nutrition.

STRUCTURE OF NICOTINIC ACID AND NICOTINIC ACID AMIDE

Nicotinic acid ($C_6H_5NO_2$) is the beta-carboxylic acid of pyridine with the structure shown in Figure 12-4. It is a white, odorless nonhygroscopic compound which is freely soluble in boiling water and fairly stable to heat, mild acids and alkalies. Nicotinic acid amide ($C_6H_6N_2O$) is much more soluble in water, has similar biologic properties and lacks the peripheral vasodilator action of nicotinic acid. In the United States nicotinic acid and nicotinic acid amide have been named niacin and niacinamide, respectively, to distinguish the vitamin from the alkaloid nicotine.

FOOD SOURCES

The vitamin occurs in plants as the acid form and in animal tissues as the physiologically active amide form. The richest food sources are lean meats, liver, yeast, whole grain products and peanuts. Small quantities are produced by the intestinal bacteria. The body also has the capacity to convert the amino acid tryptophan into nicotinic acid.

The heat stable and oxidative and light resistant properties of nicotinic acid prevent any large losses during the cooking and processing of foods. Losses that do occur usually result from diffusion of the vitamin into the cooking water.

ABSORPTION, STORAGE AND EXCRETION

Food nicotinic acid occurs mainly in the coenzyme form, from which it is readily released by the digestive processes. It is not known whether the coenzymes are absorbed directly or are first hydrolyzed to nicotinic acid and nicotinic acid amide, both of which are absorbed unchanged. Nicotinic acid is converted into the amide after absorption into the blood stream from the upper part of the small intestine. The nicotinic acid level of the blood ranges from 260 to 280 μg. per 100 ml.

Nicotinic acid is distributed in practically all tissues, predominantly as the coenzyme. The greatest quantity is contained in the liver. A period of several months is required to deplete the body of the stores of nicotinic acid to a degree sufficient to produce pellagra.

Nicotinic acid and the amide are excreted principally in the urine, both in the free forms and as the metabolic derivatives nicoturinic acid, N'-methylnicotinamide and N'-methyl-6-pyridone-3-carboxylamide. Small amounts are also excreted in the feces, sweat and milk of lactating women.

METABOLIC FUNCTIONS

Nicotinic acid is an essential part of codehydrogenase I (diphosphopyridine nucleotide) and codehydrogenase II (triphosphopyridine nucleotide) formerly known as coenzymes I and II. Nicotinamide is the functional group of the two coenzymes which, when combined with specific proteins participate in intracellular oxidation-reduction reactions involved in the utilization of all major nutrients. They act in series with the flavoprotein enzymes as hydrogen acceptors and hydrogen donors. Diphosphopyridine nucleotide is also instrumental in the conversion of vitamin A into retinene.

METABOLIC INTERRELATIONSHIPS

A metabolic relationship between tryptophan and nicotinic acid has been demonstrated in certain microorganisms and in such mammals as the rat, pig, horse and man. Pyridoxal phosphate, the coenzyme form of vitamin B_6, is essential for the conversion of tryptophan to nicotinic acid in the tissues. The accepted conversion ratio (niacin equivalent) is 60 mg. tryptophan to 1 mg. nicotinic acid. Clinically, remissions of acute pellagrous lesions have been obtained by the oral administration of tryptophan (Bean, et al., 1951).

NICOTINIC ACID DEFICIENCY

Systemic Manifestations. The end result of a prolonged deficiency of nicotinic acid in the human is pellagra, a word derived from the Italian *pelle agra,* meaning "rough skin." The disease characteristically affects the skin, gastrointestinal tract and nervous system, which has given rise to the mnemonic "dermatitis, diarrhea and dementia." The onset passes through a long period of vague prodromal symptoms, notably: insomnia, loss of appetite, weight and strength; indigestion, diarrhea, abdominal pain, burning sensations in various parts of the body, vertigo, headache, numbness, nervousness, palpitation, mental depression, irritability, distractibility, flights of ideas, apprehension, mental confusion and forgetfulness. These gradually increase in intensity as the disease progresses through the latent, subclinical and clinical phases.

The dermatitis may appear on any part of the body but is found most often at sites of irritation such as the dorsum of the hands, wrists, elbows, face, neck, under the breasts, knees, feet and perineum. The areas of dermatitis are usually bilaterally symmetrical and are separated from healthy skin by a sharp line of demarcation. In the early stages the lesions are erythematous and resemble sunburn. They are often accompanied by severe itching and burning. Subsequently, the sites become swollen, tense and more reddened in color. Vescicles and bullae may form and rupture. After an indefinite period the swelling diminishes, the color becomes reddish brown, and desquamation begins in the center of the lesions either as large flakes or fine brownish white scales. Pigmentation at the margins of the lesions and around the hair follicles may be prominent.

The gastrointestinal manifestations involve the entire alimentary tract from mouth to anus and emanate from the degenerative and inflammatory changes in the lining mucosa. Extraorally, they take the form of a pharyngitis, esophagitis, gastritis and enteritis. Severe persistent diarrhea occurs principally in the advanced cases. The stools may be hard, soft or watery and are always foul smelling. Nausea, vomiting and abdominal pain, discomfort and distention after meals is common.

The mental aberrations arise from cerebral involvement, degeneration of the spinal cord and peripheral neuritis. The initial symptoms are multiple in quantity and neurasthenic in quality. Ultimately the mental manifestations become psychotic in type. The most frequent form is memory loss, disorientation, confusion and confabulation. Others are excitement, mania, depression, delirium and paranoia.

Oral Manifestations. Pellagrous glossitis and stomatitis appear early in the course of the disease and may constitute the presenting complaint. In mild cases the tongue may be merely swollen and somewhat painful, with nondescript atrophic changes on the dorsum. In the more severe cases, hyperesthesia of the tongue frequently precedes the objective signs. Initially, only the tip and lateral margins of the tongue are swollen and reddened. Later, the swelling increases to produce tooth indentations along the margins and the reddening becomes more intense. As the disease progresses, desquamation of the superficial epithelium leaves a scarlet, smooth, dry, beefy looking dorsum. There is a diminution in the sense of taste. Deep penetrating ulcers develop along the sides and at the tip but rarely on the dorsum. During the desquamative phase secondary infection with Vincent's

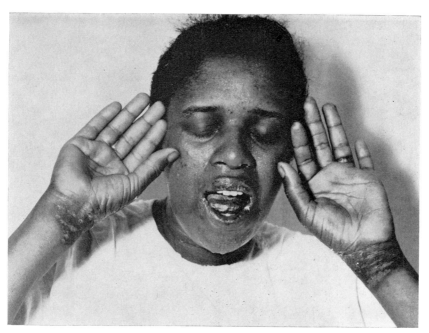

FIGURE 12-5. *Pellagrous stomatitis and dermatitis.*

organisms and monilia produces a thick whitish coating on the dorsum and under-surface which is eventually shed (Figure 12-5). The tongue is extremely painful and sensitive to food and drink.

The gingivae, lips, buccal mucosa, pharynx and floor of the mouth become reddened and ulcerated. The devitalized tissues are extraordinarily susceptible to superimposed infection, particularly with Vincent's organisms. The frequency with which ulceromembranous or acute necrotizing gingivitis occurs in pellagra does not signify that a deficiency of nicotinic acid is the precipitating factor in all cases of Vincent's infection. Only when it is pellagrous in origin does correction of the deficiency lead to a restoration of the gingival integrity.

Sialorrhea or ptyalism occurs in the acute stage of pellagrous stomatitis. The flow may be so great as to result in drooling, with an attendant irritative cheilitis and circumoral dermatitis. With time, the mouth becomes dry and reflects the dehydration resulting from a reduction in water intake and excessive water losses via the intestinal tract.

There is a strikingly low incidence of dental caries among those with chronic endemic pellagra as documented by reports from various parts of the world (Mann, et al., 1947). The low caries incidence has been associated with a deficiency of nicotinic acid which acts both as an essential growth requirement for the oral acidogenic flora and as part of the enzyme systems concerned with the degradation of fermentable carbohydrates to organic acids in amounts sufficient to alter tooth structure (Dreizen and Spies, 1953).

REQUIREMENTS

The latest recommended daily dietary allowance for nicotinic acid has been set at 6.6 mg. of niacin equivalents per 1000 calories for infants, children and adults. Although there is an increased transformation of tryptophan to nicotinic acid during pregnancy, an additional 3 mg. per day is allotted for the last two trimesters in recognition of the increased caloric intake. For lactating women an additional daily allowance of 7 mg. of niacin equivalents is recommended. The "average"

American diet contains 500 to 1000 mg. of tryptophan and 8 to 17 mg. of preformed nicotinic acid for a total niacin equivalent of 16 to 33 mg. per day.

TOXICITY OF NICOTINIC ACID AND NICOTINIC ACID AMIDE

The untoward side effects of nicotinic acid are attributable to its action as a vasodilator. These include reddening and flushing of the skin, increased skin temperature, dizziness, headache, nausea, vomiting and abdominal pain. The symptoms are transitory and harmless and may be used to measure the efficacy of absorption of water soluble vitamins. The vasodilator action of nicotinic acid is of value in conditions where it is desirable to increase the peripheral blood flow. Nicotinic acid amide does not have any vasodilator activity.

There is a huge margin of tolerance between the effective therapeutic and toxic doses of nicotinic acid and nicotinic acid amide, so much so that these compounds are regarded as essentially nontoxic.

FOLIC ACID (FOLACIN)

In 1941, Mitchell, Snell and Williams discovered that spinach leaves contain an essential growth factor for certain fastidious microorganisms; they named this factor folic acid. The vitamin was synthesized by Angier and associates in 1945. Shortly thereafter it was shown to have a beneficial hemopoietic effect in pernicious anemia, nutritional macrocytic anemia, sprue, macrocytic anemia of pregnancy, megaloblastic anemia of infancy and achrestic anemia (Spies et al., 1945; 1946).

STRUCTURE

Folic acid ($C_{19}H_{19}N_7O_6$) is comprised of an amino residue (glutamic acid) attached to a B vitamin (para-aminobenzoic acid) linked in turn to a pteridine group. The entire molecule constitutes pteroylglutamic acid with the structure shown in Figure 12-6. Folic acid is yellow-orange in color, only slightly soluble

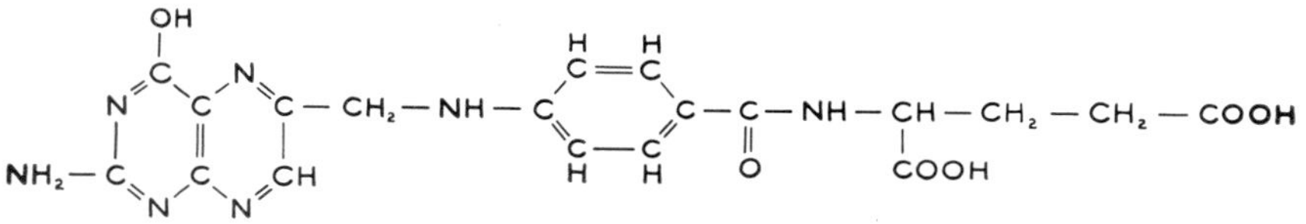

FOLIC ACID (PTEROYLGLUTAMIC ACID)
FIGURE 12-6.

in water and relatively stable to heat within the pH range 4 to 12. It is destroyed by heat in highly acid media and deteriorates when exposed to sunlight or when stored at room temperature for long periods.

FOOD SOURCES

Folic acid and folinic acid (a formyl derivative of reduced folic acid) are present in many foods as conjugates with two to six molecules of glutamic acid. Folic acid is particularly abundant in deep green leafy vegetables, liver, kidney and yeast. Moderate amounts are contained in lean beef and wheat cereals. Folic acid losses on cooking and on storage of food at room temperatures are fairly high. Refrigerated foods retain considerable folic acid for periods up to two weeks.

ABSORPTION, STORAGE AND EXCRETION

Folic acid is readily absorbed from both the upper and lower intestinal tract. Folinic acid taken orally is converted into folic acid in the upper portion of the tract before absorption. In general, the ability to utilize folic acid conjugates and hydrolyze them to free folic acid depends on the amount of conjugase-inhibiting substances present in food. In the absence of conjugase inhibitors the absorption of conjugated folic acid is comparable to that of free folic acid.

Absorbed folic acid is converted into the biologically active folinic acid by reduction and addition of a formyl group, probably in the liver. The reaction is aided by ascorbic acid which protects the folic acid reductase system.

Both folic acid and folinic acid are excreted in the urine. The fecal excretion is much greater, containing folic acid synthesized by intestinal bacteria as well as that derived from food. Small amounts are excreted in the sweat.

METABOLIC FUNCTIONS

As an integral part of folinic acid coenzyme, folic acid participates in the transfer of single carbon units to compounds undergoing metabolic synthesis. This property of folinic acid coenzyme is utilized in amino acid metabolism and in the production of the purine and pyrimidine compounds required for the formation of nucleoproteins. A folic acid deficiency leads to megaloblastic arrest of red blood cell maturation in the bone marrow. Folinic acid is also essential for mammalian cell division. Without it, mitosis is halted in the metaphase. This characteristic has led to the use of folic acid antagonists in the treatment of certain forms of acute leukemia.

METABOLIC INTERRELATIONSHIPS

The metabolic relationships between folic acid and other nutrients have not yet been definitely established. There is evidence that folic acid is required for the utilization of pantothenic acid, which is a constituent of coenzyme A. Vitamin B_{12} affects the metabolism of folic acid either directly or indirectly and may be essential for the formation of folinic acid coenzyme. Administration of ascorbic acid to anemic scorbutic patients results in increased bone marrow activity which is believed to be mediated through folic acid metabolism.

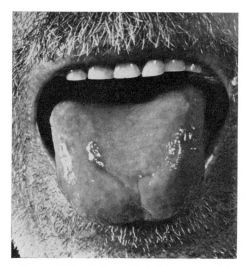

FIGURE 12-7. *Folic acid deficiency glossitis.*

FOLIC ACID DEFICIENCY

Systemic Manifestations. The most notable effect of human folic acid deficiency is a macrocytic anemia associated with a megaloblastic erythropoiesis. In addition, most folic acid deficiency states have the common clinical symptoms of glossitis, stomatitis, gastrointestinal lesions, diarrhea and intestinal malabsorption. The gastrointestinal defects are particularly prominent in sprue, a syndrome in which steatorrhea represented by foul-smelling, copious, greasy-appearing, light, liquid to semiliquid stools is a conspicuous feature.

Oral Manifestations. In sprue and other folic acid deficiency states the stomatitis is characterized by painful, hyperemic and occasionally ulcerated areas throughout the oral mucosa as well as by glossitis, cheilitis and cheilosis. The glossitis of sprue is occasionally the presenting complaint but it usually develops after the onset of steatorrhea. Initially it appears as a swelling and redness at the tip and lateral margins of the dorsum. Later crops of minute, whitish, aphthous ulcers with fiery red borders may emerge. The filiform papillae are the first to disappear, the fungiform papillae remaining for a short time as prominent spots. The papillary changes recede and recrudesce in untreated cases. In advanced cases the fungiform papillae are lost and the dorsum becomes slick, smooth and often fiery red in color (Figure 12-7). Glossodynia and sialorrhea are early and persistent manifestations. The cheilitis is usually ulcerative in type and the cheilosis is indistinguishable from that caused by riboflavin deficiency.

An ulcerative stomatitis is one of the earliest indications of toxicity when folic acid antagonists are used in the treatment of leukemia.

REQUIREMENTS

The minimum daily requirement for folic acid has not yet been established. Average well balanced diets in the United States selected from a wide variety of foods furnish from 0.5 to 1.0 mg. of folic acid daily, mostly in the conjugated form. Studies of experimental folic acid deficiency in man and the response of patients with folic acid deficiency anemia suggest that 50 μg. of folic acid will fulfill daily needs. The requirements are slightly higher during pregnancy and lactation.

TOXICITY

Folic acid is nontoxic in the human in amounts which far exceed the therapeutic range. Quantities greater than 0.1 mg. per day may suppress the development of hematologic manifestations in patients with pernicious anemia but have no effect on the neurologic complications of this disease. Consequently, the sale of mixed vitamin preparations which provide more than 0.1 mg. folic acid daily without a prescription has been prohibited.

VITAMIN B$_{12}$ (CYANOCOBALAMIN)

In 1947 Shorb reported that the growth of *Lactobacillus lactis Dorner* was related in linear fashion to the amount of antipernicious anemia factor present in liver extract. The next year, Rickes and associates (1948) in the United States and Smith (1948) in England independently isolated the extrinsic factor (vitamin B$_{12}$) in pure crystalline form from liver concentrates. Within months West (1948), Ungley (1949), and Spies and co-

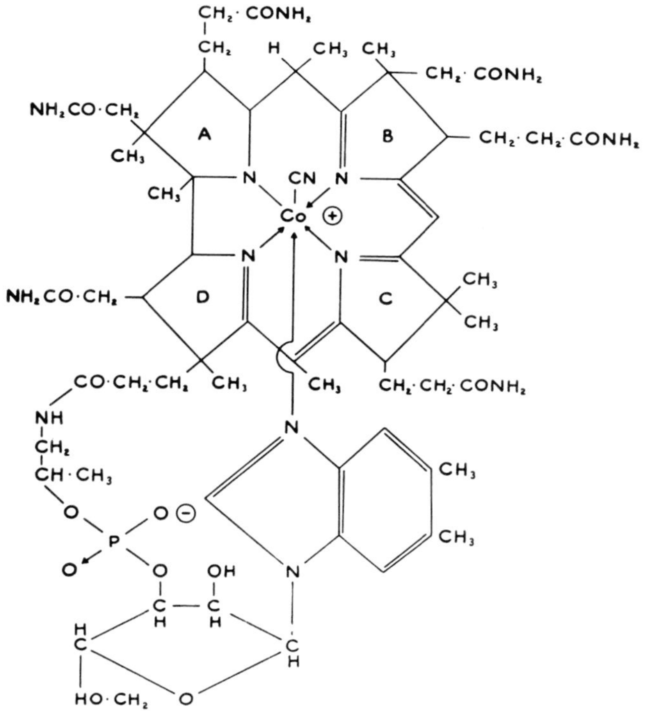

VITAMIN B₁₂
(CYANOCOBALAMIN)
FIGURE 12-8.

workers (1948b) demonstrated that vitamin B$_{12}$ was not only hematologically active in pernicious anemia in relapse but that it also prevented the subacute combined degeneration of the spinal cord induced by this disease. Spies and his group (1948a) also found that vitamin B$_{12}$ was curative for the macrocytic anemias of pellagra, tropical and nontropical sprue and nutritional macrocytic anemia.

STRUCTURE

Vitamin B$_{12}$ (C$_{63}$H$_{90}$CoN$_{14}$O$_{14}$P) is the heaviest and most complicated nonprotein food factor necessary for human health. The chemical structure shown in Figure 12-8 was first established by Hodgkin et al. (1955) and Bonnett et al. (1955). It is a water soluble, heat stable, red crystalline material which is labile to strong acids, alkalies and light. Vitamin

B$_{12}$ is unique in being the only vitamin to contain metal (cobalt) as an integral part of its structure. The vitamin is one of the most potent of all catalysts, as relatively few molecules are sufficient to maintain life. Vitamin B$_{12}$ may exist in several different forms which have been collectively named cobalamine.

FOOD SOURCES

Vitamin B$_{12}$ is found mainly in foods of animal origin in the form of a protein complex. The vitamin is released from the complex by proteolytic enzymes or by heat. It is most abundant in liver, kidney, muscle and foods derived from milk. Herbivorous animals obtain vitamin B$_{12}$ from bacterial synthesis in the rumen. In man, microbial synthesis occurs in the colon but relatively little, if any, is available from this source under ordinary circumstances.

ABSORPTION, STORAGE AND EXCRETION

Vitamin B_{12} is poorly absorbed from the intestinal tract in the absence of intrinsic factor, a heat labile mucoprotein enzyme secreted by the gastric mucosa. The intrinsic factor facilitates absorption by removing vitamin B_{12} from its combination with protein in the presence of divalent ions and links it to the wall of the lower ileum. An enzyme present in intestinal secretions, the so-called releasing factor, facilitates the passage of vitamin B_{12} across the ileal mucosa. Once absorbed, vitamin B_{12} recombines with vitamin B_{12}-binding globulins in the blood for transport to actively proliferating bone marrow.

The human body may be capable of storing relatively large amounts of vitamin B_{12} as there is a long period between the restriction of the external supply and the appearance of the earliest deficiency manifestations. In man, the major portion of vitamin B_{12} is excreted in the feces, with very little appearing in the urine. When administered parenterally in large doses, the amount excreted in the urine is proportional to the dose injected.

METABOLIC FUNCTIONS

Vitamin B_{12} participates in biochemical reactions in the form of B_{12} coenzymes which are today recognized as a distinct group of corrinoid compounds. There are at least seven different enzymatic reactions or processes catalyzed by cell-free systems in which coenzyme B_{12} or its methyl analog are active. Like folic acid, vitamin B_{12} functions in the transfer of single carbon units (methyl groups) and is concerned with nucleic acid synthesis. It influences folic acid metabolism either by effecting the release of free folic acid from the conjugated forms or by catalyzing the formation of folinic acid coenzyme.

VITAMIN B_{12} DEFICIENCY

Systemic Manifestations. In man, the most severe form of vitamin B_{12} deficiency is pernicious anemia. Other vitamin B_{12} responsive anemias are thought to be mild forms of vitamin B_{12} deficiency complicated by varying degrees of folic acid deficiency. In pernicious anemia there is apparently a genetic defect which impairs the secretion of intrinsic factor by the gastric glands. Intrinsic factor and free hydrochloric acid in the gastric juice are invariably absent or diminished in pernicious anemia patients.

The disease is characterized by a diagnostic triad of weakness, glossitis, and numbness and tingling of the extremities, in conjunction with a macrocytic anemia. Pallor, anorexia and gastrointestinal and cardiovascular dis-

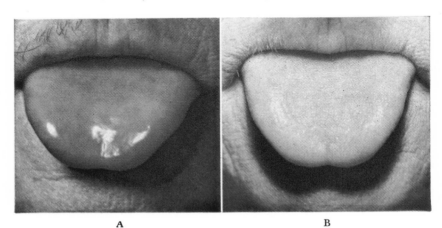

A   B

FIGURE 12-9. *A, Slick, denuded tongue of a patient with pernicious anemia in relapse, showing striking atrophy of the lingual papillae and discoloration of the tongue. B, Same patient after two weeks' treatment with vitamin B_{12}, now shows regrowth of lingual papillae and disappearance of the abnormal color of the tongue. (Dreizen, S., Stone, R. E., and Spies, T. D.: Dent. Clin. N. Amer., July, 1958.)*

turbances are frequently present. The neurologic signs are suggestive of both lateral and posterior spinal cord involvement and peripheral nerve degeneration. They are denoted by a loss of position and vibratory sense, incoordination of the lower extremities and lack of fine coordination in finger movements.

Oral Manifestations. Approximately two-thirds of all patients with pernicious anemia have recurrent attacks of sore tongue which may be the presenting complaint but which usually appears after the disease is well established (Fig. 12-9). The soreness lasts for several weeks and disappears spontaneously only to recur. When the glossitis is at its peak, the tongue is exceedingly painful and fiery red in color. The entire dorsum or only well demarcated areas may be involved. Occasional cases develop inflamed vesicles on the tongue surface; others, shallow white ulcers. The entire mouth may pain and burn on the introduction and swallowing of food. Frequently, the burning and soreness are confined to the anterior half of the tongue. Between attacks of soreness there is an atrophy of the lingual papillae commonly progressing to a complete disappearance, leaving a bald, glazed dorsal surface and a loss or diminution in the taste sense.

Cheilosis, cheilitis and fiery red, excruciatingly painful mucous membrane lesions affecting the buccal mucosa, gingivae and pharynx in addition to the tongue may develop in untreated patients. In contrast, microscopic cellular changes in the superficial oral epithelium may constitute the earliest manifestation of developing pernicious anemia. In such instances, more than 25 per cent of stained exfoliated epithelial cells obtained by oral washings show giant nuclei, nuclear polymorphism and over-all cellular enlargement. Similar microscopic changes have been observed in sprue.

REQUIREMENTS

The human requirements for vitamin B_{12} have not been conclusively established. Available studies indicate that a diet containing 3 to 5 μg. of vitamin B_{12} daily which provides 1 to 1.5 μg. of absorbed vitamin will satisfy the need for this nutrient. A diet containing 15 μg. per day will gradually replenish depleted tissue stores. Dietary deficiencies of vitamin B_{12} are extremely rare and have been found only in long-term vegetarians.

TOXICITY

Vitamin B_{12} is nontoxic to man and laboratory animals even when administered orally or parenterally in amounts thousands of times greater than the effective therapeutic dose.

VITAMIN B_6

In 1935 Birch and associates found that a pellagra-like dermatitis in rats (rat acrodynia) was cured by a food factor previously designated vitamin B_6 by György (1934). By 1938 vitamin B_6 had been isolated in crystalline form from various natural sources. One year later, chemical identification and synthesis was achieved (Harris and Folkers, 1939).

STRUCTURE

Vitamin B_6 occurs in nature in three forms: pyridoxine ($C_8H_{11}NO_3$), which is most plentiful in plants, and pyridoxal ($C_8H_9NO_3$) and pyridoxamine ($C_8H_{12}N_2O_2$), which predominate in animal tissues. Like nicotinic acid, all three are pyridine derivatives, with the structures shown in Figure 12-10. They owe

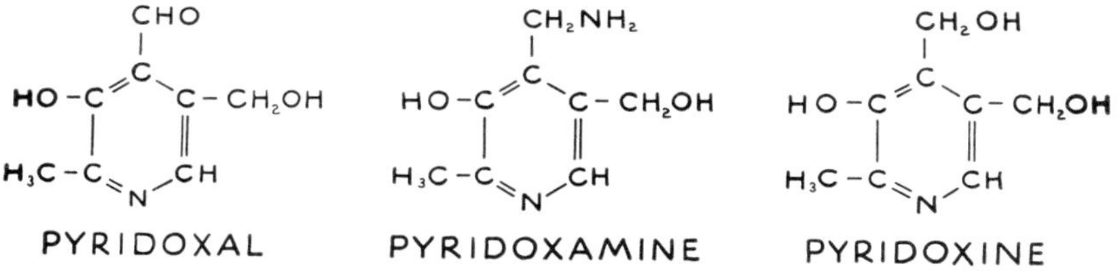

PYRIDOXAL   PYRIDOXAMINE   PYRIDOXINE

FIGURE 12-10.

their vitamin activity to the ability of the organism to convert them into the enzymatically active form, pyridoxal-5-phosphate.

Pyridoxine is a water soluble, white crystalline compound stable to heat and strong acids and alkalies but sensitive to ultraviolet light and air. Aqueous solutions of pyridoxal are heat labile.

FOOD SOURCES

The best food sources of vitamin B_6 are wheat germ, whole grain cereals, milk, yeast, legumes, meat (particularly liver and kidney) and certain vegetable fats. Milling of grain in the production of flour results in a loss of approximately half the vitamin B_6 present in the whole grain. Other losses are incurred during cooking of food in open vessels as a result of oxidation, diffusion and photosensitivity.

ABSORPTION, STORAGE AND EXCRETION

Vitamin B_6 is easily absorbed in any of its three forms from the upper intestinal tract. Absorption of pyridoxal is more rapid and complete than that of pyridoxine or pyridoxamine. Although the intestinal flora manufacture considerable quantities of vitamin B_6, little, if any, is utilized by the human from this source.

In the blood and tissues, vitamin B_6 is primarily in the bound form as pyridoxal phosphate and pyridoxamine phosphate. It is excreted in the urine chiefly as 4-pyridoxic acid along with small amounts of pyridoxal and pyridoxamine. Practically none is excreted in the urine as pyridoxine. Fecal vitamin B_6 is both microbial and dietary in origin. Negligible amounts appear in the sweat.

METABOLIC FUNCTIONS

Vitamin B_6 is an essential constituent of pyridoxal-5-phosphate, which serves as a cofactor for numerous enzyme systems concerned with the metabolism of amino acids. Practically all enzymatic reactions involving the

nonoxidative degradation and interconversion of amino acids (decarboxylation, transamination and desulfuration) require pyridoxal-5-phosphate. Vitamin B_6 also participates in the conversion of tryptophan into nicotinic acid, the utilization of essential fatty acids, the synthesis of hemoglobin, and the maintenance of proper neuronal function and activity.

VITAMIN B_6 DEFICIENCY

Systemic Manifestations. Naturally occurring vitamin B_6 deficiency has been described in infants and young children restricted to a formula of autoclaved cow's milk lacking in pyridoxine (pyridoxine deficiency) and in those whose pyridoxine needs are materially increased by a suspected genetic error of vitamin B_6 metabolism (pyridoxine dependency). All cases showed evidence of central nervous system involvement. The deficient children have a clinical triad of convulsions, hyperirritability and hyperacusis within six weeks to 4 months after birth. In the dependent children, convulsions begin 3 hours to 7 days after birth. Unless treated with high doses of vitamin B_6 each day, pyridoxine dependency children may become mentally retarded. Instances of pyridoxine-responsive microcytic and macrocytic anemia in adults are being reported with increasing frequency.

Experimentally induced vitamin B_6 deficiency has been created in adults by the administration of desoxypyridoxine, an antimetabolite of pyridoxine (Mueller and Vilter, 1950). The deficiency state is manifested by anorexia, weight loss, lassitude, depression, conjunctivitis, cheilosis, glossitis, peripheral neuritis, seborrheic dermatitis of the face and a pellagrous-like dermatitis of the extremities and neck.

Oral Manifestations. Oral lesions attributable to vitamin B_6 deficiency include bilateral angular cheilosis, and a glossitis represented by slight pain, edema, papillary atrophy and a peculiar purplish hue (Rosenblum and Jolliffe, 1941). In experimentally produced cases of vitamin B_6 deficiency the glossitis began with a scalding sensation of the tongue followed in 24 hours by reddening and hypertrophy of the filiform papillae at the tip, margins and dorsal surface. Later the tongue became swollen, blotting out the structure of the filiform papillae, the fungiform papillae standing out as hypertrophied red knobs.

Pyridoxine has been proposed as an antidental caries agent on the hypothetical ground that it selectively alters the oral flora by promoting the growth of noncariogenic organisms while suppressing the cariogenic forms. The studies to date, however, are much too limited to provide convincing evidence that pyridoxine is a caries inhibitor.

REQUIREMENTS

The human requirements for vitamin B_6 are still under consideration. They are known to increase with the protein and caloric content of the diet. The average adult need for vitamin B_6 may be in the range of 675 to 750 μg. per day when the diet supplies 2600 to 2900 calories if data from other species are applicable to man. To allow for a daily protein intake of 100 gm. or more and for conditions of stress, a tentative adult allowance of 1.5 to 2.0 mg. per day has been suggested. For artificially fed infants, 400 μg. per day probably provides a sufficient safety factor.

TOXICITY

Vitamin B_6 has an extremely low toxicity and no ill effects from the administration of large doses have been observed in man. Intramuscular injection of pyridoxine is painful because the preparations are acid in reaction.

PANTOTHENIC ACID

The discovery of pantothenic acid stemmed from investigations of a yeast growth factor (bios), a liver filtrate factor and a chick antidermatitis factor which began in 1901 and culminated in 1933 when Williams and coworkers named the fraction pantothenic acid because of its ubiquitous distribution. The compound was isolated by Williams in 1939 and structurally identified one year later by Williams and Major (1940).

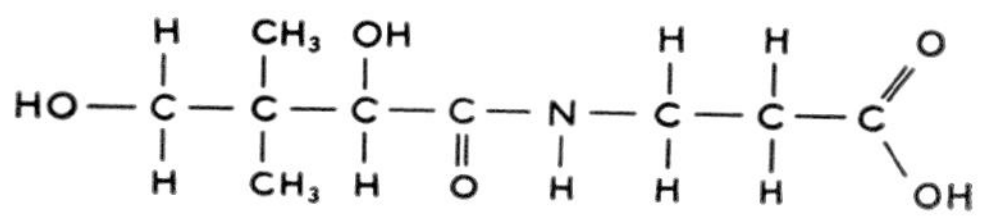

PANTOTHENIC ACID

FIGURE 12-11.

STRUCTURE

Pantothenic acid ($C_9H_{17}NO_5$) is a combination of pantoic acid and beta-alanine joined by an acid-amide linkage and has the structure shown in Figure 12-11. The dextrorotatory calcium salt which is the commercially available form of the vitamin is a white crystalline compound which is less sensitive to heat.

FOOD SOURCES

Pantothenic acid is widely distributed in foods of both plant and animal origin. The largest amounts are found in liver, kidney, yeast, wheat bran and peas. Milk, meat, eggs and fruit contain moderate quantities. Most food pantothenic acid exists in the bound form, from which it is liberated by enzymatic digestion or by hydrolysis with acids or alkalies. Significant losses occur during the milling of grains but cooking losses are not excessive.

ABSORPTION, STORAGE AND EXCRETION

Pantothenic acid is absorbed from the intestinal tract. In the blood it appears as the free vitamin in the plasma and as coenzyme A in the red blood cells. Coenzyme A is presumably synthesized intracellularly whenever needed from pantothenic acid and other components. The body distribution of pantothenic acid is uneven, with the greatest concentrations being found in the liver, kidneys and spleen. Organs with the highest coenzyme A content in order of decreasing concentration are the liver, adrenals, kidneys, brain, heart and testes. It is not known what degradation products of pantothenic acid are formed in vivo. The vitamin is excreted in the urine, feces, sweat and milk as the free acid.

METABOLIC FUNCTIONS

Pantothenic acid is an integral part of co-enzyme A which together with a specific pro-tein apoenzyme functions in many reversible acetylation reactions in carbohydrate, fat and amino acid metabolism. Acetyl-coenzyme A, the active molecule, acts as an acetyl donor or acetyl acceptor. It participates in the synthesis of fatty acids, steroids and cholesterol, in the Krebs' cycle and in the utilization of the B vitamins choline and para-aminobenzoic acid. Other B vitamins such as biotin and folic acid appear to be required for pantothenic acid metabolism. In a sense, pantothenic acid bears a relationship to the two-carbon unit similar to that of folic acid and vitamin B_{12} to the one-carbon unit and as such is extremely im-portant in cellular activity.

PANTOTHENIC ACID DEFICIENCY

Systemic Manifestations. In laboratory animals the major lesions of pantothenic acid deficiency are confined to the nervous system, adrenal cortex and skin. The pantothenic acid content of common foods is so large that a spontaneous human pantothenic acid defi-ciency has not yet been described, with the possible exception of the "burning feet syn-drome." Experimentally induced pantothenic acid deficiency in man is denoted by a clinical state of lethargy, weakness, burning, paresthe-sias, muscle cramps and gastrointestinal dis-turbances accompanied by evidence of adrenal cortical overactivity, defective carbohydrate metabolism and electrolyte imbalance (Bean and Hodges, 1954).

Oral Manifestations. There are no known specific oral manifestations of pantothenic acid deficiency in man. Isolated reports have ap-peared of the curative value of pantothenic acid in the treatment of glossitis and cheilosis in cases where other members of the vitamin B complex are ineffective (Field et al., 1945; Brown, 1949).

Pantothenic acid has been found to be an essential growth factor for the oral acidogenic flora (Dreizen and Spies, 1953). Deletion of pantothenic acid from an otherwise nutrition-ally adequate chemically defined synthetic medium or addition of sufficient quantities of the antimetabolite pantoyltaurine to such media completely inhibits growth and acid production by these organisms.

REQUIREMENTS

The exact pantothenic acid requirement for the human is still a matter of conjecture. Available evidence indicates that approxi-mately 10 mg. per day meets all human needs. This amount is easily provided by most United States diets.

TOXICITY

Pantothenic acid is relatively nontoxic and well tolerated by man.

BIOTIN

The term biotin was first applied to a yeast growth substance isolated in crystalline form from egg yolk by Kögl and Tönnis in 1936. It proved to be identical with vitamin H (egg white injury protective factor) and coenzyme R (rhizobium growth factor). Elucidation of the chemical structure of biotin was achieved

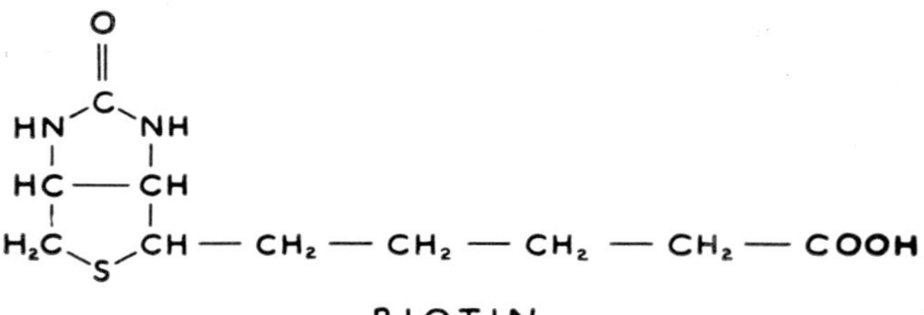

BIOTIN

FIGURE 12-12.

by du Vigneaud in 1942. One year later the compound was synthesized by Harris and co-workers (1943).

STRUCTURE

Biotin ($C_{10}H_{16}N_2O_3S$) is a bicyclic urea derivative which contains sulfur in a thiophene ring as shown in Figure 12-12. It is water soluble, heat stable, relatively resistant to acids and alkalies but susceptible to oxidation.

FOOD SOURCES

Most foods contain biotin in the free and protein bound forms. It is especially abundant in liver, kidney and yeast. Peanuts, chocolate and mushrooms are good sources. The vitamin is synthesized by the intestinal bacteria to an extent sufficient to meet requirements under normal circumstances.

ABSORPTION, STORAGE AND EXCRETION

When ingested in the bound form, biotin is liberated in the intestine by enzymatic hydrolysis provided it is not combined with avidin, a glycoprotein found in raw egg white. The avidin-biotin complex is resistant to proteolytic digestion, is nonabsorbable and is nutritionally unavailable. Free biotin is absorbed from both the upper and lower intestinal tract. It is excreted mainly in the urine and feces, with small amounts appearing in mothers' milk. The precise nature of the excretory products of biotin has not been determined but it is known that some of the biotin excreted in human urine does not combine with avidin.

METABOLIC FUNCTIONS

Biotin is a component of biotin-enzymes which act as cofactors in carboxylation, decarboxylation and deamination reactions in microorganisms. In rats it serves in the synthesis of fatty acids from acetate by carboxy-lating acetyl-coenzyme A. Its role in human metabolism has not been delineated.

BIOTIN DEFICIENCY

Systemic Manifestations. Human biotin deficiency has been produced by ingestion of a biotin deficient diet which included large quantities of raw egg white (Sydenstricker et al., 1942). By the tenth week the subjects developed a scaly dermatitis, grayish pallor of the skin and mucosa, muscle pains, depression, lassitude, insomnia, anorexia, nausea, mild anemia, precordial distress and changes in the color and topography of the tongue. All manifestations were promptly relieved by the administration of biotin.

Spontaneous human biotin deficiency has been encountered in infants treated with sulfa drugs and in adults with a passion for raw eggs. In each instance, the predominant symptom was an exfoliative dermatitis which responded to treatment with biotin.

Oral Manifestations. All cases of induced biotin deficiency studied by Sydenstricker and colleagues (1942) eventually developed a pallor of the tongue and a patchy or generalized atrophy of the lingual papillae resembling lingua geographica.

Biotin has been shown to be an essential nutrient for all oral acid-producing microorganisms tested and as such may play a role in the microbiology of human dental caries (Dreizen and Spies, 1953).

REQUIREMENTS

The biotin requirements for the human are unknown. Amounts ranging from 150 to 300 μg. per day are furnished by most United States diets which, together with that synthesized by the intestinal bacteria, is apparently sufficient to fulfill all needs.

TOXICITY

Biotin has very little, if any, toxicity. It is well tolerated by animals given large doses over prolonged periods.

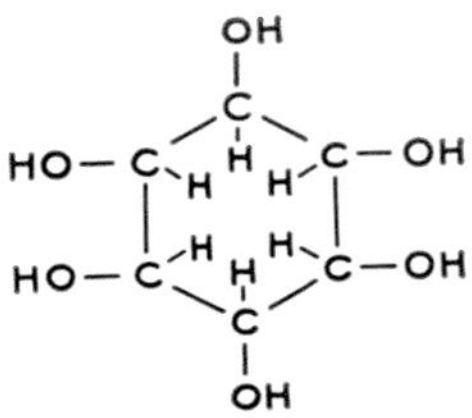

OH
|
C
HO—C   H   H   C—OH
    H   H
HO—C   H   H   C—OH
    H   H
        C
        |
        OH

INOSITOL

FIGURE 12-13.

INOSITOL

Inositol was found in muscle in 1840, identified as a cyclic hexahydroxyalcohol in 1887, synthesized in 1914 and demonstrated to be a growth factor for animals in 1940 (Robinson, 1951). Although conventionally included in the B group of vitamins, its vitamin status is somewhat doubtful as it appears to be a structural component of living tissue rather than a catalyst for metabolic reactions.

STRUCTURE

Inositol ($C_6H_{12}O_6$) is a sweet tasting, water soluble, colorless crystalline solid that melts at 247° C. It is a carbohydrate-like, nonreducing, saturated cyclic compound. Theoretically, it may exist in nine isomeric forms. Meso-inositol or myo-inositol, however, is the only biologically active isomer (Fig. 12-13).

FOOD SOURCES

Considerable quantities of inositol are contained in most plant and animal tissues. Fruits, fruit juices and cereals are good food sources. In animals, inositol occurs in combination with a protein; in plants, as phosphoric esters, the most common of which is the hexaphosphate or phytic acid. Phytic acid combines with calcium and iron in the gastrointestinal tract to form poorly absorbable calcium and iron salts which may substantially reduce the availability of these nutrients.

ABSORPTION, STORAGE AND EXCRETION

Inositol is slowly absorbed from the intestinal tract and widely distributed throughout the body. The greatest concentrations occur in the heart, muscles, brain and liver. About 7 per cent of ingested inositol is converted to glucose. Normally, most is oxidized or otherwise destroyed in the body since only small amounts are excreted in the urine and in the sweat. Renal excretion is high in patients with uncontrolled diabetes mellitus since inositol apparently competes with glucose for reabsorption by the kidney tubules.

METABOLIC FUNCTIONS

The metabolic action of inositol in the human is unknown. Studies of the nutrient requirements of human cells in tissue culture indicate that meso-inositol is necessary for growth, but there is no evidence as yet that it is a component of some essential enzyme system. When given to animals restricted to an inositol deficient diet it has a lipotropic effect.

INOSITOL DEFICIENCY

Systemic and Oral Manifestations. There have been no reports of either spontaneous or artificially induced uncomplicated inositol deficiency in man.

REQUIREMENTS

The average balanced diet of 2500 calories contains about 1 gm. of inositol, which is presumably more than sufficient to meet all necessary requirements.

TOXICITY

Inositol is nontoxic for man. No harmful effects have been noted following ingestion of doses as large as 50 gm.

CHOLINE

Choline was first isolated from bile by Strecker in 1862 and synthesized by Wurtz in 1867. Its nutritional importance was demonstrated in 1932 by Best and Huntsman when it was shown to prevent fatty infiltration of the liver in rats on a choline deficient diet. Since then fatty livers resulting from a lack of choline have been observed in at least nine species of animals ranging from ducklings to monkeys. Like inositol, choline is regarded as a member of the vitamin B complex although it appears to be a structural component of tissues instead of a metabolic catalyst.

STRUCTURE

Choline $(C_5H_{15}NO_2)$ is a viscid, colorless strongly alkaline liquid which is usually crystallized as the hydrochloride. It is water soluble, stable in acid solutions and split by strong alkalies. The chemical structure is shown in Figure 12-14.

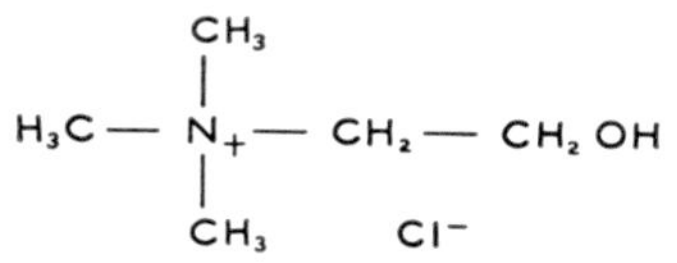

FOOD SOURCES

Large amounts of choline are present in most foods. Egg yolk, meat, fish, cereals and cereal products are particularly rich sources. Green leafy vegetables and legumes contain moderate amounts. In animal products it occurs as a constituent of lecithin and other phospholipids and as the ester acetylcholine.

ABSORPTION, STORAGE AND EXCRETION

Choline is absorbed from the gastrointestinal tract. Substantial amounts of ingested choline are broken down to trimethylamine by the intestinal flora prior to absorption. The greatest concentrations of choline are found in the brain, liver, pancreas and kidneys. Very little choline is excreted as such in the urine, feces and sweat. Most is passed out in the urine as trimethylamine and its oxide.

$$H_3C - \underset{\underset{CH_3}{|}}{\overset{\overset{CH_3}{|}}{N_+}} - CH_2 - CH_2\,OH$$

$$Cl^-$$

CHOLINE CHLORIDE

FIGURE 12-14.

METABOLIC FUNCTIONS

Choline acts as a precursor of acetylcholine and as a donor of labile methyl groups in transmethylation reactions. These reactions involve the transfer of a methyl group from one substance to another, which occurs in the metabolism of nitrogen, sulfur, proteins, fats and carbohydrates. There is apparently an interrelationship between choline and vitamin B_{12}, as the latter reduces the incidence and severity of liver and kidney lesions in choline deficient dogs and rats. It has not been shown that choline is a part of any essential enzyme system in man.

CHOLINE DEFICIENCY

Systemic and Oral Manifestations. Clearcut choline deficiency has not been encountered in the human, and the deficiency manifestations, if any, are unknown. Animal studies indicate that the omission of choline from the diet leads to a cessation of growth, fatty infiltration of the liver, hemorrhagic lesions in the kidney tubules, medial arteriosclerosis and ultimate death.

REQUIREMENTS

The choline requirement for man has not been determined. It appears to be inversely related to the quantity of lipotropic agents in the diet. Dietary components or drugs which are methylated in the body may drain off methyl groups, thereby increasing the requirement for choline. Choline can be synthesized endogenously from such precursors as protein, methionine and betaine.

TOXICITY

The toxic effects of large doses of choline are nausea and a slight fall in blood pressure. Individuals vary considerably in their tolerance of this compound.

PARA-AMINOBENZOIC ACID

Para-aminobenzoic acid was originally synthesized in 1863 and shown to be an essential metabolite for the growth of bacteria by Woods in 1940. It is an integral part of the folic acid molecule. There is considerable doubt as to whether it is required nutritionally in addition to folic acid or that it has any biocatalytic action in man independent of folic acid.

STRUCTURE

Para-aminobenzoic acid ($C_7H_7NO_2$) is a colorless, water soluble, easily oxidized crystalline compound which is stable in dilute acids and alkalies. The chemical formula is shown in Figure 12-15.

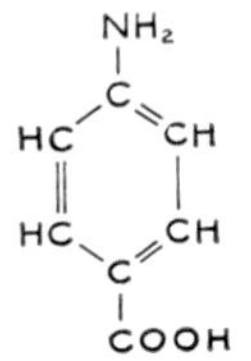

FOOD SOURCES

Para-aminobenzoic acid occurs in foods in the free state, as part of folic acid, and in combination with proteins, amino acids and polypeptides. The best food sources are yeast, liver, cereals and vegetables. It is also synthesized by the intestinal bacteria.

ABSORPTION,
STORAGE AND EXCRETION

Para-aminobenzoic acid is rapidly absorbed from the intestinal tract. There is apparently little storage or utilization of this compound. Very small amounts are excreted in the urine in the free form; the bulk appears as the glycine conjugate or as the glucuronide with a slight amount as p-acetylaminobenzoic acid. Fecal excretion is substantially in excess of dietary intake, indicative of intestinal syn-

para - AMINOBENZOIC ACID

FIGURE 12-15.

thesis. Small amounts are excreted in the sweat.

METABOLIC FUNCTIONS

The function of p-aminobenzoic acid in human metabolism is unknown. It is a nutritional requirement for bacteria that are inhibited by the sulfonamide drugs and acts as an anti-gray hair factor in rats and mice and as a growth promoting agent in chicks.

PARA-AMINO-
BENZOIC ACID DEFICIENCY

Systemic and Oral Manifestations. Para-aminobenzoic acid deficiency has not been observed in man, and the deficiency manifestations, if any, are yet to be determined.

REQUIREMENTS

Whether the human requires p-aminobenzoic acid is still a matter of speculation.

TOXICITY

It has been generally accepted that p-aminobenzoic acid is essentially nontoxic for man, but fatty changes in the liver, heart and kidneys have been found in children given large doses for the treatment of acute rheumatic fever and arthritis (Cruickshank and Mitchell, 1951).

ANIMAL STUDIES OF B VITAMINS

Animal studies permit a degree of dietary

control and the use of experimental techniques which are either unattainable or not feasible for man. Studies with monkeys, dogs, rats and mice have demonstrated the importance of various B vitamins in preserving the health and integrity of the oral structures. In monkeys, niacin deficiency produces a syndrome similar to pellagra in man typified by marked inflammatory and degenerative changes in the gingivae and periodontium; ariboflavinosis causes necrosis of the gingivae, supporting bone and periodontal membrane; folic acid deficiency leads to severe gingival breakdown and some destruction of the underlying bone (Shaw, 1962). In dogs, deficiencies of either niacin, riboflavin, pyridoxine, pantothenic acid or folic acid precipitate nonspecific gingival and periodontal lesions and a noninflammatory atrophic glossitis represented by degeneration and loss of the filiform and fungiform papillae (Afonsky, 1955). In rats, lack of either riboflavin or folic acid in the maternal diet during certain critical times in the gestation period results in a gamut of dentofacial anomalies in the offspring ranging from malocclusion to severe cleft palate (Deuschle et al., 1961; Asling et al., 1960). In mice, ariboflavinosis retards the growth of the mandibular condyles (Levy, 1949a); pantothenic acid deficiency inhibits ossification of the condylar cartilage and promotes a continuing resorption of the alveolar bone, a narrowing of the interdental septum, a broadening of the periodontal membrane and a lowering of the alveolar crests (Levy, 1949b).

REFERENCES

Afonsky, D.: Oral lesions in niacin, riboflavin, pyridoxine, folic acid and pantothenic acid deficiencies in adult dogs. Oral Surg., 8:206, 1955.

Angier, R. B., Boothe, J. H., Hutchings, B. L., Mowat, J. H., Semb, J., Stokstad, E. L. R., SubbaRow, Y., Waller, C. W., Consulich, D. B., Fahrenbach, M. J., Hultquist, M. E., Kuh, E., Northey, E. H., Seeger, D. R., Sickels, J. P., and Smith, J. M., Jr.: The structure and synthesis of the liver L. casei factor. Science, 103:667, 1946.

Asling, C. W., Nelson, M. M., Dougherty, H. D., Wright, H. V., and Evans, H. M.: The development of cleft palate from maternal pteroylglutamic (folic) acid deficiency during the latter half of gestation in rats. Surg., Gynec., Obstet., 111:19, 1960.

Bean, W. B., Franklin, M., and Daum, K.: A note on

tryptophane and pellagrous glossitis. J. Lab. Clin. Med., 38:167, 1951.

Bean, W. B., and Hodges, R. E.: Pantothenic acid deficiency induced in human subjects. Proc. Soc. Exp. Biol. Med., 86:693, 1954.

Best, C. H., and Huntsman, M. E.: The effect of components of lecithin upon deposition of fat in the liver. J. Physiol., 75:405, 1932.

Birch, T. W., György, P., and Harris, L. J.: The vitamin B_2 complex. Differentiation of the antiblacktongue and the "P.-P." factors from lactoflavin and vitamin B_6 (so-called "rat pellagra" factor). Parts I-VI. Biochem. J., 29:2830, 1935.

Bonnett, R., Cannon, J. R., Johnson, A. W., Sutherland, I., Todd, A. R., and Smith, E. L.: The structure of vitamin B_{12} and its hexacarboxylic acid degradation product. Nature, 176:328, 1955.

Brown, A.: Glossitis in Addisonian pernicious anemia. Brit. M. J., 1:704, 1949.

Casal, G.: Historia natural y medica de el principado de Asturias, Obra posthuma del Doctor D. G. Casal, Medico de Su Magnestad y su Protomedicin de Castilla, Madrid, 1762.

Cruickshank, H. A., and Mitchell, G. W., Jr.: Myocardial, hepatic and renal damage resulting from para-aminobenzoic acid therapy. Bull. Johns Hopkins Hosp., 88:211, 1951.

Deuschle, F. M., Takacs, E., and Warkany, J.: Postnatal dentofacial changes induced in rats by prenatal riboflavin deficiency. J. Dent. Res., 40:366, 1961.

Dreizen, S., and Spies, T. D.: Observations on the relationship between selected B vitamins and acid production by microorganisms associated with human dental caries. J. Dent. Res., 32:65, 1953.

du Vigneaud, V.: The structure of biotin. Science, 96:455, 1942.

Eijkman, C.: Eine beriberiähnliche Krankheit der Hühner. Arch. Path. Anat. (Virchow's), 148:523, 1897.

Elvehjem, C. A., Madden, R. J., Strong, F. M., and Woolley, D. W.: The isolation and identification of the anti-blacktongue factor. J. Biol. Chem., 123:137, 1938.

Emmett, A. D., and Luros, G. O.: Are the antineuritic and growth promoting water soluble vitamins the same? J. Biol. Chem., 43:265, 1920.

Field, H., Green, M. E., and Wilkinson, C.: Glossitis and cheilosis healed following the use of calcium pantothenate. Amer. J. Dig. Dis., 12:246, 1945.

György, P.: Vitamin B_2 and the pellagra-like dermatitis in rats. Nature, 133:498, 1934.

Harris, S. A., and Folkers, K.: Synthetic vitamin B_6. Science, 89:347, 1939.

Harris, S. A., Wolf, D. E., Mozingo, R., and Folkers, K.: Synthetic biotin. Science, 97:447, 1943.

Hodgkin, D. C., Pickworth, J., Robertson, J. H., Trueblood, K. N., Prosen, R. J., and White, J. G.: Structure of vitamin B_{12}. Nature, 176:325, 1955.

Huber, C.: Vorläufige Notiz über einege Derivate des Nicotins. Ann. Chem., 141:271, 1867.

Jansen, B. C. P., and Donath, W. F.: Isolation of antiberiberi vitamin. Mededeel. Dienst Volkgezondheid Nederland-Indie, 16:186, 1927.

Karrer, P., Schopp, K., and Benz, F.: Synthesen von Flavinen IV. Helvet. Chim. Acta, 18:426, 1935.

Kögl, F., and Tönnis, B.: Über das Bios-Problem: Darstellung von krystallisierten Biotin aus Eigelb. Ztsch. Physiol. Chem., 242:43, 1936.

Kuhn, R., György, P., and Wagner-Jauregg, T.: Ueber Ovoflavin den Farbstoff des Eiklars. Ber. deutsch. chem. Gesellsch., 68:576, 1933.

Kuhn, R., Reinemund, K., Weygand, F., and Strobele, R.: Ueber die synthese des lactoflavins (vitamin B₂). Ber. deutsch. chem. Gesellsch., 68:1785, 1935.

Levy, B. M.: The effect of riboflavin deficiency on the growth of the mandibular condyles of mice. Oral Surg., 2:89, 1949a.

Levy, B. M.: Effects of pantothenic acid deficiency on the mandibular joints and periodontal structures of mice. J. Amer. Dent. Assn., 38:215, 1949b.

Mann, A. W., Dreizen, S., Spies, T. D., and Hunt, F. M.: A comparison of dental caries activity in malnourished and well-nourished patients. J. Amer. Dent. Assn., 34:244, 1947.

McCollum, E. V., and Davis, M.: The nature of dietary deficiencies of rice. J. Biol. Chem., 23:181, 1915.

Mitchell, H. K., Snell, E. E., and Williams, R. J.: The concentration of "folic acid." J. Amer. Chem. Soc., 63:2284, 1941.

Mueller, J. F., and Vilter, R. W.: Pyridoxine deficiency in human beings induced with desoxypyridoxine. J. Clin. Invest., 29:193, 1950.

Racker, E., de La Haba, G., and Leder, I. G.: Thiamine pyrophosphate—a coenzyme of transketolase. J. Amer. Chem. Soc., 75:1010, 1953.

Recommended Dietary Allowances. Rev. 1963. Washington, D. C., National Academy of Sciences–National Research Council, Pub. No. 1146, 1964.

Rickes, E. L., Brink, N. G., Koniuszy, F. R., Wood, T. R., and Folkers, K.: Crystalline vitamin B₁₂. Science, 107:396, 1948.

Robinson, F. A.: The Vitamin B Complex. New York, John Wiley & Sons, Inc., 1951.

Rosenblum, L. A., and Jolliffe, N.: The oral manifestations of vitamin deficiencies. J. A. M. A., 117:2245, 1941.

Sebrell, W. H., and Butler, R. E.: Riboflavin deficiency in man (ariboflavinosis). Pub. Health Rep., 54:2121, 1939.

Shaw, J. H.: The relation of nutrition to periodontal disease. J. Dent. Res., 41:264, 1962.

Shorb, M. S.: Unidentified essential growth factor for Lactobacillus lactis found in refined liver extracts and in certain natural materials. J. Bact., 53:669, 1947.

Smith, E. L.: Purification of anti-pernicious anaemia factors from liver. Nature, 161:638, 1948.

Spies, T. D., Cooper, C., and Blankenhorn, M. A.: The use of nicotinic acid in the treatment of pellagra. J. A. M. A., 110:622, 1938.

Spies, T. D., Garcia Lopez, G., Menendez, J. A., Minnich, V., and Koch, M. B.: The effect of folic acid on sprue. Southern Med. J., 39:30, 1946.

Spies, T. D., Stone, R. E., and Aramburu, T.: Observations on the anti-anemic properties of vitamin B₁₂. Southern Med. J., 41:522, 1948a.

Spies, T. D., Stone, R. E., Kartus, S., and Aramburu, T.: The treatment of subacute combined degeneration of the spinal cord with vitamin B₁₂. Southern Med. J., 41:1030, 1948b.

Spies, T. D., Vilter, C. F., Koch, M. B., and Caldwell, M. H.: Observations on the anti-anemic properties of synthetic folic acid. Southern Med. J., 38:707, 1945.

Strecker, A.: Über einige neue Bestandtheile der Schweingalle. Ann. chem., 123:353, 1862.

Suzuki, U., Shimamura, T., and Odake, S.: Über Oryzanin, ein Bestandteil der Reiskleie und seine physiologische Bedeutung. Biochem. Ztschr., 43:89, 1912.

Sydenstricker, V. P., Singal, S. A., Briggs, A. D., DeVaughn, N. M., and Isbell, H.: Observations on the "egg white injury" in man. J.A.M.A., 118:1199, 1942.

Takaki, K.: Health of the Japanese Navy. Lancet, 2:86, 1887.

Ungley, C. C.: Vitamin B₁₂ in pernicious anemia: Parenteral administration. Brit. M. J., 2:1370, 1949.

Warburg, O., and Christian, W.: Ueber ein neues Oxydationsferment und sein Absorptionsspektrum. Biochem. Ztscr., 254:438, 1932.

Warburg, O., and Christian, W.: Co-ferment problem. Biochem. Ztschr., 275:464, 1935.

West, R.: Activity of vitamin B₁₂ in Addisonian pernicious anemia. Science, 107:398, 1948.

Williams, R. J.: Pantothenic acid, a vitamin. Science, 89:486, 1939.

Williams, R. J., Eakin, R. E., Beerstecher, E., Jr., and Shive, W.: The Biochemistry of the B Vitamins. New York, Reinhold Publishing Corporation, 1950.

Williams, R. J., Lyman, C. M., Goodyear, G. H., Truesdail, J. H., and Holiday, D.: "Pantothenic acid" a growth determinant of universal biological occurrence. J. Amer. Chem. Soc., 55:2912, 1933.

Williams, R. J., and Major, R. T.: The structure of pantothenic acid. Science, 91:246, 1940.

Williams, R. R., and Cline, J. K.: Synthesis of vitamin B₁. J. Amer. Chem. Soc., 58:1504, 1936.

Woods, D. D.: Relationship of p-aminobenzoic acid to mechanism of action of sulphanilamide. Brit. J. Exp. Path., 21:74, 1940.

Wurtz, A.: Synthèse de la névrine. Comptes rendus, 65:1015, 1867.

Chapter Thirteen

Ascorbic Acid (Vitamin C) and Its Role in Wound Healing and Collagen Formation

BERNARD S. GOULD, PH.D.

The involvement of an accessory food factor associated with certain green vegetation, particularly citrus fruits, in preventing scurvy and curing those suffering from the disease was recognized as early as the 16th century by the explorer Jacques Cartier. Its relationship to the proper healing of wounds, as well as to the apparent maintenance of wounds which have already healed—which we now know is largely related to the formation and maintenance of collagen—was clearly described by Walter in 1746.

With the first demonstration that experimental scurvy could be produced in the guinea pig, by Holst and Froelich (1907), the scene was set for detailed studies of the deficiency condition, but it was not until 1916 that detailed histological studies showed that the fundamental defect in the scorbutic process involved a functional defect in connective tissue, and particularly in collagen formation.

Zilva and Wells (1919) carried out what were probably the first microscopic examinations of the tooth structure in scorbutic patients and noted fibrosis and osteoid-like deposits in the tooth pulp (Fig. 13-1). Aschoff and Koch (1919) suggested that the defect in scurvy involved the inability to produce extracellular substance. Höjer (1924) was of the opinion that the defect was actually the result of impaired function of the connective tissue cell. Of historical interest is that in 1926 Wolbach and Howe proposed that in scurvy the connective tissue cells produce intercellular materials and collagen precursors normally, but that a factor is missing from the extracellular fluid which causes the collagen fiber precursor to gel and form fibrils. This factor, they felt, was not involved in the formation of fibroblasts. This concept is no longer entirely acceptable.

CHEMICAL CHARACTERISTICS

Vitamin C isolated from natural sources has been named L-ascorbic acid. It is an ene-diol lactone of an acid whose chemical configuration is analogous to that of L-glucose. Dissociation of the enolic hydrogen on C_3 gives it its acidic properties. The D-form of ascorbic acid and many closely related analogs which have been synthesized show very little if any antiscorbutic activity.

L-Ascorbic acid is very water soluble, slightly soluble in alcohol but insoluble in fats and oils. It is readily oxidized by mild oxidizing agents and, in fact, its most outstanding chemical property is its strong reducing activity, which depends upon the loss of hydrogen atoms from the ene-diol hydroxyl groups at the second and third carbons. The oxidation to dehydroascorbic acid can be effected by air, peroxide, iodine, quinones, silver nitrate and 2,6-dichlorophenol-indophenol among other oxidizing agents. These substances have been used in many instances for the qualitative detection and quantitative de-

169

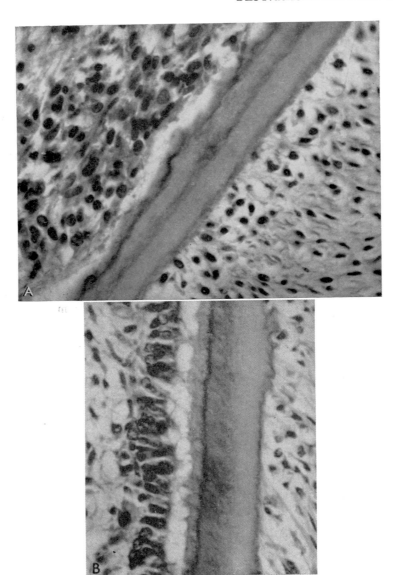

FIGURE 13-1. *Photomicrograph showing the status of the odontoblastic layer and dentin formation and the periodontal ligament in (A) a scorbutic guinea pig, and (B) a guinea pig maintained on a scorbutigenic basal diet supplemented by 0.5 mg. of ascorbic acid daily.*
A comparison of (A) and (B) shows that in ascorbic acid deficiency the odontoblastic layer is disrupted and there is cessation of dentin production. The periodontal ligament shows very little collagen and there is no organized fiber apparatus. (Courtesy of Dr. H. M. Goldman.)

termination of ascorbic acid. They are, of course. nonspecific. The dehydroascorbic acid formed has essentially full antiscorbutic activity. By the action of proper reducing agents such as H_2S or glutathione the dehydro compound can be reduced to L-ascorbic acid. It has been postulated that glutathione may, in fact, be involved in maintaining the vitamin in the reduced form under physiological conditions. However, further oxidative action yields first diketogulonic acid and other degradation products which are biologically inactive. Light, alkaline pH and certain metal ions, particularly copper and silver (but not aluminum), accelerate these reactions, especially at elevated temperatures. These facts are the bases upon which the methods have been developed that are now used in preserving the vitamin in the course of processing ascorbic acid-containing foodstuffs.

METHODS OF ASSAY

CHEMICAL

Ascorbic acid can be determined chemically and, more specifically, by bioassay. The chemical methods are relatively nonspecific since they depend in many instances upon the reducing power of ascorbic acid or upon the reaction of dehydroascorbic acid with 2,4-dinitrophenylhydrazine, a reaction also given by the biologically inactive diketogulonic acid. On the basis of the relative ease with which the chemical assays can be carried out they have largely superseded methods of bioassay.

Among the reducing methods the reduction of the blue dye 2,6-dichlorophenol-indophenol is most common. The reduced form of the dye is colorless. By controlling conditions and by converting all of the biologically active antiscorbutic substances to the reduced form (i.e., dehydroascorbic acid is converted to L-ascorbic acid by H_2S treatment), it is possible to titrate the reducing power of the vitamin using a standardized dye solution with a minimum of reduction due to nonspecific reducing compounds. A more rigorous quantitative method is that of Roe and Kuether (1943), which depends on the coupling of dehydroascorbic acid with 2,4-dinitrophenylhydrazine.

BIOASSAY

Among the common laboratory animals the guinea pig is the only animal practical for the bioassay of vitamin C. Apart from the primates and the guinea pig, none of the other animals has a dietary requirement for the vitamin. The young guinea pig (ca 250 to 300 gm.) develops signs of vitamin C deficiency rapidly (2 to 3 weeks) and with a high degree of reproducibility. Assays are based on either the protective or curative action of the vitamin with respect to a variety of manifestations of the deficiency disease such as growth, hemorrhages, bone structure as seen roentgenographically, histological examination of bones and costochondral junctions and, particularly, modifications in the incisor tooth structure, which is perhaps one of the most sensitive assays. The lowering of the blood phosphatase, which is a reflection of the impaired osteoblastic activity, may also be used as a bioassay in ascorbic acid deficiency.

DISTRIBUTION OF THE VITAMIN

As a class the citrus fruits—oranges, lemons, limes and grapefruit—rank very high as good sources of ascorbic acid. However, additional good sources are the actively growing parts of practically all plants, particularly the so-called leafy ones such as parsley, broccoli, cabbage and spinach. Peas, beans, peppers, tomatoes, strawberries, cantaloupes, carrots, bananas, turnips and potatoes are all good sources. Seeds are devoid of the vitamin but begin to synthesize it upon germination. Cow's milk is relatively low in ascorbic acid, and after pasteurization its effective concentration becomes almost negligible; it would take 4 to 5 quarts to supply the daily requirement. Eggs and meat are also poor sources of the vitamin. It has generally been assumed that almost 80 to 85 per cent of the vitamin C content of foods is lost in the course of preparation.

In animal tissue the vitamin is widely distributed in a manner suggesting a parallelism between concentration and metabolic activity. In roughly decreasing order with respect to local concentration are the pituitary, corpus luteum, adrenal cortex, liver, brain, spleen, salivary glands, lung, kidney, heart, muscle and blood. It has been found that the plasma contains about 0.6 to 1.5 mg. ascorbic acid per 100 ml.; the red blood cells contain, on the average, about twice that amount but the white blood cells are relatively rich, containing as much as 20 to 40 times the plasma level.

Despite the large total amount in the liver and the relatively high concentration in the adrenal cortex, these organs do not appear to act as storage centers for the vitamin. Apparently all body sources are called upon when the dietary intake is inadequate. In the absence of a reservoir it is possible to explain the fact that the ingestion of excessive amounts of the vitamin (amounts over and above that needed to saturate the tissues) is accompanied by prompt urinary excretion. This means that, unlike the fat soluble vitamins, which can be stored in the liver, ascorbic acid must be supplied to the body regularly. The excretion level has been used as an index of ascorbic acid nutrition. The white blood cell level is an index of the incipient scurvy as the ascorbic acid content falls only after the plasma level has reached an extremely low concentration.

FUNCTION

The specific metabolic function of ascorbic acid is unknown. The fact that it undergoes reversible oxidation-reduction reactions and appears to be involved in aromatic hydroxylation reactions at the microsomal level strengthens the feeling which has been prevalent since even before the vitamin was isolated that it serves as an electron (hydrogen) transport agent. However, the evidence is not too strong that it plays a general role of this type. It appears that it may also be involved in the specific hydroxylation reactions of proline to hydroxyproline and lysine to hydroxylysine in collagen biosynthesis.

It is known that premature infants and ascorbic acid deficient guinea pigs maintained on a vitamin C free diet *rich in aromatic amino acids* such as tyrosine and phenylalanine excrete larger amounts of p-hydroxyphenylpyruvic acid and p-hydroxyphenylacetic acid in the urine because of their inability to carry the oxidation further. The administration of ascorbic acid corrects the defect. This is not a defect due to ascorbic acid deficiency but rather to the effect of the high dosage of tyrosine and is apparent only when the body is less than saturated with respect to ascorbic acid. The tyrosine dosage produces an enzymatic imbalance consisting of an oxidative increase in the activity of tyrosine transaminase which forms p-hydroxyphenylpyruvic acid (p-HPP), and a reversible inactivation of p-hydroxyphenylpyruvic oxidase that removes the p-HPP. Therefore, p-HPP accumulates and is excreted unless sufficient ascorbic acid is available to prevent the inactivation of p-HPP oxidase.

It has been shown that ascorbic acid is important in the prevention of the megaloblastic anemia of infancy and perhaps is important in other types of macrocytic anemia.

The vitamin also plays a role in iron absorption from the intestinal tract, in the transfer of plasma-bound iron to the liver and its incorporation into ferritin.

Even though the specific metabolism interactions of ascorbic acid are unknown it is abundantly clear that it is intimately involved in the formation of the intercellular substances of connective tissue, especially collagen, particularly in the course of wound healing.

COLLAGEN FORMATION AND WOUND HEALING

Studies aimed at establishing a relationship

between ascorbic acid and collagen formation stem from the area of wound healing. It has been established that the failure to heal normally is due primarily to the failure to produce collagen, and that the tensile strength of a wound is proportional to the collagen content which, in turn, is related to the plasma vitamin C level and to the vitamin C intake.

These relationships have been substantiated not only on a histological basis but by the quantitative chemical determination of collagen in newly formed tissue, either by its isolation as gelatin or, more commonly, by the determination of the hydroxyproline content (hydroxyproline being essentially uniquely associated with collagen) of the gelatin.

THE COLLAGEN
FORMING CELL

The term fibroblast is perhaps more generic than specific. It has been suggested that, functionally, different kinds of fibroblasts exist. All fibroblasts may, however, be considered to be of mesenchymal origin with the potentiality to produce collagen, acid mucopolysaccharide, or both, when functionally differentiated. The nature and amount of these substances secreted is characteristic for each of the different kinds or modulations of the fibroblast, and such modulations of function may be induced by environmental influences. The fibroblast, the osteoblast and the odontoblast are profoundly influenced by ascorbic acid deficiency. In the absence of the vitamin these cells characteristically fail to produce their respective fibrous proteins: collagen, osteoid and dentin. It has been suggested that in most respects the cytochemical properties of these cells are analogous (except for the extremely high phosphatase production by the osteoblast) and that in all three cases ascorbic acid deficiency brings about what appears to be a reversal of the differentiated cells to more immature cell types. The administration of ascorbic acid promptly corrects the defect.

These observations point to the possibility that ascorbic acid may exert its effect on collagen biosynthesis indirectly by participation in mechanisms that effect the proper maturation of the collagen-forming cell.

Ross and Benditt (1961) have observed in their study of normal wound healing that the most striking feature of the fibroblasts was found to be the extensive and dilated endoplasmic reticulum. In many cells the profiles of the cisternae form a long anastamosing network of channels whose exterior spaces are lined with numerous closely spaced ribosomes. The cisternae appear not to have any connection to the extracellular space. The only cells which show extensive development of their endoplasmic reticulum early in the healing process are the macrophages, which suggests a transition from these cells to fibroblasts. In addition, fibroblasts are occasionally found to contain intracytoplasmic bodies characteristic of macrophages. They present some evidence that the shedding of cytoplasm, as in apocrine secretion, is a means by which collagen precursors can be extruded from the cell. Immediately following ascorbic acid administration to scorbutic guinea pigs, structures identical with individual cisternae of endoplasmic reticulum associated with numerous collagen fibrils appear, lying free in the extracellular spaces.

The endoplasmic reticulum of the fibroblasts appears to be altered in scurvy. It appears as rounded, often dilated, but rarely interconnected, profiles and not so extensively developed as in the normal fibroblasts. There is a marked decrease, but not a total absence, of collagen fibrils in the intercellular space; at the same time, a large amount of a somewhat amorphous dense matrix is present.

Studies of wound healing indicate that there is no apparent decrease in the proliferation of fibroblasts in scurvy; in some instances, increased numbers have been observed.

GENERAL CHARAC-
TERISTICS OF COLLAGEN

Collagen, which constitutes about 25 to 30 per cent of the total protein of the mammalial body, is characterized by a unique amino acid composition in which about two thirds of all the amino acid residues are made up of glycine, proline and hydroxyproline. In addition to the high concentration (approximately 14 per cent) of hydroxyproline, collagen contains a small amount (about 1 per cent) of hydroxylysine. Apart from a small amount of hydroxyproline (1.6 per cent) found in elastin, the hydroxyproline and hydroxylysine of animal tissue have been shown to be almost entirely associated with collagen and have been used as a measure of the latter. The amino acid composition is also characterized by very low content of the aromatic amino acids, tryptophan and sulfur amino acids. Collagen from a variety of sources, when examined with the electron microscope, shows a characteristic 640 Å. periodicity with a number of interband periods. Naturally occurring fibrous collagen is essentially insoluble in aqueous solvents, but suitable heating in water will convert it to soluble gelatin.

The possibility that there might be smaller units of collagen that could pass out through the cell and that precursor forms of the insoluble fibers might be more readily soluble led Orekhovich et al. (1948 a, b) to fractionate skin proteins with mildly acidic citrate buffers. They were able to remove a fraction which they called "procollagen." Further studies using labeled glycine-1-C¹⁴ suggested that this fraction is an early form of collagen.

Highberger et al. (1961) were able to extract a soluble component with mildly alkaline buffer solutions which they call "tropocollagen" and which is capable of being reconstituted in vitro into normal fibers. Soluble fractions have been isolated using 0.45 M NaCl and 0.2 M NaCl. When these soluble fractions are kept at 37°C. they form gels consisting of networks of the fine fibrils having the typical 640 Å. periodicity of native collagen.

The tropocollagen particle, a triple helical structure, found to have a length of approximately 3000 Å. with a diameter of 14 Å. and molecular weight of about 340,000, is the basic building block for the formation of collagen fibers in vivo. The amino acid composition of neutral salt-extracted collagen, citrate-extracted collagen, and insoluble collagen are essentially the same. These substances must be considered to be morphological precursors rather than biochemical precursors of collagen.

Much less procollagen was found in the skin of old than of young animals, and only about one half as much was found in the skin of scorbutic animals as compared with normal. The skin of severely scorbutic animals contains no detectable salt-extractable collagen. It appears that the defect in collagen formation associated with ascorbic acid deficiency involves not merely the inability to fibrillate a tropocollagen precursor but also one involving the biosynthesis of the soluble precursor itself.

RELATIONSHIP OF TISSUE ASCORBIC ACID TO COLLAGEN FORMATION

The relationship between ascorbic acid and collagen formation has been studied quantitatively in regenerating skin (Fig. 13-2) and

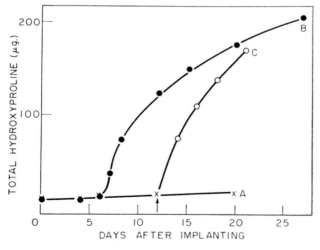

FIGURE 13-2. *Collagen formation in polyvinyl sponges implanted subcutaneously in guinea pigs maintained on: A, a scorbutigenic diet; B, a diet supplemented by 10 mg. of ascorbic acid daily; and C, a scorbutigenic diet for 12 days followed by the diet supplemented by 10 mg. of ascorbic acid daily. The hydroxyproline value × 7.46 is a measure of the apparent collagen. (Gould, 1958.)*

it has been demonstrated that very little hydroxyproline synthesis occurred in wounds made in ascorbic acid-depleted animals if they were maintained on the scorbutigenic diet. When ascorbic acid is administered to animals kept on such a diet for several days, collagen formation begins within 36 to 48 hours. If animals were wounded without prior depletion (Fig. 13-3), the tissues apparently contain sufficient ascorbic acid to insure almost normal hydroxyproline formation even though no ascorbic acid is given subsequent to wounding. If guinea pigs are wounded after varying periods of ascorbic acid depletion, the collagen synthesis parallels the metabolic decay curve for ascorbic acid. Most striking is the fact that within 5 to 6 days after the withdrawal of ascorbic acid young guinea pigs can no longer produce collagen in regenerating skin wounds, suggesting that impaired collagen formation is not a function of the scorbutic process but is related to the available tissue concentration of the vitamin and that in repair ascorbic acid plays a direct role in collagen synthesis.

Such a concept of the relationship of tissue concentration of ascorbic acid to collagen synthesis was not new. It had been the basis for clinical attempts to accelerate or improve healing by the application of ascorbic acid directly to wounds and it also emanated from studies which show that there is an apparent sequestration of ascorbic acid at sites of tissue regeneration or collagen fiber formation since wound areas had higher levels of the vitamin than adjacent normal tissue.

Ascorbic acid accumulates at the wound site shortly following wounding and before connective tissue is formed. Tissues of scorbutic guinea pigs take up a greater percentage of radioactive ascorbic acid following a single injection than do tissues of undepleted animals; and after wounding the scar tissue contains a greater amount of ascorbic acid than tissue distant from the wound.

DIRECT ACTION OF ASCORBIC ACID ON COLLAGEN FORMATION

It has been demonstrated both in vivo and in vitro with tissue minces and in tissue and

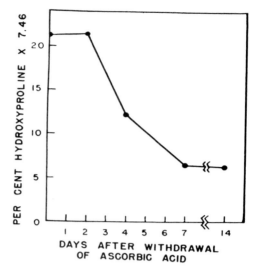

FIGURE 13-3. *Effect of ascorbic acid depletion on collagen formation. This shows the collagen content of 12-day granulation tissue of guinea pig skin wounds made at various times after the withdrawal of ascorbic acid from the diet. The hydroxyproline per cent × 7.46 is a measure of the per cent apparent collagen in the dry granulation tissue. (Gould and Woessner, 1957.)*

organ culture that ascorbic acid plays a direct role at the site of collagen synthesis rather than indirectly through some systemic effect. Figure 13-4 shows the effect of injecting very small amounts of ascorbic acid into one of a pair of polyvinyl sponges subcutaneously implanted in a vitamin C depleted guinea pig. The second sponge was injected with saline or an inactive compound. Only the ascorbic acid injected sponge formed collagen in normal amounts.

COLLAGEN SYNTHESIS

Collagen uniquely contains two amino acids, hydroxyproline and hydroxylysine, which cannot be directly incorporated into protein. It has been shown that when labeled proline is fed it is incorporated into collagen as both proline and hydroxyproline and further that if labeled hydroxyproline is fed it is not used to any significant extent in the biosynthesis of collagen. It was concluded that the hydroxyproline of collagen is derived from proline and that the conversion occurs after the pro-

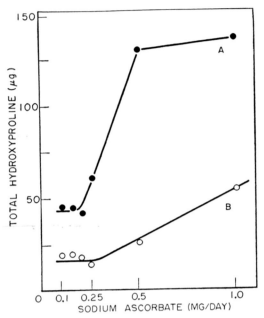

FIGURE 13-4. *The local effect of ascorbic acid. The curves indicate the collagen content (hydroxyproline × 7.46) of polyvinyl sponges implanted bilaterally under the skin of guinea pigs 7 days after withdrawal of ascorbic acid, at which time the animals were continued on the scorbutigenic diet for 12 days and thereafter injected for 4 days with varying amounts of sodium ascorbate in 0.9 per cent sodium chloride solution into one sponge, A; or with 0.9 per cent sodium chloride solution alone, B. The total amounts indicated on the abscissa were administered daily over an 8-hour period in five equal doses. (Gould, 1958.)*

line has been incorporated into a peptide or larger molecule. A similar situation has been found with respect to hydroxylysine formation.

One of the major working hypotheses with respect to the role of ascorbic acid in collagen formation has been that it is involved in some manner with the specific hydroxylation of proline and lysine, either directly or indirectly.

Numerous workers have attempted, with no success, to isolate an accumulated proline-rich precursor material from granulation tissue, from subcutaneously implanted polyvinyl sponges and from granulomas in guinea pigs deprived of ascorbic acid, which might be expected to contain such presumptive precursors. On the other hand, recent work (Peterkofsky

and Udenfriend, 1965; Prockop et al., 1965) has refocused attention on the hypothesis that the hydroxylation of proline and lysine does, in fact, occur after the peptide has been formed; in contradistinction to the idea that an activated hydroxyproline, such as s-RNA-hydroxyproline, derived from proline, is incorporated at the ribosomal level in a manner common to the assembly of most proteins.

In some of these studies it has been shown that ascorbic acid appears to be an essential component for the hydroxylation reaction. Under certain circumstances the ascorbic acid was found to be nonspecific and could be replaced to varying degrees by other substances such as reduced triphosphopyridine nucleotide in the enzymic hydroxylation of the presumptive collagen precursor. This suggests the possibility of the replacement of ascorbic acid by other substances in the biosynthetic mechanism.

There are several observations that point to the possibility that there may be both ascorbic acid-independent as well as ascorbic acid-dependent collagen forming mechanisms. Studies of collagen formation in healing wounds and in polyvinyl sponge implants indicate that, even in severe scurvy, collagen formation is inhibited about 80 to 85 per cent but rarely, if ever, completely inhibited. Electron microscopic studies of wounds show fibril formation even when the animals are scorbutic. It is not possible to conclude whether this is due to a population of fibroblasts whose activity has been impaired quantitatively by the withdrawal of ascorbic acid or whether there is a mixed population of cells, one type of which is relatively independent of ascorbic acid.

It is amply recognized that even after the withdrawal of ascorbic acid from the diet young guinea pigs continue to gain weight for about 12 to 14 days, yet it has been shown that collagen formation in wounds is markedly inhibited beginning about the fourth day after ascorbic acid withdrawal. By studying collagen increases in skin and carcasses of very young guinea pigs maintained on ascorbic acid-supplemented and -deficient diets compared with collagen formation in skin wounds and in implanted sponges in the same animals it appears that despite the absence of ascorbic acid collagen formation associated with growth continued while no appreciable collagen formation occurred in the wounded area. It has been proposed (Chmuchalova and Chvapil, 1964) that the dependence on ascorbic acid for collagen biosynthesis in various organs is related to the turnover rate of the collagen. In tissues such as liver with a high turnover rate there was great dependence, while the lung and muscle with a slow turnover showed practically no dependence.

ROLE OF ASCORBIC ACID IN THE MAINTENANCE OF COLLAGEN

Although there is little question that a relationship exists between ascorbic acid and collagen production, it is not amply clear what the relationship is between ascorbic acid and the maintenance of collagen. On the basis of experiments using labeled amino acid incorporated into collagen, Neuberger et al. (1951) and Robertson (1952) have shown that collagen is metabolically inert compared to most other proteins. However, data exist indicating that collagen is not completely inert but that there is a slow synthesis and degradation even in adult skin and tendon. In certain tissues, such as liver, bone and periodontal membrane, the process can be relatively rapid. In the case of the periodontal membrane, collagen fibers of the side of the membrane in apposition to bone are very labile while the fibers in apposition to the tooth are quite stable under conditions of ascorbic acid withdrawal.

Some of the earliest reports of scurvy contain statements to the effect that wounds which had been healed for years would break down if the individual became scorbutic (Walter, 1748). Hunt (1941) studied the effect of the withdrawal of vitamin C on healed wounds; his conclusions, based on a few animals, were that mature collagen in scars may retrogress if ascorbic acid is withdrawn from the diet for prolonged periods and that the new collagen had reverted to an argyrophilic precollagenous state very different from that found in normal control wounds. Pirani and Levenson (1953), in a carefully controlled series of experiments, also demonstrated that ascorbic acid is necessary for both the healing

of wounds and the maintenance of the collagen which had formed. In these animals the connective tissue was loose, and numerous fibroblasts and immature mesenchymal cells were present. There were numerous capillaries, many of which were defective, and several areas of hemorrhage. The changes observed were of the same type as those that occur in healing wounds in scorbutic animals.

Robertson (1952) concluded from his studies that ascorbic acid is not essential for the maintenance of preformed collagen and that possibly only certain recently formed collagen requires ascorbic acid for maintenance. Since this constitutes only a small fraction of the total, it is difficult to demonstrate the relationship. He suggested that the major portion of induced new collagen, like organ collagen, does not require ascorbic acid for maintenance.

Gould et al. (1960) implanted polyvinyl sponges subcutaneously in ascorbic acid-depleted and in normal guinea pigs and then permitted new fibrous tissue to accumulate. After varying periods of collagen formation the animals were placed on a scorbutigenic diet and the rate of collagen "resorption" from the sponges was determined chemically. It was found that almost directly after the withdrawal of ascorbic acid there was complete disappearance of salt-soluble collagen and this was followed by the slower disappearance of insoluble collagen. The rate of resorption was greater from sponges with newly formed collagen than from those which had been implanted for a prolonged period, but the latter also showed considerable resorption.

The mechanisms involved in collagen "resorption" in vivo are little, if at all, understood. It is not clear whether the process involves collagenolysis by specific or nonspecific enzymes or merely removal of the particulate material by phagocytosis.

ASCORBIC ACID, THE GROUND SUBSTANCE, AND FIBROGENESIS

Considerable attention has been directed toward the influence of ascorbic acid on the ground substance and the possible role of the latter in collagen fiber formation. The ground

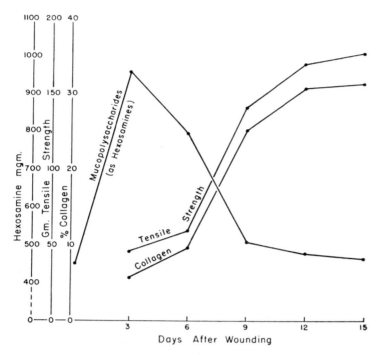

FIGURE 13-5. *Normal pattern of wound healing. (Courtesy of Dunphy, J. E., and Udupa, K. N.: Chemical and histochemical sequences in the normal healing wound. New England J. Med., 253: 847, 1955.)*

substance is the amorphous interfibrillary material thought of histologically as being the metachromatic-staining (toluidine blue) and periodic acid-Schiff (PAS)-staining material and chemically as being composed largely of mucopolysaccharides such as chondroitin sulfate and hyaluronic acid.

In healing wounds, shortly after fibroblastic proliferation but before fibrogenesis begins, there is an accumulation of material showing toluidine blue metachromasia which supposedly is a stain specific for mucopolysaccharide sulfuric acid esters. The periodic acid-Schiff stain, Hale's colloidal iron stain, and the hexosamine content parallel the metachromasia, substantiating the presence of mucopolysaccharide. This accumulation continues for a few days and then declines rapidly, the decline occurring with the onset of fiber formation (Fig. 13-5).

Bradfield and Kodicek (1951) in a study of wound sections from normal and scorbutic guinea pigs found that wounds of normal animals stained by the periodic acid-Schiff method showed a decrease of stainable material as healing progressed, whereas in the scorbutic animal it increased. They did not, however, find any difference in metachro-

matic staining of skin wounds of both normal and scorbutic animals and interpreted this as indicating that the abnormal abundance of polysaccharide material in scurvy probably was not sulfated. Scorbutic wounds appeared to be composed of thick, chaotically arranged precollagenous fibers, suggesting the accumulation of a mucopolysaccharide sheath around a precollagen core. The mucopolysaccharide supposedly interferes with the subsequent maturation of the collagen (Fig. 13-6).

Studies based on the incorporation of S^{35} into mucopolysaccharides indicate a reduction in the rate of metabolism of chondroitin sulfate in scurvy. When Kodicek and Loewi (1955) administered $Na_2S^{35}O_4$ to normal and scorbutic guinea pigs and studied the incorporation of S^{35} into regenerating tendon, they found it to be decreased in the scorbutic animals while the hexosamine content was normal. From these results it would appear that ascorbic acid may be involved in the sulfation of mucopolysaccharides.

There is no proof that mucopolysaccharides are specifically involved in the formation of collagen. They may have to do with other functions of these tissues and relatively little if anything to do with the formation of collagen

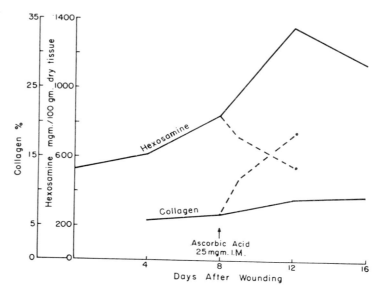

FIGURE 13-6. *Wound-healing pattern in scorbutic animals. (Courtesy of Dunphy, J. E., Udupa, K. N., and Edwards, L. C.: Wound healing: A new perspective with particular reference to ascorbic acid. Ann. Surg., 144:307, 1956.)*

fibrils. However, it must be kept in mind that it is possible that although they may not play an active role in fibrogenesis in vivo, they may play an important passive or regulatory role.

ASCORBIC ACID DEFICIENCY

SYSTEMIC MANIFESTATIONS

It has long been known that scurvy arises as a result of the continued deficiency in the diet of a factor, called vitamin C, common in fresh vegetables and fruits, particularly citrus fruits. Only man, other primates and guinea pigs exhibit this dietary requirement since they are unable to synthesize L-*ascorbic acid,* which was isolated from adrenal cortex, cabbage, paprika and lemon juice and shown to be the essential dietary antiscorbutic factor.

The deficiency disease resulting from the prolonged withdrawal of ascorbic acid from the diet of man (in 4 to 5 months) and of the guinea pig, which is the commonly used experimental animal (2 to 3 weeks), is characterized by loss of appetite and weight. Joints become swollen and tender and there is considerable subperiosteal hemorrhage. There is also hemorrhage into the mucous membranes, skin, joints, limbs and bone marrow. There is swelling at the ends of the long bones and

of the costochondral junctions. Bones become brittle and fracture easily. Capillary fragility is commonly observed due to the defective character of the connective tissue of the blood vessel walls. This defect appears to be largely related to the inability to form and maintain the intercellular ground substance of connective tissue; the matrix of bone, the dentin of teeth as well as the collagen fibers. Histologic changes show alteration in the osteoblasts, odontoblasts and fibroblasts. As a result, there is delayed or impaired wound healing as well as defective bone and tooth formation in the young when ascorbic acid is withheld from the diet.

The fact that the onset of scurvy in man is slow and that it is rare that the diet will be completely deficient in ascorbic acid, makes frank scurvy relatively uncommon. On the other hand, subacute, latent or incipient scurvy are of much greater significance. Here, too, the changes in connective tissue, the bones, teeth and capillaries can be observed microscopically and the hemorrhagic tendency is noticeable.

Scurvy is seen in infants who are fed exclusively on cow's milk, which is naturally low in vitamin C, and whose diets are not supplemented with fruit juices or pharmaceutical preparations of vitamin C. It is manifest by symptoms of irritability, crying, loss of appetite with consequent failure to gain weight, and an aversion to moving the extremities because of pain caused by hemorrhages in the joints. Clinical signs of petechial hemorrhages due to increased capillary fragility are common. There are also typical bone changes such as enlargement or beading of the costochondral joints of the rib cage and thickening of the epiphyseal plate at the end of the long bones, which appears as a characteristic white line.

In adults, development of scurvy takes longer than in infants because there is no active growth impairment and there is some body reserve of vitamin C. It has been shown that scurvy develops only after six months on a well controlled vitamin C deficiency diet, at which time there are weight loss, weakness, lethargy, irritability and nonspecific aches and pains in joints and muscles. The hair loses its luster and the skin becomes rough and coarse, due to follicular keratosis. Under the skin, because of increased capillary fragility, there are numerous petechial and ecchymotic spots. Easy and ready hemorrhage everywhere is a characteristic finding. The complexion is sallow due to loss of blood through hemorrhages.

ORAL MANIFESTATIONS

Gingival and Periodontal Tissues. The characteristic oral manifestation of scurvy in man is an enlargement of the marginal gingivae, enveloping and almost completely concealing the teeth. The gingivae are bluish red, soft, friable, hemorrhaging spontaneously or on the slightest provocation. In edentulous areas there are no mucosal changes. The gingivae become secondarily infected by anaerobic organisms, producing an acute necrotizing ulcerative gingivitis with its characteristic punched out membranous interdental papillae and fetid breath. There is lack of periodontal support, which may make the teeth loose to the point of exfoliation. On x-ray there is evidence of interruption of the lamina dura, suggesting disturbance of periodontal collagen.

Histologically, the gingivae show a chronic inflammatory cellular infiltration, with engorged capillaries and a noteworthy lack of fibroblasts and collagen fibrils. In experimental vitamin C-deficient animals, there are destruction of the periodontal fibers, disturbance in alveolar bone formation, increased osteoclastic resorption and engorged capillaries (Boyle et al., 1937).

"Gingivitis with hemorrhagic, enlarged bluish red gingiva is considered to be a classical sign of scurvy, but gingivitis is not caused by vitamin C deficiency per se. Nor do all vitamin C deficient individuals necessarily have gingivitis. If gingivitis is present in a vitamin C deficient individual, it is caused by local irritants plus the conditioning effect of the deficiency upon the gingival response to local irritation. Gingivitis does not occur in vitamin C deficient individuals in the absence of local irritation" (Glickman, 1964).

Furthermore, there are several investigators who found no correlation between ascorbic acid plasma levels and gingivitis (Radusch, 1942; Restarski, 1944; Perlitsh et al., 1960; Glickman and Dines, 1963). The opposite viewpoint has been expressed by Kruse (1942), Stuhl (1943), Blockley and Baenziger (1942) and El-Ashery et al. (1964), who independently have suggested that gingivitis is a manifestation of a latent or subclinical vitamin C deficiency and can be improved by supplementation with this vitamin.

Teeth. The effects of vitamin C deficiency have been demonstrated histologically on the guinea pig incisor tooth, which grows 2 mm. each week. Odontoblasts atrophy, producing a haphazard irregularity in their usual orderly palisade arrangement which results in an irregular dentin or no dentin at all (Fig. 13-1). The pulp is engorged and dilated from the increased blood. A few of the odontoblasts form some isolated dentin, which becomes entrapped in the pulp. In fact, the rate of dentin formation is so closely related to vitamin C levels in the diet, it is used as a bioassay criterion, as previously mentioned (Boyle et al., 1940). There has been a claim that vitamin C levels are inversely related to dental caries (Hanke, 1932), but this has not been confirmed by either survey (Russell, 1963) or clinical trials (McBeath, 1932; or Grandison et al., 1942).

ASCORBIC ACID REQUIREMENTS

In adults, an intake of 60 to 100 mg. is

needed for saturation or near saturation of the tissues and an intake of less than 40 mg. results in rapid depletion of the tissue stores. There is some evidence that a daily intake of at least 30 mg. of the vitamin is necessary for the maintenance of healthy gums. The minimal quantity that will prevent scurvy is about 10 mg. daily for infants and slightly less for adults. However, the recommended minimal daily intakes are 30 mg. for infants, 80 mg. for adolescents and 75 mg. for adults. During pregnancy the intake should be increased to 100 mg. and during lactation to 150 mg. The usual requirement appears to be increased by trauma, infections and other stress conditions.

TOXICITY

Massive intakes of ascorbic acid by ingestion by mouth are harmless. This is undoubtedly due to its apparently limited storage by the body and its rapid excretion in the urine.

REFERENCES

General

Aschoff, L. A., and Koch, W.: Eine pathologische-anatomische Studie. Fischer, Jena, 1919.

Boyle, P. E., Bessey, O. A., and Howe, P. R.: Rate of dentin formation in incisor teeth of guinea pigs on normal and on ascorbic acid-deficient diets. Arch. of Path., *30*:90, 1940.

Bradfield, J. R. G., and Kodiek, E.: Abnormal mucopolysaccharide and "precollagen" in vitamin C-deficient skin wounds. J. Biol. Chem. (Proceedings of the Biochemical Society), *49*:17, 1951.

Chmuchalova, B., and Chvapil, M.: IVth National Collagen Symp. Hlubaka, Czech., 1964.

Edwards, L. C., and Dunphy, J. E.: Wound healing. I. Injury and abnormal repair, New Eng. J. Med., *259*:224, 1958; II. Injury and abnormal repair, *259*:275, 1958.

Gould, B. S.: Biosynthesis of collagen. J. Biol. Chem., *232*:637, 1958.

Gould, B. S.: Ascorbic acid-independent and ascorbic acid-dependent collagen-forming mechanisms. Ann. New York Acad. Sci., *92*:168, 1961.

Gould, B. S., Manner, G., Goldman, H. M., and Stolman, J. M.: Some aspects of collagen formation. Ann. New York Acad. Sci., *85*:385, 1960.

Gould, B. S., and Woessner, J. F.: The influence of ascorbic acid on the proline, hydroxyproline, glycine and collagen content of regenerating guinea pig skin. J. Biol. Chem., *226*:289, 1957.

Highberger, J. H., Gross, J., and Schmitt, F. O.: In-

teraction of mucoprotein with sol. collagen. Proc. Nat. Acad. Sci., *37*:286, 1951.

Höjer, J. A.: Studies in scurvy. Acta Paediat (supp), *3*:1, 1924.

Holst, A., and Froelich, T.: Experimental studies relating to ship beri-beri and scurvy. J. Hyg., *7*:634, 1907.

Hunt, A. H.: Role of vitamin C in wound healing. Brit. J. Surg., *28*:436, 1941.

Jackson, L.: Experimental scurvy. J. Infect. Dis., *19*:478, 1916.

Kodicek, E., and Loewi, G.: Uptake of (S[35]) sulphate by mucopolysaccharides of granulating tissue. Proc. Roy. Soc. London, *144*:100, 1955.

Neuberger, A., Perrone, J. C., and Slack, H. G. B.: The relative metabolic inertia of tendon collagen in the rat. Biochem. J., *49*:199, 1951.

Orekhovich, V. N., Tustanovskiĭ, A. A., Orekhovich, K. D., and Plotnikova, N. E.: The procollagen of hide. Biokhimiya, *13*:55, 1948.

Orekhovich, V. N., Tustanovskiĭ, A. A., and Plotnikova, N. E.: Isolation of cryst. proteins of a new type (procollagen) from various organs of the vertebrates. Doklady Akad Nauk SSSR, *60*:837, 1948 b.

Peterkofsky, B., and Udenfriend, S.: Enzymatic hydroxylation of proline in microsomal polypeptide leading to formation of collagen. Proc. Nat. Acad. Sci., *53*:335, 1965.

Pirani, C. L., and Levenson, S. M.: Effect of vitamin C deficiency on healed wounds. Proc. Soc. Exp. Biol. Med., *82*:95, 1953.

Prockop, D. J., and Juva, K.: Synthesis of hydroxyproline in vitro by the hydroxylation of proline in a precursor of collagen. Proc. Nat. Acad. Sci., *53*:661, 1965.

Robertson, W. van V.: Influence of ascorbic acid on N[15] incorporation into collagen in vivo. J. Biol. Chem., *197*:495, 1952.

Roe, J. H., and Kuether, C. A.: The determination of ascorbic acid in whole blood and urine through the 2, 4-dinitrophenylhydrazine derivative of dehydroascorbic acid. J. Biol. Chem., *143*:399, 1943.

Ross, R., and Benditt, E. P.: Wound healing and collagen formation. J. Biophys. Biochem. Cytol., *11*:677, 1961.

Walter, R.: A Voyage 'Round the World. London, J & P Knapton, 1746.

Wolbach, S. B., and Howe, P. R.: Intercellular substances in experimental scorbutus. Arch. Path. Lab. Med., *1*:1, 1926.

Zilva, S. S., and Wells, F. M.: Changes in the teeth of guinea pig produced by a scorbutic diet. Proc. Roy. Soc., *90*:505, 1919.

Oral Relevance

Blockley, C. H., and Baenziger, P. E.: Investigation into the connection between the vitamin C content of blood and periodontal disturbances. Brit. Dent. J., *73*:57, 1942.

Boyle, P. E., Bessey, O. A., and Howe, P. R.: Rate of dentine formation in incisor teeth of guinea pigs on normal and on ascorbic acid-deficient diets. Arch. Path., *30*:90, 1940.

Boyle, P. E., Bessey, O. A., and Wolbach, S. B.: Experimental production of diffuse alveolar bone atrophy type of periodontal disease by diets deficient in ascorbic acid. J. Amer. Dent. Assn., 24:1768, 1937.

El-Ashery, G. M., Ringsdorf, U. M., and Cheraskin, E.: Local and systemic influences in periodontal disease: IV. Effect of prophylaxis and natural versus synthetic vitamin C upon clinical tooth mobility. Int. J. Vit. Res., 34:202, 1964.

Glickman, I.: Nutrition in the prevention and treatment of gingival and periodontal disease. J. Dent. Med., 19:179, 1964.

Glickman, I., and Dines, M. M.: Effect of increased ascorbic acid blood levels on the ascorbic acid level in treated and non-treated gingiva. J. Dent. Res., 42:1152, 1963.

Grandison, W. B., Cruickshank, D. B., and Stott, L. B.: Investigation into the influence of synthetic vitamin C as a controlling factor in the incidence of dental caries in already calcified teeth. Brit. Dent. J., 72:237, 1942.

Hanke, M. T.: Diet as a factor in the control of gingivitis: a report on nutritional studies at Moosehart, Ill. J. Dent. Res., 12:518, 1932.

Kruse, H. D.: Gingival manifestation of avitaminosis C. Milbank Mem. Fund Quart., 20:290, 1942.

McBeath, E. C.: Experiments on the dietary control of dental caries in children. J. Dent. Res., 12:723, 1932.

Perlitsh, M., Nielsen, A. G., and Stanmeyer, W. R.: Ascorbic acid levels and gingival health in personnel wintering over in Antarctica. J. Dent. Res., 40:789, 1961.

Radusch, D. F.: Vitamin C therapy in periodontal disease. J. Amer. Dent. Assn., 29:1652, 1942.

Restarski, J. S., and Pijoan, M.: Gingivitis and vitamin C. J. Amer. Dent. Assn., 31:1323, 1944.

Russell, A. L.: International Nutrition Surveys: A summary of preliminary dental findings. J. Dent. Res., 42:233, 1963.

Stuhl, F.: Vitamin C subnutrition in gingivo-stomatitis. Lancet, 1:640, 1943.

Section Two

Food

Chapter Fourteen

Standards of Dietary Adequacy in Terms of Nutrients and Foods

ABRAHAM E. NIZEL, D.M.D.

Nutrition is a science that deals with quantity as well as quality. How much of a nutrient is necessary for good health? What are minimal nutritional requirements? What are recommended allowances? How large a reserve of each nutrient is desirable?

To answer these questions the nutritional scientist has developed a set of dietary standards which can serve as a guide and as a basis of comparison to measure quantitatively the adequacy of a patient's food intake. It is important to understand that dietary standards have been developed from a study of the needs of population groups. Since the needs vary so greatly from patient to patient, not fulfilling the stipulated desirable level of food intake does not necessarily mean that the individual is nutritionally deficient. Nutritional health for an individual can be fully evaluated only if, in addition to past nutrient intake, consideration is given to clinical signs and symptoms, medical history, nutritional biochemical laboratory tests and the therapeutic response of a patient to a food or nutrient that is suspected to be the cause of the deficiency.

Several different standards of dietary adequacy have been formulated by various official bodies. For example, the World Health Organization and the Food and Agricultural Organization of the United Nations have set up international guides of dietary allowances and standards for adults. (See Table 14-1.)

Obviously, there are notable differences between each of these standards. This should be expected because the objectives and interests of each of the population groups are different since they live under different climatic conditions, have different average body weights and heights, and engage in different activities. For example, the national dietary standard for the Canadians (Table 14-2) is planned to meet minimal requirements to prevent deficiencies, whereas the British designate an average intake as adequate.

In the early 1930's a working pattern of amounts of food required daily for individuals of different ages, sexes and activities was set up by Dr. Stiebling, now Director of the Institute of Home Economics in the United States Department of Agriculture. A few years later she reported on a survey of the diets of families in the United States which must have shown them to be so inadequate that it aroused concern and spurred the President to call a national nutrition conference (Leverton, 1959).

At about the same time the National Academy of Sciences–National Research Council decided to form a Food and Nutrition Board consisting of outstanding experts and knowledgeable leaders in the science of nutrition. It was the function of this group to advise the government on matters of food and nutrition. One of their first tasks was to formulate a standard of dietary adequacy, which was designated as Recommended Dietary Allowances. By reviewing the scientific information and research available at that time, they made a judgment and a recommendation on the amounts of calories and nine nutrients that healthy individuals need for good health when subjected to the usual everyday stresses. These

TABLE 14-1.   COMPARATIVE DIETARY STANDARDS FOR ADULTS IN SELECTED COUNTRIES AND FAO WITH EXPLANATIONS AS TO THEIR MEANING

The purpose for establishing a national dietary standard is not the same in all countries. Therefore, some variation in nutrient allowances from country to country is to be expected. At the same time, it must be recognized that the "reference" individual will vary from country to country. Furthermore, even in instances when there are presumed similar objectives among countries as to the purpose and usefulness of proposed standards, it can be seen that there is by no means uniform agreement as to the nutrient allowances considered desirable as national guides. Standards are also subject to revision as newer knowledge becomes available. Particular attention should be paid to the footnotes, which explain, in brief form, the basis for nutrient allowances in the various countries and those of FAO. The original publications should be consulted for detailed explanations. The Board is indebted to Dr. L. A. Maynard for the preparation of this table.

Country	Sex	Age Years	Weight kg.	Activity	Calories	Protein gm.	Calcium gm.	Iron mg.	Vitamin A Activity I.U.	Thiamine mg.	Riboflavin mg.	Niacin Equiv. mg.	Ascorbic Acid mg.
U.S.A.[1]	M	18-35	70	Footnote[2]	2,900	70	0.8	10	5,000	1.2	1.7	19	70
	F	18-35	58	Footnote[2]	2,100	58	0.8	15	5,000	0.8	1.3	14	70
FAO[1]	M	25	65	Footnote[2]	3,200	43[3]	0.4-0.5[4]	10					
	F	25	55	Footnote[2]	2,300	36[3]	0.4-0.5[4]	12					
Australia[1]	M	25	65	Footnote[2]	2,700	65	0.7	10	2,500[3]	1.1	1.6	18[4]	30
	F	25	55	Footnote[2]	2,300	55	0.6	12	2,000[3]	0.9	1.4	15[4]	30
Canada[1]	M	25	72	Footnote[2]	2,850	50[3]	0.5	6	3,700[4]	0.9	1.4	9	30
	F	25	57	Footnote[2]	2,400	39[3]	0.5	10	3,700[4]	0.7	1.2	7	30
Central America and Panama[1]	M	25	55	Moderate Work	2,700	55	0.7	10	4,333[2]	1.4	1.4	14	50
	F	25	50	Moderate Work	2,000	50	0.7	10	4,333[2]	1.0	1.2	10	45
India[1]	M	25.4	55	Moderate Work[2]	2,800	55[3]	0.7	10					
	F	21.5	45	Moderate Work[2]	2,300	45[3]	0.7	10					
Japan[1]	M	Footnote[2]	56	Moderate Work[3]	3,000	70[4]	0.6	10	2,000[5]	1.5	1.5	15	65
	F	Footnote[2]	48.5	Moderate Work[3]	2,400	60[4]	0.6	10	2,000[5]	1.2	1.2	12	60
Netherlands[1]	M	20-29	70	Light Work	3,000	70[2]	1.0	10	5,500[3]	1.2	1.8	12	50
	F	20-29	60	Light Work	2,400	60[2]	1.0	12	5,500[3]	1.0	1.5	10	50
Norway[1]	M	25	70	None Given	3,400	70	0.8	12	2,500[2]	1.7	1.8	17	30
	F	25	60	None Given	2,500	60	0.8	12	2,500[2]	1.3	1.5	13	30
The Philippines[1]	M	None Specified	53	Moderate Work	2,600	55	0.7	6	4,000[2]	1.6	1.4[3]	16	75
	F	None Specified	45	Moderate Work	2,300	45	0.7	10	4,000[2]	1.4	1.1[3]	14	70
South Africa[1]	M	None Specified	73	Moderate Work	3,000	65	0.7	9	4,000[2]	1.0	1.6	15	40
	F	None Specified	60	Moderate Work	2,300	55	0.6	12	4,000[2]	0.8	1.4	12	40
United Kingdom[1]	M	20 up	65	Medium Work[2]	3,000	87[3]	0.8	12	5,000[4]	1.2	1.8	12	20
	F	20 up	56	Medium Work[2]	2,500	73[3]	0.8	12	5,000[4]	1.0	1.5	10	20
U.S.S.R.	M and F			Moderate Work					5,000[2]	2.0[3]	2.5	15	70[3]

U.S.A.:

[1] Source: Recommended Dietary Allowances, Revised 1963. NAS-NRC Publ. 1146. Washington (1964).

[2] Allowances are intended for persons normally active in a temperate climate.

[3] Niacin equivalents include dietary sources of the preformed vitamin and the precursor, tryptophan. 60 mg tryptophan represents 1 mg niacin.

FAO:

[1] Source: Calorie Requirements, FAO. Nutritional Studies, No. 15, Rome (1957). Protein Requirements, FAO. Nutritional Studies, No. 16, Rome (1957). Calcium Requirements, FAO. Nutrition Meetings Report Series No. 30, Rome (1962).

[2] The activity for the reference man is described as "on each working day he is employed 8 hours in an occupation which is not sedentary, but does not involve more than occasional periods of hard physical labor. When not at work, he is sedentary for about 4 hours daily and may walk for up to 1½ hours. He spends about 1½ hours on active recreations and household work." The activity of the reference woman is described as "she may be engaged either in general household duties or in light industry. Her daily activities include walking for about 1 hour and 1 hour of active recreation, such as gardening, playing with children, or non-strenuous sport."

[3] The protein value is defined as a safe practical allowance and is based on an average minimum requirement for a reference protein: increased by 50% to allow for individual variability, and by a further percentage in accordance with the estimated protein score of the protein of the diet. The values given in the table are for a diet similar to that of the USA, using a coefficient of 1.25 to allow for differences in protein quality, thus arriving at an allowance of 0.66 gm. per kilogram body weight.

[4] The value is considered a safe practical allowance. A range is given to emphasize that present knowledge does not permit any greater accuracy as to a safe allowance.

Australia:

[1] Dietary allowances for Australia, 1961 revision, Med. Journal Australia, Dec. 30, 1961. The allowances are designed to be used as a basis for planning food supplies for persons or groups.

[2] The activities specified are similar to those of the reference man and woman (Calorie Requirements, FAO Nutritional Studies No. 15, Rome). Mean annual external temperature, 18° C.

[3] Three I.U. of carotene equivalent to one I.U. of vitamin A activity.

[4] Preformed niacin plus (gms. of protein x 0.16).

Canada:

[1] "Recommended daily intakes of nutrients adequate for the maintenance of health among the majority of Canadians." Issued 1963.

[2] Five categories of activity are listed and described. The values for calories and nutrients here given apply to "most household chores," "office work," "laboratory work," "shop and mill work," "mechanical trades or crafts," various sports.

[3] Based on normal mixed Canadian diets.

[4] Based on mixed Canadian diet supplying both vitamin A and carotene. As preformed vitamin A the suggested intake would be two-thirds of amounts indicated.

Central America and Panama:

[1] Institute of Nutrition of Central America and Panama (INCAP), Boletin de la Oficina Sanitaria Panamericana, Supplemento No. 2, Noviembre, 1955, page 225. The figures for nutrients are designed to meet the needs of all individuals. Average annual temperature, 20°C., activity not defined.

[2] This figure is given in the INCAP table as 1.3 mg. of vitamin A and assumes that two-thirds of the activity is supplied from vegetable sources. International units shown were derived by converting as follows: one I.U. = 0.0003 mg. vitamin A alcohol.

India:

[1] Patwardhan, V. N., Dietary allowances for India. Calories and Protein. Indian Council of Medical Research, Special Report Series No. 35, New Delhi, 1960. The data are 1958 revisions of earlier figures.

[2] The activities corresponding to the calorie recommendations are detailed in the above publication. "Moderate" refers to activity in a "light industrial occupation."

[3] "An allowance of one gm. of protein per kilogram body weight of vegetable proteins in properly balanced diets."

Japan:

[1] Nutrition in Japan, 1962, Ministry of Health and Welfare, Tokyo, 1962. Data adopted by the Council on Nutrition in 1960. The allowances are believed to be sufficient to establish and maintain a good nutritional state in typical individuals.

[2] Age are not specified.

[3] Five categories of activity are specified for men and four for women, with corresponding intakes for calories and B-vitamins.

[4] Higher intakes are specified for heavy and very hard work.

[5] Requirement for both sexes specified as 2000 I.U. of preformed vitamin A or 6000 I.U. of carotene.

Netherlands:

[1] Recommended quantities of nutrients, Committee on Nutritional Standards of the Nutrition Council Voeding 22, 210-214, 1961. The figures for nutrients are set to cover individuals having high requirements. The figures for calories are average requirements. Figures are increased for heavy and very heavy work.

[2] Vitamin A as present in animal foods.

[3] Assumes 1500 I.U. as preformed vitamin A and 4000 I.U. of activity as carotene.

Norway:

[1] Evaluation of nutrition requirements, State Nutrition Council, 1958. Figures are "somewhat higher than average requirements."

[2] Assumes two-thirds contributed by carotene.

The Philippines:

[1] Recommended daily allowances for specific nutrients, Food and Nutrition Research Center, 1960. "Objectives toward which to aim in planning practical diets."

[2] Assumes two-thirds contributed by carotene.

[3] Grams protein x .025.

South Africa:

[1] Recommended minimum daily dietary standards, National Research Council, S. A. Med. J., 30: 108, 1956.

[2] Assumes two-thirds contributed by carotene.

United Kingdom:

[1] Report of the Committee on Nutrition, British Medical Association, 1950. The levels of nutrients recommended are believed to be sufficient to establish and maintain a good nutritional state in representative individuals or groups concerned.

[2] Values are given for 6 levels of activity for males and 5 for females. Medium work is described as 8 hours at 100 Calories per hour and traveling (130 Calories per hour).

[3] The protein allowance is increased with calories on the basis that the protein in the diet should provide not less than 11 per cent of the energy for adults not engaged in hard work.

[4] A mixed diet containing one-third vitamin A and two-thirds carotene.

U.S.S.R.:

[1] New daily vitamin supply standards in man, 1961, Yarusova, N. S., Vop, Pitan, 20: 3, 1961.

[2] I.U. is equivalent to 0.3 μg. of natural vitamin.

[3] To be increased up to 50% in far north.

allowances have been re-examined five times, in 1945, 1948, 1953, 1958 and 1963. They have served as the basis for planning adequate diets and food supplies and for checking the nutritive value of foods consumed. (See Table 14-3.)

Actually, the major objective of the recommended allowances is to serve as a yardstick in assuring the intake of the proper types and amounts of nutrients for maintaining good health and preventing disease (Food and Nutrition Board, 1964).

In the words of the report: "The allowances are designed to afford a margin of sufficiency

above average physiological requirements to cover variations among essentially all individuals in the general population. They provide a buffer against the increased needs during common stress and permit full realization of growth and reproductive potential; but they are not to be considered adequate to meet additional requirements of persons depleted by disease or traumatic stress. On the other hand, the allowances are generous with respect to temporary emergency feeding of large groups under conditions of limited food supply and physical disaster."

TABLE 14-2.   CANADIAN DIETARY STANDARD, 1963

(For Adults)

(The figures in the table give the maintenance allowances for the body weights shown plus additional needs for activity as indicated.)

WEIGHT (LBS.)	ACTIVITY[1]	CALORIES	PROTEIN[2] G.	CALCIUM G.	IRON MG.	VITAMIN A[3] I.U.	VITAMIN D I.U.	ASCORBIC ACID MG.	THIAMINE MG.	RIBOFLAVIN MG.	NIACIN MG.
MALES:											
144	Maintenance	2150	46	0.5	6	3700	—	30	0.6	1.1	6
	A	2650	46	0.5	6	3700	—	30	0.8	1.3	8
	B	3400	46	0.5	6	3700	—	30	1.0	1.7	10
	C	4000	46	0.5	6	3700	—	30	1.2	2.0	12
	D	4600	46	0.5	6	3700	—	30	1.4	2.3	14
158	Maintenance	2300	50	0.5	6	3700	—	30	0.7	1.2	7
	A	2850	50	0.5	6	3700	—	30	0.9	1.4	9
	B	3650	50	0.5	6	3700	—	30	1.1	1.8	11
	C	4250	50	0.5	6	3700	—	30	1.3	2.1	13
	D	4900	50	0.5	6	3700	—	30	1.5	2.5	15
176	Maintenance	2500	55	0.5	6	3700	—	30	0.8	1.3	8
	A	3100	55	0.5	6	3700	—	30	0.9	1.5	9
	B	3950	55	0.5	6	3700	—	30	1.2	2.0	12
	C	4600	55	0.5	6	3700	—	30	1.4	2.3	14
	D	5350	55	0.5	6	3700	—	30	1.6	2.7	16
FEMALES:											
111	Maintenance	1750	35	0.5	10	3700	—	30	0.5	0.9	5
	A	2200	35	0.5	10	3700	—	30	0.7	1.1	7
	B	2800	35	0.5	10	3700	—	30	0.8	1.4	8
	C	3300	35	0.5	10	3700	—	30	1.0	1.7	10
	D	3800	35	0.5	10	3700	—	30	1.2	1.9	12
124	Maintenance	1900	39	0.5	10	3700	—	30	0.6	1.0	6
	A	2400	39	0.5	10	3700	—	30	0.7	1.2	7
	B	3000	39	0.5	10	3700	—	30	0.9	1.5	9
	C	3550	39	0.5	10	3700	—	30	1.1	1.8	11
	D	4100	39	0.5	10	3700	—	30	1.2	2.0	12
136	Maintenance	2050	43	0.5	10	3700	—	30	0.6	1.0	6
	A	2550	43	0.5	10	3700	—	30	0.8	1.3	8
	B	3250	43	0.5	10	3700	—	30	1.0	1.6	10
	C	3800	43	0.5	10	3700	—	30	1.1	1.9	11
	D	4400	43	0.5	10	3700	—	30	1.3	2.2	13
Pregnancy—during 3rd trimester add		up to 500	10	0.7	3	500	400	10	0.15	0.25	1.5
Lactation—add		500 to 1000	10–20	0.7	3	1500	400	20	0.3	0.5	3

1. A = Sedentary; B = Moderate; C = Heavy; D = Very Heavy.

2. Protein recommendation is based on normal mixed Canadian Diet. Vegetarian diets may require a higher protein content.

3. Vitamin A is based on the mixed Canadian diet supplying both vitamin A and carotene. As the preformed vitamin A the suggested intake would be about two-thirds of that indicated.

TABLE 14-2.   CANADIAN DIETARY STANDARD, 1963 *(Continued)*

(For Girls and Boys)

(The figures in the table give the maintenance allowances for body weights
shown plus additional needs for activity.)

SEX	AGE YRS.	WEIGHT LBS.	ACTIVITY CATEGORY	CALORIES	PROTEIN[1] G.	CAL- CIUM G.	IRON MG.	VITA- MIN A[2] I.U.	VITA- MIN D I.U.	ASCOR- BIC ACID MG.	THIA- MINE MG.	RIBO- FLAVIN MG.	NIA- CIN MG.
Both	0–1	7–20	Usual	360–900	12–24	0.5	5	1000	400	20	0.3	0.5	3
Both	1–2	20–26	Usual	900–1200	25–30	0.7	5	1000	400	20	0.4	0.6	4
Both	2–3	31	Usual	1400	30	0.7	5	1000	400	20	0.4	0.7	4
Both	4–6	40	Usual	1700	30	0.7	5	1000	400	20	0.5	0.9	5
Both	7–9	57	Usual	2100	40	1.0	5	1500	400	30	0.7	1.1	7
Both	10–12	77	Usual	2500	50	1.2	12	2000	400	30	0.8	1.3	8
Boy	13–15	108	Usual	3100	75	1.2	12	2700	400	30	0.9	1.6	9
Girl	13–15	108	Usual	2600	75	1.2	12	2700	400	30	0.8	1.3	8
Boy	16–17	136	B[3]	3700	55	1.2	12	3200	400	30	1.1	1.9	11
Girl	16–17	120	A[4]	2400	50	1.2	12	3200	400	30	0.7	1.2	7
Boy	18–19	144	B[3]	3800	60	0.9	6	3200	400	30	1.1	1.9	11
Girl	18–19	124	A[4]	2450	50	0.9	10	3200	400	30	0.7	1.2	7

1. Protein recommendation is based on normal mixed Canadian diet. Vegetarian diets may require a higher protein content.
2. Vitamin A is based on the mixed Canadian diet supplying both vitamin A and carotene. As the preformed vitamin A the suggested intake would be about two-thirds of that indicated.
3. Expenditure assessed as being 113% of that of a man of same weight and engaged in same degree of activity. A = Sedentary; B = Moderate.
4. Expenditure assessed as being 104% of that of a woman of the same weight and engaged in the same degree of activity. A = Sedentary; B = Moderate.

(From "Dietary Standard for Canada," Canadian Bulletin on Nutrition, Vol. 6, No. 1. Ottawa, Canada, Department of National Health and Welfare, Nutrition Division.)

The Recommended Dietary Allowances (R.D.A.) should not be confused with Minimum Daily Requirements (M.D.R.) (1951), established for labeling purposes by the Food and Drug Administration. The former, in most instances (calories excepted), is at least twice the minimum physiological requirement to prevent symptoms of deficiency. Minimum Daily Requirements are discussed in more detail on page 191.

In the 1963 revision of the National Research Council R.D.A. there was a downward adjustment of allowances for calories, thiamine, riboflavin and niacin and an upward adjustment in the iron allowance for some segments of the population. The tabulation includes those nutrients for which sufficient quantitative information is available to permit breakdown by specified age categories. There are a number of nutrients, carbohydrates, fats, water, phosphorus, copper, vitamins B_6 and B_{12}, pantothenic acid, vitamin K and others that are not tabulated in these recommended dietary allowances either because quantitative allowances cannot be stated or because the nutrient is amply provided by an ordinary mixed diet. There is no need to give special attention to these particular nutrients if the diet includes a wide variety of foods.

CALORIE ALLOWANCES

Age, sex, body size, climate, physical activity as well as physiological stresses such as growth, pregnancy and lactation influence the recommended allowance for calories. Allowances have been calculated for a reference 70 kg. man or 58 kg. woman, moderately active, 25 years of age and living in a temperate climate with an average mean temperature of 20°C. To maintain energy balance the expenditure for the reference man should be 2900 calories and for the reference woman, 2100 calories.

TABLE 14·3.

Food and Nutrition Board, National Academy of Sciences—National Research Council

Recommended Daily Dietary Allowances,[1] Revised 1963

DESIGNED FOR THE MAINTENANCE OF GOOD NUTRITION OF PRACTICALLY ALL HEALTHY PERSONS IN THE U.S.A.

(Allowances are intended for persons normally active in a temperate climate)

	AGE[2] YEARS	WEIGHT KG. (LBS.)	HEIGHT CM. (IN.)	CALORIES	PROTEIN GM.	CALCIUM GM.	IRON MG.	VITAMIN A I.U.	THIAMINE MG.	RIBOFLAVIN MG.	NIACIN EQUIV.[3] MG.	ASCORBIC ACID MG.	VITAMIN D I.U.
Men.......	18-35	70 (154)	175 (69)	2900	70	0.8	10	5000*	1.2	1.7	19	70	
	35-55	70 (154)	175 (69)	2600	70	0.8	10	5000	1.0	1.6	17	70	
	55-75	70 (154)	175 (69)	2200	70	0.8	10	5000	0.9	1.3	15	70	
Women....	18-35	58 (128)	163 (64)	2100	58	0.8	15	5000	0.8	1.3	14	70	
	35-55	58 (128)	163 (64)	1900	58	0.8	15	5000	0.8	1.2	13	70	
	55-75	58 (128)	163 (64)	1600	58	0.8	10	5000	0.8	1.2	13	70	
	Pregnant (2nd and 3rd trimesters)			+200	+20	+0.5	+5	+1000	+0.2	+0.3	+3	+30	400
	Lactating			+1000	+40	+0.5	+5	+3000	+0.4	+0.6	+7	+30	400
Infants[4]....	0-1	8 (18)		kg. × 115 ±15	kg. × 2.5 ±0.5	0.7	kg. × 1.0	1500	0.4	0.6	6	30	400
Children...	1-3	13 (29)	87 (34)	1300	32	0.8	8	2000	0.5	0.8	9	40	400
	3-6	18 (40)	107 (42)	1600	40	0.8	10	2500	0.6	1.0	11	50	400
	6-9	24 (53)	124 (49)	2100	52	0.8	12	3500	0.8	1.3	14	60	400
Boys.......	9-12	33 (72)	140 (55)	2400	60	1.1	15	4500	1.0	1.4	16	70	400
	12-15	45 (98)	156 (61)	3000	75	1.4	15	5000	1.2	1.8	20	80	400
	15-18	61 (134)	172 (68)	3400	85	1.4	15	5000	1.4	2.0	22	80	400
Girls.......	9-12	33 (72)	140 (55)	2200	55	1.1	15	4500	0.9	1.3	15	80	400
	12-15	47 (103)	158 (62)	2500	62	1.3	15	5000	1.0	1.5	17	80	400
	15-18	53 (117)	163 (64)	2300	58	1.3	15	5000	0.9	1.3	15	70	400

[1] The allowance levels are intended to cover individual variations among most normal persons as they live in the United States under usual environmental stresses. The recommended allowances can be attained with a variety of common foods, providing other nutrients for which human requirements have been less well defined.

[2] Entries on lines for age range 18-35 years represent the 25-year age. All other entries represent allowances for the midpoint of the specified age periods, i.e., line for children 1-3 is for age 2 years (24 months); 3-6 is for age 4½ years (54 months), etc.

[3] Niacin equivalents include dietary sources of the preformed vitamin and the precursor, tryptophan. 60 mg. tryptophan represents 1 mg. niacin.

[4] The caloric and protein allowances per kg. for infants are considered to decrease progressively from birth. Allowances for calcium, thiamine, riboflavin and niacin increase proportionately with calories to the maximum values shown.

*1000 I.U. from preformed vitamin A and 4000 I.U. from beta-carotene.

TABLE 14-4.   ADJUSTMENT OF CALORIE ALLOWANCES FOR ADULT INDIVIDUALS OF VARIOUS BODY WEIGHTS AND AGES*

(At a mean environmental temperature of 20° C. [68° F.] assuming average physical activity)

| DESIRABLE WEIGHT | | CALORIE ALLOWANCE† | | |
kgs.	(lbs.)	25 years	45 years	65 years
Men		(1)	(2)	(3)
50	110	2300	2050	1750
55	121	2450	2200	1850
60	132	2600	2350	1950
65	143	2750	2500	2100
70	154	2900	2600	2200
75	165	3050	2750	2300
80	176	3200	2900	2450
85	187	3350	3050	2550
Women		(4)	(5)	(6)
40	88	1600	1450	1200
45	99	1750	1600	1300
50	110	1900	1700	1450
55	121	2000	1800	1550
58	128	2100	1900	1600
60	132	2150	1950	1650
65	143	2300	2050	1750
70	154	2400	2200	1850

Formulas

(1) $725 + 31W$ (2) $650 + 28W$ (3) $550 + 23.5W$
(4) $525 + 27W$ (5) $475 + 24.5W$ (6) $400 + 20.5W$
W = weight in kg.

* Taken from Recommended Dietary Allowances (Revised 1963), Food and Nutrition Board, National Academy of Sciences—National Research Council.
† Values have been rounded to nearest 50 calories.

(Goldsmith, G.: Med. Clin. N. Amer., *48*:1117, 1964.)

Calculation for the several variables mentioned above can be made as follows:

Age.   Reduce 5 per cent per decade between 35 and 55.

Reduce 8 per cent per decade between 55 and 75.

Reduce 10 per cent more beyond 75.

Body Size.   Take desirable weight in kilograms and apply the formula
Energy needs for man = $152 \times$ weight 0.73.
Energy needs for woman = $123.4 \times$ weight 0.73

Table 14-4 gives a simplified form for adjustment of calorie allowances for adult individuals of various body weights and ages who live at a mean environmental temperature of 20°C.

Climate.   During the cold weather a 2 to 5 per cent increase in calorie allowance will compensate for the extra energy expended in carrying warm clothing.

Recent studies indicate that, under conditions of physical activity at high temperature, calorie needs are increased; for each degree of temperature rise between 30 and 40°C., calorie allowances should be increased 0.5 per cent.

Activity.   The proper calorie allowance for an individual is that which maintains body weight and health at a level most conducive to his well-being. Unfortunately, inactive persons tend not to reduce calorie intake sufficiently to remain in energy balance; on the other hand, as a rule, physically active people do increase their caloric intake adequately.

Pregnancy and Lactation.   An increased

intake of about 200 cal./day during the second and third trimesters of pregnancy is recommended. An additional 100 cal./day is recommended during lactation, providing 120 cal. for each 100 ml. of milk produced.

Infants, Children and Adolescents. The caloric allowances for infants and children of both sexes are tabulated according to age groups. The amounts for boys and for girls after the age of nine are given separately since growth curves and levels of activity begin to differ significantly. It must be kept in mind that the allowances proposed here are averages for a group. More appropriate allowances for individual children may be derived by observation of their own growth rate, appetite and body build.

PROTEIN ALLOWANCES

Protein allowances for normal adults have been made on the basis of one gram per day for each kilogram of desirable weight. The allowances assure a diet adequate in content of calories and other essential nutrients.

The greatest demand for additional protein is during the latter half of pregnancy, when 20 gm. per day additional to the basal allowance is desirable. During lactation a supplement of 40 gm. per day is recommended.

The protein allowance for children after infancy is about 10 per cent of the caloric intake, 25 to 58 per cent of which should be animal or high quality protein. This provides approximately twice the minimal need for the average child and allows a reasonable margin of safety for the rapidly growing child.

MINERAL ALLOWANCES

About 15 mineral elements are required in the metabolism of the body. They are calcium, iron, sodium, potassium, chlorine, phosphorus, sulfur, magnesium, copper, iodine, manganese, molybdenum, cobalt, selenium and possibly chromium, bromine and fluorine. Since calcium and iron are the two minerals tabulated in the R.D.A., only these will be discussed in detail.

CALCIUM

The recommended dietary allowance for calcium for adults remains unchanged from the 1958 recommendation at 800 mg. per day. This amount was recommended after due consideration had been given to the urinary and endogenous fecal excretion of calcium and loss of calcium in sweat. These losses average 320 mg. per day. On this basis and assuming a 40 per cent absorption of calcium (because of dietary interferences by such substances as phytates, oxalates and fatty acids), 800 mg. is required to maintain equilibrium.

Calcium utilization and body requirement are influenced by previous levels of calcium intake, the phenomenon of adaptation and the physiological state of the individual. Proposals for lowered intakes as practical allowances may be valid, but for the sake of safety and optimal health the 800 mg. allowance should be striven for. Furthermore, the prevalence of osteoporosis in older persons in the United States attests to the need for adhering to the recommended allowance.

An allowance of an additional 500 mg. of calcium is recommended for the second and third trimesters of pregnancy and during lactation.

During the relatively slower growth period between the ages of one and nine, 0.8 gm. is adequate, and there is no physiological justification for increasing it. However, prepubertal and pubertal growth requires that the calcium allowance be increased from 50 to 75 per cent.

IRON

Special attention must be given to iron intake during the growth period of children and during the reproductive period of the adult woman. Adult males rarely develop iron deficiency anemia except as a result of pathological blood loss. Dietary factors which adversely affect iron absorption are phytates and phosphates; ascorbic acid tends to enhance its absorption. An absorption efficiency of about 10 per cent of food iron is a realistic value; therefore, a 10 mg. intake will maintain equilibrium, since there is a loss of about 1 mg. of iron daily.

VITAMIN ALLOWANCES

There are 13 vitamins that need to be supplied to maintain health—vitamins A, D, E, K, and thiamine, riboflavin, niacin, vitamins B_6, B_{12}, and folacin, pantothenic acid, biotin and ascorbic acid.

VITAMIN A

The requirement is 20 IU per kg. of body weight per day. Carotene is converted to vitamin A in the intestinal wall and approximately 2 units of β-carotene are equivalent to 1 unit of vitamin A. The average diet in the United States has been considered to provide about two thirds of its vitamin A activity as carotene and one third as the preformed vitamin. On this basis, the recommended allowance of 5000 IU for adults is considered to be provided by 4000 IU of β-carotene and 1000 IU of preformed vitamin A. This allowance is about twice the minimal requirements as shown by experiments in human subjects. An additional 1000 IU during the second and third trimesters of pregnancy and 3000 IU during lactation are recommended. Vitamin A is toxic when ingested in large amounts. Serious symptoms can arise if amounts in excess of 50,000 IU are ingested daily for six months.

VITAMIN D

Vitamin D is obtained either by ingestion or by activation of 7-dehydrocholesterol present in the skin through exposure to sunlight. Vitamin D is needed throughout the growth period. For maximum calcium retention during the school years and adolescence, 400 units daily is recommended.

Excessive quantities of vitamin D (1000 to 3000 units per day) are toxic and may lead to hypercalcemia.

Cow's milk is fortified with vitamin D and so are certain milk flavorings, cereals and margarines. Individuals consuming a normal diet

may therefore receive an appreciable amount of vitamin D in the absence of specified vitamin supplements.

For persons working at night, and for nuns and others whose habits shield them from sunlight, the ingestion of small amounts of vitamin D is desirable.

ASCORBIC ACID

A deficiency of ascorbic acid induces a marked but nonspecific lowering of the activity of many cellular and serum enzymes well in advance of gross indications of scurvy. Numerous studies have indicated that 10 mg. of ascorbic acid daily will prevent scurvy in infants and adults, and at least 30 mg. is necessary for maintaining gingival health. These low intakes are not satisfactory for the preservation of optimum health through long periods of time or when the body is subjected to common forms of stress. Saturation or near-saturation of the tissues of an adult can be maintained with an intake of 50 to 100 mg. daily.

The allowances recommended are 70 mg. daily for adults and 30 to 80 mg. daily for infants and growing children. These are not tissue saturation levels since higher intakes result in distinctly higher concentration in the tissues.

In conclusion, protein, calcium, iron and vitamin C have been discussed in detail because they are the food elements that are most often and most likely to be deficient. The two fat soluble vitamins A and D, when ingested in excess, can cause toxicity. For this reason, these two were discussed here. In a few of the chapters in Section I, particularly Chapter Twelve, there is detailed information about the recommended allowances of several other nutrients.

MINIMUM DAILY REQUIREMENTS

Although the National Research Council's Recommended Dietary Allowances are the basis for most of our nutrition teaching, mention should be made of another set of nutrient

TABLE 14-5. MINIMAL DAILY REQUIREMENTS COMPARED WITH
RECOMMENDED DIETARY ALLOWANCES

VITAMIN	M.D.R., ADULTS	R.D.A., MAN AGED 25	M.D.R., CHILDREN 6–11 YR.	R.D.A., CHILDREN 9–12 YR.
Vitamin A, U.S.P. units	4,000	5,000	3,000	4.500
Thiamine, mg.	1 0	1.2	0.75	0.9–1.0
Riboflavin, mg.	1.2	1.7	0.9	1.3–1.4
Vitamin C, mg.	30	70	20	70–80
Vitamin D, U.S.P. units	400	—	400	400

(Darby, W. J.: Med. Clin. N. Amer., 48:1205, 1964.)

standards in use in the United States. These are the Minimum Daily Requirements of Specific Nutrients prepared by the United States Food and Drug Administration.* These figures are minimum and "the levels chosen were such that amounts less than these would produce demonstrable deficiency signs." They are used primarily in labeling foods or preparations for "special dietary uses." The reader should be aware that if the label on a drug or food package indicates that it contains 25 per cent of the minimum daily requirements of this amount of nutrient, it probably is equal to only 12½ per cent of the recommended dietary allowance. The R.D.A. is our true guide for measuring adequacy of a diet. The differences in the two standards may be seen in the comparison table (Table 14-5).

FOOD GUIDES

To help the dentist and physician communicate clearly and readily with his patient about food and nutrition, food guides that are understandable and easy to follow have been developed. They express in a simple fashion the desirable nutrient intake requirements and serve as workable plans that help the ordinary person to select the kind and the amounts of foods that he should include in his meals each day in order to be adequately fed. Instead of prescribing x number of grams of protein and fat, the dietary prescription will read x number of ounces of chicken. Since the majority of the people eat and think in terms of food

rather than nutrients, these food guides make very good sense.

Essentially, in designing the guides an attempt has been made to group foods together on the basis of their similarity in composition and nutritive value. In the main, meat, fish and chicken all provide protein among other nutrients and therefore they are grouped together. The variety of food within each group is usually great enough to allow for a wide selection of favorite, economical and seasonably available foods.

There have been food guides devised for each culture and country. (See Figure 14-1.) In the United States we have had food guides with as few as two or three food groups and as many as 11 to 13 groups. The more popular ones have been the 7 Food Groups (National Food Guide, 1946) and the 4 Food Groups (Food for Fitness—A Daily Food Guide, 1958).

These 7 Food Groups include:

1. Leafy green and yellow vegetables—1 serving or more.
2. Citrus fruit, tomatoes, raw cabbage—1 serving or more.
3. Potatoes and other fruits and vegetables —2 servings or more.
4. Milk, cheese, ice cream—children, 3 to 4 cups milk; adults, 2 or more cups milk.
5. Meat, poultry, fish, eggs, dried peas and beans—1 or 2 servings.
6. Bread, flour, cereal (enriched or whole grain)—some every day.
7. Butter and fortified margarine—some daily.

In 1958 the United States Department of Agriculture published a somewhat simpler guide based on the "4 Food Groups." (See Figure 14-2.) In essence, the guide is similar to the basic 7 except that butter and fortified

* The 1966 suggested "standards of identity" for vitamin and mineral supplements as well as food fortification and labeling are given in Appendices VII and VIII.

FIGURE 14-1. *There are many ways to obtain an adequate diet: A sampling of daily food patterns from 5 continents—Africa, South America, North America, Asia, and Europe. (Courtesy U.S. Department of Agriculture, Agricultural Research Service; Newman Neame, Ltd., Great Britain; Ministerio de Prevision Social, Instituto Nacional de Nutricion; Daiichi Shuppan Co., Ltd.; La Scuola Editrice.) (Only upper part of Japanese poster reproduced.)*

margarine are not given group status and fruits and vegetables are grouped together. The 4 Food Groups suggested on this daily food guide are:

1. Milk—children, 3 to 4 cups; teen-agers, 4 or more cups; adults, 2 or more cups.
2. Meat—2 or more servings of beef, veal, lamb, poultry, fish, eggs, dried beans or peas and nuts.
3. Vegetable-fruit group—4 or more servings, to include a citrus fruit or vegetable important for vitamin C and a dark green leafy or deep yellow vegetable for vitamin A.
4. Bread-cereal group—4 or more servings of whole grain, enriched or restored.

No one of the above food groups provides all the recommended nutrients, but when the

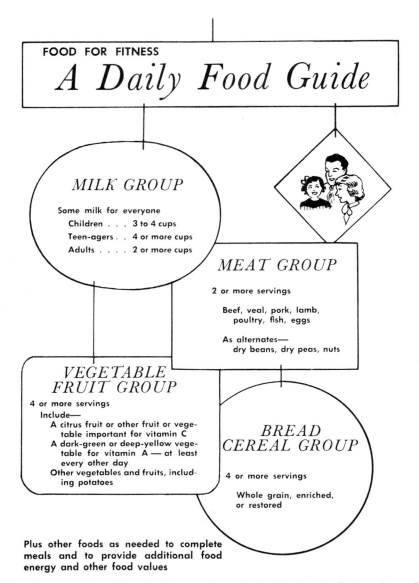

FIGURE 14-2. *The basic 4 Food Groups dietary pattern. Leaflet No. 424, Institute of Home Economics, U.S. Department of Agriculture, Washington, D.C., 1958.*

food groups, in the amounts recommended, are included in the daily diet, a person may be reasonably assured of meeting the National Research Council's Recommended Dietary Allowances for all nutrients except calories. Actually, the foods in the amounts recommended in this guide provide about 1200 calories.

These 4 Food Groups are sometimes referred to as the "protective" or "foundation"

foods. They really emphasize four key nutrients, of which three are lacking, as a rule, in the average United States diet, namely calcium, vitamin A and ascorbic acid. Protein is the fourth key nutrient which this guide stresses, not because it is found in low amounts in today's average diets, but because animal proteins tend automatically to carry with them other desirable minerals and vitamins; therefore, if the diet contains animal protein, a large share of the other essential foods will be assured.

Although the milk group is counted on for calcium, the meat group for protein, deep yellow and green vegetables for vitamin A and citrus fruits for vitamin C, each of these groups furnishes other important nutrients. For example, the milk group will provide most of the calcium, along with high quality protein, riboflavin, vitamin A and other nutrients. The meat group will supply protein, iron and the B vitamins. The vegetable-fruit group will contribute most of the vitamin C and about half of the vitamin A, as well as other minerals and vitamins. The bread-cereal group will furnish the B vitamins, iron, protein and food energy.

Since the 4 Food Groups in the amounts recommended do not meet the caloric requirement nor the taste appeal of the individual, either more of these 4 foods or "additional" (sometimes termed "other" foods) are suggested. These "additional" foods are butter, margarine, other fats, oils, sugars or unenriched refined grain products. These are often ingredients in baked goods and mixed dishes. Fats, oils and sugars are also added to foods during preparation or at the table to enhance flavor and increase appetite appeal. Even though they are expected to be part of daily meals, they are not stressed because we tend to exceed our food energy needs.

A Variety of Foods and Dishes within 4 Food Groups

MILK GROUP

Dairy products in the United States are a very important part of our dietary culture. Milk and milk products provide about two thirds of the calcium, one half of the riboflavin and one fourth of the protein in the food of families surveyed in nationwide study in 1955 (U.S.D.A. Report No. 6). It is low in ascorbic acid and iron, but it supplies more of the essential nutrients in significant amounts than any other single food. Thus, it supplements and balances incomplete foodstuffs, as, for example, using the complete protein in milk for the supplementation of the incomplete protein of cereal products

and vegetables. Whole milk and some fortified milk provide vitamins A and D. Vitamin D fortified milk is the main source of vitamin D. Milk may be taken either as whole fresh milk, evaporated milk, buttermilk, yoghurt, dry milk or cheese. Milk may be used in desserts, soups, creamed dishes or on cereals. Milk in an oyster stew has as much value as milk which is drunk from a glass. A generous slice of cheese used in a cheeseburger is equal to a glass of milk.

For the family that lives on a limited food budget, nonfat milk solids cost about one third as much as a quart of fresh milk. Nonfat milk is lacking in fat and vitamin A, but the inclusion of a dark green or deep yellow vegetable in the diet several times a week will provide an adequate supply of this vitamin. Both fat and vitamin A may be obtained economically by using fortified margarine.

Evaporated milk is also an economical food in the milk group. It may be used in cooking or in coffee or tea; not only is it less expensive than cream, it contains more calcium and protein. It is whole milk from which over half the water has been removed by an evaporating process. It differs from condensed milk in that the latter is prepared by adding about 42 per cent sugar to the milk before evaporating the water.

Those who prefer fermented milk such as buttermilk or yoghurt may choose these foods from the milk group. Buttermilk is equal to skim milk in food value; yoghurt is equal to whole milk. Despite recent propaganda, yoghurt has no special health values greater than those of whole fresh milk. Yoghurt costs more than regular milk and, unless one especially enjoys its flavor and can afford its extra cost, there is no special reason for using it. Buttermilk, on the other hand, usually costs less than regular milk. For those who enjoy it, it may well be used in place of some or all of the regular milk.

Cheese is the curd of milk separated from the whey by coagulation, and it contains most of the protein, calcium and riboflavin. Cheddar cheese is a whole milk cheese which has been cured. Processed cheese is a pasteurized cheese made by blending different cheeses and adding emulsifier. Cottage cheese is made from pasteurized skim milk. It provides high quality protein and few calories. Cream cheese is made from whole milk with cream added. Cream cheese has a high percentage of fat and vitamin A and much less protein than cottage cheese.

Since cheese and ice cream can be used to replace milk and supply calcium, it is of interest to tabulate their milk equivalent in calcium:

1 in. cube of cheddar cheese = ⅔ cup of milk

½ cup of cottage cheese = ⅓ cup of milk

2 tablespoons of cream cheese = 1 tablespoon of milk

½ cup of ice cream = ¼ cup of milk

For the amounts of milk products recommended per each age group see Figure 14-2. An average serving is one 8 oz. cup of milk or about 1 in. cube of cheddar cheese.

MEAT GROUP

The choices within the meat group are beef, lamb, veal, pork, fish, poultry, eggs, dried beans or peas and nuts.

The meat group currently supplies approximately 50 per cent of the protein and niacin, 43 per cent of the iron, and 25 per cent of the vitamin A, thiamine, riboflavin and energy intake of families in the United States.

Foods in this group are usually the most expensive item in the diet. So far as meats are concerned, cost is not related to food value. Cheaper grades and cheaper cuts are just as high in protein and iron as tenderloin of beef. In this group the organ meats (liver, heart and kidneys) deserve special mention for their high nutritional value in relation to cost. Learning to like these foods is worth the effort.

There is relatively little difference in the protein and iron content of beef, veal, lamb and pork, although pork is richer in thiamine. A family of Southern European extraction will probably choose more veal and lamb, whereas beef and pork will be used more widely by those whose background is Northern European. Fish, poultry and eggs are also complete proteins and may be used as equivalents to meat. Eggs and poultry are well accepted by many cultural groups. Fish, although not universally so well liked as eggs and chicken, is an important and economical member of the "meat" group to many people. For ex-

ample, Puerto Ricans frequently include salt codfish in their diet (Torres, 1959). Jews often use fish liberally since it is a neutral (parve) food and may be eaten with either dairy or meat meals. Only fish with scales and fins are allowed under the orthodox dietary laws (Kaufman, 1957).

Dried beans and peas are also economical foods. Although their protein is not complete, when they are used with other vegetables like corn or with milk or cheese or mixed with small amounts of meat, they may well be used to meet the complete amino acid requirements of the body. Such dishes as chili con carne, pea soup with ham, or frankfurters and beans, which combine meat and dried legumes, are commendable nutritionally and economically. In the Southwest, among the Spanish Americans, dried pinto beans are often a mainstay of the diet and may be eaten several times each day (Hacker and Miller, 1959).

To obtain full advantage from the foods in the meat group, it is recommended that a small amount of food from this group be eaten at each meal rather than a large amount at one meal. To make the most efficient use of protein foods, it is preferable to have an egg for breakfast, a fish, meat, or cheese sandwich at noon, and some meat, fish or poultry at night, rather than to have a large serving of meat, fish or poultry at only one meal and no food from this group at other meals.

Suggested amounts for a day from this group of foods are two or more servings of meat, poultry, fish or eggs, with dry beans, dry peas and nuts used occasionally as alternates.

Amounts of these foods to count as a serving are 2 to 3 ounces of lean cooked meat, poultry or fish (this amount is without bone, fat or gristle); 2 eggs; 1 cup of cooked dry beans, dry peas or lentils; 4 tablespoons of peanut butter.

VEGETABLE-FRUIT GROUP

Vegetable-fruit foods provide not only important nutrients such as 90 per cent of the ascorbic acid and 50 per cent of the vitamin A plus every other nutrient at least in trace amounts, but they also provide sensory pleasures and have laxative properties.

An assortment of fresh, canned or frozen fruits and vegetables furnishes about 215 calories for every pound eaten. Peas, lima beans, potatoes, sweet potatoes and bananas provide the highest number of calories per unit of edible material. The green-colored parts of plants, especially the leaves, tend to be highest in calcium, iron and the B vitamins. The vitamin A value of plant tissue is due to carotene and crystoxanthin, which are vitamin A precursors.

In general, the color of fruits and vegetables is the guide to food value. The deeper the yellow or the deeper the green, the better the food as a source of vitamin A; e.g., carrots are better than corn because they are yellower, or spinach is better than celery because it is greener. Specific foods that are good sources of vitamin A are apricots, broccoli, cantaloupe, carrots, chard, collards, cress, kale, mangoes, persimmons, pumpkin, spinach, sweet potatoes, turnip greens and winter squash.

Too frequently the term nutrition has been associated with eating green salads and raw vegetables. Granted these foods are excellent, nevertheless one can be well nourished and never touch a salad or raw vegetable. Fruits are an excellent substitute for vegetables. They are delicious served with a main course. For example, fried apple slices with pork, baked pears or peaches with chicken, or grilled pineapple with ham may well be used in place of a vegetable.

People who claim they do not like vegetables may change their minds when they are served fresh vegetables properly cooked in a small amount of water and delicately seasoned. Some people dislike vegetables because their only conception of a vegetable is a watery mixture of dull green peas and soft carrots.

In considering cultural patterns many adaptations may be made within the vegetable-fruit group. Scandinavians may have some of their servings from this group in the form of delicious fruit soups. Other nationalities may obtain theirs largely from vegetable soups, such as borscht (beet soup) or minestrone. Foods from this group may be used as a first course, as vegetables and salad with the main course or as dessert.

Two groups of fruits and vegetables are of particular importance: those rich in vitamin C, and those rich in vitamin A. Within these two groups, choices are wide.

Many of us rely on our daily glass of orange juice for vitamin C. However, for those in areas where the mangoes or guavas are plentiful, these may well be used. A generous serving of cabbage as a salad at lunch or dinner or a large fresh tomato may be the choice of others to fulfill the recommended serving of foods rich in vitamin C. Nor should the potato be overlooked for its contribution of ascorbic acid, especially when it is cooked in its skin. No doubt the statement that the lowly potato prevents more scurvy than the lordly orange is as true today in some areas as in former years.

When it comes to dark green or deep yellow vegetables, Southerners are more apt to use collards or mustard greens, whereas in the North people will use spinach, kale or winter squash.

From an economical standpoint the use of fresh fruits and vegetables in season is recommended. Fresh tomatoes in season are often an excellent buy, but for much of the rest of the year canned tomatoes are kinder to the food budget. This same principle applies to many other fruits and vegetables. However, there are some, like potatoes, carrots, oranges, squash, cabbage and apples, that are almost always available. Modern methods of processing, packaging and transportation have greatly increased the variety of fruits and vegetables available for year round use. Today, most of us can enjoy many frozen fruits and vegetables at all times at reasonable prices.

Many young children seem to dislike cooked vegetables but eagerly accept fruits and raw vegetables. From the standpoint of both their nutrition and their happiness, fruits and raw vegetables may well be used to provide the servings of vegetables and fruits recommended. There is no need to force cooked carrots on youngsters who will more readily accept them raw.

One half cup of a vegetable or fruit, or a portion as ordinarily served such as one medium apple, banana, orange or potato, or half a medium grapefruit or cantaloupe can count as one serving.

The food guide recommends that a dark green or deep yellow vegetable be eaten every other day. For variety see Table 14-6, which lists alternates to carrots that have an equivalent amount of vitamin A. In choosing foods

TABLE 14-6. SOURCES OF VITAMIN A
(From carotene in plants)

BEST	ABOUT HALF AS MUCH
1 raw or ⅓ cup cooked carrots	½ cup broccoli
	½ cup pumpkin
½ cup cooked greens	3 raw apricots
½ cup winter squash	1 wedge (4×8″) watermelon
½ medium sweet potato	
½ medium cantaloupe	

Smaller, but important, amounts may be obtained from asparagus, endive, tomatoes, canned apricots, yellow peaches, chili peppers.

TABLE 14-7. SOURCES OF VITAMIN C

BEST	ABOUT HALF AS MUCH
1 medium orange	1 medium tomato
½ cup orange juice	½ to ⅔ cup tomato juice
½ medium grapefruit	1 medium tangerine
½ cup grapefruit juice	1 wedge (4×8″) watermelon
½ cantaloupe	¼ raw pepper
⅔ cup fresh straw-berries	1 medium to large potato
½ cup broccoli	½ cup cooked asparagus
	½ cup cooked kale or other greens
	½ cup coleslaw

for ascorbic acid see Table 14-7 and note that it takes smaller quantities of good sources of vitamin C to fill the daily need. Therefore, it would seem expeditious to emphasize (besides orange juice) cantaloupes, grapefruit, broccoli and strawberries rather than tomato juice (unless enriched) as a source of vitamin C.

BREAD-CEREAL GROUP

Bread and cereals are the most economical source of nutrients in our daily diets. For many people they are truly the staff of life, as they provide well over 50 per cent of the food energy. A wide variety of cereal grains is available, including wheat, rice, corn, rye, oats and barley.

It should be noted that breads and cereals which are whole grain or enriched contain substantial amounts of the B vitamins and iron. The completely refined breads and cereals are much lower in these nutrients. The purpose of enrichment is to restore bread and flour to the original nutritive value of whole grain prior to refinement by adding specified amounts of iron, thiamine, riboflavin and niacin to the flour prior to baking. Thiamine is the nutrient that this food group is particularly relied upon to supply.

In milling, the bran and germ of the whole grain are removed. Bran is the hard outer layer of the grain consisting of cellulose, but more important, protein, iron, phosphorus and some B vitamins. The germ also contains some protein, vitamins and minerals. Eighty-two per cent of the whole grain kernel is endosperm, the portion of the cereal that is left after refinement and which is largely carbohydrate.

The United States Food and Drug Administration has set standards for the amounts of iron, thiamine, riboflavin and niacin which must be present if a product is to be labeled enriched. Some "dark" breads made largely of white flour with only a small amount of whole grain flour added may be of lower nutritional value than the white ones which have been enriched. Since there is no legislation controlling the enriching of "dark" bread, careful reading of labels is recommended to determine if the product is enriched or whole grain.

Cultural patterns affect choices in the bread-cereal group. In the Northern United States, wheat flour and bread are most commonly used. In the South more corn meal, grits and rice are consumed. Those of Italian background may use large amounts of spaghetti, macaroni and other pastas; those from the Far East prefer rice; and those from the Latin American countries are apt to choose corn meal and use it in tortillas.

Foods from the bread-cereal group are generally well liked; they are bland and used frequently by most people.

From a nutritional standpoint care should be taken not to use cereal foods which are unduly sweetened. Too frequently sweet rolls, doughnuts, cookies and cakes replace the recommended whole grain and enriched breads and cereals. Sweet or refined breads and cereals supply little other than carbohydrates, whereas those recommended supply iron and the B vitamins.

Today, with the current emphasis on weight control, bread is unjustly maligned as being fattening. Even for weight-conscious individuals some whole grain or enriched bread or cereals should be included in the daily diet to insure adequate amounts of the B vitamins and iron.

Counted as a serving are one ounce of ready-to-eat cereal, one half to three fourths cup of cooked cereal, corn meal, grits, macaroni, noodles, rice or spaghetti, and one slice of bread. Four servings are suggested in this group, three from bread and one from cereal. If no cereal is eaten, then five servings of bread should be eaten.

CONCLUSIONS

These 4 Food Groups suggest foods that will not only provide a balanced diet but also a varied one. Differences in taste, texture, or color of foods, as well as individual preferences, can all be dealt with nicely by using the 4 Food Groups as a foundation upon which to plan meals. Within each food group there is a wide choice of similar foods that have some individual properties that make them different enough from other members of the same group to be interesting.

To translate the basic information about foods into an actual meal plan, an arrangement such as this might be helpful:

1. Bread or cereal product at each meal; some meals might have both bread and cereal.

2. Milk for children at each meal; milk for adults usually at two of the meals.

3. One serving from vegetable-fruit group at each meal. It has become traditional, but it need not be a binding rule, to serve a citrus fruit or other source of vitamin C at breakfast. The three or more other servings from the vegetable-fruit group could be divided between the noon and evening meal.

4. A serving from the meat group is used in the evening meal and the second serving is eaten either at noon or divided between lunch and breakfast. Since we need protein at each meal, we should plan on having either meat and/or milk at each meal.

To these foundation foods, which supply only 1200 to 1300 calories, we must add more servings from the 4 Food Groups and additional foods to round out our meals and nutritional requirements.

Finally, the 4 Food Groups provide a comparatively simple guide for the dentist to use in determining the adequacy of a patient's food intake. Of equal importance is that its message can be easily communicated to the patient. The language is simple; the foods and household measures referred to in this guide are common everyday experiences. It can serve as a most effective teaching aid for improving a patient's eating habits.

REFERENCES

Food and Nutrition Board: Recommended Dietary Allowances (revised, 1963). Washington, D.C., National Academy of Sciences–National Research Council, Publication 1146.

Food for Fitness—A Daily Food Guide. Washington, D.C., U.S. Department of Agriculture, Leaflet No. 424, 1958.

Hacker, D. B., and Miller, E. S.: Food patterns of the southwest. Amer. J. Clin. Nutr., 7:224, 1959.

Kaufman, M.: Adapting therapeutic diets to Jewish food customs. Amer. J. Clin. Nutr., 5:676, 1957.

Leverton, R. M.: Recommended allowances. In Stefferud, A. (ed.). The Yearbook of Agriculture, 1959. Washington, D.C., U. S. Department of Agriculture.

Minimum Daily Requirements. Washington, D.C., U.S. Drug and Food Administration, Federal Food, Drug and Cosmetic Act. Section 403, Federal Register, June 1, 1951.

National Food Guide, Institute of Home Economics. Washington, D.C., U.S. Department of Agriculture, Leaflet No. 288, 1946.

Torres, R. M.: Dietary patterns of Puerto Rican people. Amer. J. Clin. Nutr., 7:349, 1959.

U.S. Dept. of Agriculture: Dietary Levels of Households in the United States. Household Food Consumption Survey, 1955, Report No. 6, 1957.

Chapter Fifteen

Food Patterns for Different Age Groups and Physiological States

ABRAHAM E. NIZEL, D.M.D.

All persons, throughout life, have need for the same nutrients, but in varying amounts. All persons, with the exception of infants and those on special diets for medical reasons, usually eat the same basic foods and meals. The variation in requirements between individuals is in kinds and amounts of nutrients, in food preparation and in feeding schedules. These are influenced by sex, size, activities, physiological state and age.

INFANCY

During infancy, a baby's nutritional needs are high. This is obvious when we realize that in five or six months a normal infant usually doubles his birth weight and before his first birthday has tripled it. At no other time in the life cycle is growth so rapid, unless it is during the preceding nine months of intrauterine life. Actually, the infant eats proportionately more during his first six months of life than he does in any other six months during the first two years. The baby needs approximately 45 to 50 calories for each pound each day during his first year. As a basis of comparison, the adult requirement is 16 to 20 calories.

Milk is the principal item of the infant's diet and provides the protein and calcium he needs for growing muscles and bones, plus some of the other essential nutrients. Most authorities agree that either breast milk or a carefully prepared formula will nourish an infant adequately. Breast feeding is desirable because it has both physiological and psycho-logical values for both the child and the mother. The advantage of mother's milk over commercially prepared formulas is that the former is said to be more digestible, it is always the right temperature, and it is more economical in time and energy. However, the mother should not be urged to nurse the baby if it is going to create a hardship or if the mother's nutrition and health are not optimal.

The mother who is going to nurse her child should be apprised of the fact that for the first few days the secretion from her breasts is colostrum not milk. It actually contains more protein and salts than milk, as well as certain substances which build the baby's resistance to infection. As the infant's needs increase, a milk diet alone cannot supply all the necessary nutrients, particularly iron and vitamin C. In fact, vitamin C and D supplements are usually started during the first month of life. Iron is a particularly important consideration because the hemopoietic demands of the infant increase rapidly after three months of age and they cannot be met by milk alone. The current practice is to offer a variety of specially prepared, semifluid baby foods which are rich in iron and other nutrients when the baby is three months old. Feeding iron enriched foods such as infant cereals or naturally iron rich foods such as egg yolks, meats and vegetables is highly recommended.

The infant at about nine months to one year of age, when his teeth are erupting, graduates from the puréed foods to dried bread and chopped junior foods. As the infant develops his swallowing skill and as more teeth erupt, he begins to select more solid foods. By the time he is 18 to 24 months old, he is ready to eat adult foods.

CHILDHOOD

The child after two years of age begins to have a diminution in appetite, and for good reason; his growth rate is beginning to slow down from a gallop to a trot, so to speak. Because he grows more slowly, he begins to need fewer calories in proportion to his size. This slower rate continues until he is about nine to ten years old, and then the growth rate begins to pick up speed until it reaches its climax during the midteens.

A child's requirements for minerals, calcium, phosphorus and vitamin D are greater than those of an adult but his need for protein, most vitamins and iron is less. This requirement is usually met by the use of more milk.

At this period in the life cycle new and different foods are introduced to broaden the child's food patterns. One new food at a time and in small proportions should be offered. Forcing a child to eat is unwise and should be avoided. A child will eat what he needs when he is hungry. Considerations as to flavor, texture and temperature of the food should be made because the child prefers simple, plain, smooth and tender foods which are neither too hot nor too cold. He favors foods that he can recognize and so he customarily separates mixtures of foods into their component parts, eating and enjoying the flavor of just one at a time. It is important that a sound sensible attitude about food be developed during this period because it will set the eating patterns for the rest of the child's life.

Snack eating is normal for these children, and the mother can use the proper snack foods to good advantage in influencing her child's eating habits. The real problem is to avoid sweets in general and candy in particular. Candy should not be used as a bribe or as a reward. It gives it too much importance. Candy will take away the child's appetite for the more nutritious foods that are important for his growth; furthermore, it is a proved cariogenic food.

ADOLESCENCE

With the exception of infancy, adolescence is the period of most rapid growth and development; nutritional needs are consequently among the highest in the life cycle. Actually, in boys, the nutritional requirements are the highest of any time in their lives; with girls, they are higher than ever before and will be superseded in their lifetime only when they are pregnant or nursing a baby. According to the Recommended Dietary Allowances, teen-age boys need more calories and more of every nutrient except iron and vitamin D than girls of the same age. The adolescent "growth spurt" begins in girls at about 10 or 11 years of age and in boys between 13 and 15 years and continues for about five years, which makes the teen years the period when nutrient requirements are highest.

If nutrients are interpreted in terms of foods, it can be said that the two food groups that need maximal attention are the milk group and the vegetable-fruit group. This is because they have been found to be inadequate in surveys of teen-age diets. Four cups of milk or its equivalent, plus larger portions and even second servings of each of the other three food groups should be encouraged.

Both adolescent boys and girls are interested in their personal appearance, and if the appeal is made to boys on the basis of their becoming more vigorous and athletic and to girls on the basis of their becoming more attractive, dietary advice might be followed more assiduously. One should emphasize that clear skin, good figure, pep and energy and mental alertness are all packaged in good wholesome foods.

One of the most frequent undesirable food habits encountered in teen-agers, particularly girls, is slighting or skipping breakfast. The desire to keep their weight down is the reason usually given, but actually the opposite results. Skipping breakfast causes the teen-ager to overeat at other meals or between meals to make up for the missed meal, with the result that weight is added rather than subtracted.

Fad diets are commonly followed by some teen-age girls as a means of weight control. These are dangerous to young women because they impoverish the reserve which will be called upon during the physical stresses of the reproductive period in the next few years.

Snacks must be recognized as a part of the food pattern of practically every teen-ager. Fruit juices, raw fruits (apples, peaches, etc.), vegetables (celery, carrots), cheddar cheeses and nuts are highly recommended, as are milk shakes and malteds.

In addition to good food, proper exercise and sleep should be included in any over-all program of healthful living. If the teen-ager does not respond healthwise to these simple basic rules, then the clinician should be suspicious of a systemic conditioning factor such as a low-grade infection, emotional upset or metabolic disturbance and refer the adolescent to the family physician for management.

ADULTHOOD

To maintain and to repair tissues requires a constant supply of all nutrients, even when physical growth is complete. As age increases, calorie needs decrease (providing activity is the same), but the need for protein, minerals and vitamins does not. This means that food should be chosen carefully on the basis of its nutritional content if all nutrients are to be included without an excess of calories. Since obesity is a health hazard which predisposes to many of the chronic diseases which appear in the middle years, emphasis needs to be put on the fact that energy needs decrease with age. This means that if a person is not to gain weight as he grows older, either calories must be reduced or energy expenditure must be increased through more exercise or activity. Unfortunately, instead of increasing activity, people are apt to participate in fewer sports, to ride more and to walk less. Yet programs of exercise alone for losing weight have not been very successful. What is needed is a concerted effort to combine both exercise and reduction in calorie intake. The old adage that an ounce of prevention is worth a pound of cure certainly holds in regard to the prevention of obesity in the adult years.

Although more publicity has been directed toward the hazards of obesity than toward extreme thinness, the markedly underweight adult should not be ignored nutritionally. Fatigue, proneness to infections, anemia and lack of pep and vigor may be found in the person who is undernourished or malnourished.

Although a desirable weight is not the only criterion for good nutrition, it is one of the most tangible and easily determined measures.

Pregnancy

A relationship between good nutrition during pregnancy and the health of both the mother and her infant has been proved in controlled scientific studies. During pregnancy the need for protein, minerals (especially calcium) and the vitamins is increased. Caloric needs may be slightly higher (10 to 20 per cent) than for the nonpregnant woman, but any increased need for calories is usually offset by a decrease in activity.

Most doctors allow a total gain of 15 to 20 pounds during pregnancy, depending on body build. During the first three months of pregnancy, no weight gain is desirable unless a person is underweight. An average gain of one half pound a week is usually permitted during the fourth, fifth, and sixth months. During the last three months there is an average gain of a pound or so a week.

Food needs begin to increase about the fourth month of pregnancy, and the expectant mother must eat for two only in the sense that she needs twice as much calcium as she did before pregnancy, half again as much vitamin C, and a third more protein and riboflavin and about a fourth more vitamin A, thiamine and iron. More vitamin D is also needed.

A quart of milk or its equivalent in cheese each day will provide the extra calcium and riboflavin. However, if the expectant mother is a teen-ager, a quart and a half of milk is necessary to take care of the body building and maturation needs of the mother and the needs of the baby. An extra serving of citrus fruit or other good source of vitamin C should be added. A dark green or deep yellow vegetable as a source of vitamin A is advisable every day. Vitamin D concentrates are often prescribed by the physician to insure an adequate intake of this vitamin.

Obviously, if the mother's diet is inadequate both she and the baby suffer. The diet during pregnancy is a special one in that it is increased in quality to a much greater degree than in quantity.

Lactation

The recommended dietary allowances during lactation are about the highest in adult life. The following statement (Wishik, 1959), helps us to understand why: "During the period of lactation, however, the amount of output of milk in a single month is greater than the increased mass of the whole nine months of pregnancy." A nursing mother who produces 850 cc. of milk a day requires about 1000 additional calories in order to produce this milk. She also needs extra proteins, minerals and vitamins. The contention that breast feeding is cheaper than formula feeding is a poor one to use in advocating breast feeding. The extra foods a mother needs for lactation are about equal in cost to the baby's formula. For a nursing mother, 1½ quarts of milk plus two extra servings of fruits that are good sources of vitamin C are needed, as well as a serving of dark green and deep yellow vegetables every day to meet vitamin A recommended allowances. To meet increased demands for other nutrients and calories, more foods from the bread and cereal as well as meat groups can be chosen.

It is important to remember that the quality of the breast milk will depend on the stores of the nutrients that have been built up in the mother's body over the years as a result of sound nutritional practices as well as the food that is eaten during the lactation period.

Old Age

The general food needs of the older person do not differ greatly from those in the middle years—ample amounts of protein, minerals and vitamins for maintenance and repair of his tissues. Caloric needs are lower because of a slowing down of metabolism and a decrease in physical activity.

The nutritional status of the older person is affected by the many things that have happened to him through the years. Some of the physiological factors will include a decreased ability to absorb nutrients, impaired circulation, difficulty in eating because of poor teeth or dentures, decreased sensitivity to taste and smell, and general digestive problems. Other factors which also influence eating habits of older people are loneliness, reduced income, loss of family and friends, inadequate cooking facilities, diminished interest in foods and lack of appreciation of the need for adequate nutrition.

Since protein foods, particularly meat, are apt to be eaten in too small amounts, suggestions as to how to use more eggs, milk, fish or poultry may be helpful. Nonfat milk solids (dry skim milk) are an economical food, rich in both protein and calcium. Their use in cooking or as a beverage is highly recommended and often accepted by the older person. Other protein foods are too often neglected because of the bother of cooking them. Encouragement may well be given for use of those which are easily available, such as hamburger or canned meat, chicken or fish. Since a low protein intake may be the cause of habitual fatigue, an improvement in protein intake may help the person to take a greater interest in eating better and in life in general.

In addition to protein foods from the milk and meat group, care also needs to be taken to include fruits and vegetables and whole grain or enriched breads and cereals. Frozen or canned fruits or vegetables require little effort to prepare. Since these are equal in food value to the fresh, their use may appeal to the older person who has less energy to put into food preparation. Vitamin C foods often need special attention since the foods rich in this vitamin are not too numerous and are found almost entirely in the vegetable-fruit group. Although citrus fruits are one of the best sources of ascorbic acid, sometimes tomato juice or some of the other less acid-tasting juices are more readily accepted by the older person. Today many of these, including apple, pineapple and cranberry, may be fortified with vitamin C. Labels will indicate whether vitamin C has been added. Potatoes cooked in their skins are often well accepted by older people and their use should be encouraged. Even though they contain less ascorbic acid, their daily use provides a considerable amount of this vitamin.

Breads and cereals are also generally well liked by older people. Cooked cereals such as oatmeal and other whole grain or enriched cereals served with milk not only provide B vitamins and other nutrients but are easily digested, economical and easy to eat. The use of breadstuffs such as doughnuts, sweet rolls and cakes which contain much sugar and refined flour needs to be discouraged, since these may either replace the nourishing whole grain or enriched breads and cereals or add extra, unnecessary calories.

If chewing is difficult, soft fruits and juices can be substituted for hard fruits. Chopped and puréed meats and fish are good tender sources of protein. The use of blenders, meat grinders and mixers can be very helpful in preparing the food in a form that will be acceptable to the elderly person who has chewing problems. Cheese and egg dishes, finely chopped and well cooked vegetables, puddings, soups, custards and ice creams can be added. If the sense of taste has declined and there are no medical dietary restrictions, such as a low sodium diet, then seasoning agents should be employed to make meals more appetizing.

Since today more people are living longer than ever before, medical science is focusing increased attention on ways of improving health during these added years. Adequate, but not overabundant, nutrition is one of the factors which enables older people to enjoy improved health and well-being during the later years. The benefits of good nutrition throughout the early and adult years will be reflected in a healthier, happier old age.

REFERENCES

Texts

Fleck, H., and Munves, E. D.: Introduction to Nutrition. New York, The Macmillan Company, 1962.
Martin, E. A.: Nutrition in Action. New York, Holt, Rinehart and Winston, 1963.
Stefferud, A. (ed.): Food, The Yearbook of Agriculture, 1959. Washington, D.C., United States Government Printing Office, 1959.
Wohl, M. G., and Goodhart, R. S. (ed.): Modern Nutrition in Health and Disease. 3rd ed. Philadelphia, Lea & Febiger, 1964.

Articles

Wishik, S.: Nutrition in pregnancy and lactation. Fed Proc., *18*:4, 1959.

Chapter Sixteen

Effects of Processing on the Nutritional Values of Foods

Robert S. Harris, Ph.D.

INTRODUCTION

This chapter is concerned with the changes which take place in foods as they travel from the farm to the dinner table. The discussion must be kaleidoscopic because of space limitations; the reader is referred to a book by Harris and von Loesecke (1960) which treats the subject more extensively.

A plant begins to die as soon as it is taken from the soil. The moisture and minerals which are essential to life are no longer available; as it wilts, the concentration of nutrients in the plant cells is increased, growth stops, and nutrients in the tissues are destroyed with increasing rapidity. Leafy vegetables are more sensitive to these changes because water is lost more rapidly from the extensive surfaces of the leaves. Within minutes after harvest one can measure losses in nutrients, especially in the vitamin C of vegetables.

The average truck garden vegetable in the United States travels about 1600 miles; thus, the average age of "fresh vegetables" in this country is two to four days. Losses in fresh vegetables can be minimized, but not prevented, by refrigeration under high humidity conditions. Thus, the best procedure for the shipment of a bushel of spinach is to interlard layers of the vegetable with chipped ice and ship it under refrigeration.

Most foodstuffs are unstable and must be refrigerated, frozen, canned, milled, dehydrated, pasteurized or smoked before shipment over long distances or preservation for later use. Most foods must be trimmed and cooked to increase their palatability, digestibility and safety. Therefore, in the broadest sense, most foods must be "processed," and at each step as they pass from garden to gullet they suffer losses in nutrient value.

It should be pointed out, however, that the nutritional values of foods are often *improved* by processing. Human beings cannot digest the cellulose walls of vegetable tissues, but cooking may improve the digestibility and utilization up to twentyfold. The bran layers of certain foods contain amounts of antinutrients which interfere with the absorption of vitamins, amino acids and minerals. Milling which removes these compounds improves the utilization of calcium or iron severalfold. The iron of raw meat is poorly utilized; cooking makes it available for absorption.

Food processing is essential to our sophisticated culture, and it is necessary that we devise processing methods which will minimize destruction and maximize utilization.

STABILITY OF NUTRIENTS

Man requires about four dozen nutrients (vitamins, minerals, amino acids, fatty acids) day by day. Some of these nutrients are destroyed in processing because they are sensitive to acids or alkalies; some are sensitive to oxygen and others to light. These effects are catalyzed when the temperature is raised, when catalysts (copper or iron) are present, or when enzymes are active. These effects are summarized in Table 16-1. It will be noted

207

TABLE 16-1. STABILITY OF NUTRIENTS[*]

Effect of pH

Nutrient	Neutral pH 7	Acid <pH 7	Alkaline >pH 7	Air or Oxygen	Light	Heat	Cooking Losses, range %
Vitamins							
Vitamin A	S	U	S	U	U	U	0–40
Ascorbic acid (C)	U	S	U	U	U	U	0–100
Biotin	S	S	S	S	S	U	0–60
Carotenes (pro-A)	S	U	S	U	U	U	0–30
Choline	S	S	S	U	S	S	0–5
Cobalamin (B₁₂)	S	S	S	U	U	S	0–10
Vitamin D	S		U	U	U	U	0–40
Essential fatty acids	S	S	U	U	U	S	0–10
Folic acid	U	U	S	U	U	U	0–100
Inositol	S	S	S	S	S	U	0–95
Vitamin K	S	U	U	S	U	S	0–5
Niacin (PP)	S	S	S	S	S	S	0–75
Pantothenic acid	S	U	U	S	S	U	0–50
p-Amino benzoic acid	S	S	S	U	S	S	0–5
Vitamin B₆	S	S	S	S	U	U	0–40
Riboflavin (B₂)	S	S	U	S	U	U	0–75
Thiamine (B₁)	U	S	U	U	S	U	0–80
Tocopherols (E)	S	S	S	U	U	U	0–55
Essential amino acids							
Isoleucine	S	S	S	S	S	S	0–10
Leucine	S	S	S	S	S	S	0–10
Lysine	S	S	S	S	S	U	0–40
Methionine	S	S	S	S	S	S	0–10
Phenylalanine	S	S	S	S	S	S	0–5
Threonine	S	U	U	S	S	U	0–20
Tryptophan	S	U	S	S	U	S	0–15
Valine	S	S	S	S	S	S	0–10
Mineral salts	S	S	S	S	S	S	0–3

S = stable (no important destruction).

U = unstable (significant destruction).

*From Harris, R. S., and von Loesecke, H.: Nutritional Evaluation of Food Processing. John Wiley & Sons, Inc., 1960. Used by permission.

that some nutrients (vitamin A, ascorbic acid, folic acid, etc.) are more sensitive than others (minerals, most amino acids, niacin, etc.). It will also be noted that there is no condition which is ideal for preservation; for instance, nutrients are destroyed whether the pH of the tissue is acid, neutral or alkaline. This serves to explain why no food is entirely stable during storage.

FACTORS AFFECTING NUTRIENT CONTENT OF FOODS

GENETICS

The nutrients in plant and animal tissues are mostly under genetic control. The ascorbic acid content of tomatoes was doubled to 50 mg./100 gm. by genetic selection (Lincoln et al., 1949). The protein content of corn was increased over twofold by selective plant breeding (Woodworth et al., 1952). Cravioto et al. (1945) observed an ascorbic acid content of 8.4 mg. per cent in a sample of Barbados cherries grown in Mexico while Aseñjo and de Guzman (1946) found 2960 mg. per cent of ascorbic acid in a sample collected in Puerto Rico. This 350 fold difference in the nutrient content of a food as a result of genetic influences is quite unusual. Nevertheless, it is true that the nutrient content of diets can be improved considerably when plant breeders cross high potency varieties of edible plants.

The composition of animal tissues and fluids is under genetic control, and animals are being bred to produce meat, milk or eggs of superior value. Improved feeding may increase the quantity of these foods, but to realize maximum quality, genetic control is necessary.

ENVIRONMENT

Light, temperature, season, location, soil and fertility all affect the nutrient content of foodstuffs. However, the major effect of soil improvement is to increase yield per acre rather than to enhance nutritional quality per pound.

SIZE AND MATURITY

Smaller fruits and vegetables are generally richer in nutrient content per pound, especially when these nutrients are located in the surface layers.

Some plants reach highest value while still immature, others when mature, and still others when overripe. The B-complex vitamins are highest in the youngest leaves of vegetables because metabolism is at a maximum. As the leaf matures, the vitamin content decreases and the cellulose increases.

TRIMMING

The nutritional values of most vegetables and fruits are reduced by trimming because the outer leaves are richer than the inner leaves. The ascorbic acid content of apples is about five times higher under the peel than in the flesh (Eheart, 1941). However, removal of the shell of pineapple, the rind of squash or pumpkin, the stones of fruit and the cores of apples and pears which are high in indigestible fiber will, of course, increase the nutritional values of these foods.

PASTEURIZATION AND STERILIZATION

Regardless of the method, the effect of heating on nutrients depends on time and temperature of treatment, quantity of oxygen present, pH and the presence of metal salts. For instance, when canned tomato juice was heated to sterilize it, the retention of vitamins was: ascorbic acid, 67 per cent; thiamine, 89 per cent; riboflavin, 97 per cent; niacin, 98 per cent; and carotene, 67 per cent (Cameron, 1951).

MILLING

Foods which are eaten daily in significant quantities must have a mild flavor. This explains why white wheat flour and white rice, the staple foods of mankind, are milled. Whole wheat flour has essentially the same composition as the original wheat. Sixty to 80 per cent of the minerals and vitamins in wheat are lost when the bran layers are removed. Compared with whole wheat, the flours commonly used in the United States contain approximately the following proportions of nutrients: carbohydrate 110 per cent; protein, 90 per cent; fat, 45 per cent; minerals, 25 per cent; fiber, 5 per cent; thiamine, 18 per cent; riboflavin, 33 per cent; niacin, 14 per cent (McCance et al., 1945); vitamin B_6, 40 per cent; pantothenic acid, 35 per cent (Dawbarn, 1949); vitamin E, 30 per cent (Moore et al., 1957), choline, 80 per cent (Glick, 1945); potassium, 28 per cent; magnesium, 16 per cent; and calcium, 50 per cent (McCance et al., 1945). The decrease in fat, ash and fiber reflects the removal of germ and bran, which are rich in these constituents. Removal of germ causes a small decrease in the amount and quality of protein. The rise in carbohydrate results from the fact that the endosperm consists mainly of starch. The large drop in vitamin and mineral content indicates that a major portion of these nutrients is present in the germ and bran of wheat.

Similar losses occur in the milling of rice, smaller losses result from the milling of corn, and even smaller changes take place in the milling of oats (Harris and von Loesecke, 1960).

It is to be noted that approximately 95 per cent of the phytic acid in wheat and rice is removed during milling. Phytic acid interferes with the absorption of calcium and iron from the intestinal tracts of children (Harris, 1955).

ENRICHMENT OF FOODS*

The balance of nutrients in a food may be improved by blending with a nutritionally complementary food, or by the addition of specific nutrients (vitamins, minerals or amino acids). Today most of the vitamins used in food fortification are produced by chemical synthesis. Synthetic vitamins are equally as active and equally as harmless as natural vitamins because they are chemically identical.

The toxicity of all vitamins is low; for instance, the ingestion of more than 10 times the daily requirement of one, such as vitamin A, is necessary over many days to produce evidence of even mild toxicity. Vitamins are present in fortified foods in such dilution that no adverse effects have been demonstrated, and thus there is no "toxicity" problem in relation to food supplementation.

The kinds and amounts of nutrients added to specific foods in different countries varies greatly because supplementation is carried out for many different reasons. All important types of nutrient enrichment fall into one or more of the philosophies which follow.

RESTORATION
OF NATURAL LEVELS

Significant amounts of nutrients may be removed when foodstuffs are milled or may be destroyed when foods are dehydrated, heated,

* See Harris (1959) for a fuller discussion of this subject.

canned or stored. These processing procedures are necessary if urban populations far removed from sources of food supplies are to be fed. Some processors try to restore some of the nutrients lost during manufacture. The nutrients selected for restoration are those which (a) are lost in significant amounts during manufacture, (b) tend to be deficient in population groups, and (c) are inexpensive. Usually, enough is added to raise the content of the nutrient to preprocessing levels.

This restoration philosophy has been criticized on the ground that the vitamin content of an unprocessed food bears no relation to human nutrition. The amount of thiamine in whole wheat, for instance, has no relation to the thiamine needs of a child. On the other hand, restoration can be praised because it repairs the damage done to the food during processing.

FORTIFICATION
ABOVE NATURAL LEVELS

Some food products (such as infant or geriatric foods) are designed to carry the burden of the nutrition of an individual, or to complement the nutrients of the milk with which they are to be combined. Such special purpose foods should contain optimum levels of all nutrients in available form in a daily portion. Since no one natural food can do this, supplementation is necessary. Vitamin D milk and vitamin C-enriched fruit juices are examples of foods in this category.

ENRICHMENT
WITH PUBLIC
HEALTH OBJECTIVES

An excellent example of this philosophy is enriched white flour and bread in the United States. Lane et al. (1942) reported that in the diets of the middle two-thirds to three-fourths of the United States population the thiamine, riboflavin and niacin content was 60, 80 and 95 per cent, respectively, of the amount required for health.

On the basis of (a) the Recommended Dietary Allowance, (b) the average per capita intake of flour and bread, and (c) the average per capita intake of a vitamin or mineral, officials calculated the amount of each nutrient that should be added to bread and flour to bring the average diet up to adequacy. It was reasoned that if flour and bread were enriched with these quantities of vitamins and minerals and consumed in average amounts (6 slices), they would raise the nutritive value of the average diet to desirable levels and thus eliminate the malnutrition that results from deficiencies of these vitamins. This enrichment has proved to be effective and is one of the main reasons why frank clinical vitamin deficiencies are nonexistent in areas where this public health practice has been adopted.

This public health approach is attractive because it permits improvements of diets without requiring any change in stubborn food habits, it is inexpensive, and it is effective. Harris et al. (1961) analyzed breads from a number of countries and concluded that white bread enriched according to United States standards was superior in over-all nutrient content to any other type of bread (whole wheat, rye, pumpernickel, etc.).

This enrichment philosophy has been criticized because the formula is transient and requires periodic resurvey and revision to make certain that enrichment is adequate on the one hand, and not wasteful on the other. The formula for enriched white bread in the United States today is the same as the enriched bread of 1941, the formula of which was designed to meet the deficiencies of people living on 1941 diets. Certainly, the quality of diets has changed during the past 25 years. The subject should now be reviewed to determine whether present enrichment levels should be changed and whether other nutrients should be included in the enrichment formula.

ENRICHMENT TO EQUAL THE LEVELS IN INTERCHANGEABLE FOODS

Margarine is a spread which competes with butter. The margarine manufactured in the

TABLE 16-2.   LEVELS OF ENRICHMENT OF U. S. FOODS (MG./LB.)

Enriched product	Thiamine		Riboflavin		Niacin	
	min.	max.	min.	max.	min.	max.
Bread	1.1	1.8	0.7	1.6	10.0	15.0
Flour[b]	2.0	2.5	1.2	1.5	16.0	20.0
Farina[c]	2.0	2.5	1.2	1.5	16.0	20.0
Macaroni[d]	4.0	5.0	1.7	2.2	27.0	34.0
Noodles[a]	4.0	5.0	1.7	2.2	27.0	34.0
Corn meal	2.0	3.0	1.2	1.8	16.0	24.0
Corn grits[e]	2.0	3.0	1.2	1.8	16.0	24.0
White rice	2.0	4.0	1.2[a]	2.4[a]	16.0	32.0
Whole milk	...	...	...	..	...	...

Enriched product	Vit. D[a] (U.S.P. units)		Iron		Calcium[a]	
	min.	max.	min.	max.	min.	max.
Bread	150	750	8.0	12.5	300	800
Flour[b]	250	1000	13.0	16.5	500	625
Farina[c]	250	...	13.0	...	500	...
Macaroni[d]	250	1000	13.0	16.5	500	625
Noodles[a]	250	1000	13.0	16.5	500	625
Corn meal	250	1000	13.0	26.0	500	750
Corn grits[e]	250	1000	13.0	26.0	500	750
White rice	250	1000	13.0	26.0	...	...
Whole milk	150	400	...	...	...	...

[a] Enrichment is optional; [b] in enriched self rising flour, Ca 500–1500 mg./lb. required; [c] no maximum for iron; [d] levels allow for 30–50% losses in kitchen procedures; [e] levels must not fall below 85% after washing and rinsing.

* From Harris, R. S.: Supplementation of foods with vitamins. J. Agr. Food Chem., 7:88, 1959. Reprinted by permission of the American Chemical Society.

United States today is fortified with 15,000 U.S.P. units of vitamin A per pound. This level represents the average year-round potency of butter. Because butter fluctuates between 5000 and 24,000 IU of vitamin A from winter to summer, margarine is the more reliable source of vitamin A.

Another example of this approach to food enrichment is the addition of vitamin C to fruit juices (apple, tomato, pineapple, cranberry, grape and even orange juice) to standardize them at 30 mg. per serving. This is advocated because most people use these juices interchangeably even though the natural juices differ greatly in ascorbic acid content.

The United States Food and Drug Administration has established "standards of identity" for a number of foods and food products. The kinds and amounts of enrichment ingredients are specified. Examples of enriched cereal products for which standards of identity have been established are given in Table 16-2. These products must meet the federal standards when sold in interstate commerce.

Food enrichment and fortification have been successful because they provide a relatively inexpensive way to increase the nutritional values of foods, because the synthetic nutrients are identical to the natural nutrients and are more completely absorbed and utilized, and because certain foods, especially processed foods, need improvement.

LOSSES IN NUTRIENTS DURING HOME PREPARATION

TRIMMING

Losses in the nutrient content of foods during preparation for human consumption differ, depending on the variety of food, its condition and freshness, the season of the year, cultural habits, economic status and a variety of lesser factors. When foods of plant origin are trimmed, the nutrient losses generally exceed the weight loss because vitamins and minerals are usually found in higher concentrations in the outer leaves of vegetables and outer layers of seeds, roots, tubers and fruits. On the other hand, weight losses may exceed nutrient losses when stems are trimmed from

leafy vegetables. Nutritional value per pound generally rises as one goes from a major stem to a minor stem, and from old leaf to youngest leaf in a sprig of vegetable such as broccoli. Values generally rise also as one passes from innermost layers to outermost layers of fruits and seeds and tubers. Values of meats rise as fat, bone and gristle are removed.

CHOPPING

The recent literature indicates that the losses in ascorbic acid when cabbage or salad greens are diced or minced are not so serious as first supposed. The chemical method used by earlier investigators in measuring ascorbic acid measured only reduced ascorbic acid. The reduced ascorbic acid in chopped vegetables is converted first to dehydroascorbic acid (which is nearly as active as reduced ascorbic acid) and then to diketoglutonic acid. When the assay method was improved to include the measurement of dehydroascorbic acid it was found that the loss of "vitamin C activity" was relatively small. Nevertheless, vegetables become much more unstable by mincing and chopping, and they should be frozen or stored under refrigeration and consumed promptly. The vitamin C content of potatoes drops rapidly during and following thorough mashing.

BOILING

Losses during boiling depend on the size of food particles, the volume of cooking water, the length of cooking time, etc. (Table 16-3). In general, food should be added to briskly boiling water, the water level should be sufficient only to cover the food, the cooking vessel should be covered, heating should be continued until just "done," and the food should be eaten immediately for maximum value. Rapid cooking is superior to slow. Pressure cooking is superior to the traditional method. No differences have been noted when foods are boiled in enamel, aluminum, Pyrex or stainless steel pots; copper vessels catalyze the destruction of several nutrients and should not be used.

TABLE 16-3.   AVERAGE PERCENTAGE OF
NUTRIENTS LOST DURING COOKING* †

	Thiamine	Riboflavin	Niacin	Ascorbic acid
Meats	35	20	25	—
Meats plus drippings	25	5	10	—
Eggs	25	10	0	—
Cereals	10	0	10	—
Legumes	20	0	0	—
Vegetables, leafy green and yellow	40	25	25	60
Tomatoes	5	5	5	15
Vegetables, other	25	15	25	60
Potatoes	40	25	25	60

* Based on good cooking practices with U.S. Army ration components. For individual foods in some of the food classes the loss may be considerably greater or less. These estimated average cooking losses are useful as guides, but may not necessarily apply when cooking practices and other conditions (e.g., pH) differ greatly. It is preferable to use local values where available and reliable.

† Interdepartmental Committee on Nutrition for National Defense, National Institutes of Health: Manual for Nutrition Surveys. 2nd ed. Bethesda, Maryland, 1963, p. 185.

Most of the nutrient losses during cooking are due to extraction into the cooking water. Broccoli boiled for 2, 5½ and 11 minutes lost 25 per cent, 32 per cent and 33 per cent, respectively, of the ascorbic acid content into the cooking water. Broccoli lost 18 per cent, 43 per cent and 47 per cent of ascorbic acid into the cooking water when cooked in 100 cc., 500 cc. and 1000 cc. of water, respectively.

ROASTING AND BAKING

Significant amounts of the more heat-sensitive nutrients are lost when foods are roasted or baked. The losses are influenced by temperature, time and the size and type of food being treated. Losses may be much higher in the surface layer.

FRYING

Losses during frying may be small or large, depending on temperature, exposure to light, type of pan and time of cooking. Losses in

ascorbic acid may range from 0 to 33 per cent and in riboflavin from 0 to 75 per cent.

LEFTOVER FOODS

It is common practice to refrigerate the cooked foods which have not been consumed at one meal and to serve them at a later meal. The literature indicates that leftover cooked foods lose very significant amounts of critical nutrients during cooling, refrigerator storage and reheating. Leftover foods can be avoided by slightly underestimating the amount of cooked food required at each meal and by serving plenty of enriched white or whole cereal breads and rolls. Less food will be wasted; leftovers will be avoided; the family will not have a surfeit of cooked vegetables, and yet it will be well fed.

SUMMARY AND CONCLUSION

The nutrient content of edible plants may be improved by genetic means. Fertilization improves the yield of a food per acre, rather than its nutritional quality. There is no valid evidence that organic fertilizers are superior to inorganic.

Modern sophisticated societies require foods that are processed by milling, freezing, canning, dehydration, refrigeration, etc., because foods must be transported over long distances and stored for weeks or months. Processing frequently makes foods more palatable and digestible.

Enrichment of food either through restoration to natural levels or fortification above natural levels not only overcomes nutrient losses suffered during processing, but also provides nutrients that are more completely absorbed and utilized. Public health measures such as enrichment of bread and rice have proved effective in preventing vitamin and mineral deficiencies in population groups.

In cooking, the use of small amounts of water and rapid boiling or steaming, particularly under pressure, is an effective way of reducing nutrient losses.

Optimum retention of nutrients during food preparation requires short cooking time, minimum addition of water and brief holding time before serving. Since leaching of soluble nutrients occurs during cooking in addition to thermal destruction, it is advisable to use both the solid and liquid components of the prepared food. Only the amount of food which will be consumed at each meal should be prepared, since the cooling, refrigeration, reheating and holding of leftover foods seriously reduce their vitamin content.

REFERENCES

Aseñjo, C. F., and de Guzman, F.: The ascorbic acid content of the West Indian cherry. Science, *103*: 219, 1946.

Cameron, E. J.: Use of antibiotics in preserving foods. Proc. Third Conf. on Res. Amer. Meat Inst., 1951.

Cravioto, R., Lockhart, E. E., Anderson, R. K., Miranda, F., and Harris, R. S.: Composition of typical Mexican foods. J. Nutr., *29*:317, 1945.

Dawbarn, M. C.: The effects of milling upon the nutritive value of wheaten flour and bread. Nutr. Abstr. Rev., *18*:691, 1949.

Eheart, M. S.: Factors which affect the vitamin C content of apples. Virginia Agric. Exp. Sta. Tech. Bull. No. 69, 1, 1941.

Glick, D.: The choline content of pure varieties of wheat, oats, barley, flax, soybeans and milled fractions of wheat. Cereal Chem., *22*:95, 1945.

Harris, R. S.: Phytic acid and its importance in human nutrition. Nutr. Rev., *13*:257, 1955.

Harris, R. S.: Supplementation of foods with vitamins. Agric. Food Chem., *7*:88, 1959.

Harris, R. S., Siemers, G., and Lopez, H.: Nutrients in breads from fourteen countries. J. Amer. Diet. Assn., *38*:27, 1961.

Harris, R. S., and von Loesecke, H.: Nutritional Evaluation of Food Processing. New York, John Wiley & Sons, Inc., 1960.

Lane, R. L., Johnson, E., and Williams, R. R.: Studies of the average American diet. J. Nutr., *23*:613, 1942.

Lincoln, R. E., Kohler, G. W., Silver, W., and Porter, J. W.: Breeding for increased ascorbic acid content in tomato. Bot. Gaz., *111*:343, 1949.

McCance, R. A., Widdowson, E. M., Moran, T., Pringle, W. J. S., and Macrae, T. F.: The chemical composition of wheat and rye and of flours derived therefrom. Biochem. J., *39*:213, 1945.

Moore, T., Sharman, I. M., and Ward, R. J.: The destruction of vitamin E in flour by chlorine dioxide. J. Sci. Food Agric., *8*:97, 1957.

Woodworth, C. M., Leng, E. R., and Jugenheimer, R. W.: Fifty generations of selection for protein and oil in corn. Agron. J., *44*:60, 1952.

Chapter Seventeen

Physiological Factors in Food Selection

SANFORD A. MILLER, PH.D.

How do most animals, with the possible exception of man, know how much food to eat to maintain their body weight? How do they know what foods to eat when suffering from a nutrient deficiency? In general, how does an animal regulate its food intake? These questions are part of one of the oldest, yet one of the least known, areas of physiology and nutrition and represent some of the most basic questions in terms of the animal's relationship to its environment.

In a real sense, an animal is what it eats. At the same time, what it eats determines what it is. The manner in which the animal accomplishes this task is extraordinarily complex and involves many areas including psychological and physiological factors. In each, we can describe events; in none can we completely explain the process. It is beyond the scope of this chapter to develop in detail all the information available; we can only attempt to provide a framework that will enable the reader to recognize this problem of food selection for what it is, one of the most fundamental in nutrition.

BASIC PRINCIPLES

The relationship of the animal to its nutritional environment is, like so much else in life, a cyclical process. On the basis of habit and need, an animal selects its food from its environment. Once this food is consumed, it enters the metabolic pools of the animal. The degree to which the food satisfies the needs of the animal determines his nutritional state. In turn, the cycle is completed since, in part, the nutritional state determines the food selected. It becomes evident that the selection of food is, in essence, a selection of a pattern of nutrients to satisfy body needs.

What evidence is there that the animal does, in fact, regulate its body weight by means of controlling its food intake? It has been shown, for example, that in rabbits the dry weight of their carcasses will remain at a remarkably constant level over an extended period of time. These animals, if allowed to eat ad libitum, still maintain body weight. When environmental temperature is dropped and presumably energy requirements increased, food intake increases to compensate and to maintain the energy reserves. In other studies with rats exercise was investigated for its effect on food intake and maintenance of body weight (Mayer et al., 1954). These studies indicate that, within a broad range of activity, the animal increases food intake to compensate for his additional energy needs and will maintain his body weight in this range. These studies also indicate that the short-term response is different from the long-term effect in that there is apparently an adaptation period required before the proper adjustments can be made. That this is true, to some extent, in humans is indicated in other work of Mayer (1955a). When the activities of Indian workers were plotted against caloric intake and body weight, it was found that a correlation similar to that seen in the rat could be made.

There are two components involved in the determination of the selection of nutrient patterns. The first is concerned more with habit and emotional response and has to do with attributes of food such as taste, odor, texture and consistency, all of which may be summarized in the word "flavor." Although

215

some portion of the development of the response to flavor is based upon nutritional need, it is basically a learned pattern and, if anything, is a long-term control mechanism.

The second component is, in essence, the mechanism by which the quantitative food intake is regulated and thus serves to control energy consumption and, ultimately, body weight. This mechanism is, in all probability, a physiological one and is dependent at any specific time on the nutritional state of the organism. It, too, has a long-term and short-term component, each of which may be controlled by different mechanisms.

THE REGULATION OF FOOD INTAKE

REGULATION BY APPETITE

Appetite is defined as the natural desire for satisfying some want or need. For food, flavor is perhaps the single, most important character in satisfying this want (Little, 1958). Flavor, as indicated earlier, consists of a number of different factors including taste, odor, texture and consistency. Physiologically, perception of these characteristics is dependent upon specific nerve structures located in and around the oral cavity. (See Chapter Twenty-four and Figure 24-1.) Taste is detected, for example, by receptors on the tongue. These receptors apparently have the ability to distinguish between four distinct stimuli: sweet, sour, bitter and salty. In addition, there are also receptors which respond to the temperature of the food. All tastes are apparently composed of one or more of the four primary stimuli. Odor is a function also of specific chemoreceptors in the nasal passages. For texture and consistency, there are proprioreceptors in the mouth and jaws which respond to the pressure and location of the jaws. The signals from these are interpreted in the cortical areas of the brain to give the proper impressions of texture and consistency. When food is consumed, signals from all these receptors are integrated and interpreted in the brain to result ultimately in the impression of the specific flavor of the food (Lepkovsky, 1961).

The determination of which flavors are acceptable and which are not seems to be a function of experience, although there is some evidence that sweet flavors of moderate texture are most acceptable from the earliest ages. In terms of regulation of food intake in the human, this component can cause many difficulties since the most desirable foods in terms of flavor are not necessarily the best in terms of an adequate and well-balanced diet. It is also apparently true in the human that those components having to do with the sensations of flavor play a more important role in regulating food intake than do other more fundamental mechanisms, which are more dominant in other animals.

REGULATION BY HUNGER

Hunger is defined as a strong craving for food. As differentiated from appetite, hunger is a more basic phenomenon with little or none of the nuances of flavor associated with appetite. In terms of the regulation of food intake, hunger is most often associated with energy requirements. However, there is some evidence that requirements for essential nutrients can also produce similar responses (Lepkovsky et al., 1959).

The regulation of food intake can be discussed in two parts; the first is associated with the short-term or day-to-day regulation while the second is associated with the long-term regulation of body reserves. Each of these can be defined in terms of their precision, sensitivity, reliability and rapidity of response. In general, the short-term mechanism is more precise, more sensitive, of some reliability and certainly more rapid. Moreover, it can be shown that, within a specific group of animals, there is a marked variation in the ability of individual animals to perform the tasks of short- or long-term regulation. There is, then, an individual component of food intake regulation that must be recognized.

Fundamentally, the controlling factors in both mechanisms are apparently those concerned with energy and nutrient utilization, both of which can be considered a component of metabolic regulation. Apparently, the short-term mechanism responds to the balance between the energy intake and expenditure, controlling daily food intakes so as to prevent either a loss or a gain in the size of the energy reserves. The response to changes in the utilization of other nutrients such as protein or vitamins is apparently a longer term activity and is thus considered a part of the second mechanism (Mayer, 1964).

The hypothalamus plays a central role in these control mechanisms (Anon., 1962). This organ is located at the mid-base of the brain, just above the pituitary gland. Its activity is both neural and neurohumoral and its function is concerned with regulation of temperature, food and water intake and other endocrine activities.

Regulation of food intake is apparently a function of two areas of the hypothalamus. The ventromedial area is associated with satiety, while the lateral areas are associated with appetite. Experiments in which lesions are placed in specific areas indicate that lesions in the lateral areas produce aphagia; lesions in the ventromedial area produce hyperphagia. Stimulation of these areas produces opposite responses, the lateral area stimulation resulting in increased food intake while ventromedial stimulation results in depression of food intake. More recently, data have been collected which indicate that the principal area of regulation of food intake is the ventromedial, in which there is application or release of inhibition to the constantly activated lateral area; in other words, the ventromedial area acts, in a sense, as a brake on satiety.

The next question that arises is, what are the stimuli which control the action of the ventromedial area? A number of hypotheses have been presented to answer this question. Among those which have been suggested are the glucostatic hypothesis of Mayer (1955b), the thermostatic hypothesis of Brobeck (1960) and the lipostatic hypothesis of Kennedy (1953).

Glucostatic Hypothesis. The glucostatic hypothesis states that the arteriovenous (A-V) blood glucose difference reflects the physiological and metabolic state of the animal with regard to energy supply. This difference constitutes a signal to the hypothalamus. Thus, when the A-V difference is great, food intake is reduced. On the other hand, when the difference is small, hunger is experienced. In other words, there exist in the ventromedial satiety centers glucoreceptors sensitive to the blood glucose in the sense that they utilize it.

Thus, the organism, through the utilization of glucose in the cells of the hypothalamus, can respond to the ebb and flow of its energy reserves.

The evidence for the validity of this theory is based upon the following observations: The central nervous system is dependent upon the availability of glucose for its proper function. Moreover, since the carbohydrates are rapidly utilized and not stored to any extent, their level in the body will change more rapidly than any other reserve in response to changes in metabolic energy needs. If this is considered in terms of the interrelationships of the carbohydrates with the other nutrients, it appears as if a hypothesis based upon glucose regulation could explain the regulation of food intake.

Conditions which cause hypoglycemia, such as insulin injection or decreased utilization of glucose (diabetes mellitus), produce a small A-V blood glucose difference and a feeling of hunger. In addition, when hunger is experienced, there is an increased uptake of P^{32} and glucose-C^{14} in the region of the satiety center but not in the appetite center. The opposite effect is experienced with a feeling of satiety. Finally, gold thioglucose (but not other gold-containing compounds) is selectively taken up by the cells in the region of the satiety center, leading to their destruction and ultimately to obesity of the test animal.

Thermostatic Hypothesis. The hypothesis of Brobeck, or thermostatic hypothesis, is concerned with the heat produced by ingestion of food. This occurs both as a result of the specific dynamic action of the food and as a result of the changes in metabolic rate accompanying the level of nutrients in the body. It was proposed that heat-sensitive cells exist in or ahead of the rostral hypothalamus which, by modifying the blood flow, can stimulate the appropriate hypothalamic areas. This hypothesis has been criticized by many workers. However, it still represents a significant contribution to the field.

Lipostatic Hypothesis. As an explanation of the long-term control of body reserves of energy, the lipostatic hypothesis of Kennedy (1953) has been proposed. This theory states that the ventromedial hypothalamus is sensitive to the concentration of circulating metabolites. Thus, the ability of the animal to convert ingested food to fat stores and the rate at which this occurs controls the level of circulating metabolites. The animal can, in essence, sense changes in body stores and thus be able to regulate its intake in terms of long-term changes in these stores.

The evidence for this hypothesis is based upon the following observation: The long-term control of the size of the reserves is better than the short-term control. Under conditions of a high rate of lipogenesis, as in lactation, an increased food intake can be observed. Finally, there is some evidence that the amount of endogenous fat mobilized daily in ad libitum feeding is proportional to the fat stores. That this hypothesis can be associated with the theories related to glucose utilization is probable since lipid and carbohydrate metabolism are closely interrelated.

It is apparent that there are still many unanswered questions concerning the mechanisms of long- and short-term control of food intake. It is only a reflection of our ignorance that this is true. In any case, what is known is that the animal has a remarkable ability to regulate its food intake and body weight, apparently through control centered in the ventromedial and lateral areas of the hypothalamus. Just how this is done is still an important question.

NUTRITIONAL IMPLICATIONS

The implication of the foregoing discussion for the human is apparent. In the human, more than in any other animal, control of food intake rests more in the areas of emotion and gratification of appetite than in the area of regulation of body weight and satisfaction of the needs of the nutrient reserves. The result of the greater intelligence and increase in civilization is an endemic obesity. Whether this obesity is the result of environmental or evolutionary changes is a difficult question to answer. It has been suggested that for the accurate regulation of food intake a certain minimum amount of exercise is required. Our

modern life is marked by a pronounced trend toward a sedentary existence. It is apparent that the body regulatory mechanisms have been unable to adapt to this change. If it is impossible to increase the use of exercise, then it must become the function of the diet to re-establish this control of food intake. It becomes more and more important that our diets be designed to take into account our present faulty regulation. This must be done in spite of a lack of specific knowledge of the mechanisms in this area. Although data have been accumulated in recent years, as indicated earlier, one must never forget that the description of an event is not an explanation of its mechanism. It therefore becomes the function of the nutritionist to be concerned with more than the facts of dietary needs; he or she must be aware of the effect of flavor on food-intake regulation. Only then can reasonably rational diets be designed.

REFERENCES

Anon.: Hunger motivation and basic feeding systems. Nutr. Rev., 20:116, 1962.

Brobeck, J. R.: Food and Temperature. Rec. Prog. Horm. Res., 16:439, 1960.

Kennedy, G. C.: The role of depot fat in the hypothalamic control of food intake in the rat. Proc. Roy. Soc. London (Series B), 140:578, 1953.

Lepkovsky, S.: Potential pathways in nutritional progress. Proc. Nutr. Soc., 21:65, 1961.

Lepkovsky, S., Feldman, S., and Sharon, I.: Some basic mechanisms of hunger and satiety. Food Tech., 13:421, 1959.

Little, Arthur D., Co. Flavor Research and Food Acceptance. New York, Reinhold Publishing Co., 1958.

Mayer, J., Marshall, N. B., Vitale, J., Christensen, J. H., Mashayeihi, M. B., and Stare, F. J.: Exercise, food intake and body weight in normal rats and genetically obese adult mice. Am. J. Physiol., 177:544, 1954.

Mayer, J.: Regulation of energy intake and the body weight; The glucostatic theory and the lipostatic hypothesis. Ann. New York Acad. Sci., 63:15, 1955 a.

Mayer, J.: The physiological basis of obesity and leanness. Part I. Nutr. Abs. Rev., 25:597, 1955 b.

Mayer, J.: Regulation of food intake. In Beaton, G. H., and McHenry, E. W. (ed.): Nutrition: A Comprehensive Treatise. New York, Academic Press Inc., 1964. Vol. 1, pp. 1-40.

Chapter Eighteen

Food Habits and Faddism

CHARLOTTE M. YOUNG, PH.D.

Food habits are one of the most complex aspects of human behavior. The complexity of the problem is well stated by Wallen (1943), "Feeding activity ranks with sexual behavior as a demonstration of that peculiar and delicate interaction of biological, psychological and cultural influences so often found in the study of human wants." Studies of determinants of food habits have been made primarily by psychologists interested in motivation. But anyone concerned with giving nutritional advice or in influencing what a patient eats has reason to be interested in the determinants of food habits.

At best, food habits in the adult are not easy to change. There is more hope of doing so if one has some realization of the manifold factors which have determined his patient's present food patterns. An understanding of the complexity of the situation may lead to patience in the process of education and will save some feelings of complete frustration when a patient does not eat what he has been "told to eat." As Bruch (1961) has said, "If human behavior always functioned reasonably there would be no problem in applying new scientific knowledge about nutrition to our daily meals." Or as Harris (1961) points out, "People do not eat foods because they are good for them—rather because they appeal to their appetite, to their emotions, to their soul."

If the dentist is going to try to influence his patient's food habits it behooves him to understand all he can about what determines the foods his patient eats and also that these foods serve many uses for his patient other than nutrition. In fact, food habits are one of the most intimate facets of an individual, which explains why they are so difficult to change.

Before considering some illustrations from the writings of psychologists, psychiatrists, cultural anthropologists and sociologists on the subject of the non-nutritive uses of food, let us look at some experiences around us. Have you ever heard what it means to a homesick Oriental to find rice available when he is going through a cafeteria line? Have you listened to a homesick college freshman's reactions to good dormitory food, the only fault of which is that at any single meal what he was most accustomed to at home might not be available? Have you been present at a luncheon of a group of dairy farmers meeting in concern over a market for their product and been the *only* person in the group who drank milk for lunch? Have you ever eaten at a vegetarian sanitarium where the soy-bean substitute for meat was made up in the form of a chop or steak? More than intellect is involved in why people eat what they eat.

How about the ersatz gum, candies and puddings made especially for diabetic patients, or the diabetic patient who longs to have his prescription interpreted so that he may have apple pie and cheese as an occasional treat? The same nutritive value could be obtained in other forms of food, but not the same emotional or cultural values. Most of us have experienced the "buttering up" dinners of high prestige foods which precede financial campaigns. Have you ever been part of an experimental diet squad fed a perfectly adequate diet of good quality, well prepared natural foods, where the only limitation was that everything served must be eaten, no more and no less? If you have not,

219

you should; for then you would never question that food has non-nutritional uses and significance. Some individuals cannot take control of their food intake, even when they have been tested psychologically to eliminate the potentially neurotic. Some of the testing squad may become grumpy, irritable and fussy, displaying anger over minor things which ordinarily would never concern them. Ultimately, they probably will break the diet. Others on the squad will not be affected at all. The curious part is that those who are affected most really had no idea that they would react in this fashion under circumstances of controlled eating.

Thus, human *food* behavior (like most human behavior) is very complex, being determined by multiple motives and directed and controlled by multiple stimuli. The old axiom, "You can lead a horse to water but you can't make him drink," certainly applies to feeding him as well. Eppright (1947) has written: "Food acceptance is a complex reaction influenced by biochemical, physiological, psychological, social and educational factors. Metabolic conditions play a part. Age, sex and mental state are factors of importance. People differ greatly in their sensory response to foods. . . . The food likes and dislikes of the individual move in a framework of race, tradition, economic status and environmental conditions."

STUDIES OF FOOD LIKES AND DISLIKES

The subject of food likes and dislikes has been widely treated historically and anthropologically, but there has been relatively little scientific attempt to study the food likes and dislikes of present-day man in the United States and, even more important, in learning what determines his food likes and dislikes or his food acceptance.

As indicated earlier, psychologists have studied food acceptance as a means of studying motivation and personality maladjustment; child psychologists have made some contributions on the relation of various family, developmental and maturity factors to food acceptance. But much of the interest in food acceptance has come from the military services,

both in terms of reviewing the literature for what is known concerning food acceptance of various population groups and factors affecting it and in the initiation of new studies along these lines, including techniques for the study of food acceptance (Peryam et al., 1960; Committee on Food Research, 1946). World War II gave stimulus to work by a Committee on Food Habits of the National Research Council, which resulted in two reports, "Manual for the Study of Food Habits" (1945) and "The Problem of Changing Food Habits" (1943). Recently, Mead (1964) has brought research of trends in food habits up to date.

There have been limited studies of food acceptance by different population groups in different communities, such as those in the military service, preschool and school children, and college students. A variety of techniques has been used: oral or written attitude studies; sensory tests either of discriminative judgment or affective reactions; and measurement of the actual consumption of food under either normal or experimentally modified conditions (Pilgrim, 1957). Probably the most extensive studies of food acceptance have been made of men in the U. S. Armed Forces. On the basis of these studies it has been concluded that food consumption is predictable and that one of the important factors is food preference, that is, the degree of food like or dislike. Evidence indicates that food preferences of men in the army correspond generally to those of the general population of the United States.

What foods were best liked? Fresh milk stood at the top of the list, in a class by itself, with hot rolls and biscuits, grilled steak, ice cream, strawberry shortcake, fried chicken, french fried potatoes and roast turkey coming close behind as well liked foods. What foods were least liked? Almost all those listed were vegetables, particularly turnips, broccoli, squash, parsnips, asparagus and cauliflower, although iced coffee was also listed. The method of preparation had a definite effect on preference for vegetables. Within a given class of foods there were distinct preferences. For example, most meats were well liked, but lamb, fish and organ meats were low preference items.

In general, well liked foods were restricted

to those that are simply prepared. The more one does to a food, the less well liked it seems to be. It is characteristic of the United States male (and, for that matter, most of the population groups, with the possible exception of homemakers) that he likes his food plain and simple. When foods are combined, the preference rating is likely to be similar to that of the least liked ingredient in the combination. Many foods have not been tried, and this fact points to the problem of familiarizing individuals with a variety of foods early in life.

The evidence seems to be that there is relatively little spontaneous change in food preferences once they are established in spite of advertising and marketing techniques. For men, satiety or the "fillingness" property of a food is one of the better predictors of food consumption. Preference for certain food items seems to be strongly related to the geographical region of origin of the individual, though there are certain staples of the United States diet, such as bread and white potatoes, for which there is almost complete acceptance among various regions of the country. There are many more dishes which are typically southern and are preferred in southern areas than there are dishes which appear to be specialties of other areas.

Another factor strongly related to food preferences in the Army was age. Preference for soups and vegetables tended to increase with age, while preferences for beverages, cereals, desserts and fruits tended to decrease with age. In general, one might say that young people like sweet foods better, and they also like frankfurters and ground meats better; older people like "hot" condiments better.

Relatively few food group preferences correlated with educational differences. In general, however, vegetable preferences decreased with increasing education, as did preferences for fish, certain meat combinations, frankfurters, hot condiments and sauces.

A statement taken from the retrospective comments concerning food preferences of men in the U. S. Armed Forces seems particularly appropriate: "Man eats not just what his system needs nutritionally and what suits him physiologically; the range of food stuffs which fulfill these purposes is broad. He eats what is available, what he likes, what his culture defines as food, what his personal history dic-

tates, and what society and his peers say he should eat. This listing could be extended much further. It is important to recognize that no single factor has predominant control, but that they interact in complex ways to determine final acceptance and usage" (Peryam et al., 1960).

Experience would seem to dictate that if one were engaged in mass feeding in this country of individuals whose particular preferences one either did not know or in a practical sense could not cater to individually, one should serve foods in as natural a form as possible without added seasoning or sauces except salt, singly and not in combination, and in a form in which the food is readily identifiable. A variety of condiments, seasonings and sauces could be made available for the consumer to apply at his own discretion.

BASES OF FOOD LIKES AND DISLIKES

Various studies have tried to analyze the bases of particular food habits. Many of these have been summarized by Torrance (1955) as: the influences of personality maladjustment; psychophysical factors; learning, motivation and beliefs; and psychosocial and psychocultural factors, including group influences, family factors, regional and national factors, cultural traditions and social structure and status. One thing seems certain, food habits are formed early in life and are influenced by all forces which mold personality and behavior.

EMOTIONAL FACTORS

Emotional factors have a tremendous role in the formation of food habits and the use of food. We learn about the emotional uses of food primarily from three sources: observations of normal eating habits and behavior in the study of normal development of human behavior, observations of abnormal eating habits and behavior especially in young children and the solution of these eating disturbances and, finally, contributions from psychoanalysis.

Babcock (1948), Simon (1960), Hamburger (1960) and Rabinovitch and Fischhoff (1952) all discuss the significance of early experiences in feeding in relation to personality development. As Simon points out, the mouth has a psychology all its own, and the mouth and eating become involved in most of the hierarchy of developmental influences, drives and conflicts beyond the early infancy phase. Later conflicts over bowel and bladder training are fought out in food habits, and still later sexual problems and conflicts become manifest through eating habits and practices. Rabinovitch and Fischhoff (1952) describe the development of food habits in children and how they reflect emotional needs.

Babcock (1948) discusses in some detail emotional uses to which food is put: (1) as an instrument to relieve anxiety; (2) negatively to deny one's needs (food may be rejected because the taking of it for one's pleasure produces anxiety); (3) to increase one's feeling of acceptableness and security in society; and (4) to affect the response and behavior of another person—to cajole, to discipline, to punish, to threaten, to deprive or to establish a non-verbal means of communication. In connection with the use of food to relieve anxiety, Dr. Babcock makes the point that the need of some patients for the continuance of their patterns of eating in order to accommodate their psychological equilibrium may be greater than any possible rational and nutritive goal seen for them by their physician or health educator. Food from the day of birth is associated with intimacy and carries not only the feeling of security, protection, love and developing strength but also the sense of pain, rejection, deprivation and the potential terror of starvation (Babcock, 1961).

The range of attitudes about food is tremendously wide and complex (Moore, 1952). Moore discusses food as a symbol, as a substitute for other kinds of satisfactions the patient is unable to acquire, as a weapon turned against oneself, as a technique for punishing and rewarding (an unspoken language) and as a symbol of social prestige. She also has an interesting discussion of social class differences in food attitudes.

Several investigators have shown that the neurotic individual has a greater number of food aversions than the non-neurotic (Wallen, 1945; Gough, 1946). Adolescents are particularly prone to food dislikes, and Hellersberg (1946) has shown that the adolescent's food habits are definitely related to his general adjustment and level of maturity or immaturity.

CULTURAL FACTORS

Harris (1961) has indicated that for most people food is cultural, not nutritional. A plant or animal may be considered edible in one society and inedible in another. Probably one of the most important things to remember in connection with the cultural factors involved in food habits is that there are many combinations of foods which will give the same nutritional results. The combinations which may be best in the United States may not be best for other areas of the world. The combinations which are best for the middle class may not be the best for the poorer classes. Health workers are often accused of being culture bound and tending to reject concepts and programs of behavior (including eating patterns) which are different from their own.

Culture consists of values, attitudes, habits and customs, acquired by learning which starts with the earliest experiences of childhood, much of which is not deliberately taught by anyone and which is so thoroughly internalized that it is unconscious but "goes deep" (Fathauer, 1960). Food habits are among the oldest and most deeply entrenched aspects of many cultures and cannot, therefore, be easily changed, or if forceably changed, can produce a series of unexpected and unwelcome reactions. Food and food habits as a basic part of culture serve as a focus of emotional association, a channel of interpersonal relations, a channel of love, discrimination and disapproval and usually have a symbolic reference. The sharing of food symbolizes a high degree of social intimacy and acceptance.

Many factors in a culture must be known before one can attempt to change food habits. Fathauer notes that in the United States there are numerous distinct subcultures which, for effective nutrition education, need to be carefully defined so that the educational program can be designed to fit specifically the particular subcultural group. Harris (1961) has indi-

TABLE 18-1. RECOMMENDED DIETARY ADAPTATIONS

U. S. Dept. of Agriculture Guide (1958)	Cultural Background			
	Irish	Italian	Jewish	Spanish
Milk Children, 3–4 cups Teen-agers, 4 or more cups Adults, 2 or more cups	Milk Buttermilk	Milk Cheese	Milk Cottage cheese Sour cream	Tortillas (calcium) Peppers (vitamin A)
Meat 2 or more servings of beef, veal, pork, lamb, poultry, fish, eggs, dried beans or peas, nuts	Beef stew	Veal cacciatore	Pot roast	Arroz con pollo (rice with chicken)
Vegetable-fruit group 4 or more servings to include a citrus fruit or vegetable important for vitamin C, and a dark green leafy or deep yellow vegetable for vitamin A	Kale Turnip greens	Tomatoes Greens	Carrots Various greens	Mango Avocado Citrus
Bread-cereal group 4 or more servings of whole grain, enriched or restored	Potato	Spaghetti	Rye bread Pumpernickel bread	Rice

cated how the 1958 U. S. Department of Agriculture simple guide to good eating may be adapted for certain cultural groups in the United States (Table 18-1).

Wherever society makes social distinctions between groups there are foods considered appropriate to one's own class and others for different classes. Food use is patterned in relation to physical status, age, sex, state of health, pregnancy, etc. Certain foods are labeled "baby foods," "female foods" or "convalescent food." In many cultures food has a social or ceremonial role. Certain foods are highly prized; others are reserved for special holidays or religious feasts; still others are a mark of social position. There are cultural classifications of food such as "inedible," "edible by animals," "edible by human beings but not by one's own kind of human beings," "edible by human beings such as self," "edible by self." In different cultures certain foods are considered "heavy"; some, "light"; some, "foods for strength"; some, "luxury," etc.

The more one knows of the particular meanings of food in any culture or subculture the less errors one is likely to make in suggesting dietary modifications. The more one knows concerning the induction of children into a food pattern the more one may be able to influence the development of food habits in a desired direction.

GEOGRAPHY, CLIMATE, AND TRANSPORTATION

The influences of geography, climate and transportation are far more important in determining food patterns in other parts of the world than in the United States. In most countries food patterns are built around foods which are indigenous to the particular locality. Thus, the principal cereal and legume used varies with the geography of the country. Nonindigenous foods ordinarily should not be introduced into a diet unless the change is necessary, the food can be produced locally, and it is acceptable to the people. Too often people accustomed to the United States dietary make the mistake of trying to transport it to other parts of the world without due regard to local food customs, supplies and needs.

ROLE OF NUTRITION EDUCATION

What role does nutrition education play in determining what people eat? The actual part is to a large extent unknown. Some people probably are too optimistic about what can be done. Some feel that education plays a relatively small role and lay greater stress on emotional, cultural and economic factors.

Olson (1958) has said that no program of nutrition education can be effective if the emotional aspects of food and diet are ignored. There is a real need for well conceived and well controlled studies on the effects of nutrition education, both in terms of increase in nutritional knowledge and more especially in terms of changes in nutritional practices.

NUTRITION EDUCATION IN SCHOOLS

Nutrition education may be of the more formal variety, such as is taught in school rooms or in adult or extracurricular classes such as extension groups, Red Cross, etc.; it may be of the more informal sort through mass media of communication such as radio, television, magazines, newspapers, booklets; or it may be by face-to-face contacts among family, friends or neighbors.

At what grade level should the more formal classroom approach to nutrition be used? Should it be taught by a specialist or by the regular teacher? Can one effectively teach nutrition by words or must the lessons incorporate activities? Should it be taught as a separate subject, in health or science courses or as an integrated part of social studies, history, geography, etc.?

To date there has been little opportunity to know just what role nutrition education could play in changing food habits and food acceptance. Few teachers have any training in nutrition and relatively few efforts have been made in any systematic fashion to incorporate nutrition education into the school curriculum. We look to the day when at least the elements of nutrition are incorporated in the curricula of teacher training institutions so that hopefully it may become an integral part of the knowledge each teacher has to share with pupils.

The primary role of the school in the formation of food habits probably should be teaching what constitutes an adequate diet in the early grades as an integrated part of other learning experiences. Good school lunches and personable school lunch personnel can be important factors, whose value should not be underestimated, especially for children who at home have not known good menu planning, good food preparation or a variety of foods.

Limited studies have been made in some schools of Canada and the United States of the effect of brief periods of nutrition education directed toward the apparent weaknesses in nutrition practices in the school on subsequent nutrition practices. In addition, several projects have developed materials or activities suitable for nutrition education at various school levels (Lockwood 1947, 1948, 1950; Jacobson et al., 1959; Nutrition Education Research Project, 1961). Two national conferences on nutrition education have been held (Proceedings Nutrition Education Conference, 1957; 1962).

NUTRITION EDUCATION IN HOMES

Lewin (1943) has called the homemaker the "gate keeper" of the food available to the family. A number of studies have been undertaken to determine the adequacy of homemakers' knowledge of what to feed their families, where they appear to have gotten their information and how their nutritional knowledge relates to their actual practices in feeding their families (Young et al., 1956). Results indicate that those who have ever studied "about what to eat" had a better knowledge of nutrition than those who had not, and the principal source of information seemed to be the schools. The homemakers with the best nutritional knowledge appeared to do a better job of feeding their families. Formal educational attainment of the homemaker seemed to be the single most important factor in gauging her nutritional knowledge. The homemakers' greatest need for more nutritional knowledge was with regard to ascorbic acid-rich fruits and vegetables, carotene-rich fruits and vegetables, the adult need for milk and the nutritional value of breadstuffs and cereals and of butter and fortified margarine. Although magazines and papers were important sources of information, there is no evidence of the relative influence of editorial content and advertising.

Parents have the primary responsibility and influence in the formation of children's food habits. Eating habits are largely an expression of child-parent relationships, and much child-parent conflict may be reflected in a child's food habits. Frequently, arguments about food are really hidden ways of arguing about other things such as love, jealousy and hostility. Good advice to parents with regard to feeding young children might run something like this: "You have the responsibility to know what constitutes an adequate diet and some of the many ways by which it might be achieved. You have the responsibility to do your best to provide an adequate diet in an attractive well prepared form and in a considerable variety so that your child may become acquainted with many kinds of food. And, then, you have the responsibility to be quiet. Food should not become an emotional football. Few children in this country suffer from severe nutritional problems; many suffer from emotional conflicts which center around the feeding process. Nonverbal example on the part of parents contributes far more constructively and far less destructively than a verbal tirade on the subject of food."

CHANGING FOOD HABITS

As previously explained, food habits are extremely difficult to change; hence, one should never attempt to change them for the mere sake of change. If changes must be made for therapeutic purposes, they should be limited to the minimum necessary to accomplish any critical change in the diet. It is well to remember that many different combinations of food may be used to attain an adequate diet.

A food prejudice may sometimes be overcome by a change in the form of preparation of the food. Changes in diet may be accomplished by a change in the frame of reference, in an individual's perception of a food through social or prestige suggestions, in terms of valuation or in an individual's feeling of belonging (Lewin, 1943). Other methods may include offering a choice of foods, changing the name of a food, psychotherapy and group decision techniques. It has been pointed out frequently that, especially for growing children, adolescents and young people who are well, health is not a primary consideration in the formation of food habits. As Rowntree

(1949) has said, "Food is eaten because it provides something more gratifying than health." She also states that the child who eats well has been blessed with an understanding mother.

FOOD HABITS OF ADOLESCENTS

The appeal that will be effective in changing food habits once they have been established will vary with different age groups. Adolescents or teen-agers, for example, have special psychological characteristics which must be taken into consideration (Gallagher, 1962). They are at an age of self-concern and want one to be interested in *them*, not just their problems. They are great imitators and learn more from example than from advice or lectures. They live for the moment, want quick results and are unimpressed by warnings of what may happen in years to come. They want to be accepted by others of their own age group and dislike being different. They are in the process of attempting to gain independence and, hence, authoritarian advice, warnings, orders and suggestions are likely to be rejected. All adolescents, especially girls, want to look nice. Appeals for change based on a promise of improvement in personal appearance—glossier hair, better complexion, firmer flesh, trimmer figure—have meaning to the adolescent girl. Physical fitness has tremendous appeal for adolescent boys. Anything which will give energy, build muscles or contribute to stamina and endurance has meaning for them. Health as an abstract concept has little or no meaning for adolescents, but health tied to physical fitness (endurance, being able to compete) or appearance (in concrete terms) has tremendous appeal.

Another effective approach for teen-agers may be based on social acceptance. They want to be wanted, want to be good guests, want to be accepted. One of the best means of persuading an adolescent to try new vegetables or other foods strange to him is on the basis of his learning to be a more appreciative and socially acceptable dinner guest. This is a far more powerful incentive than any number of nutrition facts.

1. Give the teen-ager credit for knowing something.
2. Give him a chance to talk and ask questions.
3. Allow him to accept or reject an idea while talking about it. Then he won't have to reject it by action. (This does not mean giving up your stand based on your knowledge just because the teen-ager says he doesn't agree with you. You can stick to your guns and still allow him the opportunity to dissent verbally.)
4. Don't fall into the trap of "taking sides" in an argument between generations that probably is not really based on food.
5. Don't force-feed food or ideas to teen-agers. Force will make them resist or become more childish and less able to be responsible adults.
6. Don't be afraid to express your honest feelings about what is important. Adolescents don't really want to be "strictly on their own" yet. They are not ready for absolute responsibility as long as they are living in the family home.
7. The best teaching is by example, not by talking. How many of us have fruits and vegetables at our coffee breaks instead of doughnuts or a candy bar?
8. Good nutrition is a long-term thing, not a meal-by-meal or even a day-by-day emergency.
9. There is no perfect food. Yet there is some food value even in the snacks we worry about (milkshakes and hamburgers do contain proteins as well as carbohydrates and fats).
10. Aim for acceptance of one's self instead of going along with an adolescent's dream/wish. Time and energy are often wasted wishing for, or investing in, magic promises to achieve ideal health, ideal nutrition, an ideal mate or an ideal world. That time and energy could be much better spent utilizing to the fullest what is available in this problem-filled world.

* Roth, A.: J. Amer. Diet. Assn., *36*:27, 1960.

A teen-ager is interested in himself and how he as an individual "stacks up" with regard to some attribute. Many teen-agers do a far better job of eating balanced meals than one might think. The combinations of foods may seem peculiar to the adult, but the net result may be good. Teen-agers want straight nutrition information. They are interested, and the information does not have to be sugar coated. Above all, they do not want to be "talked down to" or "preached at." They want to share information. Given independence and the opportunity to make their own decisions, they welcome helpful unemotional advice on which to base *their* decisions. Like most of us, they cannot stand to have their ideas made

the subject of scoffing or ridicule no matter how peculiar or far-fetched they may seem. Often one finds the teen-ager is quite willing to accept with respect a dispassionate, uncritical evaluation of the accuracy or inaccuracy of his ideas.

Dr. Arthur Roth, Director of the Teen-age Clinic at the Permanente Medical Group of California, has, on the basis of his experiences, made ten excellent suggestions to parents who wish to help educate the teen-ager concerning proper nutrition. Dentists might find them very helpful in working with their teen-age patients (Roth, 1960) (Table 18-2).

In a study of food practices of 12 to 14 year old girls in Iowa, knowledge of nutrition, social status, psychological adjustment and family relations, and food enjoyment all had a highly significant relationship to diet adequacy; health as a value was significantly related. Those who fell within the following categories were inclined to have poor diets: those concerned about their weight; those who were obese or underweight per age and state of maturation; those who placed a high value on sociability, independence, status or enjoyment of food (Hinton et al., 1963). These findings are of interest since the teen-age girl is usually considered the most poorly fed member of the family group.

A Minnesota study of adolescents' views on their foods indicated that it is essentially a mother's world as far as the sanctioning of food habits is concerned. Apparently other persons, such as physicians, dentists and contemporaries, play only a peripheral role in this area, while teachers are virtually ignored. The authors conclude that it is possible that the value of these other groups may lie in their influence as a catalyst rather than as an active agent (Litman et al., 1964). Such findings may have important indications for dentists.

FOOD HABITS
OF THE ELDERLY

Savitsky and Zetterstrom (1959) point out some of the particular problems involved in feeding the elderly and in attempting in any way to change their eating patterns. They stress disturbances in appetite as sensitive indicators of anxiety, and food prejudice as a ritual which is an adaptive technique designed to avoid anxiety. Security is achieved through the maintenance of rigidly held attitudes and rituals in which food acceptances play a part. In the elderly, the attempt to wean the individual from food prejudices must be done cautiously in view of his diminished capacities for protection against anxiety.

FOOD FADS AND FALLACIES

In an article on food faddism Beeuwkes (1954) quotes from a book on "Fads and Feeding" which was published in London in 1908. The quotations are as applicable today as when they were written. Food faddism has been with us for generations and probably will continue in some form or degree as long as man exists. Fads come and go and are replaced by others. Sometimes a fad persists so long that it becomes incorporated into the culture and is then very difficult to remove.

Why do food fads persist in spite of incontrovertible scientific evidence that denies their validity? Olson (1958) believes it is because food has emotional rather than intellectual value to the average person and that food faddists capitalize on this fact, appealing to the emotional drives of their believers rather than to their intellects. The food faddist appreciates the symbolic value of food much better than does the professional concerned with scientific nutrition. The faddist plays on the hopes, fears and superstitions of the individual. With the expansion of fundamental knowledge of nutrition, the public has become conscious of the vital importance of nutrition and wants to know more about it in order to provide the best nutrition for self and family. Sebrell (1954) believes it is this very receptiveness on the part of the public which now provides such a good opportunity for food faddism to flourish.

Milstead, of the Food and Drug Administration, U. S. Department of Health, Education, and Welfare (1959) expands this point by presenting several facts which are basic to his group's efforts in combating nutritional quackery:

1. Quackery is not new. It has been with us since ancient times. It will continue to be with us because it feeds and grows on powerful incentives of human nature—the desire for health; the fear of pain, disease and death; the desire to make easy money, and the desire to save money even at the risk of health. These incentives are strongest in the prejudiced, the ignorant, and the uninformed. They are strengthened through misinformation—they are weakened and destroyed through education and enlightenment.

2. The American people are health-conscious and eager for knowledge about health in all of its aspects. They want to know about the developments in medicine, about new medicines and new treatments. They are hungry for information about nutrition and our foods—how they are prepared and if they are safe and nutritious. Never in the history of science has the public been more interested in the developments in the scientific field, and especially in the fields of health and medicine.

3. The increased interest in science and particularly in health and medicine is good and the dissemination of accurate and truthful knowledge in this field can be of great benefit. But the concern of the public about health has created a favorable climate for quackery and is no doubt one of the most significant factors in the increase in medical and nutritional quackery in recent years. . . .

4. The most widespread kind of quackery in the United States today is quackery in the field of nutrition. This racket is based on misstatement of the facts of nutritional science. . . . Food faddism and other forms of medical quackery are big business in the United States. . . .

5. A vast amount of misinformation about diet, nutrition and health is reaching the public by many channels. Through modern media of mass communication—newspapers, magazines, books, radio, television, person-to-person and even through educational institutions and libraries—the American people are being propagandized in a way that tends to undermine their confidence in the purity and nutritional value of their food supply. Misrepresentation from a multitude of sources is designed to encourage the purchase and consumption of a wide variety of special dietary preparations ranging from simple vitamin preparations to complex mixtures containing such things as royal jelly, unsaturated fats, dried alfalfa, water cress, rose hips, seaweed, etc. Many of these products are being offered as cure-alls for serious diseases. . . .

6. Vitamin and mineral products have a recognized place in modern preventive medicine, when for some special reason the diet requires this kind of supplement. But they are not cure-alls, and it is dangerous for anyone to assume that they are. People who have serious medical problems are misled by these false claims to rely on products which do not actually help them, and thus fail to get proper medical attention until it is too late. If symptoms of disease or ill health are present, it is a matter for the physician—not the food quack or food faddist.

What are food fads? How do they arise? Many arise from both modern and ancient myths and superstitions about food. Usually they are a matter of misinformation which

TABLE 18-3.   EXAMPLES OF POPULAR FOOD FADS AND FALLACIES

Type of Fallacy	Specific Examples	Fact
Fad—special power or wonder foods	Black strap molasses	Crude molasses has more vitamin B and iron, along with impurities, than refined molasses but is not eaten in sufficient quantities to add appreciably to iron and B vitamins in normal diet.
Combination of foods harmful or poisonous	Milk and fish eaten together are harmful or poisonous	No clinical or experimental evidence. Fish chowders and milk at fish meals have been eaten by many for years. Fallacy may have had origin in days before refrigeration when fish may not have been fresh.
"Natural" foods only	Enriched flour should not be used because of dangerous chemicals added or Only whole grain flour should be used because refined flour has been "impoverished" with the "removal of vital elements" by "overprocessing"	No dangerous chemicals have been added to enriched flour in restoring elements removed during milling. A great deal of flour is refined to increase the keeping quality of the flour and because it is preferred by the American people. Enrichment restores the elements removed by preservation.
Only organic fertilizers should be used	Food is nutritionally inferior because it is grown on impoverished soil Food grown on soil treated with chemical fertilizers is poisoned	Not true. Depleted soil yields very poor crops; hence, little of the food consumed comes from such soils. Research shows that the nutritional value of crops is not affected by the soil or fertilizers used. The only disease in man known to be associated with any deficiency of water or soil is simple goiter due to lack of iodine in certain areas. This deficiency is easily remedied by the use of iodized salt.
Need for extensive food fortification or supplementation	Foods need to be fortified or supplemented with proteins, minerals and/or vitamins because of "subclinical" deficiencies from which many people suffer	Scientific surveys of the nutritional status of people in the United States have shown few indications of any nutritional deficiencies.
Cure or prevention of disease by special foods	Faulty nutrition is the basis of all disease Disease is caused by chemical imbalance in the body, which in turn is due to faulty diet	There are certain well known diseases which are caused by specific dietary deficiency such as scurvy, due to a lack of vitamin C. These diseases are rarely encountered in the United States and are readily recognized by competent physicians.

may arise through honest error, misguided zeal or deliberate deception. Fads and fallacies are numerous, but many of the modern ones tend to fall into five or six general types (Table 18-3). One of the oldest of food fads has to do with special virtues or powers of some particular food or group of foods out of all proportion to the actual nutrient content. Certain fruits, yoghurt, black strap molasses, seaweed, wheat germ oil and royal jelly have been promoted in this fashion. Still other fads or fallacies would have one believe that particular foods or groups of foods are harmful either generally or to certain individuals or that certain combinations of foods, such as fish and milk, must never be eaten together. Another

fallacy centers about the use of only "natural" foods because of the supposed great dangers to health of materials added to foods during processing, on the one hand, or the supposed great "impoverishment" of the food, on the other hand, as a result of the "removal of vital elements" in "overprocessing." Some promote the idea that there are "nutritionally inferior" foods grown on "impoverished soil" or soil "poisoned" by chemical fertilizers; according to this fallacy only "organic" or "natural" fertilizers should be used.

Some advocate the fortification of the diet with various vitamins, trace minerals, etc., on the grounds that these will improve health or correct the supposed "subclinical" nutritional deficiencies, which are supposed to be the cause of one's tiredness, aches and pains. Still more disastrous fallacies have to do with all the diseases that diet is supposed to cure and that are assumed to have grown out of a "chemical imbalance" in the body due to "faulty diet."

Many of these food fads and fallacies are so ridiculous that perhaps some feel we should be able to look on them with good humor. But they are so malicious and dangerous that they cannot be ignored. Why? The economic loss alone is tremendous. A few years ago the American Medical Association estimated that quackery involving false and misleading claims for a great variety of food supplements was costing 10 million Americans over $500,-000,000 a year. Far more serious than this is the threat to health and life posed by false claims for special products which have caused people with serious medical problems to be misled and to fail to get proper medical attention until it is too late. A third, more subtle, danger from food faddism is the general undermining of sound nutrition knowledge and practice and of the confidence of the public in the value of legitimate nutrition practices.

Examples of specific food fads and fallacies and the facts to counteract them may be found in two excellent source materials: the packet of materials on "Food Faddism and False Claims," published by the American Medical Association; and the booklet, "Food Facts Talk Back," of the American Dietetic Association. A popular paperback, Deutsch's "The Nuts Among the Berries," is a rich source of information on food fallacies over the years.

IMPROVEMENT THROUGH EDUCATION

What can be done in a constructive way to combat food fads and fallacies? There seem to be two general answers: education and law enforcement to protect the public. In the long run, a less gullible public must be the answer. However before the general public can be educated, the professionals (physicians, dentists, nurses, dietitians, nutritionists, biochemists and teachers) on whom they depend for facts must be informed. Sound nutrition

instruction should be a part of the background of all these professionals if they are to be a reliable source of information for the public. It is sad indeed that, on occasion, the professional himself propagates food faddism either through ignorance and lack of scientific background, carelessness, lack of discrimination in information passed on, or even indifference. The family physician and dentist, each in his respective field, should be the court of appeals in all matters of health and should be able to give discriminating answers or be able to direct their patients to a source of reliable information.

Sebrell (1954) stresses the role of the scientist in relation to food fads. He sees the scientist's responsibility as being threefold: "First, he must constantly guard against initiating fads—a hazard in speculating upon experimental results. Moreover, in the pursuit of knowledge, in the acceptance of hypotheses and sometimes conclusions, the scientist himself is not immune to faddism. He should be careful not to impart his unfounded enthusiasms to the public. Secondly, the scientist can attack faddism by subjecting current fads to experimental evaluation. With definitive answers derived from research, he can state not only that the faddists' claims are groundless or implausible, but also untrue. Third, the scientist is obligated to be a leader in nutrition education, and to lend his authority in combating falsehood. The best ammunition against food fads has come from the research laboratory, and the public, by and large, will continue to turn there for guidance."

Sebrell also has said that there is no substitute for an enlightened consumer. The incorporation of nutrition education into the curricula of teacher training institutions would mean the ultimate education in nutrition of all children in the schools—an ideal but long-range solution to the problem. Responsible use of mass media such as magazines, newspapers, television, radio, etc., to inform the public about sound nutrition practices would help. This could be handled either directly by reputable scientists or by scientific writers whose materials could be checked by appropriate competent scientists.

The Food and Drug Administration now has a Division of Public Information which has undertaken an aggressive educational drive against nutritional quackery. They have joined with the A.M.A., other professional organizations, and the National Better Business Bureau in campaigns against food faddism and quackery. Personnel from the group are talking to civic and consumer groups. Educational materials such as the circular "Food Facts vs. Food Fallacies" have been prepared and are being distributed. The Administration is working with writers in preparing articles on nutritional quackery for newspaper syndicates and national magazines. Legal actions are being publicized.

IMPROVEMENT THROUGH LAWS

Law enforcement is another means of attacking food faddism and nutritional quackery. Under the Federal Food, Drug and Cosmetic Act, the Food and Drug Administration may proceed by seizure, prosecution and injunction in the Federal courts against all forms of nutritional quackery in which labeling is involved. Their enforcement program is directed against three types of operators: (1) shippers and distributors of food supplements and other special dietary preparations which bear exaggerated and unwarranted claims on their labels; (2) peddlers of "shotgun" mixtures of vitamins and minerals, plus some "secret" factor, who make a "doorstep diagnosis" and prescribe their preparations not only as an answer to all health problems, but all too frequently for the treatment of serious diseases; and (3) "nutritional educators," "nutritional advisors" and "health food lecturers" who rent halls and deliver pseudoscientific lectures about the cause and treatment of practically all the diseases known to medical science and incidentally prescribe their own special preparations for the treatment of whatever ails those who come to hear and believe.

Furthermore, the Federal Trade Commission under the Federal Trade Commission Act of 1914 and the subsequent Wheeler-Lea Amendment of 1938 can attack nutritional quackery by procedures against false and misleading advertising, either as a form of "unfair competition" or as "unfair and deceptive acts and practices," if either the product ad-

vertised or the advertisement itself moves in interstate commerce. False advertising includes not only direct falsehoods but also failure to reveal material facts respecting consequences resulting from use of the product.

Unfortunately, the jurisdiction of Federal law enforcement agencies is limited to matters of interstate commerce. Effective state and local laws and enforcement are necessary as well.

In the long run, the most effective answer to food faddisms will be a nutritionally well educated professional person who, in turn, can educate the public.

REFERENCES

Babcock, C. G.: Food and its emotional significance. J. Amer. Diet. Assn., 24:390, 1948.

Babcock, C. G.: Attitudes and the use of food. J. Amer. Diet. Assn., 38:546, 1961.

Beeuwkes, A. M.: Food faddism and the consumer. Fed. Proc., 13:785, 1954.

Bruch, H.: Social and emotional factors in diet changes. J. Amer. Diet. Assn., 63:461, 1961.

Campaign Kit to Combat Food Faddism and False Claims. Chicago, Ill. American Medical Association, 1962 (includes reading list).

Committee on Food Habits: Manual for the Study of Food Habits. Bull. No. 111. Washington, D. C., National Academy of Sciences—National Research Council, 1945.

Committee on Food Habits: The Problem of Changing Food Habits. Bull. No. 108. Washington, D. C., National Academy of Sciences—National Research Council, 1943.

Committee on Food Research: Conference on Food Acceptance Research. Quartermaster Corps List QMC 17-9. Washington, D. C., Office of Quartermaster General, War Dept., 1946.

Deutsch, R. M.: The Nuts Among the Berries. New York, Ballantine Books, 1961.

Eppright, E. S.: Factors affecting food acceptance. J. Amer. Diet. Assn., 23:579, 1947.

Fathauer, G. H.: Food habits—an anthropologist's view. J. Amer. Diet. Assn., 37:335, 1960.

Food Facts Talk Back. Chicago, Ill., American Dietetic Association, 1957.

Gallagher, J. R.: Weight control in adolescence. J. Amer. Diet. Assn 40:519, 1962.

Gough, H G.: An additional study of food aversions. J. Abnormal Soc. Psychol., 41:86, 1946.

Hamburger, W. W.: Appetite in man. Amer. J. Clin. Nutr., 8:569, 1960.

Harris, R. S.: Cultural, geographical and technological influences on diet. J. Amer. Dent. Assn., 63:465, 1961.

Health Frauds and Quackery. Hearings before the Subcommittee on Frauds and Misrepresentations Affecting the Elderly, of the Special Committee on Aging, U. S. Senate, 88th Congress, 2nd Session. Washington, D. C., Gov't Printing Office, 1964.

Hellersberg, E. F.: Food habits of adolescents in relation to family training and present adjustment. Amer. J. Orthopsychiat., 16:34, 1946.

Hinton, M. A., Eppright, E. S., Chadderdon, H., and Wolins, L.: Eating behavior and dietary intake of girls 12 to 14 years old. Psychologic, sociologic and physiologic factors. J. Amer. Diet. Assn., 43:223, 1963.

Institute of Home Economics, Agricultural Research Service: Food for Fitness, Daily Food Guide. U. S. Dept. of Agriculture Leaflet No. 424. Washington, D. C., Gov't Printing Office, 1958.

Jacobson, W. J., Boyd, F. L., and Hill, M. M.: Promising Practices in Nutrition Education in the Elementary School. New York, Bureau of Publications, Teachers College, Columbia Univ., 1959.

Lee, D.: Cultural factors in dietary choice. Amer. J. Clin. Nutr., 5:166, 1957.

Lewin, K.: Forces Behind Food Habits and Methods of Change. In The Problem of Changing Food Habits. Bull. No. 108. National Academy of Sciences —National Research Council, Washington, District of Columbia, 1943, pp. 35-65.

Litman, T. J., Cooney, J. P., Jr., and Stief, R.: The views of Minnesota school children on food. J. Amer. Diet. Assn., 45:433, 1964.

Lockwood, E.: Goals for Nutrition Education for Elementary and Secondary Schools. New York, The Nutrition Foundation, Inc., 1947.

Lockwood, E.: Activities in Nutrition Education for Kindergarten through Sixth Grade. New York, The Nutrition Foundation, Inc., 1948.

Lockwood, E.: Activities in Nutrition Education— a Unit for High School Classes. New York, The Nutrition Foundation, Inc., 1950.

Mead, M.: Food Habits Research: Problems of the 1960's. Publ. 1225. Washington, D. C., National Academy of Sciences—National Research Council, 1964.

Milstead, K. L.: The Food and Drug Administration's Attitude on Food Fads and Nutritional Quackery. In The Role of Nutrition Education in Combating Food Fads. New York, The Nutrition Foundation, Inc., 1962, pp. 16-26.

Moore, H. B.: Psychologic facts and dietary fancies. J. Amer. Diet. Assn., 28:789, 1952.

Nutrition Education Research Project: Selected List of Nutrition Education Materials for Teachers. New York, Teachers College, Columbia University, 1961.

Olson, R. E.: Food faddism—why? Nutr. Rev., 16:97, 1958.

Peryam, D. R., Polemis, B. W., Kamen, J. M., Eindhoven, J., and Pilgrim, F. J.: Food Preferences of Men in the U. S. Armed Forces. Dept. Army, Quartermaster Research & Engineering Command. Chicago, Ill., Quartermaster Food and Container Institute for the Armed Forces, 1960.

Pilgrim, F. J.: The components of food acceptance and their measurement. Amer. J. Clin. Nutr., 5:171, 1957.

Proceedings of Nutrition Education Conference, April 1-3, 1957: Increasing the Effectiveness of Nutrition Education. U. S. Dept of Agriculture Misc. Publ. No. 745. Washington, D. C., Gov't Printing Office, 1957.

Proceedings of Nutrition Education Conference, Jan. 29-31, 1962: Improving Nutrition Education for Children. U. S. Dept. of Agriculture Misc. Publ. No. 913. Washington, D. C., Gov't Printing Office, 1962.

Rabinovitch, R. D., and Fischhoff, J.: Feeding children to meet their emotional needs. A survey of the psychologic implications of eating. J. Amer. Diet. Assn., 28:614, 1952.

Roth, A.: The teen-age clinic. A resource for health service and education. J. Amer. Diet. Assn., 36:27, 1960.

Rowntree, J. I.: Techniques in changing food habits. J. Amer. Diet. Assn., 25:1016, 1949.

Savitsky, E., and Zetterstrom, M.: Group feeding for the elderly. Psychologic factors of importance. J. Amer. Diet. Assn., 35:938, 1959.

Sebrell, W. H., Jr.: Food faddism and public health. Fed. Proc., 13:780, 1954.

Simon, J.: Psychologic factors in dietary restriction. J. Amer. Diet. Assn., 37:109, 1960.

Spindler, E. B.: Motivating teen-agers to improve nutrition. J. Home Econ., 55:28, 1963.

Torrance, E. P.: Food Prejudices and Survival. Note C. R. L. 55-LN-210, U. S. Armed Forces. Reno, Nevada, Survival Training School, Stead Air Force Base, 1955.

Wallen, R.: Sex differences in food aversions. J. Appl. Psychol., 27:288, 1943.

Wallen, R.: Food aversions of normal and neurotic males. J. Abnormal Soc. Psychol., 40:77, 1945.

Young, C. M., Berresford, K., and Waldner, B. G.: What the homemaker knows about nutrition. I to IV. J. Amer. Diet. Assn., 32:214, 218, 321, 429, 1956.

Young, C. M.: Eating problems in adolescence. New York J. Med., 61:939, 1961.

Section Three

Clinical Nutrition

Chapter Nineteen

The Clinical
Evaluation of Nutritional Status

NEVIN S. SCRIMSHAW, M.D., PH.D., AND WERNER ASCOLI, M.D.

Dentists, in general, have not availed themselves sufficiently of the simple but valuable tool of clinical inspection to obtain insight into the adequacy of the nutritional state of their patients and to identify the relationship of nutrition to some types of oral pathology. Although more time-consuming, the taking of a brief dietary history when a patient is first seen or when suspicious clinical signs appear also will be rewarding.

The dentist should become particularly familiar with the clinical signs suggestive of nutritional deficiency and should look for them routinely in his patients. It is convenient to do so, as he has ample opportunity to examine closely and under good light those body areas in which such signs are most likely to be detected: the hair, eyes, lips, mouth and skin of the face, neck and arms. No special equipment is required, and the patient need not disrobe. It should be very clear, however, that these tissues can respond to stimuli in only a limited number of ways; hence, the signs are seldom diagnostic of nutritional deficiency, merely suggestive of the problem. Suspicions aroused by the clinical examination must be verified by medical and dietary history and biochemical tests.

Clinical examination of the individual for signs suggestive of nutritional deficiency is an integral part of all nutrition surveys. The U.S. Interdepartmental Committee on Nutrition for National Development (ICNND) (1963) has had worldwide experience in this field and has developed a manual of procedures. A more concise document on the subject has been prepared by a committee of experts convened in 1962 by the World Health Organization (1963). Both include methods of collection and evaluation of dietary, biochemical and anthropometric data, as well as clinical data. Much of the information compiled in population surveys can be used in the appraisal of the nutritional status of patients, but interpretations may not be fully reliable.

MEDICAL HISTORY

A detailed medical history for diagnostic purposes is outside the role of the dentist, but he should inquire of the patient about the presence of diseases which may influence the interpretation of clinical signs of nutritional deficiency or be pertinent to the dentist's handling of oral problems of the patient. These will include acute and chronic infections and disorders such as diabetes, hypertension, heart disease, and allergy to foods or drugs. Other diseases that may condition secondary malnutrition are shown in Table 21-1, Chapter Twenty-one.

CLINICAL EXAMINATION

Each patient should be looked at as a whole before examination of specific areas, for general evaluation will enable the dentist to ascertain whether a person is grossly overweight or underweight, or has excessive pal-

235

FIGURE 19-1. *Flag sign of poor nutrition.*

lor, generalized skin lesions or other indications of unsatisfactory health that may possibly be related to diet. At this stage the height and weight of the patient should be obtained and possibly skinfold thickness measured at selected sites.

Once the patient is seated in the dental chair, the following order of examination will be convenient:

    General inspection
    Hair
    Eyes
    Face
    Neck
    Arms
    Lips
    Tongue
    Teeth and gums

(The only additional examinations made routinely by physicians when they are making clinical nutrition surveys are of the skin of the legs and feet, inspection or palpation of the ankles for edema, with determination of the presence of normal knee and ankle jerk reflexes, and a tuning fork test for vibratory sense. Rarely do these add any pertinent information except among rice-eating peoples or alcoholics, in whom altered reflexes due

to thiamine deficiency may be encountered; very mild cases of kwashiorkor also show lesions in the lower extremities.)

The following signs have proved useful in the assessment of nutritional status, although it must be emphasized again that they are largely nonspecific, and most are more reliable indicators when their frequency is assessed in population groups than in a single individual. Excellent color illustrations of most of the lesions described will be found in Jolliffe's recent book (1962) and will supplement the black and white illustrations of this chapter.

HEIGHT, WEIGHT, AND SUBCUTANEOUS TISSUE

Although much can be told about the status of grossly underweight or overweight individuals by simple inspection, standard weight for height is a more reliable means of discrimination (Appendix VI). Similarly, a gross estimate of subcutaneous fat can be obtained by pinching a double fold of skin over the outer surface of the upper arm (midtriceps region), although skin calipers provide a more precise estimate.

HAIR

Protein Deficiency. Protein malnutrition causes the hair to become light in color, fine, dry and brittle. Characteristically, the hair of persons with severe protein deficiency can be pulled out of the scalp without discomfort; whole tufts of hair come out readily by the roots. Since the hair is brittle, the tips also break off easily. In persons known to have brown or black hair, the color change is readily detected; without this knowledge it could be thought to be of genetic origin. The hair of Negroes becomes reddish in cases of protein deficiency. In blond persons with normally fine hair, the change is difficult to detect with certainty.

Color Changes. In assessing hair changes, the bleaching action of the sun and the care given the hair must be taken into consideration. Color change resulting from exposure to the sun generally will involve the upper layers of hair, while that caused by protein deficiency will affect deeper as well as superficial layers. Use of grease and oil and frequent washing or lack of it complicate interpretation. Individuals exposed to alternate periods of very poor and good nutrition may develop transverse light and dark hair bands; this is known as the flag sign. Children recovering from the severe protein malnutrition of kwashiorkor often develop this sign as normal hair grows out (Fig. 19-1).

EYES

Xerophthalmia. Both vitamin A and riboflavin deficiencies are known to affect the eyes. The xerophthalmia due to avitaminosis A begins with a dryness of the bulbar conjunctiva, as evidenced by loss of the light reflex, lack of luster and decreased lacrimation. It may proceed to keratomalacia, which is a softening of the cornea leading to ulceration, perforation, rupture and destruction of the cornea. The final result is a scarred, opaque cornea and a sightless eye.

Bitot Spots. Bitot spots often are associated with vitamin A deficiency, although they cannot be regarded as specific to it (Fig. 19-2). They appear as frothy, irregular, white or light yellow spots from one to several mm. in diameter, most often on the conjunctiva lateral to the cornea. They look as if they could be wiped away but are beneath conjunctival epithelium. A small Bitot spot may consist of only a few tiny air bubbles visible in the triangles, especially the outer ones. Both photophobia and the inability to see in dim light may be due to vitamin A deficiency.

Circumcorneal Injection. The area in the eye where the sclera changes to the cornea is called the limbus. The circumcorneal injection which is seen in riboflavin deficiency consists of penetration of the corneal limbus and branching of the subconjunctival arterioles that normally terminate within 0.5 mg. of the limbus (Fig. 19-3). This proliferation and congestion of the blood vessels in the sclera and their extension into the clear corneal tissue is by no means pathognomonic of riboflavin deficiency. Moreover, excessive exposure to sunlight, smoke, dust and other irritants is a recognized conditioning factor even when riboflavin deficiency is the underlying cause. Riboflavin deficiency may also

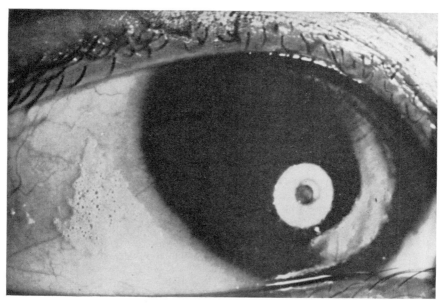

FIGURE 19-2. *Bitot spot.*

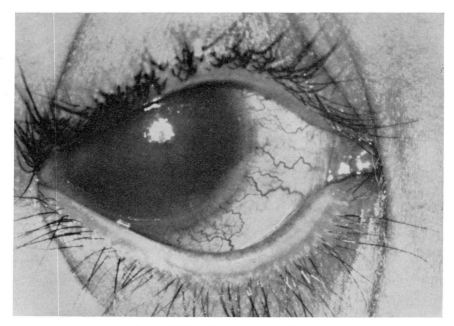

FIGURE 19-3. *Circumcorneal injection.*

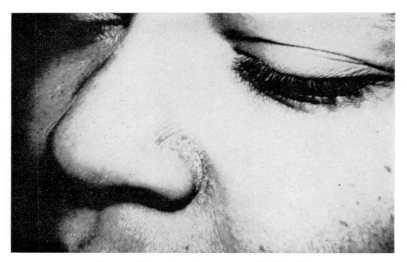

FIGURE 19-4. *Dyssebacea of nasolabial fold.*

produce a moist, red lesion of the external angle of both eyes.

Signs of Non-nutritional Diseases. There are a number of other signs in the eye which may be confusing to the examiner but which have no nutritional significance. Corneal scars, usually gray in color, are generally due to infectious conjunctivitis. Tiny hemorrhages at the end of the subconjunctival arterioles, as well as an increased vascularization and pigmentation of the exposed part of the conjunctivae, are very likely due to irritating environmental factors. Pinguecula are very small, white or yellow subconjunctival cholesterol or lipid deposits which may be confused with Bitot spots. A pterygium consists of a fleshy, red growth which encroaches from either angle of the eye upon the cornea and eventually may cover the pupil. Its cause is unknown.

SKIN

The areas of the skin most affected by nutritional deficiency usually are accessible to examination because they are also those most exposed to the environment. Thus, the skin of the face, neck, arms and legs, and the skin over pressure points such as the elbows, knees and ankles are most likely to show positive findings in nutritional deficiencies.

Dyssebacea. Dyssebacea is the clinical term used to designate a series of disturbances of the sebaceous glands characterized by increased oiliness and dermatitis, fissuring and exfoliation. The lesion of this type already mentioned as seen at the external angle of the eye in riboflavin deficiency is also found at the nasolabial fold (Fig. 19-4) and behind the ears, and may occur in other skin folds of the body. The nasolabial lesions should not be confused with acne in adolescents or with the irritation from nasal discharge in small children. For positive identification, the lesion must be red and humid.

Xerosis. Xerosis or generalized dryness of the skin is characteristic of vitamin A deficiency but difficult to evaluate in the individual because the appearance of the skin is influenced so much by bathing habits and exposure to dust and sunlight. Follicular hy-

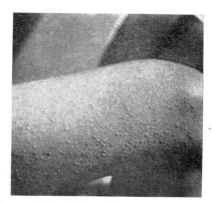

FIGURE 19-5. *Early follicular hyperkeratosis resembling "goose flesh." (Jolliffe, N., Tisdall, F. F., and Cannon, P. R.: Clinical Nutrition. Paul B. Hoeber, Inc.)*

perkeratosis is characterized by keratinization of the hair follicles to give a goose flesh appearance which is not altered by brisk rubbing. Fully developed lesions consist of symmetrically distributed, rough, horny papules formed by keratotic plugs projecting from hypertrophied hair follicles; they make the skin look and feel like coarse sandpaper (Fig. 19-5). Traditionally, the disease is associated with vitamin A deficiency and more recently in India with essential fatty acid deficiency. However, in Central America where it is exceedingly prevalent among schoolchildren, investigators have been unable to demonstrate any relationship between this lesion and either dietary vitamin A or fat.

Ascorbic Acid Deficiency. It is important to differentiate follicular hyperkeratosis from the perifolliculitis of ascorbic acid deficiency (Fig. 19-6). The latter consists of perifollicular congestion, swelling and, eventually, hypertrophy of the follicles. Only in this late stage is the skin rough to the touch. Other skin changes caused by severe ascorbic acid deficiency include petechiae, purpura, hematomas and increased capillary fragility with trauma easily caused by the blood pressure cuff, shoes or an accidental bruising. Again, it must be emphasized that there are many other causes of the phenomenon.

Pellagra. Cutaneous lesions of pellagra are variable because of differences in the degree of deficiency, original skin color and conditioning factors but are sufficiently char-

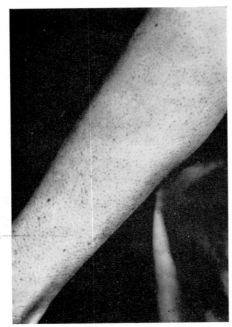

FIGURE 19-6. *Perifollicular hemorrhages on the leg of a boy, age 16, with scurvy. (From the Merck Report, May, 1956, Merck and Co., Inc., Rahway, N. J.)*

acteristic to be diagnostic in many instances. The most common chronic form is thickening, inelastic, fissured and deeply pigmented skin in areas especially exposed to sunlight or over pressure points. Eventually, the skin in affected areas becomes dry, scaly and atrophic.

Erythema upon exposure to sunlight is an acute manifestation, with subsequent vascularization, crusting and desquamation out of proportion to the precipitating exposure. Also, a common acute but highly nonspecific reaction is redness, maceration, abrasion and superimposed infection in intertriginous areas. Heat friction and poor personal hygiene are major conditioning factors. The symmetrical distribution of pellagrous skin lesions over exposed areas and pressure points is the most characteristic feature.

Edema. In severe protein deficiency the skin of the legs, arms and thighs shows marked pitting edema (which is a sine qua non of the diagnosis) and a dermatosis characterized by hyperkeratosis, hyperpigmentation and desquamation. Although often similar in appearance to the lesions of pellagra,

they are not limited to the exposed areas and characteristically extend to the thighs and lower trunk.

NECK

Casal's Necklace. Casal's necklace is a pellagrous dermatitis following the neckline when a low-necked dress or open shirt collar is habitually worn outdoors by persons on a niacin deficient diet.

Goiter. Endemic goiter resulting from a deficiency of iodine available to the thyroid gland usually can be detected by inspection of the neck with the head thrown back, and in case of doubt evaluated by simple palpation. If the goiter is visible with the head in the normal position, its pathological significance is beyond doubt, and it is Grade II according to the classification of the World Health Organization (Fig. 19-7). Grade III goiter is one easily recognizable at a distance. The difficulty in this superficial examination lies in the fact that in thin individuals not all thyroid glands that become visible with the head thrown back can be classified as being goiterous. As a rough guide, the size of a lobe of the normal thyroid can be taken as that of a lima bean about the size of the thumbnail of the person being examined. When the volume of the lobe exceeds this by 4 or 5 times, the goiter is classified as a Grade I or higher.

MOUTH

The mouth is one of the areas most sensitive to nutritional deficiencies, but the changes are nonspecific, confusing and difficult to evaluate. Pallor of the lips and mucous membranes, like pallor of the skin and fingernails, may be a consequence of anemia, but its clinical appraisal is so subjective as to be of little value except in severe cases. The angular stomatitis which may be a consequence of riboflavin deficiency also has been mentioned (Fig. 19-8). Angular scars may be the result of past episodes of acute ariboflavinosis but also may be wholly non-nutritional in origin. Even loss of teeth which produces closed vertical dimension or closed bite may

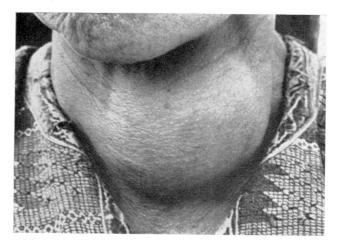

FIGURE 19-7. *Goiter.*

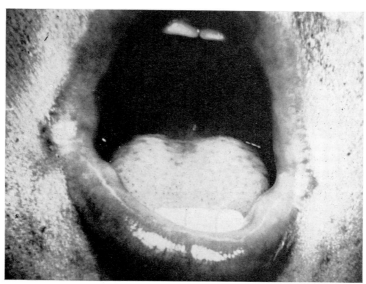

FIGURE 19-8. *Angular stomatitis.*

FIGURE 19-9. *Scorbutic gingivitis. (Courtesy, Dr. G. Shklar.)*

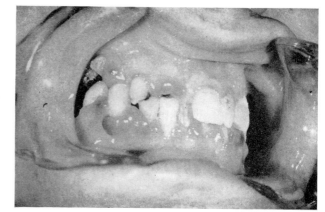

FIGURE 19-10. *Hypertrophy of tongue suggestive of vitamin B-complex deficiency.*

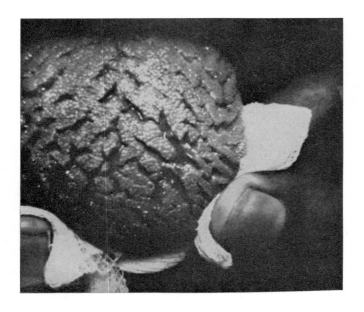

FIGURE 19-11. *Scrotal tongue*

result in maceration of the lips and corners of the mouth.

Gingivae. The gums may reflect any of a variety of nutritional deficiencies, but even the dentist will have difficulty differentiating the changes from those resulting from accompanying local oral irritants. This is especially true because neglect of diet and of oral hygiene are common in the same individual. Gingivitis is reported to be common in people whose diets are deficient in ascorbic acid, but most attempts to prove such a relationship by ascorbic acid therapy have been failures. On the other hand, engorged, dark red and bleeding gums are almost pathognomonic of scurvy (Fig. 19-9). (See Chapter Thirteen.)

Both marginal and generalized gingivitis have been stated to be associated with various other nutritional deficiencies, including vitamin A, niacin and riboflavin, but without proof. The problem is that many factors may play a causal role simultaneously so that the experimental correction of a single factor does not necessarily effect a cure.

Tongue. Marked hypertrophy of the filiform and fungiform papillae of the tongue is more frequently seen in populations in which B-complex deficiencies, particularly of thiamine and riboflavin, are present (Fig. 19-10). Papillar atrophy is sometimes related to niacin or iron deficiency, although any disease causing severe anemia will have this effect. It is for this reason that nutritional megaloblastic anemia results in a smooth tongue. A markedly furrowed, so-called scrotal tongue is a possible manifestation of vitamin A deficiency (Fig. 19-11).

Impression of the teeth on the borders of the tongue may be due to the edema resulting from protein deficiency as well as to defective dentures and a variety of other causes. When edema is the cause, there is also likely to be evidence of edema of the extremities.

Nutritional changes also occur in the color of the tongue, although they are usually difficult to evaluate. A reddening of the tongue which is generalized or limited mainly to the distal third may be associated with the deficiency caused by sprue. It affects first the tip and lateral margins but may progress to include not only the entire tongue but all

TABLE 19-1. CLINICAL SIGNS MORE OR LESS SUGGESTIVE OF NUTRITIONAL DEFICIENCIES

Deficiency	Suggestive	Possibly related
Vitamin A	Xerosis of skin	Bitot spots
	Xerophthalmia	Follicular
	Keratomalacia	hyperkeratosis
	Nonpurulent	Conjunctival
	conjunctival	thickening
	injection	Fungiform papil-
	Photophobia	lary hypertrophy
	Night blindness	of tongue
		Marginal gingivitis
		Scrotal tongue
		Circumcorneal
		injection
Riboflavin	Angular stomatitis	Circumcorneal
	External angle	injection
	lesion of eye	Papillary hypertro-
	Nasolabial	phy of tongue
	seborrhea	Red tip of tongue
	Canthal fissures	Magenta tongue
		Angular scars
Niacin	Pellagrous	Gingivitis
	dermatitis	
	Red tongue	
Thiamine	Absence of ankle	Papillary hypertro-
	jerk	phy of tongue
Ascorbic	Gingival injection	
acid	and swelling	
	Gingival bleeding	
	Subcutaneous	
	petechiae,	
	folliculitis	
Iodine	Endemic goiter	
Iron	Papillary atrophy	Pallor of mucous
	of tongue	membranes
Protein	Hair changes:	Retarded eruption
	color, texture,	of primary teeth
	easily pluckable	Malposition of teeth
	Edema (children)	
	Lingual edema	

of the oral mucous membranes as well. A beefy red glossitis resembling raw beefsteak probably represents most frequently a deficiency of niacin, but other B-complex vitamin deficiencies may be involved. Riboflavin deficiency is probably most often responsible for the purplish discoloration known as magenta tongue. Patchy areas of paler discoloration give rise to the name geographic tongue, which has no nutritional significance.

Teeth. Although discussed in more detail in Chapter Twenty-nine, the relationship between nutritional deficiency in infancy and early childhood and caries susceptibility, retarded eruption and malposition of teeth should be mentioned here. Malposition of primary and secondary teeth resulting from retarded bone development is sometimes the result of early protein deficiency.

In many underdeveloped countries a so-called hypoplastic line across the upper primary incisors has been described, into which, in time, a yellow or brown pigment is deposited, followed by the development of caries on the labial surface of the teeth. Subsequently, the teeth break off along this line. A nutritional or febrile insult in the neonatal period is a probable cause.

Fluorosis is characterized by mottled enamel and chalky white patches distributed over the surface of the teeth. In some cases, the entire tooth surface may be dull white, giving an unglazed appearance.

The signs suggesting nutritional deficiency are easier to evaluate and more specific in children than in adults. In adults lesions from other causes, such as aging and repeated trauma, and confusing signs such as pinguecula, pterygium, and geographic tongue are more likely to be present. The clinical signs which suggest more or less strongly the possibility of a nutritional deficiency are summarized in Table 19-1. It should be clear from the foregoing that clinical examination alone will rarely be sufficient to establish a definite diagnosis. A medical history is useful and so are dietary and biochemical data.

DIETARY HISTORY

The techniques of obtaining quantitative data on nutrient intakes by interview or diet records methods, although beyond the scope of this discussion, have been described in Chapter Twenty-eight and also delineated elsewhere (*FAO, 1953*). The dentist can and should obtain qualitative dietary information which could shed light on the probability that signs of nutritional deficiency will be encountered and on the cariogenicity of the habitual diet. For example, an individual who drinks milk regularly is not likely to be deficient in riboflavin, and one consuming abundant citrus and other fruit and juice will not be deficient in vitamin C. In fact, a person consuming a diet which is normal and balanced is not likely to be deficient in

any nutrient. In such a case it is quite wrong to try to interpret minimal changes in the skin, eyes or mouth as a consequence of some nutritional deficiency and to suggest vitamin supplements. On the other hand, if questioning reveals that a patient is an alcoholic or food faddist, signs of nutritional deficiency may well be present. In the first case, the signs are likely to be those associated with B-complex vitamin deficiencies, and in the second, nutritional anemia. Similarly, the questioning may reveal a heavy consumption of candy and sweet desserts which are exacerbating the problem of dental caries.

The dietary questioning need not be involved or time-consuming when the diet is relatively good, but when it is poor, time and skill are required to obtain reliable information and to draw proper conclusions. For more detailed work on quantitative dietary estimates, the services of a dietitian or nutritionist should be enlisted by dentist and physician alike.

BIOCHEMICAL TESTS

Just as a single picture may be worth more than a thousand words, so a single biochemical determination, if accurate, may decisively confirm or deny the nutritional origin of an uncertain complex of clinical signs. Here again, biochemical measures vary in their reliability and specificity. Serum levels of vitamin A and ascorbic acid are quite useful, but levels approaching deficiency may be present for some time before lesions appear. Total serum protein and serum albumin are decreased markedly in protein deficiency but only when the deficiency has progressed to the point at which it is clinically evident. The ratio of urea to creatine has been recommended recently as a measure of the loss of lean body mass in protein-calorie malnutrition. For the B-complex vitamins, blood serum levels are relatively insensitive, and measurement of thiamine, riboflavin, and niacin excretion in 24 hours or in timed urine samples is recommended. Anemia is best diagnosed by determination of plasma hemoglobin and hematocrit accompanied by examination of a thin blood smear for red cell size. It should be noted that most biochemical

TABLE 19-2.   TENTATIVE GUIDE TO THE INTERPRETATION OF SELECTED BIOCHEMICAL DATA USEFUL IN THE APPRAISAL OF NUTRITIONAL STATUS*

	Deficient	Low	Acceptable	High
Plasma protein: gm./100 ml.	< 6.0	6.0 – 6.4	6.5 – 6.9	≧ 7.0
Serum albumin (electrophoretic method): gm./100 ml.	< 2.80	2.80– 3.51	3.52– 4.24	≧ 4.25
Hemoglobin: gm./100 ml.				
Men	<12.0	12.0 –13.9	14.0 –14.9	≧ 15.0
Women (nonpregnant, nonlactating; ≧13 yr.)	<10.0	10.0 –10.9	11.0 –14.4	≧ 14.5
Children (3-12 yr.)	<10.0	10.0 –10.9	11.0 –12.4	≧ 12.5
Hematocrit (PCV): per cent				
Men	<36	36–41	42–44	≧ 45
Women (nonpregnant, nonlactating: ≧13 yr.)	<30	30–37	38–42	≧ 43
Children (3-12 yr.)	<30.0	30.0 –33.9	34.0 –36.9	≧ 37.0
Plasma ascorbic acid: mg./100 ml.	< 0.10	0.10– 0.19	0.20– 0.39	≧ 0.40
Plasma vitamin A: μg./100 ml.	<10	10–19	20–49	≧ 50
Urinary thiamine: μg./gm. creatinine	<27	27–65	66–129	≧130
Urinary riboflavin: μg./gm. creatinine	<27	27–79	80–269	≧270
Urinary N-methylnicotinamide: mg./gm. creatinine	< 0.5	0.5 – 1.59	1.6 – 4.29	≧ 4.3

* From Manual for Nutrition Surveys. 2nd ed. Bethesda, Md., Interdepartmental Committee on Nutrition for National Defense, National Institutes of Health, 1963.

measures show only the present situation so that chronic nutritional lesions are possible even in patients with normal nutritional biochemistry.

Table 19-2 provides a guide to the interpretation of biochemical measures of particular value in appraising nutritional status. It should be noted that a vitamin deficiency lesion could be present as the result of a previous deficiency even when biochemical examination at the time of an office visit gives normal values. A short period of improved nutrient intake could obscure biochemical values before a lesion caused by a deficiency has had time to heal.

The Dentist's Responsibility with Respect to Nutritional Advice

The use made by the dentist of information regarding the nutritional status of the patient necessarily varies greatly with the situation. When anemia is suspected, the patient should be urged to see his physician for more detailed diagnostic procedures and treatment. This should be the case also when severe deficiency disease of any kind is present. On the other hand, advice can be given appropriately by the dentist when there is obvious excessive use of cariogenic foods, evidence of imbalanced diets likely to lead to

difficulty, or minimal suggestive clinical signs coupled with compatibly poor dietary habits. In between lies a spectrum of conditions for which the need for referral will depend upon the nature of the situation, the training and experience of the dentist, and the diagnostic and consultant facilities at his disposal.

In mild cases involving oral lesions under his direct observation, the dentist may be justified in recommending dietary changes or vitamin supplements as a therapeutic trial before deciding on the need for outside consultation. Because of the systemic ramifications of any frank nutritional disease, however, and the importance of precise diagnosis of the etiology of anemias, the dentist always should obtain medical consultation or be sure that the patient sees his physician when these entities are identified or strongly suspected.

SUMMARY

The dentist is in a favorable position to observe, without inconvenience to his patients or disruption of his routine, nearly all the areas in which signs suggestive of nutritional deficiency are commonly encountered. These areas are the hair, face, mouth, neck and skin of the arms and legs. The lesions found are generally not diagnostic but only suggestive of possible nutritional deficiency; they are seldom cues to nutritional deficiency alone but instead to a variety of conditioning factors acting upon weakened tissue.

Clinical lesions in these areas, however, usually serve to alert the dentist to nutritional factors which may be influencing adversely the oral tissues or the general health of his patient. Qualitative information on dietary habits, as well as a medical history, and laboratory examinations made at the same time may sufficiently confirm the clinical impression to justify dietary counseling or medical referral, depending upon the nature and severity of the suspected condition.

REFERENCES

Brozek, J.: Body Measurements and Human Nutrition. Detroit, Wayne University Press, 1956.

Expert Committee on Medical Assessment of Nutritional Status, World Health Organization: World Health Organization Technical Report, Series No. 258. Geneva, 1963.

Food and Agriculture Organization: Dietary Surveys, Their Technique and Interpretation. Nutritional Studies No. 4. Second Printing. Rome, 1953.

Interdepartmental Committee on Nutrition for National Defense, National Institutes of Health: Manual for Nutrition Surveys. 2nd ed. 1963.

Jolliffe, N. (ed.): Clinical Nutrition. 2nd ed. New York, Harper & Brothers, 1962.

World Health Organization: Endemic Goitre. Monograph Series No. 44. Geneva, 1960.

Chapter Twenty

Nutritional Status and Infection

NEVIN S. SCRIMSHAW, M.D., PH.D.

The interaction of nutritional status and infection has dual significance for the dentist. Stress from any cause, including oral trauma, anxiety or fear of dental manipulation, as well as both localized and generalized infections, has an adverse influence on nutritional status. The consequences of this stress depend on the prior and continuing diet of the individual.

Of even greater importance in ordinary dental practice is the fact that individuals with significant deficiency of almost any essential nutrient are more susceptible to infections and their complications. Furthermore, a vicious cycle is easily established whereby the infection is more severe because of the malnutrition, and the malnutrition becomes worse because of the infection. In fact, most frank nutritional disease is the result of a combination of dietary deficiency and conditioning factors, of which infection is by far the most common.

Although relatively few patients in the United States and in more privileged socioeconomic groups of developing countries are likely to be sufficiently malnourished for this interaction to be of practical concern, identification by the dentist of the occasional individual in private practice whose oral infections are secondary to poor nutritional status is very important. In clinic practice in the United States and in the developing countries, nutritional status is much more likely to be a factor in oral infections. Similarly, under such circumstances stress arising from pathologic conditions in the oral cavity may actually precipitate clinical nutritional disease.

EFFECT OF INFECTION AND OTHER CAUSES OF STRESS ON NUTRITIONAL STATUS

When an individual is deprived of all food, a series of metabolic adjustments occurs which lead to the mobilization of amino acids from skeletal muscle and other tissues from which they can be spared for gluconeogenesis in the liver. This is necessary because liver reserves of glycogen from which blood glucose is maintained are soon exhausted, yet glucose is essential to the functioning of the central nervous system and the heart. This process also makes available the amino acids required for necessary synthesis of enzymes and other proteins in the liver.

The reaction described involves increased secretion of glucocorticoid hormone by the adrenal cortex, also a common feature of stress of any origin. It maintains the individual ready to fight or flee regardless of the timing of meals, a feature undoubtedly of crucial importance to the survival of early man. Part of the amino acids thus mobilized are deaminated to provide the carbon skeleton of glucose; their nitrogen is discarded and excreted in the urine, largely as urea. Serum levels of ascorbic acid, vitamin A, and probably of certain other nutrients also are lowered by the stress reaction and may appear in the urine in increased amounts.

There is good evidence that even such mild infections as vaccination against smallpox and immunization with the 17-D strain of yellow fever vaccine, as well as small localized abscesses, will provoke this stress reaction. Trauma has a similar effect, although studies have not been made of the minimum

247

trauma required to cause a detectable stress reaction. Pain and anxiety, even concern about a college examination, have been shown to measurably increase urinary nitrogen excretion. It can be presumed that anxiety prior to and during dental treatments can have the same effect, since increased glucocorticoid hormone excretion has been demonstrated under these circumstances.

The duration of the metabolic effect of stress depends upon the nature and extent of the precipitating cause. If it is of short duration, it may be balanced by increased retention during the same or a succeeding 24 hours. The stress arising from infection, however, is likely to be sufficiently long-lived as to be followed by a recovery period of increased nitrogen retention of similar or greater duration. During this time amino acids are returned to the tissues from which they have been mobilized during the time of stress.

The duration of nutritional effect of an infection or any prolonged cause of stress is thus much greater than would be suspected from the length of the acute episode. Though the magnitude of the effect is, of course, influenced somewhat by the intensity of the stimulus, much more important is its duration. Moreover, in the case of nitrogen, at least, some adaptation occurs at lower levels of protein intake so that the urinary loss measured is less than would be expected from observation of well nourished individuals.

SPECIFIC EFFECT OF INFECTION

One of the most constant and significant consequences of infection is decreased appetite and, along with this, even decreased tolerance for food. As a result, the ingestion of protein and other nutrients is often decreased just at a time when metabolic losses are increased. In addition, changes in the diet to less solid and more liquid food are common therapeutic measures which may further depreciate the nutritional value of the diet during infection.

Among preschool children in developing countries, these changes are probably the most important factors in the chain of events lead-ing from infection in an already malnourished child to the development of the severe protein malnutrition of kwashiorkor, of the xerophthalmia and keratomalacia of avitaminosis A, of scurvy, and of other nutritional diseases. Diarrhea either due to enteric infection or secondary to infection elsewhere in the body may also reduce the absorption of nutrients.

When the combination of metabolic and cultural factors leads to widespread malnutrition, infection will be present to make it worse and to become the actual precipitating cause of most nutritional disease. It is probable that psychological factors such as maternal deprivation of the child, anxiety over supposed witchcraft, as well as trauma, sometimes also have this effect.

EFFECT OF MALNUTRITION ON INFECTION

Most types of nutritional deficiency, if sufficiently severe, reduce the resistance of the human host to infection. On the other hand, if the nutritional deficiency is relatively mild but the infecting agent particularly virulent or host resistance excessively low, improving nutritional status may have no effect on the severity of infectious disease. However, virulence and susceptibility vary so widely in population groups that for many individuals adequacy of past dietary intake may be the factor determining the consequences or even the presence of infectious disease.

The effect of malnutrition on resistance to infections has been extensively studied in experimental animals. Under appropriate experimental conditions, nutritional deficiencies are almost always found to be synergistic with bacterial, rickettsial, helminthic, or intestinal protozoal infections, so that the consequences are more serious than in well nourished animals. Sometimes a severe deficiency of a specific vitamin or other nutrient, especially when worsened by the feeding of a metabolic antagonist to the deficient vitamin, will be antagonistic to the infectious agent. In this case, the deficiency interferes more with the reproduction of the agent than with the resistance of the host. Most instances of antagonism in experimental animals are limited to viral and systemic protozoal infections.

There are no known examples of nutritional deficiencies limiting the spread of an infection in man. The kind of malnutrition found or induced in man, when sufficient to produce an observed effect, seems always to interfere with resistance mechanisms without being severe and specific enough to inhibit the infectious agent itself. Even if a type of nutritional deficiency which would interfere with viral replication could be induced in man, it would still be highly dangerous and detrimental since it would lower resistance to the secondary bacterial complications which are the usual cause of death in viral diseases.

Direct evidence of the synergistic interaction of malnutrition and infection in man is less extensive than for experimental animals and comes mainly from epidemiological rather than experimental studies. The increased susceptibility of protein deficient children to infectious disease is well documented, as are the effects of avitaminosis A and scurvy. The common infectious diseases of childhood, particularly measles, and diarrheal disease are far more frequently fatal in the poorly nourished preschool children of developing areas than in well nourished children of the same age. Furthermore, nutritional improvement decreases the severity of these infections even when an unsanitary environment and lack of medical care persist. It has been recognized since early in the century that children with xerophthalmia or with scurvy are much more likely to develop bronchitis, bronchopneumonia, otitis, pyelonephritis and other septic complications.

There have been few systematic studies of the direct effect of malnutrition on oral infections in man. Secondary infections of the mouth have been reported to be more common in riboflavin deficiency and to benefit from correction of the deficiency. More detailed documentation of evidence for the effect of malnutrition and infection will be found in the bibliographical material at the end of the chapter. Published studies of the influence of malnutrition on infection in man are summarized in Table 20-1 according to the type of infection and nutrient deficiency present.

MECHANISMS OF RESISTANCE INFLUENCED BY MALNUTRITION

Antibody Formation. Although other mechanisms may be of more importance in resistance to oral as well as various other infections, the ability to form specific antibodies to infectious agents or their toxins has received the most attention. This is due in part to the excellent techniques for their measurement and also to the success of immunization procedures in protecting against many infections.

In experimental animals severe deficiencies of protein, vitamin A, ascorbic acid, riboflavin, thiamine, pantothenic acid, biotin, and niacin-tryptophan have all been shown to interfere with antibody formation under some circumstances. There are also a number of studies indicating that protein deficiency interferes with antibody formation in man. False negative tuberculin reactions in malnourished persons with tuberculosis have also been reported several times. Severe pellagra, moderately severe deficiency of several of the B-complex vitamins, and highly specific deficiencies of pantothenic acid or pyridoxine induced by feeding the corresponding antimetabolites desoxypyridoxine and omega methyl pyridoxine have also been shown to have this effect in human subjects.

Phagocytic Activity. The macrophages and microphages of the reticuloendothelial system are also important factors in the resistance of the body to infectious diseases, particularly those of bacterial origin. In theory, nutritional deficiency could interfere

TABLE 20-1.   SYNERGISM OF INFECTION AND
        MALNUTRITION IN MAN*

Deficiency	Bacteria and Rickettsia	Viruses	Protozoa and Helminths
Multiple	11	2	1
Protein	7	4	6
Vitamin A	7	—	—
Vitamin C	4	1	—
B-complex	3	—	—

* Summarized from 47 studies. (Scrimshaw et al., 1966.)

with either the number or the phagocytic capacity of these cells. Both consequences have been demonstrated. Protein deficient diets regularly impair leukocytosis in rats, and children with kwashiorkor show little or no leukocyte response to superimposed infection. Similarly, in protein deficient rabbits and roosters the polymorphonuclear leukocytes appear unable to phagocytize bacteria normally.

Decreased macrophage activity also has been reported in deficiencies of vitamin A, ascorbic acid, thiamine and riboflavin. In both experimental and clinical observations, folic acid deficiency has been shown to interfere with the production of phagocytes in mammalian bone marrow to such an extent as to nullify the effect of protective antibodies. The relationship of these observations to secondary infections of the oral tissues is obvious.

Tissue Integrity. Antibody formation and phagocytosis are mechanisms which become involved only when the infectious agent has already entered the host. Strictly speaking, they are not mechanisms which prevent infection, but ones which prevent or minimize the possible pathological consequences of infection. One means of defense which is specifically involved in preventing infection per se is the integrity of the skin, mucous membranes and other tissues.

There are a variety of pathological changes known to occur in tissue, depending on the type and the severity of nutritional deficiencies. Such changes as increased permeability of mucosal surfaces, reduction or absence of mucous secretions, alterations in intercellular substance, epithelial metaplasia, tissue edema and accumulation of cellular debris to form a more favorable culture media are all changes which may be observed in the oral cavity as a result of various deficiencies of essential nutrients and which have been stated to reduce local resistance to infection. Deficiencies of vitamin A, ascorbic acid, thiamine, riboflavin, niacin and protein are particularly likely to produce tissue changes leading to decreased tissue resistance as the oral manifestations of these deficiencies discussed in Chapter Nineteen might suggest.

Wound Healing and Collagen Formation. The effect of nutritional deficiencies on wound healing, fibroblastic response to local trauma, walling off of abscesses, and collagen formation is closely related to their influence on tissue integrity. In terms of the impact of any infection on the total period of disability, the most important consideration is often the rapidity with which the infection can be localized and contained. This is particularly true of some types of oral infections.

The walls of sterile subcutaneous abscesses induced in protein deficient rats are much thinner than in well nourished animals, and when such abscesses are spontaneously or experimentally infected, there is much less fibroblastic response. As a consequence, fatal septicemia develops with greater frequency.

Protein deficiency, particularly when methionine is limited, interferes not only with the conversion of procollagen to collagen but also with the tensile strength of the collagen fibers. Fasting for a week also has been shown to reduce the procollagen content of guinea pig skin and the synthesis of skin collagen as measured by the uptake of radioactively labeled glycine.

Ascorbic acid is also essential to the synthesis of the amino acids from which collagen is formed, particularly hydroxyproline and hydroxylysine, which are almost unique to collagen. The skin at the edges of the wounds of scorbutic guinea pigs has been shown to contain no hydroxyproline. When ascorbic acid was restored to the diet, this substance was rapidly produced. The importance of these observations for the dentist obviously transcends that of their possible relation to infections.

Nonspecific Resistance Factors. The relative importance of the various nonspecific mechanism of resistance to infection is difficult to assess, and little is known of the effects of nutritional deficiency upon most of them. Enzymes which help to destroy pathogenic microorganisms have long been recognized to occur in body fluids and to be excreted in tears, sweat and saliva. The activity of these so-called lysozymes has been reported reduced in human subjects by vitamin A deficiency and by a poor nutritional state in general.

Properdin is a euglobulin found in the serum of all normal animals thus far tested and is associated with natural resistance to many diseases of bacterial, viral and even protozoal origin. Under conditions which include the presence of magnesium, it appears to function by combining selectively with polysaccharides of high molecular weight. The properdin system is vulnerable to nutritional deficiency at several points, at the formation of the compound itself, as well as in the presence of an appropriate complement of magnesium. The only published evidence of this to date is the report of a marked reduction of properdin in the serum of pantothenic acid deficient rats.

Interferon is a natural product of animal cells which supplements other mechanisms of resistance to viral and some other infections. It appears to act by uncoupling oxidation from phosphorylation, so that the production of adenosine triphosphate is inadequate for replication of the infectious agent. Since interferon is a protein molecule, it seems likely that its formation is depressed in deficiency states in which protein synthesis is impaired.

Destruction of Bacterial Toxins. Most of the destruction of bacterial toxins or resistance to them in the animal host occurs as the result of the toxins combining with specific antibodies. There is some evidence that other mechanisms may exist as well. Rats suffering from deficiencies of B-complex vitamins or vitamin A are more susceptible than controls to diphtheria toxin even though antitoxin titers and rate of disappearance of injected toxin are similar in the two groups. Fasting mice also are rendered more susceptible to *Klebsiella pneumoniae* endotoxin, an effect seemingly independent of antibody formation and readily reversed by a good diet. Probably because of the ethical difficulties of investigating the problem directly in man, there are no clinical observations of this phenomenon.

Intestinal Flora. There is evidence that changes in the intestinal flora induced by diet can influence susceptibility to some intestinal pathogens. In kwashiorkor, there is a tendency for the bacteria of the lower intestinal tract to appear at a higher level. In milder degrees of protein malnutrition it seems likely that intestinal organisms which are not normally pathogenic may become so in the well nourished host and cause diarrhea.

Not only is no known pathogen identified in most diarrheal episodes in malnourished children, but also the frequency and severity of these episodes are decreased when dietary supplements are provided. Changes in gastrointestinal motility secondary to malnutrition may also play a role in the severity of protozoal and helminthic infections. Little is known about the effect of nutritional deficiencies either on the normal flora of the mouth or their relationship to oral infections.

Endocrine Balance. Endocrine factors are intimately involved in a number of the mechanisms of resistance to infection already mentioned, although little is known of their precise role. Examples include the loss of resistance to infection of adrenalectomized animals and of patients with Addison's disease and the serious problem of secondary infections in poorly controlled diabetics. Prolonged adrenocorticotrophic hormone (ACTH) or cortisone therapy, however, increases the spread of many infections, presumably because these hormones inhibit local inflammatory reactions at the site of bacterial proliferation. In experimental animals cortisone increases resistance to a number of bacterial endotoxins. Rabbits with alloxan diabetes develop the lesions of nasal, pulmonary and cerebral mucormycosis when inoculated with the responsible fungus, *Rhizopus oryzae*. These observations have a bearing on the question of how malnutrition influences resistance to infection, since protein deficiency, caloric deprivation and a number of other nutritional deficiencies influence endocrine balance. Because of the possible relationship to local resistance to infection, the dentist should be informed of any endocrine disorder in his patient and whether the patient is receiving any kind of endocrine therapy.

Practical Importance to Dentists

The dentist should take into routine consideration the fact that malnutrition in his patient is likely not only to interfere with the healing of oral lesions but also to be synergistic with oral as with other infections. Although very few individuals seen by the dentist in private practice are likely to have a degree of malnutrition sufficiently severe to have this significance, it is more likely to be the case for clinic and charity patients. Moreover, alert recognition of malnutrition in the occasional private patient with infections or other problems of the soft tissues of the mouth can be an important contribution to the effectiveness of the dentist's therapy and to the welfare of his patient. In his private practice the dentist should look for such cases particularly among the food faddists, elderly patients with chronic disease or too debilitated to eat properly, and teen-agers with poor dietary habits. The techniques for the evaluation of nutritional status presented in Chapter Nineteen are for everyday routine use and are not just background information.

SUMMARY

Whenever the dentist encounters a patient who is already malnourished, he should recognize that nutritional disease can be precipitated by trauma, infection or even excessive anxiety. Infection exerts an unfavorable effect on nutritional status by reducing appetite and tolerance to food. At the same time the diet is likely to be altered for the worse by reducing the amount of solid food and giving more liquid and carbohydrate. Home remedies and prescribed medicines may interfere with gastrointestinal absorption of nutrients or synthesis of the B-complex vitamins. With infection as with stress caused by trauma or anxiety, there are increased metabolic losses of nitrogen, ascorbic acid and probably of other nutrients.

Conversely, severe deficiency of almost any nutrient interferes with antibody formation and leukocyte activity. Tissue barriers to possible infection are also weakened by most nutritional disease. In addition, some deficiencies affect nonspecific resistance to infectious agents and to bacterial toxins. The effect of these deficiencies on intestinal flora and on endocrine balance may also have unfavorable repercussions on resistance to infection. Malnutrition may be responsible, therefore, for increased severity of infectious diseases and their secondary complications; and infections of the oral cavity are no exception.

Recognition by the dentist of the synergism between malnutrition and infection in man is of vital importance, for each may contribute to the severity of the other. When infection is present and clinical signs suggestive of nutritional deficiency described in Chapter Nineteen are detected, the dentist should give a high priority to confirmation of the presence of malnutrition and to its correction.

REFERENCES

Hodges, R. E.: Nutrition in relation to infection. Med. Clin. N. Amer., *48*:1153, 1964.

Scrimshaw, N. S., Taylor, C. E., and Gordon, J. E.: Interactions of nutrition and infection. Amer. J. Med. Sci., *237*:367, 1959.

Scrimshaw, N. S., Taylor, C. E., and Gordon, J. E.: Interactions of Nutrition and Infection. 2 volumes. WHO Monograph Series. Geneva, World Health Organization, 1966.

Viteri, F., Behar, M., Arroyave, G., and Scrimshaw, N. S.: Clinical aspects of protein malnutrition. *In* Munro, H. N., and Allison, J. B. (ed.): Mammalian Protein Metabolism. Vol. II. New York, Academic Press, Inc., 1964, Chap. 22.

Chapter Twenty-one

The Nutritional Implications of Obesity, Coronary Heart Disease and Secondary Malnutrition

GEORGE CHRISTAKIS, M.D., AND THEODORE B. VAN ITALLIE, M.D.

The dentist as a member of the health team can and, in fact, is expected to impart sound nutritional information to his patients, particularly if it has an oral relevance. For several reasons he should be informed on the more common clinical nutritional problems that his medical colleagues deal with daily. First, he will be able to educate and thereby resolve his patient's nutritional misinformation and misconceptions. Secondly, he will recognize his own professional limitations in the area of nutritional guidance. Thirdly, he will be wary of giving any nutritional advice for oral problems that might, in fact, be contraindicated for the general well-being of any patient who may be subject to concurrent medical problems. Fourthly, he will better understand the factors and mechanisms that contribute to oral changes which sometimes accompany common clinical nutritional problems. Thus, the purpose of this chapter is to acquaint the dentist with some newer knowledge on the nature and management of obesity, coronary heart disease and secondary malnutrition—three of the medical problems with important nutritional implications that prevail in people living in the Western hemisphere. Other diseases, such as the anemias, diabetes, allergic states, etc., also fall in this category but are discussed more appropriately in texts that are addressed primarily to the medical profession.

OBESITY

Incidence and General Concepts

Obesity is one of the most prevalent public health problems affecting the United States and many other Western countries; it has been estimated that 20 per cent of the population may be considered obese. Moreover, both the incidence of obesity among men in the United States and the magnitude of weight gain appear to be increasing. If "best weight," the term referring to that weight associated with the lowest mortality, is applied as an actuarial standard, three of every five American men aged 50 to 59 years are overweight by at least 10 per cent, and one out of three is at least 20 per cent above "best weight."

While not strictly a disease, per se, there is epidemiological and laboratory evidence that obesity may act as a "risk factor" associated with and thereby predisposing to such diseases as hypertension, diabetes mellitus, and coronary heart disease. Ultimately, obesity may contribute not only to morbidity but to decreased longevity as well. An individual who is 10 per cent over standard weight, as determined by actuarial height-weight tables, is subjected to a 13 per cent degree of excess mortality; if he is 20 per cent overweight, the degree of excess mortality is 25 per cent; at

253

30 per cent overweight, excess mortality is 40 per cent. Obesity, therefore, poses a great public health challenge by virtue of the numbers of persons afflicted, the diseases associated with it, and its possible independent influence on longevity. Obesity is no less a clinical challenge because its correction is frustrating to patient and physician despite the availability of theoretically effective therapy.

It is important to distinguish between obesity and overweight, since excess weight, in contradistinction to excess fat, may result from water retention, as in hypothyroidism or congestive heart failure, massive tissue formation, as in certain pelvic tumors, and, indeed, extreme muscular development. Obesity may occur in association with Cushing's syndrome, insulinoma or hypothalamic brain tumors.

Obesity appears most often to result from a complex interplay of host and environmental factors, ranging from possible genetic determinants to psychosocial and physical activity factors. Family and ethnic eating habits, the use of food as a demulcent for psychosocial stresses or as a recreational modality as evidenced by ever-present coffee breaks and cocktail and dinner parties, illustrate several causal variables. The decreased calorie need which occurs with advancing age must be considered within this type of environmental framework. The importance of physical activity is illustrated by the observation that obese adolescent girls actually consume fewer calories than girls of normal weight of the same age.

Definition and Diagnosis

Obesity is a state of excessive fatness. The diagnosis of obesity may be as easy as looking in the mirror to behold, unhappily, folds of facial or abdominal fat, or it may require body densitometry or isotope dilution techniques to determine that the overweight is, in fact, due to excessive fat accumulation and not to large bone structure or increased muscularity. "Normal weight" has also escaped a satisfactory definition since height-weight tables change with time, and some authorities advocate large–medium–small frame categories

without providing the anthropometric or clinical criteria required to classify subjects. The concept of "ideal" or "desirable" weight was introduced in 1942, when "ideal" weight was related to low mortality. The limitations of weight tables become apparent when it is recognized that excess weight may be due to large bone structure and muscularity and have little relation to excess body fat. Total body weight is comprised of the sum of all tissue weight such as muscle, bone, internal organs, body fluids, fat or neural tissue. Body fat may be estimated by a calibrated skin caliper. This instrument has yielded valuable data relating body fat accumulation to age and sex in various population groups.

Etiological Factors

What makes man eat? What hunger and satiety mechanisms regulate his food intake? Recent investigations indicate that central nervous system, gastrointestinal, hormonal, nutritional, and metabolic factors operate to influence appetite, hunger and satiety. Psychological and social conditioning are also important. For example, a child with one obese parent has a 40 per cent chance of becoming obese; if both parents are obese, this doubles to 80 per cent. It is both a fascinating and difficult problem to determine the extent to which genetic or environmental factors operate to explain these observations. There is, however, growing evidence suggesting that genetic factors are more important than was hitherto supposed. A large survey of obese middle-aged women disclosed that 20 per cent became obese in association with their first pregnancy. Once again, endocrinologic as well as a number of other factors may be involved.

In still other patients, obesity is related to profound psychological factors and may operate as an important buffer to psychological and social problems which the obese individual cannot resolve. Although obesity can be said to exist as a *sign* of serious physical or mental disease in a minority of patients, it is more often related to milder psychological needs or to sociological and physiologic processes of the sort cited previously. Once the obese state is initiated, the obese individual appears to behave metabolically somewhat differently from the normal; however, to date no predictive metabolic factors have been identified in the pre-obese state. It has been demonstrated that certain obese subjects mobilize fatty acids more sluggishly than normal during an acute fast. Such individuals also do not develop the same degree of hyperketonemia as lean subjects deprived of all food. Elevated urinary 17-ketosteroids have been found in some obese female patients.

Prognosis

Obesity is a paradox to both patient and physician. Few diseases are less challenging diagnostically and more frustrating therapeutically. Before a mode of therapy is initiated, the therapist must consider the "differential diagnosis" of obesity. Although excess weight can be attributed to an endocrine disorder in less than 1 per cent of obese persons, a high index of suspicion for hypothalamic damage, hypogonadism in the male, insulinemia and hyperadrenocorticism should be maintained by the physician. If it is determined that serious psychological factors are associated with the obesity, weight reduction should not be attempted until the underlying psychopathology is controlled.

Once organic and functional diseases have been ruled out in the obese patient, two therapeutic approaches (not mutually exclusive) exist: dietary and pharmacologic. The patient most likely to respond favorably to dietary treatment exhibits the following characteristics: (1) is slightly or moderately overweight, (2) became overweight as an adult, (3) never attempted to lose weight previously, and (4) is well adjusted emotionally and accepts weight reduction as a realistic goal.

The inclusion of anorexigenic drugs in weight reduction regimens usually is indicated only when dietary therapy alone has been unsuccessful. However, drug tolerance, side effects and addiction limit their usefulness. *Moreover, unless the causal factors which initiated and maintain the obese state are identified and amended and faulty diet patterns of the patient are permanently corrected*

by appropriate dietary counseling, weight loss will be transient.

Indications for Weight Reduction

Valuable clues to factors underlying the obesity can be obtained from carefully elicited dietary, medical and psychiatric histories. Since overeating and obesity can be interpreted as providing a buffer to the environment of children or adolescents unprepared to accept social and psychological challenges, parental or professional pressure to lose weight without prior psychotherapy may induce further untoward psychological effects. Since many preadolescent boys and girls who are active, healthy and moderately obese lose their obesity in their early teens, specific efforts at weight reduction during childhood may be inappropriate. However, guidance in the establishment of sound dietary patterns should be an integral part of health counseling. Since it has been shown that the obese adolescent may have low serum iron values, weight reduction in these instances should be accompanied by specific nutritional therapy. This illustrates the important concept that, contrary to popular opinion, the obese individual is not necessarily a repository of nutrients but may actually be malnourished.

Because obesity can be the result of specific diseases, the need for weight reduction and appropriate therapeutic measures should be determined by a physician.

The course of such chronic diseases as diabetes, osteoarthritis and hypertension are favorably influenced by weight reduction. Weight reduction prior to elective surgery is advantageous from the standpoint of both surgeon and patient and makes postoperative complications less likely.

Methods of Weight Reduction

SPECIFIC DIETARY REGIMENS

A logical dietary approach to weight reduction must provide all the essential nutrients while limiting calories. The reducing diet must introduce the patient to a "new nutritional way of life" which corrects defective diet patterns and becomes the foundation for the weight maintenance diet.

The menu plan of the reducing diet should include a large variety of the foods available in any supermarket and contain no food fad characteristics. The reducing diet for most female patients should contain no less than 1200 calories daily in order to supply all essential nutrients. For most obese males, a weight reduction diet of 1600 calories usually is adequate. At a caloric deficit of approximately 1000 calories per day, a weekly loss of weight of between two and two and one-half pounds should result.

The reducing diet should take into account cultural influences and methods of food preparation. Within the daily caloric allowances, eating need not be restricted to three meal periods. Two snack periods may be allowed. A nutritionist or dietitian, working with the physician, can contribute much toward modifying the patient's food habits and menu planning.

Lean meats should replace those which are especially fatty. Most vegetables may be eaten freely. Fruit and skim milk provide good snacks.

Diets varying in the relative proportions of protein, fat and carbohydrate have been designed in quest of a diet which will result in optimal weight loss. High protein diets seek to take advantage of the specific dynamic action of protein, the lower caloric content per gram and the prevention of excessive urinary nitrogen loss occurring with negative caloric balance. In order to maintain nitrogen equilibrium on a reducing diet, at least 60 gm. of protein and a minimum of 1200 calories per day must be provided.

High fat diets have also been proposed on the grounds that they provide a special satiety value. There is no evidence that high fat diets affect the metabolic equilibrium between lipogenesis and lipolysis. In studies in which obese women consumed diets varying in proportions of fat, carbohydrate and proteins, no relationship was shown between weight loss and composition of the diet consumed for a given caloric level.

FORMULA DIETS

During the last decade liquid formula diets have emerged from the metabolic ward and become available to the public. These are usually ingested in four equal 225 calorie portions throughout the day. The advantages of using a formula diet in reducing regimens lie in the accuracy of estimating daily nutrient intake, convenience of preparation, and perhaps in the limitation of food choices so that temptation is minimized. Prolonged consumption is untenable, not only because many persons find formula diets monotonous and unpalatable but also because a switch to conventional sources of food ultimately must be made. A return to pretreatment dietary patterns often will result in a reaccumulation of lost weight.

Numerous reports of experience with formula diets in the treatment of obesity are to be found in the medical literature. While all studies report successful weight loss during the first two to four weeks of use, even in recalcitrant obese patients, long-term results are discouraging.

ANOREXIGENIC AGENTS

The amphetamines induce habituation and tolerance after about six weeks of use. In addition, the following side effects have been reported: insomnia, irritability, restlessness, palpitations, tension, sweating, dry mouth and alterations in gastrointestinal function. Adrenergic effects in the cardiovascular system, such as increase in cardiac work, heart rate, blood pressure and potentiation of cardiac arrhythmias make the use of anorexigenic agents especially hazardous in patients with cardiac disease.

Results with anorexigenic agents in weight reduction regimens vary greatly. Opinions concerning their value range from complete rejection to enthusiasm. In some instances, results are inferior to those obtained on diets without simultaneous drug therapy. Results have been best during the early weeks of therapy and in obesity of recent onset. Newer anorexigenic drugs, such as phenmetrazine (Preludin), diethylpropion (Tenuate), chlorphentermine (Lucofen), and benzphetamine (Didrex), do not appear to be better than amphetamine since they are equally liable to produce habituation, side effects and tolerance.

Since anorexigenic agents are effective only with simultaneous dietary restriction, they should play a minor role in the treatment of obesity. It has proved difficult to evaluate these drugs in double blind studies since patients and doctors can easily distinguish the drugs from placebos because of their excitatory side effects. Moreover, individual responses to the effects of these drugs are variable.

Thyroid Preparations. Thyroid hormone has been used as an adjunct to weight reduction on the grounds that obese patients are in a hypometabolic state and can lose weight more easily if metabolic stimulation is accomplished; however, most obese patients are euthyroid. Besides the lack of rationale in the use of thyroid preparations in euthyroid obese patients, it has been shown that thyroid hormone has no effect on the success of weight reduction. Moreover, thyroid preparations can induce a variety of side effects including insomnia, nervousness, sweating and tachycardia.

Diuretics. The indiscriminate use of diuretics in weight reduction regimens can delude both patient and physician into believing that weight loss represents loss of triglyceride rather than loss of water. There is evidence, however, that some obese persons have a propensity for sodium and water retention. During the first few days of fasting, weight loss may be the result primarily of sodium and water excretion. It would appear that the use of diuretics may occasionally be indicated in obese persons who have signs of water retention or who, on restricted caloric intake, reach a plateau in weight loss which might be the result of water retention. Routine use of diuretics in weight reducing regimens is undesirable.

Glucagon. The parenteral administration of glucagon depresses food intake in healthy subjects. Glucagon may cause anorexia by decreasing gastric motility. The potential use of this hormone as an anorexegenic agent for weight reduction requires further evaluation.

Chorionic Gonadotrophin. The effectiveness of chorionic gonadotrophin as an adjunct to dieting is doubtful. It has been suggested that chorionic gonadotrophin promotes mobilization of fat from adipose tissue, making it available for metabolic use. Only one study has confirmed the original observation with this hormone, while five others failed to do so.

Bulk Producing Agents. Indigestible, hydrophilic substances such as methylcellulose have been included in weight reduction regimens on the grounds that they produce anorexia by expanding in the stomach and reducing gastric hunger contractions. The use of hydrophilic compounds has not been successful; these substances are unpalatable and may contain up to 20 calories per dose.

EXTREME MEASURES USED FOR WEIGHT REDUCTION

The treatment of obesity presents one of the most exasperating experiences in clinical and preventive medicine. It is not surprising, therefore, that extreme therapeutic measures have recently been applied to the grossly obese, including fasting and surgical excision of the ileum. Both of these procedures frequently are unreasonably drastic solutions to the obesity problem. Undesirable side effects occur in the application of both of these procedures, and the possibility of residual harmful effects has not been adequately assessed. Fasting induces certain metabolic alterations which may potentiate weight loss in the obese. These include: ketonemia, which is associated with and may cause anorexia; hyperuricemia, sometimes associated with exacerbation of clinical gout; qualitative changes in plasma amino acids; and urinary nitrogen excretion of 4 to 12 gm. daily. Potassium and multiple vitamin deficiencies may occur if supplements of these substances are withheld. Orthostatic hypotension, anemia unresponsive to vitamin B_{12}, iron or folic acid, decreased stamina and increased sensitivity to cold have also been observed as complications.

Surgical procedures which temporarily bypass the small intestine, such as the jejunocolic shunt, have been used as treatment for severe intractable obesity. In the largest series reported (10 patients), 50 per cent of the patients regained their former weight, only 30 per cent were able to maintain their ideal weight and only one of these had an adequate period of followup after intestinal continuity was restored. All patients required oral potassium and calcium supplements, five patients required intravenous potassium to restore their depleted potassium reserves, four patients had symptomatic hypocalcemia and in two of these, the shunt had to be terminated because hypocalcemia could not be controlled, even with intravenous therapy. All patients developed malabsorption syndromes associated with frequent frothy, bulky and foul-smelling stools and decreased serum carotene levels. Other potential harmful side effects noted in patients who have had ileal excision for reasons other than obesity are megaloblastic anemia, iron deficiency and impaired vitamin B_{12} absorption.

EXERCISE

While it is true that the energy equivalent of a pound of adipose tissue is approximately 3500 calories, and 35 miles of walking are required to expend this energy, an extra mile of exercise daily for 360 days will result in the equivalent total expenditure of a 10-pound supply of calories. It is the preventive long-term effects of small increments of exercise which may well deserve emphasis rather than attempts to lose weight by isolated attempts at excessive activity. Passive exercises such as massage and external manipulation with rollers, vibrators and chin straps are of no value.

In the United States there has been a gradual decrease in the amount of physical activity required to perform daily tasks. This applies particularly to the adult population in whom the gradual onset of obesity in middle age is a problem. It has been noted that obese women are far less active than nonobese women. As mentioned previously, obese adolescent girls also were found to be less active than nonobese adolescent girls, yet the obese actually ate less.

While dietary patterns are being revised

in the course of weight reduction, an assessment and adjustment of daily exercise patterns also can be carried out. Regular exercise contributes to a rather intangible physical and perhaps psychological state of well-being. The effect of exercise on circulatory dynamics and metabolic processes has yet to be fully investigated. The benefits of exercise may prove to be much greater than those related to caloric expenditure alone.

Evaluating the Results of Weight Reduction Programs

The evaluation of weight reduction experiences as recorded in the medical literature is difficult because: (1) existing criteria for measuring weight loss are not standardized and universally used; (2) relatively small numbers of subjects are sometimes used in weight reduction studies, and data on dropouts are often lacking; (3) the follow-up period is often short; and (4) the criteria of a valid "double blind" study have seldom been fulfilled in studies evaluating antiobesity drugs.

CORONARY HEART DISEASE

Early Observations

Populations subjected to nutritional deprivation had experienced a decreased incidence of heart attacks, according to reports made after World War II. These observations provided strong support for the concept of a relationship between diet and coronary heart disease. Thus, in the postwar years, teams of investigators representing a variety of disciplines have studied the relationships between diet, serum cholesterol, and coronary heart disease incidence in different parts of the world. It had already been known that persons with diabetes mellitus, hypothyroidism and familial hypercholesterolemia* carried a greater risk of coronary heart disease.

Epidemiologists studying populations of many nations have observed that the level of cholesterol appears to be related to the type

* A condition in which the cholesterol level may be extremely high because of a hereditary metabolic defect.

of diet consumed, although there has been disagreement concerning the specific elements of the diet responsible for influencing the cholesterol level. Some investigators implicated the type of dietary fat, stating that in those nations in which a diet high in saturated fat was consumed, the level of serum cholesterol also was high. Others believed that the total quantity of fat in the diet regardless of its nature was the governing factor. Still other workers related the amount of protein or sugar in the diet to the level of serum cholesterol.

Tenets about Coronary Heart Disease

Despite disagreement over the relative importance of certain nutritional factors in relation to coronary heart disease, the following tenets appear to be well founded:

1. Many pathogenic factors appear to operate in the causality of coronary heart disease. These include heredity, sex, diet, smoking, hypertension, level of physical activity, blood coagulation, obesity, psychological stress, and presumably others yet to be discovered.

2. The Framingham study, a prospective epidemiological study now in its twelfth year, has identified some of the risk factors associated with coronary heart disease: these include hypercholesterolemia, hypertension, smoking and obesity.

3. The nexus between dietary fats and coronary heart disease has been thought of in terms of the following relationships:

    a. The type and amount of dietary fat are related to the level of serum cholesterol.

    b. The level of serum cholesterol is related to atherogenesis.

    c. Atherogenesis is related to coronary heart disease.

4. Diets have been developed (such as the one utilized in the Anti-Coronary Club Study Project of the Bureau of Nutrition of the New York City Department of Health) that are palatable, nutritionally complete, safe, and effective in lowering the serum cholesterol level.

Experimental evidence as it relates to coronary heart disease in man derives from three major sources: the metabolic research ward, clinical observations (including pathological), and epidemiological studies. Metabolic ward research can yield direct and definitive results, but is necessarily carried out in limited numbers of selected hospitalized subjects. Clinical and epidemiological observations tend to be circumstantial and obtained from self-selected populations for which control groups are difficult to find. However, clinical and epidemiological studies frequently provide the initial observations which lead to the discovery of causal factors operating in disease. Tenets 1 and 2 are based on epidemiological evidence: the data supporting 3a have been derived from controlled metabolic research studies; 3b and 3c are based on clinical and pathological studies; and 4 is based on an epidemiological approach. The most difficult of these statements to defend would be 3b, i.e., the proposition that high serum cholesterol levels and atherogenesis are directly and causally related.

SERUM CHOLESTEROL LEVELS AND ATHEROGENESIS

At present we do not have detailed knowledge of the intimate mechanism whereby an elevated serum cholesterol level promotes atherogenesis. Moreover, we do not now know whether induced lowering of an elevated serum cholesterol level will retard or arrest the atherogenic process. The variables associated with coronary heart disease are difficult to measure and more difficult to standardize.

When a multifactor entity as complicated as coronary heart disease is approached by the epidemiological method, single variables cannot be easily manipulated while the other reacting variables are kept constant.

Not only is it unlikely that randomly selected subjects will fit neatly into the various cells of a classic experimental design, but the variables active in coronary heart disease are themselves interdependent. For example, the serum cholesterol level is affected by diet, lack of exercise and even psychological stress. Separation and assessment of the effects of these variables on serum cholesterol is indeed difficult. Moreover, the therapeutic means by which the serum cholesterol level is lowered may be crucial in adequately testing the hypothesis that a lowering of serum cholesterol level will be attended by a decreased morbidity and mortality due to coronary heart disease.

It may be that the genetic characteristics of the host determine the degree of individual susceptibility or resistance to these various factors. The environment may then inflict its effects through diet, psychological stress, sedentariness, etc., in a critical sequence and dosage. Then, when thrombosis is superimposed upon an arterial intima already damaged by an atherosclerotic lesion, the stage is set for myocardial infarction and its often catastrophic consequences.

Atherosclerosis may be considered in terms of two kinds of causal factors: (1) those difficult or impossible to influence, for example, factors deriving from the process of aging, and (2) those environmental and host factors associated with coronary heart disease which are potentially more susceptible to control. Since such manipulative factors may be of great etiologic importance, a public health approach to the problem of atherosclerosis appears feasible. The hazards of obesity can be influenced by educational and clinical programs. Carefully designed physical fitness programs based on age, weight, occupation and medical status may be promoted. If destructive psychological stresses operating in our society can be identified, it may be possible to reduce them through psychiatric practice and community health programs. The physician can today effectively treat hypertension. Antismoking campaigns and clinics are already in operation. Target populations subject to high serum cholesterol levels can be delineated and advised to adopt a serum cholesterol-lowering regimen.

"DESIRABLE" RANGE OF SERUM CHOLESTEROL LEVEL

What constitutes an elevated serum cholesterol level? Can "desirable" or "normal" ranges of serum cholesterol be validly defined for various age groups? On the basis of the Framingham and Albany studies and the experience obtained from two decades of international epidemiological investigations into the relationship between diet, serum cholesterol level and the incidence of coronary heart disease, a "desirable" maximum level for adults would appear to be 220 mg./100 ml. However, the serum cholesterol level of the male populace of the United States is above 220 mg./100 ml. by 27 years of age and above 250 mg./100 ml. by age 43. The serum level of the average woman may be over 200 mg./100 ml. by her fortieth birthday.

The mean serum cholesterol levels of a large cooperative study of male subjects age 40 to 59 indicate that about 32 per cent are below 220 mg./100 ml., 36 per cent fall within the 220 to 259 mg. range, and 32 per cent are 260 mg./100 ml. or higher. A recent survey indicated that among male New York City clerical workers age 40 to 59, 23 per cent were under 220 mg. and 40 per cent were above 260 mg./100 ml. The levels of 220 and 260 mg./100 ml. were chosen as demarcation lines since the Framingham study has revealed that subjects with serum cholesterol levels above 220 mg./100 ml. exhibited three times the normal risk of developing coronary heart disease, and those above 260 mg. per cent exhibited six times the normal risk. It can, therefore, be postulated that from 30 to 40 per cent of all United States males age 40 to 59 have serum cholesterol levels above 260 mg/100 ml. and carry a high risk of developing coronary heart disease.

Is this target population large enough to justify recommending a diet modification for all males in the United States in the age group 40 to 59, conceding that 20 per cent of this same group may not respond to a serum cholesterol lowering diet? And what about the 30 to 35 per cent of the males age 25 to 40 having serum cholesterol levels ranging from 220 to 259? For the younger male, should not the serum cholesterol level be prevented from rising to the high risk range? Do all men in the United States with a cholesterol level above 220 carry a substantially marked risk of coronary heart disease? Perhaps not, if they are not subject to the other potent variables active in the causality of this disease. However, a change of 1 per cent in the average serum cholesterol of population groups is believed to be associated with a corresponding change of 2.66 per cent in risk at all levels of serum cholesterol.

DIET AND SERUM CHOLESTEROL LEVELS

What conditions in the United States contribute to levels of serum cholesterol which are associated with a high incidence of coronary heart disease? Approximately 20 to 25 per cent of our population can be considered obese. Sedentariness is also prevalent. Perhaps of greater significance is the dietary basis for our high serum cholesterol levels. Approximately 40 to 44 per cent of total daily calories in the diet are derived from fat, one half to two thirds of which is saturated. In the past several decades, there has been a significant increase in per capita consumption of fats and refined carbohydrates, with no recommendation as to where this trend should stop. It is logical to assert that levels of serum cholesterol associated with an increased risk of developing a disease as serious and as prevalent as coronary heart disease are neither "normal" nor "desirable." It seems equally logical to regard a diet responsible for elevated serum cholesterol levels as being unsatisfactory. These are some of the vital questions and issues challenging physicians, nutritionists and public health workers today.

One diet that has been found successful in lowering serum cholesterol levels effects two major departures from the usual United States pattern. It (1) decreases the total dietary fat from 40 to 44 per cent of total daily calories to 30 to 33 per cent; and (2) reverses the usual predominance of saturated fat by providing approximately equal amounts of the three types of fat, saturated, polyunsaturated and monounsaturated.

The first change attempts to decrease the caloric density of the diet, thereby helping to avoid obesity; the second is directed toward establishing moderation in saturated fat consumption, thereby correcting what is believed by some to be an excessive intake of saturated fat.

The constituents of such a regimen are available from food markets everywhere, are palatable, fulfill the nutrient recommendations of the National Research Council and are free of "food fad" characteristics. The essence of the diet is simply an equal representation of beef, poultry and fish in the diet pattern; the usually recommended amounts and types of vegetables and cereals and fruits; and a substitution of low fat dairy products for those high in fat. No untoward clinical effects have been observed in seven years of studying 850 men who have voluntarily adopted this kind of regimen. The Framingham study has identified the risk factors operative in the development of coronary heart disease, and the New York Health Department study is demonstrating the feasibility of reversing two of these risk factors—hypercholesterolemia and obesity.

Conclusions and Recommendations

At present three choices appear to lie open to public health practitioners in approaching the problem of diet and heart disease:

1. Consider the links in the diet and coronary heart disease hypothesis as weak and inconclusive and requiring further substantiation.

2. Consider the diet and coronary heart disease hypothesis as sound and the identification of the Framingham risk factors as of public health significance and immediately institute public health programs to reduce them in the population in general.

3. Accept the coronary heart disease risk factor concept as of potential public health significance and institute programs only after it has been demonstrated that successful reversal of risk factors are, in fact, feasible and effective, *and associated with a definite and significant decrease in coronary heart disease incidence.*

Approach number 3 appears most reasonable. A public health recommendation instituting a change in the diet pattern of the adult population in general should be based not only on the ability of such changes to lower the serum cholesterol but also on the proved efficacy of the lowered serum cholesterol to significantly decrease coronary heart disease morbidity and mortality.

It is certain, moreover, that the physiologic needs and metabolic attributes of infants, children, adolescents and pregnant and lactating mothers are sufficiently specific to warrant extensive investigation in relation to particular nutrient needs before dietary changes are considered for these populations.

Although it is realized that a nutritional approach to coronary heart disease will not erase the problem since many factors operate in this disease, there is hope that if current research trends are confirmed, public health programs may blunt the onslaught of the most insidious and catastrophic disease afflicting Western civilization.

SECONDARY MALNUTRITION

Basic Concepts

Traditionally, nutrition has been thought of in terms of such classic deficiency diseases as beriberi, rickets, scurvy and pellagra. These disorders exemplify the primary deficiency state, a state in which one or more essential nutrients is absent from the diet. While primary malnutrition is still the principal nutritional problem in many parts of the world, particularly in the developing areas, secondary or "conditioned" malnutrition is the major type of deficiency disorder encountered in the United States. Although a comprehensive exposition of the principles and practice involved in the management of conditioned malnutrition is beyond the scope of this brief discussion, an attempt will be made to indicate what is involved in such management and its importance.

In secondary or conditioned malnutrition a nutritional deficiency exists because of a disorder *in the patient* that interferes with the ingestion, digestion or absorption of food. There may also be abnormalities of transport, cellular uptake and nutrient utilization. Finally, excessive losses of important metabolites may occur. Examples of such forms of

conditioned malnutrition are given in Table 21-1.

Furthermore, these conditioning factors can lead to a sequence of events which may result in death of tissue or cells. When bodily reserves of nutrients such as the water soluble vitamins are exhausted, tissue depletion results; if this exhaustion continues, biochemical lesions, functional and morphologic changes follow (Figure 21-1).

Management of Conditioned Malnutrition

In some types of conditioned malnutrition the basic disorder is reversible and treatment is directed principally toward the underlying condition. For example, hyperthyroidism often can be managed by surgery. However, the individual who has become emaciated because of hyperthyroidism needs acute nutritional therapy even though his hypermetabolic state has been corrected. On the other hand, the patient with pernicious anemia requires what might be called "compensatory" therapy for the rest of his life. Hence, extra vitamin B_{12} must be given by mouth or the gastrointestinal tract circumvented and the vitamin administered parenterally.

If one considers the many expressions that conditioned malnutrition can have, it becomes obvious that the treatment of this form of malnutrition must be highly individualized. The proper management of such conditions requires a special understanding of the principles of nutrition in addition to knowledge of the physiology of the disorder under treatment.

In certain patients who cannot take sufficient food by mouth, other forms of alimenta-

TABLE 21-1. EXAMPLES OF CONDITIONED MALNUTRITION

Function	Condition	Deficiency State
Appetite	Anorexia nervosa	Semistarvation
Ingestion	Esophageal carcinoma	Semistarvation
Displacement	Alcoholism	Protein and thiamine deficiencies
Digestion	Pancreatic insufficiency	Multiple deficiencies (calories, protein, calcium, fat soluble vitamins)
Absorption	Pernicious anemia	Vitamin B_{12} deficiency
Transport	Acanthocytosis	β-Lipoprotein deficiency
Translocation	Diabetes mellitus	Cellular carbohydrate deficiency
Utilization	Hyperthyroidism	Calorie deficiency
Excretion	Addison's disease	Sodium deficiency

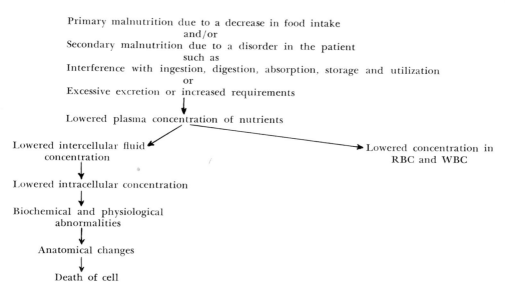

Primary malnutrition due to a decrease in food intake

and/or

Secondary malnutrition due to a disorder in the patient

such as

Interference with ingestion, digestion, absorption, storage and utilization

or

Excessive excretion or increased requirements

Lowered plasma concentration of nutrients

Lowered intercellular fluid concentration

Lowered concentration in RBC and WBC

Lowered intracellular concentration

Biochemical and physiological abnormalities

Anatomical changes

Death of cell

FIGURE 21-1. *Pathogenesis of a nutritional deficiency disease. (Adapted from schematic representation of factors which contribute to the various stages of deficiency disease. In (Follis, R. H.: Deficiency Disease. Charles C Thomas, 1958.)*

tion may become necessary. These range from simple gavage by means of a nasogastric tube to gastrostomy or jejunostomy feeding. Frequently, parenteral nutrition is necessary. When nutrients are administered by intravenous infusion, it is usually impossible to provide optimal quantities of protein and calories. Thus, a nutritional "compromise" is necessary, and a judgment must be made concerning the most desirable ratio of calories to protein to be provided under a given set of circumstances. Such a judgment must be based on an understanding of nutritional priorities. For example, it has been shown that 100 gm. of glucose fed to a subject during the acute phase of starvation will cut losses of nitrogen, water and sodium by one half over a six-day period. At the same time, ketosis is suppressed and losses of valuable cations in the urine are prevented. This effect of glucose cannot be duplicated by a similar quantity of protein or fat since the body has a specific need for carbohydrate apart from its role as a source of calories. In the management of individuals who are exhibiting acute secondary malnutrition, such as postsurgical patients, an awareness of the importance of carbohydrate in the nutritional hierarchy is most valuable. The roles of other nutrients such as potassium,

magnesium and calcium must also be appreciated.

ROLE OF NUTRITION IN SUPPORTIVE CARE

In recent years, surgery and anesthesia have become increasingly safe and effective. At the same time, as a result of the development of a variety of antibiotic agents, infections have lost much of their menace. It is frequently the less dramatic aspects of patient care that will determine the severity, duration and even the outcome of an illness. These aspects of day-to-day management can be collectively designated "supportive care" and, under this broad heading, the role of nutritional management is usually of major importance. There is a relationship between nutriture (nutritional status) and the resistance to infection and the proper healing of wounds (see Chapter Twenty). The debilitating effects of an illness or its therapy can be compounded by neglect of nutritional considerations. Examples of such neglect might be the patient with severe liver damage who is given too much protein or the patient being treated with high doses of diuril who is given too little potassium.

Conclusions

From this brief summary of what is involved in secondary malnutrition, it may be appreciated that conditioned deficiencies can and usually do occur as by-products of a variety of illnesses. The ability to manage such secondary disorders successfully is not a simple matter of treating the underlying disorder; frequently the underlying disorder may not be treatable. The patient with chronic destructive pancreatitis cannot be given a new pancreas; however, his diet can be adjusted and his diabetes treated with insulin. Thus, compensatory and supportive nutritional management are essential components of patient care just as preventive nutrition is an essential part of preventive medicine.

REFERENCES

Texts

Duncan, G. G.: Diseases of Metabolism. 5th ed. Philadelphia, W. B. Saunders Company, 1964.

Jolliffe, N. (ed.): Clinical Nutrition. 2nd ed. New York, Harper & Brothers, 1962.

Wohl, M. G., and Goodhart, R. S. (ed.): Modern Nutrition in Health and Disease. 3rd ed. Philadelphia, Lea & Febiger, 1964.

Articles

Christakis, G.: Treatment at New York City's Obesity Clinics. Columbia-Presbyterian Therapeutic Talks. New York, The Macmillan Company, 1963, pp. 177-181.

Christakis, G., Rinzler, S. H., Archer, M., Winslow, A., Jampel, S., Stephenson, J., Friedman, G., Fein, H., Kraus, A., and James, G.: The Anti-Coronary Club: A dietary approach to the prevention of coronary heart disease—A seven year report. Amer. J. Public Health, *56*:299, 1966.

Doyle, J., Heslin, A. S., Hilleboe, H., Formel, P., and Korns, R.: A prospective study of degenerative cardiovascular diseases in Albany. Report of 3 years' experience. Amer. J. Public Health, *25*: (part 2), 1957.

Kannel, W. B., Dawber, T. R., Kagan, A., Revotskie, N., and Stokes, J.: Factors of risk in the development of coronary heart disease—6 year follow-up experience. Ann. Int. Med., *55*:33, 1961.

Schilling, F. J., Christakis, G. J., Bennett, N. J., and Coyle, J. F.: Studies of serum cholesterol in 4,244 men and women: an epidemiological and pathogenic interpretation. Amer. J. Public Health, *54*: 461, 1964.

Van Itallie, T. B.: Obesity. Amer. J. Med. *19*:111, 1955.

Chapter Twenty-two

Hereditary Molecular Diseases
(Inborn Errors of Metabolism)

HOWARD M. MYERS, D.D.S., PH.D.

It has long been considered a truism that the seeming diversity of science conceals what is, in actual fact, a basic unity. Only as knowledge grows in breadth and depth does this unity emerge from an apparent disorder. Nowhere is this better illustrated than in the group of diseases known as "inborn errors of metabolism" or "hereditary molecular disease," the latter, more modern label being a less restrictive definition of these often disabling conditions. First introduced by Sir Archibald Garrod in 1908, this concept originally applied to only four conditions (alkaptonuria, albinism, cystinuria and pentosuria). Today, as a result of a fusion of biochemistry and genetics, there are some 70 entities which also fit into this general grouping and there no doubt will be many more.

Although the number of people affected by the conditions classified under this heading is not large, the diseases have a significance well out of proportion to their occurrence. Distinct and often precisely defined biochemical flaws can be related to clinical findings and to hereditary background. Such close corroboration of biochemical concepts with actual clinical situations, and vice versa, is rarely achieved.

GENERAL CONCEPT OF
HEREDITARY MOLECULAR DISEASES

There is a growing emphasis on the relationship between genes and protein structure. The realization that specific minor variations in the amino acid sequence of proteins are not only tolerated in nature but are, in fact, found as a result of genetic mutation, has had an enormous impact on biology. The growing evidence that genes have a decisive role in determining the amino acid sequence of proteins has led to the hypothesis that these same molecular variations of proteins are also the first direct demonstrable action of genes. This is strongly supported by studies of the relationship between DNA, RNA and proteins.

We should expect, then, that to explain inherited errors of metabolism we must seek first to identify a specific error in protein structure and relate this to a specific mutation of a gene. From this primary protein error, most of the other findings in the disease entity can be understood. Thus, there is a logical sequence by means of which we may pass from a given clinical finding to a directly determined biochemical error in protein assemblage. We can then gain broader insight into the genetic origin of the error by examination of the family pedigree. It may be concluded, therefore, that protein structure serves as a link between the genetic basis and the clinical pattern of a disease.

In a few instances, the protein defect has been very precisely identified, while in many others it can only be inferred from the confirmed absence of a specific enzyme activity. In still others, even the absence of the enzyme must be presumed from indirect data. The best characterized defects are those having to do with hemoglobinopathies, such as sickle cell anemia. In this condition a single valine substitution for one glutamic acid residue in

265

a polypeptide of 280 amino acids is responsible for a change in solubility of hemoglobin which results in gel formation in the red cell, causing the latter to distort into a sickle shape. This, in turn, is responsible for an increased viscosity of the blood, which leads to the clogging of small blood vessels and produces the infarction that characterized the "crises" of sickle cell anemia.

In many inborn errors the defect is not clearly known to be a single error in amino acid sequence. In these situations, the presence of a noncatalytic protein immunologically resembling an enzyme may be identified. From this it is inferred that the genetic code for assembling the enzymatically active protein is defective and that the improper coding results in a slightly altered protein which is unable to serve as a catalyst. Some instances have also been reported in which the altered protein is still catalytic but different in its properties from the normal enzyme. There are also instances in which no related active or inactive protein has been demonstrated and it is presumed that the defect is severe enough to block entirely the synthesis of the specific protein.

It should be understood that the presumption of an inborn error of metabolism can best be proved by the demonstration of a precise defect in a protein. When the pinpointing of the specific biochemical lesion is not yet possible, the necessity for doing so should not be overlooked, for it is the genetic control of a single specific biochemical event that is the basis of the concept of hereditary molecular disease. Thus, missing enzyme activities do not establish that a condition belongs in the above category. Such phenomena can just as well be caused by inhibitors or a lack of cofactors. It is only when a protein is isolated and shown to be composed of amino acids differing from the normal state that the final link in the chain of suppositions can be forged.

"ONE GENE–ONE ENZYME THEORY"

Very significant support for the concept that single genetic defects can give rise to specific biochemical lesions has appeared (Beadle and Tatum, 1941). These authors were able to show that mutant strains of Neurospora lacked the ability to perform certain specific chemical transformations and, as a result, failed to grow. That each of these blocked reactions was due to a single missing chemical event could be shown by the fact that the entire complex of faulty growth could be corrected by the addition of a single pure chemical compound, such as arginine, citrulline or ornithine. Since in vivo production of these amino acids is normally under enzymic control, the absence of a single enzyme could well be responsible for the inability to transform ornithine to citrulline to arginine. That the missing enzyme is under the control of a single gene could be shown by a simple genetic testing of the asexual spores of Neurospora.

Thus, analyzing the behavior of the various mutant strains, Beadle and his co-workers founded the science of "chemical genetics." They proposed the "one gene–one enzyme theory"—that is, every gene governs the formation of a single enzyme and, correspondingly, a single mutation governs a single biochemical reaction.

Recent findings have emphasized that more than one gene can influence the production of a single enzyme or protein, as, for instance, the occurrence of several distinct repressor genes which can modify the effect of a single gene mutation (Yanofsky, 1956). In addition, it is now clear that a gene which may be considered a functional portion of a chromosome can be represented by a variety of different sized units, depending on the method of study being used. In the case of hemoglobin, the evidence is quite strong that a single gene controls a single polypeptide rather than the entire protein. In general, however, the one gene–one enzyme theory is still a highly useful concept both for the planning of experiments and for an understanding of the bases of inherited metabolic disorders (Horowitz and Leupold, 1951).

HETEROZYGOSITY AND HOMOZYGOSITY

In considering the genetic pattern of inborn errors it is necessary to take into account the various ways in which inherited disease is passed along from parents to offspring. Some traits are inherited in such a way that they appear in each successive generation. Thus, grandparents, parents and children will all exhibit the trait. Mutations or traits which follow such a pattern are said to be dominant. This means that a single gene carrying this trait is able to cause its overt appearance in any individuals who receive it. An individual carrying such a gene not only exhibits the trait himself but passes it along to half of his progeny who, in turn, also have the trait. Since those who exhibit the traits governed by dominant genes have only one mutated gene they are heterozygotes. When both genes are mutated, the individual is homozygous for that trait and will have a more severe form of the disease.

A rather large number of inborn errors are inherited as recessive traits. In this pattern of inheritance the trait, in a great majority of instances, does not appear in successive generations. If children show the overt trait, their parents do not. If one of these same children later has offspring of his own, they will not show overt signs of the mutation, unless the other parent is at least heterozygous for the trait. Recessive genes, therefore, can give rise to overt signs only if the individual carrying them is homozygous for the gene. It is also obvious that both parents must be at least heterozygotes if the trait is to appear in a child. In the above event, 25 per cent of the offspring of this mating will be homozygotes and exhibit the trait, and 50 per cent will be heterozygotes and will not show overt signs of the trait. The remaining 25 per cent will be entirely normal, having inherited a single normal gene from each parent.

It is not an easy matter to determine the presence of the heterozygous state when recessive mutations are involved. In studying inborn errors it is often necessary to apply special tests to uncover heterozygosity. These tests usually involve loading or challenging the patient with a large amount of the critical metabolite in order to demonstrate a deficiency or abnormality of intermediate severity.

The so-called sex-linked mutations are those seen overwhelmingly in one sex only. If a trait is evenly distributed between the sexes its gene is not likely to be a part of the X- and Y-chromosomes and it is said to be autosomal.

It is not possible to discuss in a single chapter all the known or suspected conditions which are classified as inborn errors of metabolism. Several books are currently available which discuss fully the history and pathogenesis of these conditions (Stanbury et al., 1960; Hsia, 1960). This chapter will be limited to those conditions which have either nutritional implications or dental aspects which make them particularly interesting to the intended users of this book.

GLYCOGEN DEPOSITION DISEASE

Glycogen, a normal constituent of liver and muscle, represents a rather limited reserve of carbohydrate fuel which can be mobilized when needed. The glycogen polymer has a molecular weight between 2.5 and 4.5 million and is made up entirely of units of α-D-glucose. The glycogen molecule is organized as a tree-like structure with a single reducing terminus at one end and with a number of nonreducing branches increasing progressively toward the other end and periphery.

Synthesis and Breakdown. Glycogen is synthesized by the combined action of synthetase and the branching enzyme upon uridine diphosphoglucose. Two types of linkage are known to occur, the α-1,4, which is responsible for the straight chain portion, and the α-1,6, responsible for the branch points.

The enzymatic breakdown of the branched structure of glycogen to glucose requires a series of reactions with four enzymes. First, phosphorylase cleaves the α-1,4 linkage to produce glucose-1-phosphate and this, in turn, is transformed into glucose-6-phosphate by phosphoglucomutase. A specific phosphatase dephosphorylates glucose-6-phosphate to yield glucose and inorganic phosphate. Branch points which cannot be cleaved by phosphorylase are hydrolyzed to glucose by the so-called debranching enzyme, amylo 1,6 glucosidase.

If there are errors due to an enzyme defect in either the synthetic pathway, which is rare, or in the catabolic scheme, which is more probable, inborn errors of glycogen metabolism will result. Six forms of glycogen storage disease, based on the nature of the enzyme defect and the corresponding effect on the

TABLE 22-1.   GLYCOGEN STORAGE DISEASES

Descriptive Name	Missing Enzyme	Glycogen Structure
Hepatorenal glycogenosis	Glucose-6-phosphatase	Normal
Limit dextrinosis	Amylo 1,6 glucosidase (debrancher)	Increased number of branch points
Muscle glycogenosis	Muscle phosphorylase	Normal
Liver glycogenosis	Liver phosphorylase	Normal
Amylopectinosis	Amylo (1,4 ⟶ 1.6) transglucosidase	Longer straight chains, decreased branch points
Idiopathic generalized glycogenesis	Unknown	Normal

structural architecture of the glycogen, are listed in Table 22-1.

GLYCOGEN STORAGE DISEASE DUE TO GLUCOSE-6-PHOSPHATASE DEFICIENCY

In glucose-6-phosphatase deficiency, also called von Gierke's disease, the enzyme is either missing or is present in very small amounts in both the liver and the kidney. Both organs are the sites of massive deposition of normal structured glycogen because of the inability to mobilize glucose from glucose-6-phosphate. The disease is not sex linked and is, therefore, said to be autosomal. Heterozygotes can be detected by the presence of elevated blood levels of various phosphorylated intermediates of glycolysis. There appears to be no difference in these levels for hetero- and homozygotes.

Signs and Symptoms. The clinical signs and symptoms are related directly to the deficiency of glucose: namely, enlarged liver, hypoglycemia and ketosis when food is withheld, and subnormal response of blood sugar to epinephrine. In the classic pattern of von Gierke's disease children fail to develop properly, have progressive inanition, develop large protuberant abdomens because of the hepatorenomegaly and frequently suffer from hypoglycemic convulsions and coma. Death usually occurs before adulthood as a result of intercurrent infection.

Laboratory Findings. A routine finding is elevation of the blood pyruvate and lactate levels. This is thought to be due to an increased glycolysis imposed by the lack of direct accessibility of the glycogen stores. There is usually a noteworthy degree of lipemia. The hexose monophosphate pathway is also stimulated by the surplus of glucose-6-phosphate, and this latter fact serves to explain the occurrence of marked storage of lipid in spite of the accelerated lipolysis necessary for energy requirements. Muscle which ordinarily does not have glucose-6-phosphatase shows only a slight storage of glycogen in this deficiency.

Treatment. Therapy of the condition must take into account the need to maintain the blood glucose by allowing for only minimal periods of fasting. Frequent feeding of small carbohydrate meals would appear to be a rational dietary scheme. Promoting gluconeogenesis would seem of little value since the glucose formed from protein would still require the glucose-6-phosphatase step for final mobilization of glucose from the necessary end-product, the 6-phosphate ester. Some success has been reported from use of L-thyroxine, and presumably this is effective by virtue of its ability to uncouple oxidative phosphorylation, causing a greater rate of activity in the Krebs' tricarboxylic acid cycle. The wasteful nature of this mechanism appears to have a beneficial effect on the surplus of carbohydrate stores which characterize this condition.

GLYCOGEN STORAGE DISEASE DUE TO AMYLO 1,6 GLUCOSIDASE DEFICIENCY

Second in frequency to the glucose-phosphatase deficiency is a form of glycogen storage disease due to a faulty debranching enzyme (Andersen, 1964). Lack of this protein brings about a molecular distortion of glycogen in the form of an increase of branch points from a normal of 7 or 8 per cent to 10 to 13 per cent. Since branch points contribute a relatively minor fraction of glucose from glycogen, the result is a more mobilizable molecule, which leads to milder symptoms than those of the phosphatase deficiency. There is, in addition, a greater tendency for glycogen storage in the skeletal and cardiac muscle, leading to a somewhat different set of clinical manifestations. The muscle type of deficiency generally shows symptoms relating to pain, weakness and stiffness in muscles after variable periods of exercise, the masseter muscle often being specifically involved. The accumulation of abnormally structured glycogen, although not massive in amount, is significant—about 2.5 to 4.0 per cent compared to normal levels of 0.2 to 0.9 per cent.

Since the missing enzyme is normally found in a greater variety of tissues, the glycogen storage defect is also more widespread than in the case of deficient liver glucose-6-phosphatase. Inasmuch as glucose-6-phosphatase activity is normal in such individuals, gluconeogenesis can be of use in maintaining blood sugar levels. For the same reason, these patients also show normal responses to galactose and fructose infusions. A more varied diet is possible in such cases, although some tendency toward ketosis remains and high fat diets seem ill advised. The most probable inheritance is by way of an autosomal recessive gene.

GENERALIZED GLYCOGEN STORAGE DISEASE WITH NO SPECIFIC ENZYME DEFICIENCY

In the generalized type of glycogen storage disease no specific enzyme deficiency has yet been uncovered. The clinical picture is dominated by the effects of massive accumulations of glycogen in all muscles. The cardiac effects are the most significant, although generalized muscle weakness is present. The tongue is often enlarged and protruding because of massive glycogen stores. Mental deficiency is usually marked.

GALACTOSEMIA

The hexose galactose is a part of the human diet by virtue of its presence in milk sugar. Under normal circumstances, a sequence of reactions readily converts galactose to glucose. An inherited condition has been described in which this usual pathway is unable to function because of a missing step and, as a result, galactose and galactose-1-phosphate accumulate in the red blood cells and plasma.

The usual sequence of metabolism of galactose requires an initial phosphorylation followed by an incorporation of galactose-1-phosphate into a nucleotide structure in the form of uridine diphosphate galactose. Another enzyme converts this form of nucleotide-bound galactose to glucose. Subsequently, the newly formed glucose, still in the nucleotide linkage, is utilized just as are glucose molecules obtained directly from the diet.

In galactosemia the vital step of transferring galactose-1-phosphate to the uridine nucleotide is blocked by the absence of the enzyme galactose-1-phosphate-uridyl-transferase. The accumulation of galactose-1-phosphate and galactose is, therefore, due to an inability to metabolize the ester phosphate via its chief normal pathway.

The condition is an autosomal recessive and quite rare since only about 100 cases have been reported in the literature. Heterozygotes can be detected by their inability to tolerate relatively large doses of galactose.

Signs and Symptoms. A triad of symptoms consisting of hepatosplenomegaly, cataracts and mental retardation are the dominant clinical findings.

Treatment. Virtual elimination of galactose from the diet can be effectively accomplished by the use of milk substitutes and will prevent all the sequelae of this rather severe condition. If begun during the first five weeks of life, dietary elimination of galactose may even cause complete regression of signs and symptoms already present. No special need for dietary galactose exists since all of this hexose necessary for galactolipid formation in the central nervous system can be synthesized from glucose. Tolerance for galactose has been shown to increase with age with no corresponding increase in the deficient enzyme. This is apparently due to a compensatory increase in the activity of a pyrophosphorylase which utilizes galactose-1-phosphate and uridine-triphosphate in a different mechanism to make uridine-diphosphate galactose along with pyrophosphate.

PHENYLKETONURIA

The enzyme system phenylalanine hydroxylase, found in the liver and responsible for oxidizing phenylalanine to tyrosine, has been shown to be missing in a condition called phenylketonuria, so named because of the phenylketones found in urine by means of the ferric chloride test. The urine becomes olive green in color because of the presence of phenylpyruvic acid, the typical metabolite of this disease. The important feature of this molecular hereditary disease is that dietary treatment can prevent the serious consequences of mental deficiency, which is one of the main characteristics of this disease

Metabolism of Phenylalanine. In the ordinary course of events dietary phenylalanine is converted to tyrosine by utilizing O_2, one atom becoming part of the hydroxyl group of tyrosine and the other forming water with

reduced NADP (niacinamide adenine dinucleotide phosphate) or TNPH (triphosphopyridine nucleotide). The reaction is carried out in liver microsomes by a labile protein called Fraction I. A second protein more widely distributed and more stable, called Fraction II, is necessary, as is ferrous iron for complete enzymatic activity. The inborn disease is due only to the absence of Fraction I from the liver. This metabolic block leads to an accumulation of phenylalanine which then may enter other existing but usually less important metabolic schemes. One of these, oxidative deamination, leads to phenylpyruvic acid accumulation. This compound itself is readily reduced to phenyllactic acid or may be decarboxylated to phenylacetic acid. This latter product usually is rapidly conjugated with glutamine and excreted by the kidney. The metabolic scheme is presented in Figure 22-1.

The genetic nature of this disease is now well established. An autosomal recessive trait should give an incidence of 25 per cent in the offspring of those families carrying the defect,

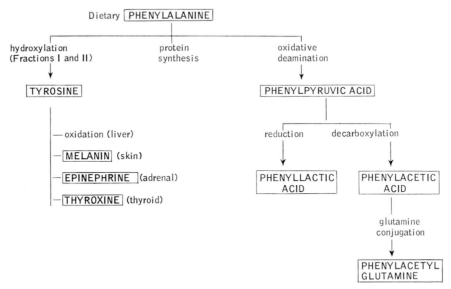

FIGURE 22-1. *The hydroxylation pathway for phenylalanine is blocked in phenylketonuria with resultant effects on all the biochemical derivatives of tyrosine.*

Protein synthesis is not blocked except by the limitation imposed by the deficiency of tyrosine. This requirement must be met during therapy with restricted diets.

Oxidative deamination ordinarily a minor pathway, becomes overly active, and three new deratives of phenylpyruvic acid accumulate.

In this scheme products are listed in capital letters and reactions in small letters.

which has been confirmed. Heterozygotes tend to have higher (1.5 to 2.5 times) phenylalanine plasma levels than normals when tested either in the fasting state or after a test dose of the amino acid.

SIGNS AND SYMPTOMS

Phenylketonuria is associated with profound mental deficiency, about 1 per cent of institutional cases being so classified. The great majority are idiots, a few are imbeciles, and only occasional individuals can be classified as morons. It is thought that the massive accumulation of metabolites in blood and cerebrospinal fluid is responsible for the mental deterioration. However, it has not been possible to correlate severity of mental deficiency and plasma levels of phenylalanine. There appears to be a delay in myelinization in the central nervous system, with secondary overgrowth of neuroglial tissue.

Lessened hair and skin pigmentation is also a prominent feature of this disease. This may be traced to faulty melanin formation due, in turn, to inhibition of the enzyme tyrosinase by the phenylalanine excess. Increasing tyrosine or lowering phenylalanine intake will rapidly bring about darkening of new grown hair.

Patients with phenylketonuria have been reported to be overly sensitive to epinephrine and to have lower than normal plasma concentrations of this hormone. Patients with this condition give a blood pressure response half again as much as a control group.

Delayed eruption of teeth until after 11 months has been reported in this condition. Affected children also often exhibit eroded and broken teeth, and this has been interpreted as being due to an abundance of hypoplastic defects.

LABORATORY FINDINGS

Elevated levels of all the products of phenylalanine mentioned previously are found in the urine. Blood levels of phenylalanine are high, but those of the other metabolic derivatives are found in excess only in urine because of the efficiency of tubular secretion.

TREATMENT

Treatment of this condition consists of a careful restriction of dietary phenylalanine. A balance must be struck, inasmuch as the growth requirement for this indispensable amino acid cannot be ignored. Insufficient intake results in the release of excessive amounts of phenylalanine derived from tissue breakdown. There is considerable variation of the requirement for phenylalanine during growth, and it is clear that the reported value of 15 mg. per kg. per day cannot be consistently depended on, especially for younger children. Plasma phenylalanine levels (normally 1 mg. per 100 ml.) must be repeatedly used as a guide to dietary control.

The basis of the diet used in phenylketonuria is a casein hydrolysate from which phenylalanine is removed by charcoal absorption. Tyrosine, cystine and tryptophan are also eliminated by this treatment. Addition of appropriate amounts of phenylalanine in the form of some protein such as milk may then be made. Caloric requirements of the patients are high, apparently because of the less efficient utilization of the free amino acids of the hydrolysate as opposed to whole proteins (Blainey and Gulliford, 1965). Some reversal of almost all the effects of phenylketonuria can be accomplished by dietary control of phenylalanine. The amelioration of the mental deficiency depends on the age at which therapy is begun. However, even when not ideal, the degree of progress reported justifies the time, expense and difficulty utilizing these diets.

BLOOD CLOTTING DEFECTS

Since the turn of the century it has been known that blood clotting is an enzymatic action of thrombin on the protein fibrinogen. The active enzyme thrombin obviously does not exist in blood except during the brief periods when it is actually producing the insoluble fibrin of the blood clot. The inactive form of thrombin is called prothrombin and this, in turn, must be activated by a complex group of presumed substances called thromboplastin.

Two pathways of prothrombin activation

TABLE 22-2.   A LIST OF COAGULATION FACTORS AND PROPERTIES

Name		Synonyms	Chemical Nature	Disappears During Coagulation	Intrinsic or Extrinsic System
Factor	I	Fibrinogen	Globulin	Yes	———
	II	Prothrombin	α_2 Globulin	Yes	———
	III	Thromboplastin, thrombokinase	Complex of substances	No	———
	IV	Ca++	Calcium ion	No	Both
	V	ac globulin		Yes	Both
	VI	Same as V; term no longer used			
	VII	Serum prothrombin conversion accelerator (SPCA)	β Globulin	No	Extr.
	VIII	Antihemophilic globulin (AHG)	β_2 Globulin	Yes	Intr.
	IX	Christmas factor plasma thromboplastin component (PTC)	β Globulin	No	Intr.
	X	Stuart-Prower factor	α Globulin	No	Both
	XI	Plasma thromboplastin antecedent (PTA)	α Globulin between β and α	No	Intr.
	XII	Hageman factor	Between β and α globulins	No	Intr.

are now known, an extrinsic one which requires a contribution derived from tissues other than blood and an intrinsic one in which all the activators are derived from plasma. In the extrinsic system either a tissue extract or Russell's viper venom is the external factor, the two activations being slightly different. Three other activators present in blood are also required, factors V, VII and X. In the intrinsic system six factors all present in plasma are required, V, VIII, IX, X, XI and XII. Both the extrinsic and the intrinsic systems thus require factors V and X, and it has long been known that both require calcium ions. Table 22-2 summarizes the factors thought to be involved in blood coagulation.

FIBRINOGEN

Fibrinogen is a globulin type of protein present in plasma to an extent of 0.2 to 0.4 per cent. The action of thrombin on this molecule is proteolytic, cleaving a glycine-arginine bond and releasing a peptide which contains the terminal glutamic acid residue of the original fibrinogen. Its removal from the parent molecule allows a reaggregation of the remainder, which leads to polymerization and precipitation as fibrin. The polymerizing action is spontaneous and does not require thrombin. The final stabilization of the fibrin clot appears to involve S-S bonding.

The protein fibrin consists of a network of fibers about 0.1 μ or less in diameter. A marked cross striation with major bonds at 230 Å is present. It is possible to account for the striation pattern by aggregating fibrinogen molecules in a particular alignment (Hall and Slayter, 1959). Fibrin molecules readily undergo a stretching from the folded (α) form to the unfolded (β) form, which the original fibrinogen cannot do. This supports the idea of a rather specific cross linking being involved in the final conversion of the molecules to an insoluble fibrous network.

Fibrinogen Deficiency. Fibrinogen deficiency is a very rare condition transmitted as an autosomal recessive trait. Surprisingly, the hemorrhagic complications from this defect are rather mild, even after surgery. This has given rise to some doubts about the role of the fibrin clot in actually producing hemostasis (Biggs and Gason, 1960). A vascular contraction and plugging by platelets may in fact play a role in hemostasis earlier than the formation of the fibrin clot and this may account for the survival of patients with afibrinoginemia.

PROTHROMBIN

Prothrombin has been shown to be converted slowly to thrombin in the absence of any added activators. The preparation, made in 25 per cent citrate, contained only traces of other substances. Although the observed generation of thrombin cannot be said to rule out the role of activators which might not have been completely removed, it is highly probable that the amount of contaminant could not have been a physical precursor of the observed thrombin. Thus, prothrombin is almost certainly the single precursor of thrombin.

The intrinsic system of prothrombin activation, in common with the extrinsic system, requires calcium as well as factors V and X. Other β and α globulins also required are factors VIII, IX, XI and XII.

Hereditary defects involving the absence of VIII and IX are the most important, in terms both of severity and incidence. Both conditions have similar clinical manifestations, but may be distinguished by the fact that equal parts of blood from patients with each of these conditions clots normally when compared with unmixed samples of each. These conditions are said to be sex-linked, the recessive gene being located on the X-chromosome. Since males inherit only one X-chromosome, they may exhibit the overt disease even though they carry a single gene for the trait. Female heterozygotes, since they possess two X-chromosomes, may act as transmitters of the gene with no overt sign of the disease and thus would be classified as carriers. The female will exhibit the disease only when homozygous, but this is quite rare since it requires possession of the mutation by both parents.

Prothrombin Deficiency. Pure prothrombin deficiency is extremely rare among the inborn errors in coagulation. Those patients who have been observed usually exhibit severe symptoms. The activation of prothrombin is also thought to be degradative in nature. In both intrinsic and extrinsic systems factors V and X are necessary. Factor X is a α globulin not fully consumed during the coagulation process. Inherited deficiencies of this factor are known but are not clinically distinctive. Factor V, which is consumed during clotting, has not been characterized.

FACTOR VII

In the tissue extract dependent extrinsic system, one other factor besides V and X is needed to accelerate the conversion of prothrombin to thrombin—factor VII—a β globulin not consumed in the coagulation process. Factor VII can be distinguished from factor X which it closely resembles by not being required for either the intrinsic system or for that part of the extrinsic system which is effective with Russell's viper venom.

FACTOR VIII

Factor VIII globulin is a very unstable protein. Efforts to isolate it and separate it from fibrinogen have not been successful without gross loss of activity. Stored human blood will lose 70 per cent of its factor VIII in 24 hours. Somewhat better preservation is obtained if plasma is separated quickly and frozen at −20 C, but even under these conditions some loss of activity occurs.

Antihemophilic globulin (AHG), or factor VIII, disappears during coagulation, and a part of this loss is believed to be due to adsorption of the protein on the fibrin clot. The close physical association of fibrinogen and AHG has already been mentioned, so it is of interest that afibrinogenemia patients do have a normal component of factor VIII.

Hemophilia. Hemophilia is a severe disorder of blood coagulation, the life expectancy of the average patient being only 22 years. The disease is virtually limited to males and is transmitted as a sex-linked recessive mendelian trait. Presumably, the gene harboring the defect is contained in the X-chromosome and is transmitted by an affected male through an unaffected daughter to a grandson.

Signs and Symptoms. Episodes of frequent massive hemorrhages are common. Often the hemorrhage involves the joints and requires confinement to bed. Wasting of muscle follows from the immobility imposed during the long and frequent periods of convalescence. The slightest trauma may trigger a prolonged oozing of blood, which often can be remedied only by transfusion.

In dentistry the frequent minor cuts and

bruises of the mouth may initiate periods of severe bleeding (Webster and Graham, 1961), and dental extractions can become hazardous to life. Not infrequently hemophilic bleeding is delayed until 6 to 8 hours after completion of surgery. At this point, a persistent flow of blood may begin which cannot easily be arrested by pressure. The oozing can continue as long as 4 to 6 weeks from a single tooth socket if no specific treatment is given. Less severe forms of the disease occur, however, in which the defect is uncovered only after serious injury or when unexpected bleeding stimulates a laboratory study. In these milder cases the factor VIII levels may not be so severely reduced (about 25 per cent of normal) as in the more profound cases in which the level is close to zero. Heterozygous female carriers often show depressed levels of factor VIII.

Treatment. Treatment of hemophilia usually requires the local application of cold and pressure. In tooth extraction, careful packing of the socket plus protection of the wound site by mechanical means has given good results (Lucas and Tocantins, 1960). Transfusion with fresh whole blood or fresh frozen or lyophilized plasma is often resorted to as a systemic measure. Human AHG is available but is in limited supply. Bovine or porcine factor VIII has been used in severe cases to raise the blood level of the factor 30 to 50 per cent of normal for a limited period in order to allow for major surgery. These are antigenic, however, and the duration of the administration must be limited.

Von Willebrand's Disease. A number of minor hemorrhagic disturbances have been identified recently which often involve bleeding from the tooth sockets or gingivae as their chief manifestations. Von Willebrand's disease is an autosomal dominant condition involving a vascular defect of uncertain nature. In more severe cases a diminished level of factor VIII has been identified, leading to the theory that the vascular and circulating factors are separately heritable.

FACTOR IX

Christmas Disease. The Christmas factor is so named for the person who first was diagnosed with a bleeding condition very similar to hemophilia (Hsia, 1960). Sometimes called hemophilia B, this condition is somewhat rarer than the A form, by a factor of about 10. Also a sex-linked recessive trait, this deficiency results from the absence of factor IX, one of the necessary components of the intrinsic system of prothrombin activation. Little is known about the missing component except that, unlike AHG, it is present in serum. The use of fresh plasma or whole blood is effective in controlling this condition.

FACTOR XI

PTA Deficiency. Plasma thromboplastin antecedent (PTA), or factor XI, deficiency is a mild condition inherited as an autosomal dominant trait which appears to have a predilection for those of Jewish ancestry. The manifestations are usually a tendency to bruise easily, menorrhagia and postoperative bleeding. The missing factor appears to be necessary for the initial reaction between a glass surface and blood. This reaction can be prevented by siliconizing the glass container. PTA is therefore a part of the intrinsic system of prothrombin activation.

FACTOR XII

Factor XII deficiency can be demonstrated in vitro only. No known clinical defect has been found in individuals with this deficiency, although the clotting time in glass tubes is prolonged. Factor XII is believed to be another of the required contact factors which initiate the intrinsic system of prothrombin activation.

HYPOPHOSPHATASIA

There are at least two immunologically distinguishable proteins which contribute to the serum enzymatic activity known as alkaline phosphatase. Twenty-seven per cent of the activity is derived from the intestinal cells, the remainder mostly from bone (Schlamowitz, 1958). Long suspected of having a role in the formation of calcified tissues, this enzyme activity is useful in the diagnosis of rickets and metastatic tumors of bone. However, its exact role in bone formation has never been clarified. The alkaline phosphatase derived from bone-forming cells apparently contributes to elevated enzyme levels in skeletal diseases such as Paget's disease. Intestinal alkaline phosphatase is thought to be largely responsible for the marked elevation of the enzyme in obstructive jaundice. Congenital absence of the enzyme from serum is associated with a condition of skeletal inadequacy which closely resembles rickets histologically.

SIGNS AND SYMPTOMS

The severity of this condition correlates with the onset of appearance. In the very young it is quite often fatal, but when onset is delayed, the symptoms are usually milder. In adults there may be asymptomatic episodes alternating with skeletal defects following the appearance of fractures. Most of the clinical findings can be related to defective bone formation, which usually manifests itself as uncalcified osteoid.

The long bones show typical rachitic changes. The cranial sutures appear wide in infancy, but this is misleading since this appearance is due to the presence of uncalcified osteoid. Subsequent calcification of this osteoid without the presence of fibrous septa results in premature synostosis and its attendant increased intracranial pressure. Serious brain damage and even death may result if surgical relief is not provided.

Oral Manifestations. Premature loss of deciduous teeth is a common feature of this condition, the effect usually being confined to single rooted members. An incomplete or defective cementum is seen in such teeth, along with a higher incidence of localized resorption. It is not clear what relationship, if any, exists between the exfoliation of the tooth and the apparent defect in the suspending apparatus. Similar defects have been reported without attendant loss of teeth.

LABORATORY FINDINGS

Alkaline Phosphatase Deficiency. In hy-pophosphatasia the enzyme alkaline phosphatase is missing in both serum and urine. Intestinal alkaline phosphatase has, in some cases, been assayed at normal levels in liver and duodenal juice. Since the two forms of the enzyme are different proteins, this is to be expected. Several independent efforts to identify an inhibitor of the enzyme have been made with no success. Thus, there is apparently a true absence of the circulating enzyme, which is presumably derived from bone-forming cells.

A perplexing characteristic of the deficiency as observed in adults is its periodic nature. Periods of absence of the enzyme alternate with intervals of normal or slightly subnormal levels. It is not known whether the enzyme observed during these periods is identical with the one found in normal individuals. It is difficult to reconcile the inability to synthesize the enzyme with its appearance even on a fluctuating basis.

Hypercalcemia. The inability to deposit mineral apatite into the excess of osteoid or cartilage probably explains the frequent appearance of hypercalcemia in these cases. Even with normal amounts of calcium absorption, the failure to remove calcium from the circulating fluids in the form of bone mineral could explain the elevated levels of this cation. Hypercalciuria is also a logical consequence of the nonparticipation of the skeleton in normal calcium metabolism.

Uncalcified Cartilage. Of interest is the fact that uncalcified cartilage from hypophosphatasia patients unlike that from patients with rickets does not calcify when placed in normal serum. Thus, the role of phosphatase cannot be to elevate phosphate levels in order to allow precipitation of bone salt. The suggestion that phosphatase may be necessary to remove an inhibitor of calcification is worthy of further study (Neuman et al., 1951).

Presence of Phosphoethanolamine. The observation that hypophosphatasia is accompanied by the presence of phosphoethanolamine in the urine seems to suggest a relationship between this substance and the enzyme. Although alkaline phosphatase is capable of hydrolyzing this substrate, no rational relationship has yet been established to link one with the other. Phosphoethanolamine has no apparent role in mineralization, being absent from serum and urine under normal circumstances.

Adding confusion to the picture are the facts that phosphoethanolamine is a normal constituent of brain tissue and that it is also found in the urine in liver disease, celiac disease and erythroblastosis fetalis. A relationship between the appearance of phosphoethanolamine in the urine and the onset of periods of reduced enzyme activity has been reported (Eisenberg and Silverman, 1965). In addition, there is reason to believe that this substance is found in higher levels in the urine of heterozygotes than in normals. This, too, coincides with the lower than normal levels of alkaline phosphatase usually found in these individuals.

The combined use of tests for phosphoethanolamine and alkaline phosphatase has produced a detectable heterozygosity in 81 per cent of the parents of children exhibiting the condition. A sibship incidence of the disease close to the theoretical 25 per cent has also been observed, all of which is consistent with the premise that the condition is an autosomal recessive trait.

TREATMENT

No adequate therapy is known for this condition. Vitamin D often produces hypercalcemia, and intoxication from such vitamin therapy has even been reported. The superficial similarity between this condition and rickets should not be considered a justification for the use of vitamin D.

ACATALASIA

The congenital absence of catalase, ordinarily one of the most ubiquitous enzymes, is responsible for a rather rare condition known as acatalasia. First identified in Japan, this condition has so far been found to produce no deleterious effects with the exception of a curious gangrene of the oral cavity in about 70 per cent of those who are afflicted.

The condition is governed by an autosomal recessive gene. Heterozygotes have lower catalase levels than normal, while those homozygous for the condition have a virtual complete absence of the enzyme.

Catalase promotes the destruction of hydrogen peroxide, a product formed by the aerobic dehydrogenase group of oxidative enzymes. Though rather toxic by virtue of its strong oxidizing tendency, hydrogen peroxide does not accumulate sufficiently to present a toxicity problem. This may be due to the widespread occurrence of the enzyme catalase in virtually all tissues. However, even in the absence of this enzyme from the tissues, hydrogen peroxide does not ordinarily accumulate.

The apparent ability of patients with acatalasia to handle hydrogen peroxide in most instances has been explained as being due to the presence in tissue of peroxidases. These enzymes are capable of utilizing such peroxide as normally occurs for oxidative functions. The consumption of peroxide by these enzymes apparently is sufficient even in the absence of catalase unless rather special circumstances intervene.

ORAL MANIFESTATIONS

In the oral cavity, the gums are constantly subjected to masticatory injuries, and these sites are highly suitable for invasion by such bacteria as hemolytic streptococci and Type I pneumococci. The organisms are not abundant but produce hydrogen peroxide without being able to synthesize catalase. The combination of hydrogen peroxide production without the presence of catalase in either the tissue or the bacterial cells appears to provide a unique situation for an accumulation of toxic levels of hydrogen peroxide. The nascent oxygen released has a severe effect on the hemoglobin present in the tissues adjacent to it, and the resulting oxidation to methomoglobin deprives the area of its oxygen supply.

A necrosis extending to gangrene develops and requires surgical intervention and antibiotic therapy. The inflammatory response is usually very slight because of the local anoxia, and the predominant clinical picture is one of necrosis, gangrene and sequestration (Takahara et al., 1960).

The oral condition is usually encountered before the age of 10 and multiple loss of teeth is a common result of the surgical intervention for the recurring infectious episodes. The oral affliction is the only symptom that has been found in these acatalasia cases (Wyngaarden and Howell, 1960).

LABORATORY FINDINGS

The enzyme has been shown to be missing in a variety of tissues as well as the blood. If hydrogen peroxide is added to the blood of these patients, the usual release of gas bubbles does not occur. Instead, the blood turns a brown-black color because of formation of the methemoglobin. Subsequently, it is possible to demonstrate the formation of further products of hemoglobin oxidation down to the colorless dipyrrole called propentdyopent. This is possible only when hydrogen peroxide is able to exert a prolonged attack on the hemoglobin molecule undisturbed by catalase activity.

TREATMENT

Treatment of the condition is largely dental care to prevent the establishment of infections. The use of oxidizing mouthwashes would appear to be unwise in view of the accepted pathogenesis of this condition.

REFERENCES

Andersen, D. H.: Miscellaneous disorders of metabolism: V. Glycogen storage diseases and galactosemia. *In* Thompson, R. H. S., and King, E. J. (ed.): Biochemical Disorders in Human Disease. New York, Academic Press, Inc., 1964.

Beadle, G. W., and Tatum, E. L.: Genetic control of biochemical reactions in Neurospora. Proc. Nat'l Acad. Sci., 27:499, 1941.

Biggs, R., and Gaston, L. W.: The blood clotting factors. *In* Stanbury, J. B., Wyngaarden, J. B., and Fredrickson, D. S. (ed.): The Metabolic Basis of Inherited Disease. New York, McGraw-Hill Book Co., Inc., 1960, pp. 1167-1171.

Blainey, J. D., and Gulliford, R.: Phenylalanine restricted diets in the treatment of phenylketonuria. Arch. Dis. Child., 31:452, 1956.

Eisenberg, E., and Silverman, S., Jr.: Hypophosphatasia. Gordon Research Conference Presentation, Kimball Union Academy, Meriden, New Hamphire, July 13, 1965.

Hall, C. E., and Slayter, H. S.: The fibrinogen molecule: Its size, shape, and mode of polymerization. J. Biophys. Biochem. Cytol., 5:11, 1959.

Horowitz, N. H., and Leupold, W.: Some recent studies bearing on the one gene-one enzyme hypothesis. Cold Spr. Harb. Symp., 16:65, 1951.

Hsia, D. Y.: Inborn Errors of Metabolism. Chicago, Year Book Publishers, Inc., 1960.

Lucas, O. N., and Tocantins, L. M.: Problems in hemostasis in hemophilic patients undergoing dental extractions. Ann. N. Y. Acad. Sci., 151:470, 1964.

Neuman, W. F., Distefano, V., and Mulryan, B. J.: Surface chemistry of bone. III. Observations on the role of phosphatase. J. Biol. Chem., 193:227, 1951.

Schlamowitz, M.: Immunochemical studies on alkaline phosphatase. Ann. N. Y. Acad. Sci., 75:373, 1958.

Stanbury, J. B., Wyngaarden, J. B., and Frederickson, D. S. (ed.): The Metabolic Basis of Inherited Disease. New York, McGraw-Hill Book Co., Inc., 1960.

Takahara, S., Hamilton, H. B., Neel, J. V., Kobara, T. Y., Ogura, Y., and Nishimura, E. T.: Hypocatalasemia: A new genetic carrier state. J. Clin. Invest., 39:610, 1960.

Webster, W. P., and Graham, J. B.: Physiologic and genetic aspects of hemostasis. *In* (Witkop, C. J.) (ed.): Genetics and Dental Health. New York, McGraw-Hill Book Co., Inc., 1961, pp. 135-160.

Wyngaarden, J. B., and Howell, R. R.: Acatalasia. *In* Stanbury, J. B., Wyngaarden, J. B., and Frederickson, D. S. (ed.): The Metabolic Basis of Inherited Disease. New York, McGraw-Hill Book Co., Inc., 1960.

Yanofsky, C.: Gene interactions in enzyme synthesis. *In* Gaebler, O. H. (ed.): Enzymes: Units of Biological Structure and Function. New York, Academic Press, Inc., 1956.

Section Four

Oral Biology
and Nutrition

Chapter Twenty-three

The Chemical Composition
of Bones, Teeth, Calculus, Saliva
and the Periodontium of the Human*

ISADORE ZIPKIN, PH.D.

There is unquestionably some truth in the statement, "We are what we eat." But, what are we made of? More specifically, what are the structures unique to the oral cavity made of? It is essential to have some knowledge of the chemistry of the hard and soft tissues, salivary fluids and concretions on the teeth in order to better understand how nutritional biochemistry affects the development and preservation of these oral structures and tissues. In fact, changes in the composition of the oral tissues may be related to nutritional problems concerned with dental caries, periodontal disease and oral wound healing.

This chapter will present the current consensus of information on the chemical composition of a number of hard and soft tissues in the human oral environment. The discussion of hard tissues will include bone in general, alveolar bone in particular, the tooth (enamel, dentin, cementum, pulp) and concretions (salivary stones and supra- and subgingival calculus). The discussion of soft tissues will be concerned with the epithelial and connective tissues of the gingivae. The chemistry of the whole mixed secretion of saliva and the differences in the various discrete secretions need to be discussed, too, because they are an important part of the oral environment.

* This article was written by Dr. Zipkin in his private capacity. No official support or endorsement by the National Institute of Dental Research, National Institutes of Health, Public Health Service, is intended or should be inferred.

BONE

The inorganic portion of hard cortical bone makes up about 60 per cent of its total weight; the organic portion, about 25 per cent; and water, about 15 per cent. A small portion of the water of compact bone is so tightly bound to the inorganic phase that even high centrifugal forces cannot remove it. Heating to 100° C. removes that water believed to be associated with the organic portion of bone, which is chiefly collagen (Stack, 1955) associated with small amounts of polysaccharides (Meyer, 1956).

INORGANIC COMPOSITION

The inorganic portion of bone is made up largely of calcium and phosphate, which give a chemical structure $Ca_{10}(PO_4)_6X_2$, generically called apatite. When X is hydroxyl or fluoride the compound is called, respectively, hydroxyapatite or fluorapatite. It should be emphasized that the crystals of apatite of bone are very small, in the order of 1500 Å in length in the adult, compared to the apatite of enamel, which may exceed 10,000 Å in length. Bone crystals give a poorly resolved pattern on x-ray diffraction analysis compared to the well crystallized pattern in dental enamel. In addition, x-ray diffraction analysis does not identify the presence of small quantities of

281

TABLE 23-1.   INORGANIC COMPOSITION OF ADULT HUMAN BONE
Dry, Fat-free-Basis (%)*

Constituent	Calvarium (8)[†]	Sternum (8,9)	Rib (1,2,5,8,9)	Iliac Crest (5,8,9)	Vertebra (3,8,9)	Femur (1,2,5,7)	Skull (3,5)	"Bone" (4,10)	Mean (1-10)
Ash	65.7	52.1	58.5	56.4	52.6				57.1
Ca	24.0	19.3	22.6	21.9	19.4	25.6	25.0		22.5
P	10.7	9.0	10.3	9.8	9.3	11.8	11.1		10.3
Ca/P[‡]	2.24	2.14	2.19	2.23	2.09	2.17	2.25		2.18
Mg		0.22	0.28	0.28	0.25			0.34	0.26
Na		0.33	0.54	0.51	0.43	0.69	0.61		0.52
K		0.029	0.080	0.085	0.090		0.050	0.20	0.089
CO_2		2.3	3.6	4.0	2.6		5.1		3.5
Cl			0.13	0.15			0.10	0.050	0.11
F[§]	0.056	0.062	0.050	0.047	0.056				0.054

* Means are unweighted and are expressed on a wt./wt. basis.
† Numbers in parentheses refer to citations listed below.
‡ Ca/P given as weight ratio.
§ Fluoride values are taken from bones of individuals residing in an essentially fluoride-free area.

(1) Follis, R. H., Jr., 1952; (2) Davies, R. E., Kornberg, H. L., and Wilson, G. M., 1952; (3) Edelman, I. S., et al., 1954; (4) Duckworth, R., and Hill, R., 1953; (5) Agna, J. W., Knowles, H. D., and Alverson, G., 1958; (6) Rogers, H. J., Weinmann, S. M., and Parkinson, A., 1952; (7) Dickerson, J. W. T., 1962; (8) Call, R. A., et al., 1965; (9) Zipkin, I., McClure, F. J., and Lee, W. A., 1960; and (10) Long, C., 1961.

other ions associated with the apatite such as sodium, potassium, magnesium, chloride, carbonate and citrate. It is possible, therefore, that more than one apatitic type compound* may exist in the mineral phase of bone, and also that one form such as octacalcium phosphate, $Ca_8H_2(PO_4)_6 \cdot 5 H_2O$, may be converted to fluorapatite under the influence of fluoride.

The inorganic composition of human bone, based on ten selected references, is presented in Table 23-1. It is generally believed that only fluoride can substitute for hydroxyl to give $Ca_{10}(PO_4)_6F_2$, although partial substitutions may exist to give mixed crystals of $Ca_{10}(PO_4)_6$ (F, OH). The other ions are presumed to be oriented somehow on the surface of the crystals. Recent data indicate that carbonate may substitute for phosphate (LeGeros et al., 1965), but this issue is still not resolved. Ions such as calcium, phosphate and fluoride which make up the apatite lattice are also presumed to be absorbed to some extent on the surface (Armstrong and Singer, 1965).

It is not clear how the inorganic phase is associated with the organic matrix. Current thinking indicates that collagen, the chief protein of bone, provides the template for the

* Apatitic compounds are structures giving patterns similar to well crystallized hydroxyapatite. A number of calcium phosphate compounds belong to the apatite series but have Ca/P ratios other than the theoretical 2.15 on a weight basis.

nucleation of apatite somewhere along its fiber length. It should be mentioned, also, that other biological polymers besides collagen may provide the sites for the nucleation of calcium and phosphate to produce apatitic or apatitic-like compounds. For example, the elastin of the aorta may preferentially provide nucleation sites in vitro for apatite even in the presence of collagen (Martin et al., 1963).

As shown in Table 23-1, the main constituents of the inorganic phase of bone are calcium and phosphorus, which account for nearly all the ash. With the exception of carbonate and magnesium, all the other inorganic constituents together comprise about 1 per cent of the dry, fat-free weight of bone. Other elements reported in bone are iron, 0.014 per cent; copper, 0.0025 per cent; lead, 0.006 per cent; strontium, 0.010 per cent; and barium 0.0004 per cent. Spectrographically, a number of other elements have been detected in traces such as silver, arsenic, bismuth, lithium, molybdenum, nickel, selenium, silicon and zinc (Long, 1961).

Fluoride has been shown to alter the concentration of some of the inorganic components of bone (Zipkin et al., 1960). No data appear to be available, however, on the effect of fluoride on the organic fraction in human bone which may affect the nucleation and precipitation of the mineral phase of bone.

It has been demonstrated that different

parts of the same bone may vary in their inorganic composition and that bones from the same individual may differ slightly (Agna et al., 1958). In addition, species differences exist and age is a factor, but sex apparently is not.

ORGANIC COMPOSITION

The dry, fat-free adult human femur contains 4.7 per cent nitrogen (Rogers et al., 1952) and 25.8 per cent protein. About 23 per cent of the bone is made up of collagen, comprising 90 to 95 per cent of the protein of bone, with an amino acid composition characteristic of most collagens from other species (Eastoe, 1956). Approximately 0.3 per cent of the protein is combined with a mucopolysaccharide. The composition of this particular protein is quite different from that of collagen and approaches albumin in amino acid make-up.

No comprehensive review appears to be available on the mucopolysaccharide or lipid content of human compact bone. However, ox shaft bone appears to contain chondroitin sulfate A (Long, 1961; Meyer, 1956), hyaluronic acid, keratosulfate and other ill-defined sulfated fractions. Hydrolysis of the protein-polysaccharide complexes yields glucosamine, galactosamine, glucose, galactose, mannose and glucuronic and perhaps other uronic acids. The compact bone of ox femurs contains about 0.1 per cent total lipids.

The location of citrate in bone is not known but is generally presumed to be associated with the inorganic matrix. The dry, fat-free iliac crest, rib and vertebrae of humans contain 1.2, 1.1 and 1.0 per cent, respectively (Zipkin et al., 1960); Thunberg, in 1948, recorded values of 0.89 to 1.89 per cent.

Alveolar Bone. Few comparative studies on skeletal tissues have included the alveolar bone, which is of paramount importance to the maintenance of the integrity of the dental apparatus. Alveolar bone is highly vascularized and indeed has been shown to incorporate more fluoride than the incisal portion, the ramus or the body of the mandible of the rat (Ezra and Gedalia, 1964). It has been demonstrated that various insults such as hydrocortisone administration (Zipkin et al., 1965), low protein diets (Zipkin and Larson,

1963) and other dietary regimens (McClure, 1961; Likins et al., 1963; Baer and White, 1960) can produce marked alveolar bone loss in the experimental animal. In addition, alveolar bone differs from other skeletal tissue in being rich in alkaline phosphatase activity and in its ability to incorporate various bone-seeking isotopes. It is important, therefore, to extend future studies on the composition and metabolism of various skeletal components to include alveolar bone.

TEETH

Teeth of the human make up about 0.066 per cent of the body weight, or about 46 gm. About 22 per cent of a tooth by weight is enamel, and dentin and cementum comprise the remainder. The specific gravities are: enamel, 2.9 to 3.0; dentin, 2.14; and cementum, 2.03. The refractive indexes of enamel and dentin are 1.62 and 1.56, respectively, whereas that of synthetic apatite is 1.63. The chemistry of enamel and dentin will be presented separately. No review seems to have been written on the chemistry of cementum, but it has been considered to be similar in composition to dentin (Long, 1961).

ENAMEL

Inorganic Composition. About 95 per cent of the enamel is inorganic, and the major constituents, as in the case of bone, are calcium, phosphorus, magnesium and carbonate. The inorganic phase is believed to belong to the apatitic class of compounds with the general formula $Ca_{10}(PO_4)_6X_2$, in which X is either hydroxyl or fluoride, with a theoretical Ca/P weight ratio of 2.15. The apatite of enamel is considerably better crystalized than that of dentin or bone (Frank et al., 1960).

Some differences in the inorganic composition of sound and carious enamel have been reported (Johansen, 1963). It should be stressed that carious material itself was analyzed in these studies. Magnesium and carbonate showed a significant decrease in the carious material, whereas a marked increase was observed in the fluoride content. Nikiforuk and Grainger (1966) reported that

TABLE 23-2. INORGANIC COMPOSITION OF SOUND ENAMEL FROM ADULT HUMAN TEETH
Dry, Fat-free Basis (%)*

	(1)†	(2)	(3)	(4)	Mean (1-4)
Ash	97	94.9	95.2		95.7
Ca	36.0	36.4	35.8	35.5	35.9
P	17.5	17.1	16.9	16.5	17.0
Ca/P‡	2.08	2.13	2.12	2.15	2.11
Mg	0.42	0.43	0.39	0.44	0.42
Na	0.77	0.64		0.24	0.55
K		0.17	0.17		0.17
CO_2	2.50	2.49	2.5	1.9	2.35
Cl	0.25			0.29	0.27
F§	0.01	0.01	0.01	0.02	0.01

* Means are unweighted and are expressed on a wt./wt. basis. Values refer to whole body enamel. (See Brudevold, 1962.)

† Numbers in parenthesis refer to citations quoted below.

‡ Ca/P given as weight ratios.

§ Fluoride values are taken for teeth of individuals residing in an essentially fluoride-free area. For relation of F in drinking water to F in teeth, see McClure and Likins, 1951.

(1) Brudevold, F., 1962; (2) compiled from Leicester, 1949; (3) compiled from Long, 1961; (4) taken from Trautz, 1955, and calculated to dry, fat-free basis (95 per cent ash).

sound enamel from carious teeth had a lower carbonate content than sound enamel from sound teeth. In a recent review, Brudevold (1965) has interpreted his data to indicate that no relation exists between the deposition of fluoride and either carbonate or magnesium in human enamel. The evidence is equivocal, therefore, that the effect of fluoride on enamel (Johansen, 1963; Brudevold, 1965; Nikiforuk and Grainger, 1966) is similar to that observed in bone (Zipkin et al., 1960). It is generally accepted that no differences have been seen in the fluoride content of cleaned bulk enamel from sound and carious teeth.

The data which have been presented in Table 23-2 refer to the composition of bulk or whole body enamel rather than to specific surfaces which can be removed by grinding or by dissolution in acid.

Distribution of Ions in Enamel. Brudevold (1962) has shown that fluorine, zinc, lead, nitrogen and, to a lesser degree, iron, silver, manganese, silicon and tin normally occur in higher concentrations in surface than in subsurface enamel. Other ions such as carbonate, sodium and magnesium show a reverse distribution; namely, a lower concentration in the surface than in the more interior enamel. Strontium, copper, aluminum and potassium show no particular distribution pattern.

Organic Composition. The organic template (structure) on which calcium and phosphate are presumed to be nucleated in bone is collagen, but the composition of the organic matrix of enamel is in doubt (Piez, 1962). About 0.6 per cent of dry enamel is organic and 99.2 per cent is inorganic (Brudevold, 1962). As much as 98 per cent of the nitrogen occurs in protein and comprises about 0.08 per cent of the dry weight of enamel. Since enamel is of ectodermal origin, it would be expected that its protein would be keratin as in hair, feathers, nails or skin. It cannot be unequivocally assigned to the keratin group since its basic amino ratio is not characteristic of this protein and, in addition, its cystine content is much too low for eukeratin. Furthermore, the protein of enamel contains about 4 per cent of hydroxyproline, an amino acid characteristic of collagen. The fact is that proteins of enamel are unique because most proteins contain 16 per cent nitrogen, whereas 13 per cent of the acid insoluble protein of enamel (comprising about one third of the enamel protein) is nitrogen, while 10 per cent of the soluble protein is nitrogen. It should also be remembered that the protein composition is markedly affected by age (Brudevold, 1962).

Enamel also contains traces of phospholipid and cholesterol and 0.1 per cent citrate. Some of the carbohydrate is combined with the soluble protein as a mucoprotein. The organic composition of enamel is as follows: keratin, 0.15 per cent; mucoprotein, 0.10 per cent; peptides, 0.15 per cent; collagen, 0.10 per cent; and citrate, 0.10 per cent.

DENTIN

Inorganic Composition. The inorganic phase of dentin, like that of bone and enamel, is apatitic in structure and has the general formula $Ca_{10}(PO_4)_6X_2$, in which X is either hydroxyl or fluoride but is much less crystalline than enamel. It will be noted in Table 23-3 that dentin contains calcium and phosphorus in concentrations similar to those in bone but is higher in ash content and in magnesium.

TABLE 23-3. INORGANIC COMPOSITION OF
SOUND DENTIN FROM ADULT HUMAN TEETH
Dry, Fat-free Basis (%)*

	(1)†	(2)	(3)	(4)	Mean (1-4)
Ash	69	71.0	70.2		70.0
Ca	27.0	26.2	25.7	24.8	
P	13.0	12.8	12.4	12.3	12.6
Ca/P‡	2.07	2.05	2.07	2.02	2.06
Mg	0.84	0.88	0.84	0.70	0.82
Na	0.30	0.27		0.17	0.25
K		0.09	0.09		0.09
CO_2	3.30	3.44	3.15	2.80	3.17
Cl				0	0
F§	0.02	0.02	0.02	0.02	0.02

* Means are unweighted and are expressed on a
wt./wt. basis. Values refer to whole body dentin.

† Numbers in parentheses refer to citations quoted
below.

‡ Ca/P given as weight ratios.

§ Fluoride values are taken for teeth of individuals
residing in an essentially fluoride-free area.

(1) Brudevold, F., 1962; (2) compiled from Leicester,
1949; (3) compiled from Long, 1961; (4) taken from
Trautz, 1955, and calculated to dry, fat-free basis
(70 per cent ash).

Compared to enamel, dentin contains less calcium and phosphorus but more magnesium,
carbonate and fluoride.

Johansen (1963) reported that carious dentin contained more water and organic material than sound dentin. There was also a
marked increase in fluoride in soft, carious
dentin. When expressed on an ash basis, magnesium and carbonate were lower in carious
than in sound dentin. Some trace elements
such as zinc, lead, tin, manganese and iron
were reported to be elevated in carious dentin
as compared to sound dentin.

Distribution of Ions in Dentin. As in
the case of enamel, a gradient exists in the
concentration of a number of ions throughout
the dentin. It should be remembered that dentin is bounded by the dentinoenamel junction
on one aspect, and by the pulp on the other.
The concentration of fluoride and zinc, for
example, increases from the dentinoenamel
function to the pulp (Brudevold, 1962). Lead,
tin and iron and, to a lesser extent, aluminum
and silicon also appear to concentrate in the
layers of dentin adjacent to the pulp (Brudevold et al., 1960). Strontium and copper are
relatively uniformly distributed.

Organic Composition. The organic matter of dentin accounts for about 20 per cent of

its dry weight. Dentin contains about 3.5 per
cent nitrogen, virtually all of which is present
in collagen. The nitrogen content of collagen
is 18.4 per cent so that about 18 per cent of
the dentin is collagen, assuming that about
95 per cent of the protein present is collagen;
that is,

$$\frac{100 \times 3.5 \times 0.95}{18.4} = 18 \text{ per cent.}$$ The collagen

of dentin has been shown to be similar to that
of skin and bone in the proportion of amino
acids and by the fact that on heating it lique-
fies to produce gelatin. About 0.2 per cent of
the organic residue of dentin is made up of
an insoluble noncollagenous protein. Small
amounts of chondroitin sulfates can be isolated as part of the mucopolysaccharide fraction. In addition, cholesterol and aminophos-
pholipids have been identified in the lipid
fraction. Citrate has been shown to occur to
the extent of about 0.9 per cent in dentin.
An approximate composition of the total organic content of dentin may be summarized
below as presented by Brudevold (1962) and
Sognnaes (1961):

Collagen	18.0%
Insoluble protein	0.2
Mucopolysaccharide	0.2
Lipid	0.2
Citric acid	0.9

Total organic matter 19.5%

The mineral crystallites of dentin are presumed to be oriented and linked in some unknown fashion to the major organic component, collagen.

CEMENTUM

Little is known about the chemical composition of cementum except that it is presumed
to be similar to bone and dentin. Brudevold
et al. (1960) have indicated that ions concentrate in various layers of cementum in a fashion similar to dentin. It is interesting that
cementum concentrates fluoride to a greater
extent than does surface enamel (Yardeni
et al., 1963; Singer and Armstrong, 1962) or
dentin, mandible, femur and iliac crest (Singer
and Armstrong, 1962). It has a specific grav-
ity of 2.03, which is somewhat less than that
of dentin (sp. gr., 2.4) and appears with the

| | Age in Years | | | |
	< 20	20-29	30-49	> 50
ASH				
Fresh basis	2.2	1.6	5.8	19.2
WATER				
Fresh basis	8.2	8.1	6.7	5.5
Ca				
Fresh basis	0.51	0.74	1.32	4.71
Ash basis	27.4	29.1	38.2	37.8
P				
Fresh basis	0.28		0.67	2.36
Ash basis	16.9		19.2	18.5
Ca/P	1.82		1.97	2.00
F				
Fresh basis		0.011		0.068
Ash basis		0.220		0.578

dentin fraction when separated by usual flotation procedures.

The primary function of cementum is to furnish the attachment for fibers of the periodontal membrane. During eruption, secondary layers of cementum are deposited and old fibers become detached while new fibers replace them. Cementum appears to be independent of the pulp and dentin for its nutritional requirements, so that a pulpless tooth still presents a vital cementum for attachment of periodontal fibers.

PULP

The chemistry of pulp has been reviewed by Yoon et al. (1965). It has been demonstrated that the ash, calcium, phosphorus and fluoride are many times higher in pulp than in other soft tissues when examined on a fresh weight basis. The ash of human pulp contains calcium in varying amounts which increase with age from 27.4 to 37.8 per cent; phosphorus, 16.9 to 19.2 per cent; and fluorine, 0.011 to 0.068 per cent (Table 23-4). The uptake of fluoride by pulp appears to be directly related to its degree of calcification.

Pulp is a vital tissue, undergoes active respiration and has been shown to be an important storehouse of a number of enzymes.

Summary

Table 23-5 gives summary data on the inorganic components of bone, enamel and dentin which are expressed on both a dry fat-free and on an ash basis, since the mineral (ash) content varies in these tissues.

It can be seen that the concentration of the major components is similar when expressed on an ash basis. The calcium content is 37 to 39 per cent, the phosphorus about 18 per cent and the Ca/P ratio 2.1–2.2. Bone and enamel contain similar concentrations of magnesium, but dentin is appreciably higher. Sodium appears to be higher in bone than in either enamel or dentin. The carbonate content is highest in bone and lowest in enamel. More fluoride concentrates in the ash of bone than in the ash of either enamel or dentin.

It is difficult on the basis of this review to draw any general conclusions regarding the deposition of inorganic elements in the hard tissues. While cementum, dentin and bone present a presumably similar organic template

TABLE 23-5. INORGANIC COMPOSITION OF BONES AND TEETH OF THE ADULT HUMAN

| | Dry, Fat-free Basis | | | Ash Basis | | |
	Bone	Enamel	Dentin	Bone	Enamel	Dentin
Ash	57.1	95.7	70.0			
Ca	22.5	35.9	25.9	39.4	37.5	37.0
P	10.3	17.0	12.6	18.0	17.7	18.0
Ca/P	2.18	2.11	2.06	2.18	2.11	2.06
Mg	0.26	0.42	0.82	0.46	0.44	1.17
Na	0.52	0.55	0.25	0.91	0.57	0.36
K	0.089	0.17	0.09	0.16	0.18	0.13
CO_2	3.5	2.35	3.17	6.1	2.46	4.53
Cl	0.11	0.27	0	0.19	0.28	0
F	0.054	0.01	0.02	0.095	0.010	0.029

TABLE 23-6. ORGANIC COMPOSITION OF
BONES AND TEETH
Dry Basis (%)

	Bone	Enamel	Dentin
Organic matter	25.0	0.6	19.5
Collagen	23.5	0.1	18.0
Peptides		0.15	
Keratin		0.15	
Mucoprotein	0.3	0.1	0.2
Insoluble protein			0.2
Lipid	0.1	trace	0.2
Citrate	1.1	0.1	0.9

(collagen) for deposition of mineral, some differences exist in sodium, magnesium, carbonate and fluoride. These differences may be related to the degree of remodeling or perhaps to the degree of vascularity of the various tissues, which could influence ion transport. Enamel protein is different from either collagen or keratin and so it would be expected that the calcification and nucleation patterns for the mineral phase of enamel would be different from dentin, cementum or bone.

Almost all the protein matrix of bone and dentin, as shown in Table 23-6, is collagen which is similar in its amino acid composition to collagen from other sources. The protein of enamel is not clearly defined and probably contains more than one protein, all in small quantities. Lipids and polysaccharides, not yet clearly elucidated, are present in small quantities in bone, enamel and dentin.

Citrate is present to the extent of about 1 per cent in bone and dentin but its role in the calcified tissues is not known. Some investigators believe it is only adventitiously present because of its occurrence in the extracellular fluids permeating bone, and others indicate it plays an important role in calcification. It has been proposed that citrate is associated with a peptide although this finding still lacks confirmation. The thesis that citrate may play a role in bone resorption is no longer strongly supported.

Only little attention has been paid to the chemistry of alveolar bone even though it may be susceptible to changes in the composition of the diet. More information is needed in view of the important role the alveolar bone plays in the maintenance of the integrity of the dental apparatus.

CALCULUS

Calculi of dental interest may be found in the salivary glands or their ducts and are generally referred to as salivary calculi. When present as hard solid masses of calcified material adhering to the teeth or to prosthetic appliances in the oral cavity, they are usually called dental calculi. Stains, materia alba, food debris and the bacterial plaque are not included.[*] Dental calculi may be further classified as supragingival or subgingival.

Both salivary and dental calculi are heterogeneous mixtures of at least four calcium phosphates, three of which occur rarely or not at all in other types of biological calcifications. Thus, apatite or $Ca_{10}(PO_4)_6X_2$ in which X is either hydroxyl or fluoride occurs most abundantly; brushite, $CaHPO_4 \cdot 2 H_2O$; magnesium whitlockite, a type of $Ca_3(PO_4)_2$ in which some of the calcium is substituted by magnesium; and octacalcium phosphate, $Ca_8H_2(PO_4) \cdot 5 H_2O$, have also been identified in appreciable quantities (Leung and Jensen, 1958). Octacalcium phosphate has been proposed as a possible precursor of apatite during the initial stages in the calcification of bone but as yet has not been unequivocally identified in calcifying systems in vivo. The proportion of these calcium phosphates varies in supra- and subgingival dental calculi and in salivary calculi (Rowles, 1964) and is also dependent on the age of the calculus deposits (Schroeder and Bambauer, 1966).

The organic composition of salivary and dental calculi is ill defined, and no attempts have been made to separate the heterogeneous calculi into component parts prior to analysis.

SALIVARY GLAND CALCULI

Data on the composition of calculi surgically removed from the salivary glands or their ducts are quite meager. Their inorganic composition is presented in Table 23-7.

Inorganic Composition. As shown in Table 23-7, concretions removed from the salivary glands contain about 80 per cent mineral

[*] It is possible, however, that samples of calculus prepared for analysis may contain adventitious amounts of these materials.

TABLE 23-7. INORGANIC COMPOSITION OF SALIVARY CALCULI*
Dry, Fat-free Basis (%)

	(1)[†]	(2)	(3)[‡]	(4)	(1-4)
Ash	80.0		78	80.5	79.5
Ca	31.7	30.7	31	29.9	30.8
P	14.9	16.9	15.7	14.4	15.5
Ca/P	2.13	1.82	1.97	2.08	1.99
Mg.	0.45	1.03		0.48	0.65
CO_2	1.91	1.39		2.89	2.06

* Salivary calculi refer to calculi surgically removed from either the salivary gland or its ducts.

† Numbers in parenthesis refer to citations listed below.

‡ Data of (3) expressed on a dry basis, but included since values are similar to those of (1) and (2) expressed on a dry, fat-free basis.

(1) Karshan, M., and Schroff, I., 1928; (2) Glock, G. E., and Murray, M. M., 1938; (3) Harrill, J. A., King, J. S., and Boyce, W. H., 1959; (4) Hodge, H. C., and Leung, S. W., 1950.

matter (Karshan and Schroff, 1928; Glock and Murray, 1938; and Harrill et al., 1959). The inorganic or mineral material is made up largely of calcium and phosphorus with a weight ratio of 1.97. The magnesium content is less than 1 per cent, and the carbonate (CO_2) is about 1.7 per cent. No data appear to be available on the sodium, potassium, chloride and fluoride content. These calculi would be expected to accumulate appreciable quantities of fluoride in view of their high ash and calcium concentrations, as has been shown for urinary tract calculi (Zipkin et al., 1958).

Organic Composition. Only fragmentary data are available on the organic composition of salivary calculi. About 20 per cent of the weight of calculus is organic in nature. Glock and Murray (1938) gave the following composition of salivary duct calculi:

Ash	81.5%
Protein	8.3
Fat	2.7
Water and CO_2	7.5
	100.0%

Harrill et al. (1959) reported the presence of galactose, glucose, mannose, rhamnose, fucose, hexosamine and a deoxypentose in demineralized, hydrolyzed samples of salivary duct calculi. King and Boyce (1957) determined the amino acid proportions of salivary duct calculi and noted the similarity of their composition to that of urinary tract calculi.

DENTAL CALCULI

As previously noted, dental calculi in this report refer to the hard concretions which form on prosthetic devices in the mouth or on the teeth. Concretions forming on the teeth may be divided into those occurring above the gingival margin (supragingival) and those below (subgingival).

Inorganic Composition. Information on the inorganic composition of subgingival calculus is very meager. Only small differences appear to exist in the inorganic composition of supra- and subgingival calculus, as seen in

TABLE 23-8. INORGANIC COMPOSITION OF DENTAL CALCULUS
Dry Basis (%)

		Supragingival				Mean*	Subgingival		Mean	Total[†]
	(1)[‡]	(2)	(3)	(4)	(5)	(1-5)	(6)	(7)	(6-7)	(8)
Ash	78.0	77	75	74.5		76.1	77.8	72.8	75.3	76.7
Ca	26.4	25.0	25.7	28.9	26.0	26.4	32.6	28.1	30.4	26.6
P	15.4	13.9	13.3	14.2	11.6	13.7	16.9	15.3	16.1	14.3
Ca/P	1.73	1.80	1.93	2.03	2.24	1.93	1.93	1.84	1.89	1.86
Mg					0.2	0.2				
Na	2.7				1.5	2.1		1.5	1.5	
CO_2				2.8		2.8	2.79		2.8	
Cl					0.9	0.9				

* Means are unweighted.

† Total calculus refers to three samples containing both supra- and subgingival calculus.

‡ Numbers in parenthesis refer to citations listed below.

(1) Little, M. F., Casciani, C. A., and Rowley, J., 1963; (2) Schroeder, H. E., 1963; (3) compiled from other data quoted by Schroeder, H. E., 1963; (4) Hodge, H. C., and Leung, S. W., 1950; (5) Söremark, R., and Samsahl, K., 1962; (6) Glock, G. E., and Murray, M. M., 1938; (7) Little, M. F., and Hazen, S. P., 1964; (8) Zipkin, I., unpublished data.

Table 23-8. The ash content of the two types of dental calculus is similar (about 75 or 76 per cent); the mean calcium content is 26 per cent for supra- and 30 per cent for subgingival calculus; while, respectively, the phosphorus concentration is about 14 and 16 per cent. The Ca/P ratios are quite similar at about 1.9. The carbonate content on the basis of very meager data seems similar, i.e., about 2.8 per cent of the dry weight. The lipid content of dental calculus is about 0.2 per cent (Mandel and Levy, 1957), so that comparisons of the composition of supra- and subgingival calculi are valid, although the former are expressed on a dry, fat-free basis and the latter on a dry basis only.

Fluoride accumulates in both supra- and subgingival dental calculus (Yardeni et al., 1963; Gedalia et al., 1963) to give mean values exceeding those reported in enamel of individuals exposed to essentially fluoride-free water (McClure and Likins, 1951). Thus, supragingival calculus has been shown to contain 0.039 per cent fluoride (Yardeni et al., 1963) and 0.047 per cent (Gedalia et al., 1963); subgingival calculus, 0.024 per cent fluoride (Yardeni et al., 1963) compared to about 0.01 per cent fluoride in enamel (McClure and Likins, 1951). It seems possible that varieties of calcium phosphate such as brushite, whitlockite and octacalcium phosphate identified in calculus but not in enamel may incorporate greater amounts of fluoride than hydroxyapatite.

It has been shown also that no relation exists between the concentration of fluoride in saliva and in calculus (Gedalia et al., 1963).

Organic Composition. Very little quantitative data appear to be available on the organic composition of dental calculi, i.e., the hard concretions on the surface of teeth. Little et al. (1961) reported that about 22 per cent of the EDTA decalcified and dialyzed supragingival calculus was organic matrix. On ashing, this matrix lost 90 to 95 per cent of its weight. About 16 per cent by weight of the matrix was carbohydrate as determined by the anthrone reaction, and about 38 per cent was protein according to the Kjeldahl procedure. Mandel et al. (1962) found that 12 per cent of the original calculus remained after decalcification with EDTA, followed by dialysis. About 13 per cent of this matrix was soluble

TABLE 23-9.   COMPOSITION OF SOLUBLE, INSOLUBLE AND TOTAL CALCULUS MATRIX*

	Soluble	Insoluble	Total
Hexose	13.4	6.1	7.9
Methylpentose	3.3	2.2	2.5
Hexosamine	0.2	2.2	1.7
Pentose	2.0	0.9	1.2
Sialic acid	0.3	0.2	0.2
Nitrogen	2.7	10.	8.2
Ash†	13.4	4.4	6.7

* Taken from Mandel, Hampar, and Ellison, 1962. Calculated on the basis that the total matrix consists of 75 per cent insoluble and 25 per cent soluble matrix.

† Ash calculated on a dry basis.

in water, and both the soluble and insoluble fractions were analyzed for several carbohydrate components by Mandel and his associates as shown in Table 23-9.

It is apparent that differences exist in the organic composition of the soluble and insoluble portions of the calculus matrix, i.e., that portion of calculus which remains after decalcification and dialysis. No quantitative data appear to be available on the amino acid chemistry of dental calculus.

Dry samples of mixed oral dental calculus have been found to contain about 64 mg. per cent citrate.*

Summary

The composition of salivary gland and oral calculus is summarized in Table 23-10, and expressed on both a dry weight and ash basis. Calcium and phosphorus seem to be lower in supragingival calculus, but the Ca/P ratio is not markedly different. Magnesium also appears to be lower in supragingival calculus than in subgingival or salivary gland calculi, whereas the latter appear to be lower in carbonate. However, data are insufficient at present to make any rigid comparisons. Quantitative data on the composition of the organic matrix of calculi of dental interest are also too sparse for valid comparisons.

The mechanism of formation of salivary gland calculi and oral dental calculi is unknown. Calculus deposits within the salivary glands and on the surfaces of teeth appear to

* Zipkin, I. Unpublished data.

TABLE 23-10. INORGANIC COMPOSITION OF SALIVARY AND DENTAL CALCULUS

	Dry Basis			Ash Basis		
	SALIVARY	DENTAL Supragingival	Subgingival	SALIVARY	DENTAL Supragingival	Subgingival
Ash	79.5	76.1	75.3			
Ca	30.8	26.4	30.4	38.8	34.6	40.4
P	15.5	13.7	16.1	19.5	17.9	21.4
Ca/P	1.99	1.93	1.89	1.99	1.93	1.89
Mg	0.65	0.20		0.82	0.26	
CO_2	2.06	2.80	2.80	2.60	3.67	3.72

be derived from the saliva (Mandel, 1963). Subgingival calculus may not be derived from the saliva, since it has been shown that dyes fail to penetrate into periodontal pockets.

If some template or specific organic structure is necessary for calculus deposition, as has been postulated for calcification of bones and teeth, it has not been elucidated for calculus precipitation. The composition of saliva is sufficiently similar to that of salivary gland concretions and supragingival concretions to suggest that it plays a role in the formation of both the inorganic and organic phases of calculi (Mandel, 1963). It is of interest to note that calculus deposition has been observed in both germ-free mice (Baer and Newton, 1959) and in rats (Fitzgerald and McDaniels, 1960).

It is generally assumed that calculus formation is an orderly process rather than a conglomerate precipitation of a number of inorganic and organic phases. Calcification processes in bone and calculus are probably dissimilar in view of the differences in organic composition and in view of the number of species of calcium phosphates which have been identified in calculi. However, the general premise is still assumed that there is some structural and perhaps chemical relationship between the organic and inorganic structures of both salivary gland and oral dental calculus.

SALIVA

Whole mixed saliva is a composite of the secretions of the three major salivary glands, the parotid, submaxillary and sublingual. A number of minor glands of the oral mucosa also contribute small amounts to the saliva. Under paraffin stimulation, about 40 to 50 per cent of the whole mixed saliva is contributed by the parotid gland, and an approximately equal volume by the submaxillary gland. The secretion of the sublingual gland is viscous and, together with the secretions of the minor oral mucosal glands, contributes about 5 per cent of the total volume of whole saliva.

The composition of whole saliva is determined, of course, by the composition of the individual salivary secretions. In addition, saliva contains bacteria and cellular debris from the oral mucosa. An adequate rationale may be presented for either retaining such suspended elements as just mentioned or removing them prior to analysis. The induced rate of secretion alters the concentration of a number of constituents and, in most cases, the methods used for the collecting of saliva do not simulate the normal physiological situation.

It is not known to what extent many of the constituents represent an ultrafiltrate of blood and to what extent they are products of salivary gland manufacture. Little is known about the role that sex or age may play in the concentration of constituents in saliva. Large variations have been reported by various investigators, who have used different methods for stimulating saliva; some have reported only a so-called "resting saliva" or saliva obtained under minimal stimulation. For these reasons it is difficult to present unequivocal values for the constituents of saliva. The values presented here should, therefore, be regarded as consensus values taken from a number of sources; they refer to saliva obtained under paraffin stimulation.

Whole mixed human saliva contains about 99.5 per cent water, and the remaining 0.5 per cent is approximately equally divided between inorganic and organic constituents. It varies in pH from 7.2 to 7.6 and has a significant buffering effect due to phosphate and bicarbonate ions and to proteins. Saliva is secreted at an approximate rate of 1.0 to 1.5 liters per day.

INORGANIC COMPOSITION

As seen in Table 23-11, calcium in saliva is much lower than in blood serum (9 to 11 mg.). Inorganic phosphorus in saliva is about 75 per cent of the total P value quoted or about 11.7 mg. per cent, which is appreciably higher than that reported for inorganic phosphorus in serum (3 to 5 mg. per cent). Magnesium is lower in saliva (0.4 mg. per cent) than in blood (1.6 mg. per cent); sodium is much lower in saliva (67.3 mg. per cent) than in blood (330 mg. per cent), whereas the potassium content of saliva (71.0 mg. per cent) far exceeds that found in blood serum (19.0 mg. per cent). The carbon dioxide content is somewhat lower (34 vol. per cent) than that of blood serum (45 to 65 vol. per cent), and the chloride content (91.4 mg. per cent) is markedly lower than that of blood serum (355 mg. per cent).

It appears, therefore, that the inorganic constituents of saliva do not wholly represent an ultrafiltrate of blood unless appreciable resorption of some ions by the salivary duct cells occurs to give lower values than those in blood. The higher concentration of some ions in saliva as compared to blood may perhaps be explained by resorption of water by the salivary duct cells. At any rate, the phenomena responsible for the concentration of various ions in saliva must be complex and, in addition, are influenced by the degree of stimulation of the various glands and, hence, the rate of secretion.

About 15 mg. per cent thiocyanate has been found in saliva, as well as traces of copper, cobalt, iodide, bromide and fluoride (0.1 to 0.2 ppm). Small amounts of dissolved oxygen and nitrogen have also been detected. Values for the saliva of children were within the range of values for adults.

Data on the inorganic chemistry of individual paraffin-stimulated salivary secretions are very limited. Mandel et al. (1964), for example, found human parotid saliva to contain 3.5 mg. per cent calcium and 13.3 mg.

per cent phosphorus, whereas submaxillary saliva contains 8.0 mg. and 10.6 mg. per cent, respectively. No data appear to be available on the inorganic composition of human sublingual saliva.

ORGANIC CONSTITUENTS

As already indicated, 0.5 per cent of saliva is made up of solids, about half of which are organic in nature. As seen in Table 23-11, the major organic constituent is protein, much of which is present as a carbohydrate-protein complex precipitable by dilute acid and called "mucin." A number of proteins exist in whole saliva which can be separated electrophoretically. Immune sera can be produced in rabbits following injection of concentrated human saliva. Blood group substances presumably similar to those in blood can be found and can be used for typing purposes.

A large number of vitamins have also been reported in human saliva but in trace amounts.

A large number of enzymes have also been shown to be present in saliva, most of which are due to suspended bacteria (Chauncey, 1961). One of the richest sources of α-amylase has been shown to be saliva.

The organic constituents of the individual secretions have been studied to some extent.

TABLE 23-11. COMPOSITION OF WHOLE MIXED HUMAN SALIVA*
(mg./100 ml.)

Inorganic Components†		Organic Components†	
Ca	5.7	Protein	230
P‡	15.7	Carbohydrate	15
Mg	0.4	Cholesterol	25
Na	67.3	Urea	4.5
K	71.0	Uric acid	2.5
CO_2§	34.0	Citrate	1.0
Cl	91.4	Lactate	3.0
NH_3	6.5	Creatinine	<1.0
		Sialic acid	5.0

* Data refer to paraffin stimulated saliva.
† The data for the composition of whole mixed human saliva are taken from Jenkins, 1960; Afonsky, 1961; Long, 1961; Burgen and Emmelin, 1961; Chauncey et al., 1956; Zipkin and Hawkins, 1964; Niedermeier et al., 1956; Anders, 1956; Newbrun, 1961; Zipkin et al., 1957; Williams, 1961; Zipkin et al., 1961
‡ P refers to total phosphorus.
§ CO_2 values are given in vol. per cent.

The parotid saliva has received the most attention because of its relative ease of collection. Parotid saliva contains about 200 mg. per cent protein, about 40 per cent of which is precipitable with trichloracetic acid. The soluble portion is presumed to be either low molecular weight peptides or glycoproteins (Zipkin et al., 1964; Hawkins, et al., 1963). For the separation of the various proteins, the disc electrophoresis technique seems to be most suitable (Caldwell, 1965). Submaxillary saliva appears to contain less protein than the parotid secretion, or about 120 mg. per cent, most of which is not precipitated by trichloracetic acid (Dawes, 1965).

The major portion of the carbohydrates in the parotid and submaxillary secretions are bound to protein and on hydrolysis by acid yield hexosamine, fucose, sialic acid, methyl pentose, deoxyribose, glucose, galactose and mannose. These constituents account for about 35 to 40 mg. per 100 ml. of the parotid saliva and about 75 mg. per 100 ml. of the submaxillary gland secretion. The lipids of both parotid and whole saliva have been qualitatively fractionated by paper chromatography to yield cholesterol esters, fatty acids, cholesterol and mono- di- and triglycerides as well as a number of phospholipids (Dirksen, 1962).

Summary

The chemistry of whole mixed human saliva and the individual secretions of the parotid and submaxillary glands have been studied to some extent. The data found in the literature need further support and verification. The degree to which the chemistry of the individual salivary secretions may serve as an index of systemic disease is only now receiving some attention. In studying the chemistry of the salivary secretions, it must be remembered that their quantitative composition may be related to the rate of secretion. Thus, for example, it has been shown that increasing the rate of secretion will enhance the concentration of calcium, sodium, chloride, citrate and lactate. The concentration of phosphorus decreases with an increased rate of secretion, potassium seems unchanged, and equivocal data have appeared on protein. The mechanisms responsible for these relationships need further study.

PERIODONTAL TISSUES

The periodontal tissues include gingival tissue, the membranes between the alveolar bone and the teeth, the alveolar bone supporting the teeth, and the cemetum (Engel et al., 1960).

Gingival tissue is made up of epithelial tissue and connective tissue and is physically located circumferentially around the tooth above the cementoenamel junction. An irregular margin separates the epithelial tissue from the contiguous connective tissue. No chemical data appear to be available on the gingival epithelium, but it can be presumed to be similar to other cellular soft tissues, since it is essentially cellular in nature and contains no fibrillar elements. Most soft cellular tissues contain about 70 to 80 per cent water, about 20 per cent organic solids and a small proportion of inorganic matter. The organic phase is made up largely of protein-polysaccharide complexes, as yet undefined. Small amounts of lipid materials are also present.

The connective tissue underlying the epithelial layer of the gingival tissue contains cells, collagen, reticular, elastic and oxytalan fibers embedded in a so-called "ground substance." Schultz-Haudt (1965) reviewed his contributions on the organic composition of human gingival tissue. Although qualitative, it presents the most extensive work on the polysaccharide and protein components of these tissues. Two polysaccharide-protein complexes were isolated using paper electrophoresis. On hydrolysis followed by paper chromatography, both fractions yielded glucuronic acid, glucosamine, galactose, glucose, mannose and ribose. The slowest moving component also contained fucose. The associated protein contained hydroxyproline in concentrations characteristic of collagen and reticulin. The protein component of the second fraction remains unidentified. Other electrophoretically determined components showed the presence of mucoproteins containing hyaluronic acid and chondroitin sulfates. Little or no hydroxyproline was associated with these latter fractions. Little more information is available at the moment on the chemical composition of gingival tissue. Histochemical identification of the components of gingival tissue and of the periodontal structures in general has been extensively reviewed by Fullmer (1966). Glycogen has been reported to comprise about 0.2 per cent of the wet tissue, and small amounts of lipid material have also been shown to be present.

The chemistry of cementum and alveolar bone has been briefly discussed previously in this review.

Summary

Very little quantitative information on the chemistry of the periodontium (gingival tissue, periodontal membrane, cementum and alveolar bone) appears to be available. The major contribution comes from histochemical examination of these tissues as reviewed by Fullmer (1966) and Schultz-Haudt (1965). The protein-polysaccharide complexes have been partially and only quantitatively identified by Schultz-Haudt and co-workers using histochemical, chemical, chromatographic and electrophoretic means.

It is quite clear that a large void exists in our knowledge of the chemical composition of the periodontium. Virtually no information is yet available on the biosynthesis of the protein and carbohydrate moieties in the periodontal tissues in health. Of perhaps even greater import is the lack of any definitive dynamic chemical information on composition of these tissues in one of the most prevalent diseases in man, namely, periodontal disease.

REFERENCES

Agna, J. W., Knowles, H. C., Jr., and Alverson, B.: The mineral content of normal human bone. J. Clin. Invest., 37:1357, 1958.

Afonsky, D.: Saliva and Its Relation to Oral Health. Tuscaloosa, University of Alabama Press, 1961.

Anders, J. T.: Physiologic chloride level of the saliva. J. Appl. Physiol., 8:659, 1956.

Armstrong, W. D., and Singer, L.: Composition and constitution of the mineral phase of bone. Philadelphia, J. B. Lippincott Co., Clin. Orthopaed., 38:179, 1965.

Baer, P. N., and White, C. L.: Studies on periodontal disease in the mouse. I. Effect of age, sex, cage factor and diet. J. Periodont., 31:27, 1960.

Baer, P. N., and Newton, W. L.: The occurrence of periodontal disease in germ-free mice. J. Dent. Res., 38:1238, 1959.

Brudevold, F.: Chemical composition of the teeth in relation to caries. In Sognnaes, R. F., (ed.): Chemistry and Prevention of Dental Caries. Springfield, Ill., Charles C Thomas, 1962, pp. 32-88.

Brudevold, F.: Caries resistance as related to the chemistry of the enamel. In Wolstenholme, G. E. W., and O'Connor, M. (ed.): Ciba Foundation Symposium on Caries-Resistant Teeth. Boston, Little, Brown and Co., 1965, pp. 121-148.

Brudevold, F., Steadman, L. T., and Smith, F. A.: Inorganic and organic components of tooth structure. In The Metabolism of Oral Tissues. Ann. N. Y. Acad. Sci., 85:110, 1960.

Burgen, A. S. V., and Emmelin, N. G.: Physiology of the Salivary Glands. Baltimore, The Williams & Wilkins Co., 1961.

Caldwell, R. C.: Electrophoretic studies of human submaxillary saliva. I. A. D. R. Abstracts No. 218, 1965.

Call, R. A., Greenwood, D. A., LeCheminant, W. H., Shupe, J. L., Nielson, H. M., Olson, L. E., Lamborn, R. E., Mangelson, F. L., and Davis, R. V.: Histological and chemical studies in man on effects of fluoride. Pub. Health Rep., 80:529, 1965.

Chauncey, H. H.: Saliva enzymes. J. Amer. Dent. Assoc., 63:360, 1961.

Chauncey, H. H., Levine, D. M., Kass, G., Shwachman, H., Henriques, B. L., and Kulczyeki, L. L.: Composition of human saliva. Parotid gland secretory rate and electrolyte concentration in children with cystic fibrosis. Arch. Oral Biol., 7:707, 1962.

Davies, R. E., Kornberg, H. L., and Wilson, G. M.: The determination of sodium in bone. Biochem. J., 52:xv, 1952.

Dawes, C.: Some characteristics of parotid and submaxillary salivary proteins. Arch. Oral Biol., 10:269, 1965.

Dickerson, J. W. T.: Changes in the composition of the human femur during growth. Biochem. J., 82:56, 1962.

Dirksen, T.: The Lipid Constituents of Whole and Parotid Saliva. Report SAM-TDR-62-111, Brooks Air Force Base, Texas, 1962.

Duckworth, J., and Hill, R.: The storage of elements in the skeleton. Nutr. Abstr. Rev., 23:1, 1953.

Eastoe, J. E.: The organic matrix of bone. In Bourne, G. H. (ed.): The Biochemistry and Physiology of Bone. New York, Academic Press, Inc., 1956, pp. 81-105.

Edelman, I. S., James, A. A., Baden, H., and Moore, F. D.: Electrolyte composition of bone and the penetration of radiosodium and deuterium oxide into dog and human bone. J. Clin. Invest., 33:122, 1954.

Engel, M. B., Joseph, N. R., Laskin, D. M., and Catchpole, H. R.: A theory of connective tissue behavior. Its implications in periodontal disease. In The Metabolism of Oral Tissues. Ann. N. Y. Acad. Sci., 85:399, 1960.

Ezra, B., and Gedalia, I.: The fluoride content in

different parts of the mandible of the rat. J. Dent. Res., *43*:716, 1964.

Fitzgerald, R. J., and McDaniels. E. G.: Dental calculus in the germ-free rat. Arch. Oral Biol., 2:239, 1960.

Follis, R. H., Jr.: Inorganic composition of human rib with and without marrow elements. J. Biol. Chem., *194*:223, 1952.

Frank, R. M., Sognnaes, R. F., and Kern, R.: Calcification of dental tissues with special reference to enamel ultrastructure. *In* Sognnaes, R. F. (ed.): Calcification in Biological Systems. Washington, D. C., American Association for the Advancement of Science, Publ. No. 64, 1960, pp. 163-202.

Fullmer, H. M.: Organization of the dental supportive tissues. *In* Greulich, R. C., and Miles, A. E. W. (ed.): The Chemical and Structural Organization of the Teeth. Vol. II, New York, Academic Press, Inc., 1966.

Gedalia, I., Yardeni, J., and Gershon, I.: Dental caries rate and fluoride content of dental calculus and saliva. J. Amer. Dent. Assn., *66*:525, 1963.

Glock, G. E., and Murray, M. M.: Chemical investigation of salivary calculi. J. Dent. Res., *17*:257, 1938.

Harrill, J. A., King, J. S., and Boyce, W. H.: Structure and composition of salivary calculi. Laryngoscope, *69*:481, 1959.

Hawkins, G. R., Zipkin, I., and Marshall, L. M.: Determination of uric acid, tyrosine, tryptophan, and protein in whole human parotid saliva by ultraviolet absorption spectrophotometry. J. Dent. Res., *42*:1015, 1963.

Hodge, H. C., and Leung, S. W.: Calculus formation. J. Periodont., *21*:211, 1950.

Jenkins, G. N.: The Physiology of the Mouth. 2nd ed. Oxford, England, Blackwell Scientific Publication, 1960, pp. 30-57.

Johansen, E.: Ultrastructural and chemical observations on dental caries. *In* Sognnaes, R. F. (ed.): Mechanisms of Hard Tissue Destruction. Washington, D. C., Amer. Assn. Advance. Sci. Publ. No. 75, 1963, pp. 187-211.

Karshan, M., and Schroff, J.: Composition of some salivary calculi. J. Dent. Res., *8*:454, 1928.

King, J. S., and Boyce, W. H.: Amino acid and carbohydrate composition of the mucoprotein matrix in various calculi. Proc. Soc. Exp. Biol. Med., *95*:183, 1957.

LeGeros, R. Z., Trautz, O. R., and LeGeros, J. P.: Carbonate substitution in the apatite structure. Abst. No. 14, 43rd Gen. Meeting, IADR, 1965.

Leicester, H. M.: Biochemistry of Teeth. St. Louis, The C. V. Mosby Company, 1949.

Leung, S. W., and Jensen, A. T.: Factors controlling the deposition of calculus. Intern. Dent. J., *8*:613, 1958.

Likins, R. C., Pakis, G., and McClure, F. J.: Effect of fluoride and tetracycline on alveolar bone resorption in the rat. J. Dent. Res., *42*:1532, 1963.

Little, M. F., Casciani, C., and Lensky, S.: The organic matrix of dental calculus. J. Dent. Res., *40*:753, 1961. (Abstract.)

Little, M. F., Casciani, C. A., and Rowley, J.: Dental calculus composition. I. Supragingival calculus: ash,

calcium, phosphorus, sodium and density. J. Dent. Res., *42*:78, 1963.

Little, M. F., and Hazen, S. P.: Dental calculus composition. 2. Subgingival calculus: ash, calcium, phosphorus and sodium. J. Dent. Res., *43*:645, 1964.

Long, C.: Biochemists' Handbook. Princeton, N. J., D. Van Nostrand Company, Inc., 1961.

Mandel, I. D.: Histochemical and biochemical aspects of calculus formation. J. Amer. Soc. Periodont., *1*:43, 1963.

Mandel, I. D., Eisenstein, A., Ruiz, R., Thompson, R. H., Jr., and Ellison, S. A.: Calcium and phosphorus in human parotid and submaxillary saliva. Proc. Soc. Exp. Biol. Med., *115*:959, 1964.

Mandel, I. D., Hampar, B., and Ellison, S. A.: Carbohydrate components of supragingival salivary calculus. Proc. Soc. Exp. Biol. Med., *110*:301, 1962.

Mandel, I. D., and Levy, B. M.: Studies on salivary calculus. I. Histochemical and chemical investigations of supra- and subgingival calculus. Oral Surg., *10*:874, 1957.

Martin, G. R., Schiffmann, E., Bladen, H. A., and Nylen, M.: Chemical and morphological studies on the *in vitro* calcification of aorta. J. Cell Biol., *16*: 243, 1963.

McClure, F. J.: Effect of diet on alveolar bone loss. J. Dent. Res., *40*:380, 1961.

McClure, F. J., and Likins, R. C.: Fluorine in human teeth studied in relation to fluorine in the drinking water. J. Dent. Res., *30*:172, 1951.

Meyer, K.: The mucopolysaccharides of bone. From Wolstenholme, G. E. W., and O'Connor, C. M. (ed.): Ciba Foundation Symposium on Bone Structure and Metabolism. Boston, Little, Brown and Co., 1956, pp. 65-74.

Newbrun, E.: Application of atomic absorption spectroscopy to the determination of calcium in saliva. Nature, *192*:1182, 1961.

Niedermeier, W., Dreizen, S., Stone, R. E., and Spies, T. D.: Sodium and potassium concentrations in the saliva of normotensive and hypertensive individuals. Oral Surg., *9*:426, 1956.

Nikiforuk, G., and Grainger, R. M.: Fluoride-carbonate-citrate interrelations in enamel. *In* Stack, M. V., and Fernhead, R. W. (ed.): Tooth Enamel: Its Composition, Properties and Fundamental Structure. Bristol, England, J. Wright & Sons, Ltd., 1966, pp. 26-31.

Piez, K. A.: Chemistry of the protein matrix of enamel. *In* Butcher, E. O., and Sognnaes, R. F. (ed.): Fundamentals of Keratinization. Washington, D. C., American Association for the Advancement of Science, 1962.

Rogers, H. T., Weidmann, S. M., and Parkinson, A.: Studies on the skeletal tissues. II. The collagen content of bones from rabbits, oxen and humans. Biochem. J., *50*:537, 1952.

Rowles, S. L.: The inorganic composition of dental calculus. *In* Blackwood, H. J. J. (ed.): Bone and Tooth. New York, Pergamon Press, 1964, pp. 175-183.

Schroeder, H. E.: Inorganic content and histology of early dental calculus in man. Helv. Odont. Acta, *7*:17, 1963.

Schroeder, H. E., and Bambauer, H. V.: Stages of calcium phosphate crystallization during calculus formation. Arch. Oral Biol., 11:1, 1966.

Schultz-Haudt, S. D.: Connective tissue and periodontal disease. In Hall, D. A. (ed.): International Review of Connective Tissue Research. Vol 3. New York, Academic Press, 1965, pp. 77-89.

Singer, L., and Armstrong, W. D.: Comparison of fluoride contents of human dental and skeletal tissues. J. Dent. Res., 41:154, 1962.

Sognnaes, R. F.: Dental aspects of the structure and metabolism of mineralized tissues. In Comar, C. L., and Bronner, F. (ed.): Mineral Metabolism. Vol. I. Part B. New York, Academic Press, Inc., 1961, pp. 677-741.

Söremark, R., and Samsahl, K.: Analysis of inorganic constituents in dental calculus by means of neutron activation and gamma ray spectrometry. J. Dent. Res., 41:596, 1962.

Stack, M. V.: The chemical nature of the organic matrix of bone, dentin and enamel. In Recent Advances in the Study of the Structure, Composition and Growth of Mineralized Tissues. Ann. N. Y. Acad. Sci., 60:585, 1955.

Thunberg, T.: Some information on the citric acid content of bone substance. Acta Physiol. Scand., 15:38, 1948.

Trautz, O. R.: X-ray diffraction of biological and synthetic apatites. In Recent Advances in the Study of the Structure, Composition and Growth of Mineralized Tissues. Ann. N. Y. Acad. Sci., 60:696, 1955.

Williams, E. S.: Salivary electrolyte composition at high altitude. Clin. Sci., 21:37, 1961.

Yardeni, J., Gedalia, I., and Kohn, M.: Fluoride concentration of dental calculus, surface enamel and cementum. Arch. Oral Biol., 8:697, 1963.

Yoon, S. H., Brudevold, F., Smith, F. A., and Gardner, D. E.: Fluoride, calcium, phosphate, ash and water content of human dental pulp. J. Dent. Res., 44:696, 1965.

Zipkin, I., Bernick, S., and Menczel, J.: A morphological study of the effect of fluoride on the periodontium of the hydrocortisone-treated rat. Periodontics, 3:111, 1965.

Zipkin, I., Bullock, F. A., and Mantel, N.: The relation of salivary sodium, potassium, solids and ash concentration to dental caries experience in children 5 to 6 and 12 to 14 years of age. J. Dent. Res., 36:525, 1957.

Zipkin, I., and Hawkins, G. R.: Formal discussion: Species specificity in saliva biochemistry. J. Dent. Res., 43:1040, 1964.

Zipkin, I., Hawkins, G. R., and Mazzarella, M. A.: The tyrosine, tryptophan, and protein content of human saliva in oral and systemic disease. Use of ultraviolet absorption techniques. In Sreebny, L. M., and Meyer, J. (ed.): Salivary Glands and Their Secretions. New York, The Pergamon Press, 1964, pp. 331-350.

Zipkin, I., and Larson, R. A.: Effect of level of protein on caries and periodontal disease in the rat. Proc. 6th International Congress Nutr. 1963, p. 184. (Abtract.)

Zipkin, I., Lee, W. A., and Leone, N. C.: Fluoride content of urinary and biliary tract calculi. Proc. Soc. Exp. Biol. Med., 97:650, 1958.

Zipkin, I., McClure, F. J., and Lee, W. A.: Relation of the fluoride content of human bone to its chemical composition. Arch. Oral Biol., 2:190, 1960.

Zipkin, I., Mergenhagen, S. E., and Kass, R.: The siliac acid content of human saliva. Biochem. Biophys. Res. Commun., 4:76, 1961.

Chapter Twenty-four

Oral Physiological Factors Concerned with Ingestion of Food

KRISHAN K. KAPUR, D.M.D., HOWARD H. CHAUNCEY, PH.D., D.M.D.,
AND IRVING M. SHARON, D.D.S.

Man has an individualized physiological and anatomical constitution that affects his selection, perception and enjoyment of food. There are numerous human factors that produce different eating patterns, e.g., body type (endomorph, mesomorph or ectomorph), the age of the individual, as well as cultural and environmental conditions. These and other psychological, social and general physiological factors that influence food selection and food habits have been discussed in Chapters Fifteen, Seventeen and Eighteen. This chapter focuses on the influence of oral mechanisms concerned with one of the most fundamental functions of the body, eating.

Eating is the ingestion of food by chewing and swallowing. It serves to meet the physiological demands associated with body metabolism and also elicits certain oral sensations. Thus, mastication and salivation function synchronously with the special chemical sensory processes of gustation and olfaction to influence our food selection. It is incumbent upon the dentist to have a thorough knowledge of these functions because it is his responsibility to diagnose and even, in some cases, to manage impairments that may be associated with these oral sensory and motor mechanisms.

PALATABILITY OF FOOD

A palatable food is one that has properties of taste, smell, appearance, texture, tempera-ture and spiciness that are pleasing to a given individual at a given time. Food which is palatable to one person may be repulsive to another. Palatability may be judged with total disregard for nutritional values. In common usage, the word "taste" is often used interchangeably with "palatability." Taste, in its precise scientific denotation, refers to a specific chemical quality, whereas palatability connotes an agreeable or pleasing sensation. For example, when a person says a pizza "has a good taste," what he actually means is that it is palatable because it meets his sensory demands. It has an appealing aroma (smell), has a crust that is crisp and not doughy (texture), is warm but not so hot that it burns his palate (temperature), is spicy enough to produce the desired mucosal irritation (pain) and, finally, the pizza stimulates the proper combination of gustatory submodalities (taste).

Man thus receives his final gratification when the particular qualities of a palatable food are perceived through a complex set of interrelated sensory and motor activities in the oral cavity. The primary activities involved are mastication, salivation, gustation and olfaction.

MASTICATION

Mastication is the biomechanical act of comminuting food so as to prepare it for swal-

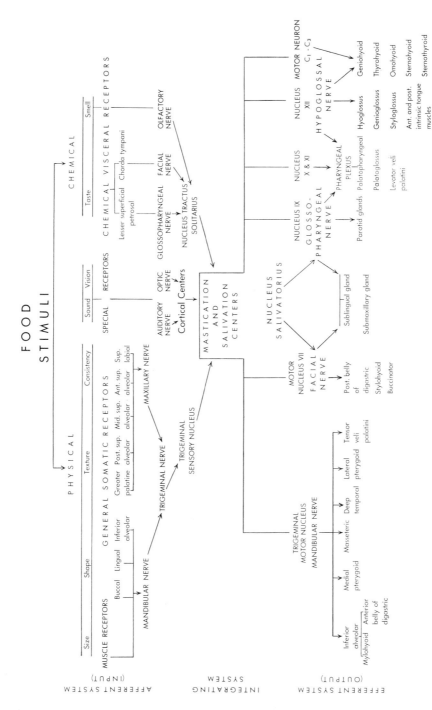

FIGURE 24-1. *The complex neuromuscular mechanism involved in the act of chewing food. The special senses of sound, vision, taste and smell participate indirectly in chewing through their influence on the salivary secretion.*

lowing. It also serves to induce salivation and excites taste and olfactory receptors through the liberation of gustatory and olfactory stimuli from the food. Mastication encompasses many intricate sensory and motor activities of the mouth. Various oral structures—the lips, cheeks, tongue, teeth and temporomandibular joints—are coordinated through a complex neuromuscular mechanism to accomplish the act of chewing (Fig. 24-1).

In man, chewing is both a voluntary and involuntary act. The incisive action, biting, is primarily voluntary. The mandible is normally maintained in rest position by the tonic contraction of the closing muscles and the teeth are held slightly apart. When the oral mucosa and teeth receive tactile stimulation, an involuntary process is set in motion which causes the mouth to open by evoking a reflex inhibition of the closing muscles. The amount of opening and lateral gliding movement are determined by the size, shape and consistency of the food morsel. Once the jaw is open, a rebound reflex causes the mandible to return to the centric occlusal position. If food is still present, the opening is again initiated. Chewing then proceeds in a rhythmic and involuntary manner.

Although some people chew on both sides of the mouth simultaneously, the majority chew on only one side. The preference for one side generally stems from the increased efficiency and/or comfort accompanying the use of that particular side. Masticatory habits are commonly developed between the ages of 7 and 12 years (Dahlberg, 1946).

For the purpose of simplification, the act of mastication can be considered to involve two unique, well synchronized activities. First, the actual grinding or subdivision of food and, second, the manipulation of food for grinding and swallowing.

GRINDING OF FOOD

The grinding of food occurs primarily between the occlusal surfaces of teeth, expressed as *masticatory units* or chewing elements. These chewing elements are brought into action by the movements of the mandible and are controlled by a complex of neuromuscu-

lar activity. Various sensory cues from the food help proprioceptors in the muscles, temporomandibular joints and the periodontal membrane of teeth to direct these movements of the mandible.

The grinding function is directly dependent upon the efficacy of the manipulating system, which constantly brings particles to the *food platform*. The efficient manipulation of food for mastication requires proper selection, transportation and confinement of coarser particles to the occlusal surfaces of teeth. If the discriminatory ability is altered and if coarser particles are not selectively carried in sufficient quantities to the food platform, even the best possible natural chewing mechanism would fail to be effective in pulverizing food.

The ability of the mouth to perceive size, form, texture and consistency of three-dimensional objects permits it to determine which particles will be chewed and which swallowed. This complex talent is called *stereognosis*. The proprioceptors in the masticatory muscles, the temporomandibular joints and the periodontal membranes of teeth all play an important role in this activity.

MANIPULATION OF FOOD

The tongue and the cheeks are the two structures chiefly engaged in food manipulation during chewing, with the tongue being the more dominant and active. The tongue is well suited for this function because of its highly developed tactile localization and *two-point discrimination* sensitivity. The latter determines the proximity of two points for which the tongue can still distinguish two separate stimuli. The tongue demonstrates a more acute tactile sensibility than any other part of the body. Greisheimer (1963) reported two-point tactile discrimination measurements of 1.4 mm. for the tongue, 2 mm. for the fingertip, and 36.2 mm. for the nape of the neck.

MASTICATORY PERFORMANCE

At present, *masticatory performance* can be quantitated by the degree of pulverization that food particles undergo during mastication. The following method of measuring masticatory performance has been described (Yurkstas and Manly, 1950): A subject chews a 3-gm. portion of peanuts for 20 strokes on one side of the mouth. The chewed food is recovered and passed through a ten-mesh screen. The ratio of the volume of the test food particles which passes through the ten-mesh screen divided by the total volume of the test food recovered (volume of particles passing through and remaining on the screen) multiplied by 100 represents the per cent masticatory performance. A masticatory performance of 80 per cent on the ten-mesh screen would be derived as follows:

$$\frac{12 \text{ ml. test food passed through 10-mesh screen}}{15 \text{ ml. total test food volume recovered}}$$

$$\times\ 100 = 80\%$$

MASTICATORY EFFICIENCY

The masticatory performance test described above cannot, however, accurately express any degree of masticatory impairment that may have occurred. The average performance of 100 test subjects with full complements of natural dentition has been found to be 78 per cent (Manly and Braley, 1950), while the average performance of 140 denture wearers was 24.5 per cent (Kapur and Soman, 1964). Although the average performance of denture wearers is about one-third that of people with natural dentition, it does not follow that the denture wearer is one-third as efficient. Efficiency must be related to some norm or standard. It should be based on the amount of effort expended by an individual to achieve a certain degree of food pulverization. Manly and Braley (1950) established a norm or basis of comparison from a group of subjects with complete natural dentition. From this norm a subject's masticatory performance may be converted into *masticatory efficiency*. The latter is defined in terms of the amount of effort expended by a subject to achieve the same degree of pulverization as that of the norm or standard person. It has been observed that the average chewing *performance* of denture wearers is one-third that of normal subjects, while the chewing *efficiency* is about one-sixth of the norm.

Integrity of both the maxillary and mandibular dental arches is necessary for effective masticatory function. Missing teeth result in loss of food platform area and reduced masticatory efficiency. Reductions of 33, 44, and 66 per cent in masticatory efficiency have been reported with the loss of first permanent molars, second and third molars, and first and third molars, respectively (Yurkstas, 1954). Persons with deficient dentition tend to swallow large particles (Manly and Vinton, 1951). The replacement of missing teeth with fixed and removable prostheses has been shown to improve chewing efficiency (Yurkstas et al., 1951).

Efficiency of Denture Wearers. Studies dealing with the effect of various denture factors on masticatory performance have revealed that the chewing ability of denture wearers is altered because the dentures make the food manipulating process of the oral cavity less effective (Kapur and Soman, 1964; Kapur et al., 1965). Furthermore, the masticatory process with artificial dentures is different from that with natural dentition. Manly and Braley (1950) found that mastication with natural dentition is selective, because as chewing progresses coarse particles are comminuted further, in preference to finer particles. On the other hand, mastication in subjects with complete dentures is a nonpreferential process, wherein particles of all sizes are ground at random (Kapur and Soman, 1964).

The presence of dentures thus alters both the form and environment of the oral cavity. In the presence of dentures and, thus, altered muscular activity, the action of the tongue and cheeks fails to carry coarser food particles to the food platform. These changes, along with blunted oral sensations resulting from the loss of teeth, are responsible for this nonselective grinding of food by denture wearers.

The implication of the foregoing is that dietary likes and dislikes are often directly dependent upon one's ability to chew and taste. This is of special significance for the older person since other sources of pleasure are frequently reduced. In dealing with the elderly person, a diet should be selected with attention to the limits set by chewing ability as well as appetite and nutritional needs.

SALIVATION

Salivation is the discharge of fluid (saliva) by various glands into the oral cavity. It helps to accomplish the following functions: (1) moisten and lubricate the oral mucosa and teeth, (2) flush away food particles and other debris, (3) dilute unpleasant and noxious substances, (4) neutralize acids, (5) facilitate chewing and swallowing by lubricating the food, (6) dissolve and transport gustatory stimuli released by chewing, and (7) initiate the first phase of digestion through the action of its enzyme (amylase) content.

SOURCE OF SALIVA

Saliva is secreted by three pairs of major glands and several minor glands. The minor glands are situated in the submucosal layer of the lips, cheeks, palate and tongue. Salivary glands are also classified as serous, mucous or mixed types, depending upon their secretion. Of the three major glands, the parotid is mainly serous; the submaxillary, mixed; and the sublingual, predominantly mucous.

All three pairs of major glands receive both sympathetic and parasympathetic innervation. The sympathetic fibers synapse in the superior cervical ganglion and reach the glands by way of the cervical sympathetic trunk. The parasympathetic innervation of the parotid glands is received by way of the glossopharyngeal nerve, while the submaxillary and sublingual glands are innervated by the chorda tympani, a branch of the facial nerve (Fig. 24-1).

SECRETION OF SALIVA

Secretion is usually considered to be a reflex response which can be either conditioned or unconditioned. The conditioned salivary response may be elicited by the stimulation of special senses other than those of mastication or taste. The conditioning of a dog to salivate with the ringing of a bell is a prime example

of this type of response (Pavlov, 1927). In establishing this conditioned response, the hungry dog is served food immediately after the ringing of a bell. The dog becomes so conditioned that he salivates at the sound of the bell. The conditioned response, in this instance, is established by the eating of food which, by itself, is considered as an unconditioned stimulus. To date, exact evaluation of the conditioned salivary secretion response in humans has been extremely difficult to determine (Winer et al., 1965).

The unconditioned secretion of saliva is evoked by the stimulation of receptors in the oral cavity and pharynx. Salivary secretion in response to chewing is an example of an unconditioned reflex (Chauncey and Shannon, 1959). Both the masticatory-salivary and gustatory-salivary reflexes are involved in the secretion of saliva during the chewing of food. The masticatory-salivary reflex excites the salivary centers through the stimulation of both the general receptors of the oral cavity and proprioceptors in the periodontal membrane of teeth, masticatory muscles and temporomandibular joints. The stimulation of the special sensory receptors of taste is responsible for the gustatory-salivary reflex.

In addition, the unconditioned reflex secretion of salivary fluid by the parotid salivary gland, following application of gustatory stimuli, can be used for a quantitative assay of human response to taste stimuli. It has been shown that a correlation exists between secretory response, stimulation frequency and stimulus concentration (Chauncey and Shannon, 1960; Chauncey et al., 1963). The effect on secretory response appears to be dependent on the molecular configuration and physical properties of the test substance (Feller et al., 1965).

Rate of Secretion. The secretion rates and composition of saliva produced by the separate glands show considerable variation and are dependent upon the type of stimulus used. The total electrolyte concentration and pH of parotid, submaxillary and sublingual saliva are the result of a complex series of physiological mechanisms (Chauncey and Shannon, 1965; Chauncey et al., 1966).

Salivary secretion rates are determined by measuring the volume of saliva that is secreted in a given time (Shannon et al., 1962).

The *stimulated secretion rate* is obtained by introducing a gustatory and/or masticatory stimulus and measuring the response. In the absence of a stimulus, the rate of output of the salivary gland is called the *unstimulated* or *resting flow rate*. Since a collecting device is present, the unstimulated flow rate actually represents the amount derived with minimal stimulatory conditions. Because salivation is influenced by a myriad of factors, secretion rates become meaningful only if they are obtained using standard techniques or if all existent conditions are recorded. Much conflict has occurred when data have been compared from studies in which the investigators used different procedures and secretion rates were not presented.

Whole saliva, the mixture of the fluid from all the salivary glands, is commonly collected by instructing a subject to accumulate the saliva in his mouth and to expectorate every minute for a given period into a graduated tube. While this fluid can be of value in determining the metabolism of the oral microbiota and leukocytes, it is of little value in studying salivary gland metabolism. Although wide variation exists in salivary secretion rates among various persons, adults secrete an average of one to two liters of whole saliva in a 24-hour period, the greater part of which is stimulated and comes from the major salivary glands.

Collection of Secretion. Secretions from the individual major salivary glands are collected by special devices. The collection of secretions from the sublingual and submaxillary glands frequently requires a specially constructed appliance for each individual subject (Henriques and Chauncey, 1961). Parotid saliva is easily collected by placing a modified Carlson and Crittendon's device over the opening of the Stenson duct (Shannon et al., 1962). Because of its relative accessibility, the parotid gland is more frequently used in clinical studies.

Factors Affecting Secretion. A variety of conditions can affect salivary secretion. Salivary flow rates are influenced by certain mental states such as fear or anxiety, by sleep and by certain physical states resulting in the loss of body fluid (diarrhea, vomiting, excessive sweating, hemorrhage, or polyuria accompanying diabetes). Endocrine disturbances

during the menopause in women, inflammation of salivary glands or irradiation in the area of the salivary glands may also result in decreased flow rates and dryness of the mouth.

A marked reduction in salivation, resulting in dry mouth, is termed *xerostomia*. Prolonged or chronic xerostomia can produce severe changes in the oral mucosa such as atrophy, inflammation, fissuring and cracking. A burning sensation, pain and dryness of the mouth are frequently the patient's chief complaints. Chewing and swallowing become difficult and taste is sometimes altered. Denture wearers are uncomfortable while using their dentures and are often bothered by denture sore spots. Food selection becomes limited to foods which are easily ingested and, thus, the enjoyment of eating is greatly lessened.

GUSTATION

Gustation is the act of perceiving taste. Four submodalities—acid, salt, sweet, and bitter—are generally acknowledged as comprising gustatory sensation. These primary sensations, in addition to the somatosensory components of pain, temperature and touch, contribute to the response elicited when receptors are activated by gustatory stimuli. Measurement of human response to gustatory stimuli is usually based on subjective evaluation. Recent studies have employed the parotid gland secretory rate for objectively measuring response (Chauncey et al., 1963; Feller et al., 1965). This technique has yielded an accurate assay of a physiological response to innumerable test conditions.

The tongue, the primary structure involved in taste perception, shows regional variations in sensitivity to the four primary tastes. The tip is most sensitive to sweet and sour, the lateral borders to sour and salt, and the base of the tongue to bitter (Moncrief, 1944). The receptors of taste stimuli lie in small barrel-shaped or ovoid bodies called taste buds. Each taste bud contains from 10 to 12 neuroepithelial cells, centrally situated and surrounded by long spindle-shaped sustentacular cells. Taste buds are present in large numbers in the epithelium of the mucous membrane of the tongue, anterior faucial pillars, palate, pharynx and larynx. They are most numerous on the outer surface of the tongue, particularly along the sides of grooves surrounding the circumvallate papillae, on foliate papillae, and on some of the fungiform papillae. Clusters of buds are also found on the laryngeal surface of the epiglottis and the lateral surfaces of the arytenoids.

Taste buds are more widely distributed in the mouths and pharynxes of fetuses, infants and children than in adults (Arey et al., 1935). The number of buds, especially in the grooves of the circumvallate papillae, decreases in older persons.

Although it was formerly held that taste receptors were specific in their sensitivity to the four primary tastes, recent electrophysiological studies have revealed that most receptors respond to more than one submodality. While some receptors respond only to acid, others respond to both acid and salt, and still others, to acid and quinine. This mixed and multiple sensitivity of receptors has been observed both in studies of the action potential of single gustatory afferent fibers (Pfaffmann, 1941) and depolarization receptor potentials of individual taste cells (Kimura and Beidler, 1961). From these findings it is now considered that portions of the receptor membrane of individual cells rather than the whole receptor show specific taste sensitivity.

Taste buds on the anterior two-thirds of the tongue are innervated by the chorda tympani, while the glossopharyngeal nerve supplies the posterior one-third of the tongue; in the pharynx and larynx, innervation is via the vagus nerve. This multiple sensory innervation of taste differs from the other special senses of vision, smell and hearing, whose receptors are supplied by a single nerve.

Nerve fibers enter the taste bud from its base and arborize around the neuroepithelial cells. The contact between the terminal fiber and neuroepithelial cell is a surface-to-surface synaptic junction (Zotterman, 1963).

TASTE THRESHOLDS

There are several methods by which taste thresholds are measured. The most common is a choice method, requiring a subjective judgment as to whether any difference in taste is perceived. This involves presenting a blindfolded subject with two beakers, one containing distilled water and the other a subliminal concentration of taste solution. The subject is instructed to taste the two liquids as often as he wishes. The concentration of the taste solution is gradually increased until the subject reports a difference. This value or concentration is recorded as the *taste discrimination threshold* for the given (primary) taste. At this point the subject usually cannot identify the taste. The concentration of the taste solution is then increased further until the subject can identify the definite taste. This concentration value is considered the *taste perception threshold*. The average taste perception thresholds for humans for the four basic submodalities are given in Table 24-1.

LOSS OF TASTE

The loss of taste, or ageusia, can result from pathological conditions such as inflammation of the geniculate ganglion or herpes orticus. *Hypoageusia,* or diminished taste sensitivity, can result from damage to the chorda tympani nerve and has often been seen in patients following total mastoidectomy. *Parageusia* refers to an altered taste condition, which is usually transient; it may sometimes be associated with influenza, hypertension, otitis media, stone formation in the submandibular gland or some head injuries. Some people are totally unable to taste specific substances such as phenylthiocarbamide and are said to be *taste blind.* Their taste blindness is considered to be an inherited characteristic that is transmitted by a simple recessive gene, as is color blindness.

Taste acuity seems to diminish with age. It is interesting to note that the presence or absence of dentures does not affect gustatory

TABLE 24-1. TASTE THRESHOLDS

Submodality	Taste Discrimination	Taste Perception
Sweet*	0.17% Sucrose	0.41% Sucrose
Salt*	0.010% Sodium chloride	0.065% Sodium chloride
Bitter*	0.0003% Phenyl-thiocarbamide	0.0003% Phenyl-thiocarbamide
Sour†	0.0012% Hydro-chloric acid	0.0022% Hydro-chloric acid

* Richter, 1942.
† Cooper et al., 1959.

sensitivity or taste thresholds (Kapur et al., 1966). Even the use of certain denture adhesives fails to affect taste discrimination and taste perception thresholds. Giddon et al. (1954) noted that denture wearers failed to perceive subtle differences in the sweetness of solids that can be appreciated with natural dentition. However, this finding does not necessarily contradict the above statement. The lack of perception of subtle differences in the sweetness of solids may well be a function of the reduced tactile threshold observed in denture wearers (Manly et al., 1952).

OLFACTION

Olfaction is the act of perceiving smell. It is a special chemical sense that is activated by stimulation of olfactory receptors situated in the upper part of the nasal cavity. The olfactory surface receptors are hair-like projections arising from the distal end of rod-shaped cells located in a yellow pigmented epithelium. In contrast to taste receptors, the nerve fibers on the proximal ends of the olfactory cells continue as unmyelinated nerve fibers which terminate in the olfactory bulb via the cribriform plate (Mullins, 1955).

Since the olfactory receptors are relatively secluded in the olfactory chamber, comparatively few stimuli reach the receptors in normal breathing. However, greater influx of stimuli to the receptors can be achieved by a quick, short inspiration, such as sniffing.

Various theories have been propounded regarding the mechanisms responsible for olfaction, but to date no satisfactory explanation is available for many of the factors involved (Patton, 1950; Pfaffmann, 1956; Beidler, 1961).

In contrast to gustation, olfaction is more acute and can be stimulated by extremely low chemical concentrations. While the taste threshold for ethyl alcohol is 3 M, the olfactory threshold is 0.000125 M (Olmstead, 1961).

The olfactory threshold, representing the minimum identifiable odor, is measured by the "blast injection test" (Elsberg, 1960). This involves pumping into both nostrils while the subject is holding his breath a standard amount of air containing a subliminal amount of the volatile substance. The concentration of the volatile substance is gradually increased

until the subject can identify the odor. This concentration value represents the *olfactory threshold*.

CONCLUSIONS

Exactly how much each of the factors mentioned contributes to the over-all development of appetite is impossible to measure; however, the role of each is important.

A person with *anodontia vera* (complete absence of teeth) and without dentures may never have had the pleasurable experience of eating a steak. Any desire to eat a steak would probably be to satisfy curiosity rather than appetite. In this way, chewing limitations may dictate the range of a patient's eating experiences.

Swallowing thresholds are influenced by salivary flow rates. Swallowing occurs when there is a certain accumulation of saliva and the food has been reduced to a suitable size. Both factors can vary tremendously from one individual to the next. The rate of salivation depends not only on the presence of food and on chewing but also upon the concentration of gustatory and olfactory stimuli released from the food by the chewing. Therefore, it is possible that persons who have higher secretion rates may tend to eat faster and/or chew for shorter periods of time prior to swallowing. Saliva can accumulate only up to a certain volume without producing discomfort. Thus, these persons may choose foods which provide less physical and chemical salivary stimulation. On the other hand, persons who have reduced flow rates will be able to chew for longer periods without being forced to swallow.

People with diminished masticatory efficiency, theoretically, should chew longer to compensate for their deficiency, although this has not been borne out by studies (Dahlberg, 1946). Unfortunately, no attention was paid to salivary secretion rates in swallowing threshold studies, i.e., these rates are more likely to determine the swallowing threshold. Although a person can regulate his chewing efficiency, its duration is still greatly dependent upon his salivary secretion rate.

It would, therefore, seem likely that people select foods whose chemical and physical properties are such that chewing efficiency and salivary secretion rates can be harmonized with the special senses to result in optimum oral gratification.

REFERENCES

Arey, L. B., Tremaine, M. J., and Monzingo, F. L.: The numerical and topographical relations of taste buds to human circumvallate papillae throughout the life span. Anat. Record, 64:9, 1935.

Beidler, L. M.: Taste receptor stimulation. In Progress in Biophysics and Biophysical Chemistry. New York, Pergamon Press, 1961.

Beidler, L. M.: The chemical senses. Ann. Rev. Psychol., 12:363, 1961.

Chauncey, H. H., Feller, R. P., and Henriques, B. L.: Comparative electrolyte composition of parotid, submaxillary, and sublingual secretions. J. Dent. Res., 1966, in press.

Chauncey, H. H., Feller, R. P., and Shannon, I. L.: Effect of acid solutions on human gustatory chemoreceptors as determined by parotid gland secretion rate. Proc. Soc. Exp. Biol. Med., 112:917, 1963.

Chauncey, H. H., and Shannon, I. L.: Parotid gland secretion rate as a method for measuring response to gustatory and masticatory stimuli in humans. School of Aviation Medicine Report 59-66, May, 1959.

Chauncey, H. H., and Shannon, I. L.: Parotid gland secretion rate as a method for measuring response to gustatory stimuli in humans. Proc. Soc. Exp. Biol. Med., 103:459, 1960.

Chauncey, H. H., and Shannon, I. L.: Glandular mechanisms regulating the electrolyte composition of human parotid saliva. Ann. N. Y. Acad. Sci., 131:830, 1965.

Cooper, R. M., Bilash, I., and Zubek, J. P.: The effect of age on taste sensitivity. J. Geront., 14:56, 1959.

Dahlberg, B.: Masticatory habits: Analysis of the number of chews when consuming food. J. Dent. Res., 25:67, 1946.

Elsberg, C. A.: In Ruch, T., and Fulton, J. (ed.): Medical Physiology and Biophysics. 18th ed., Philadelphia, W. B. Saunders Company, 1960.

Feller, R. P., Sharon, I. M., Chauncey, H. H., and Shannon, I. L.: Gustatory perception of sour, sweet, and salt mixtures using parotid gland flow rate. J. Appl. Physiol., 20:1341, 1965.

Giddon, D. B., Dreisback, M. E., Pfaffmann, C., and Manly, R. S.: Relative abilities of natural and artificial dentition patients for judging the sweetness of solid foods. J. Prosth. Dent., 4:263, 1954.

Greisheimer, E. M.: Physiology and Anatomy. 8th ed. Philadelphia, J. B. Lippincott Co., 1963.

Henriques, B. L., and Chauncey, H. H.: A modified method for the collection of human submaxillary and sublingual saliva. Oral Surg., 14:1124, 1961.

Kapur, K. K., Colistee, T., and Fischer, E. E.: The Effect of Denture Factors on the Masticatory and Gustatory Salivary Reflex Secretions of Denture Wearers. I A D R Abstracts, March, 1966.

Kapur, K. K., and Soman, S.: Masticatory performance and efficiency in denture wearers. J. Prosth. Dent., 14:687, 1964.

Kapur, K. K., and Soman, S.: The effect of denture factors on masticatory performance. Part II. Influence of the polished surface contour of denture base. J. Prosth. Dent., 15:231, 1965a.

Kapur, K. K., and Soman, S.: The effect of denture factors on masticatory performance. Part III. The location of the food platforms. J. Prosth. Dent., 15:451, 1965b.

Kapur, K. K., and Soman, S.: The effect of denture factors on masticatory performance. Part IV. Influence of occlusal patterns. J. Prosth. Dent., 15:662, 1965c.

Kapur, K. K., Soman, S., and Shapiro, S.: The effect of denture factors on masticatory performance. Part V. Food platform area and metal inserts. J. Prosth. Dent., 15:857, 1965.

Kapur, K. K., Soman, S., and Stone, K.: The effect of denture factors on masticatory performance. Part I. Influence of denture base extension. J. Prosth. Dent., 15:54, 1965.

Kimura, K., and Beidler, L. M.: Microelectrode study of taste receptors of rat and hamster. J. Cell. Comp. Physiol., 58:131, 1961.

Manly, R. S., and Braley, L. C.: Masticatory performance and efficiency. J. Dent. Res., 29:448, 1950.

Manly, R. S., Pfaffmann, C., Lathrop, D., and Keyser, J.: Oral sensory thresholds of persons with natural and artificial dentitions. J. Dent. Res., 31:305, 1952.

Manly, R. S., and Vinton, P.: Factors influencing denture function. J. Prosth. Dent., 1:578, 1951.

Moncrief, R. W.: The Chemical Senses. New York, John Wiley & Sons, Inc., 1944.

Mullins, L. J.: Olfaction. Ann. N. Y. Acad. Sci., 62:247, 1955.

Olmstead, J. M. D.: In Bard, P (ed.): Medical Physiology. 11th ed. St. Louis, The C. V. Mosby Co., 1961.

Patton, H. D.: Physiology of smell and taste. Ann. Rev. Physiol., 12:469-484, 1950.

Pavlov, I. P.: Conditioned Reflexes. New York, Oxford University Press, 1927.

Pfaffmann, C.: Gustatory afferent impulses. J. Cell Comp. Physiol., 17:243, 1941.

Pfaffmann, C.: Taste and smell. Ann. Rev. Psychol., 7:391, 1956.

Pfaffmann, C.: Taste. McGraw-Hill Encyclopedia of Science and Technology, 1960, pp. 399-401.

Richter, C. P.: Total self regulatory functions in animals and human beings. Harvey Lect., 38:63, 1942.

Shannon, I. L., Prigmore, J. R., and Chauncey, H. H.: Modified Carlson-Crittendon device for the collection of parotid fluid. J. Dent. Res., 41:778, 1962.

Winer, R. A., Chauncey, H. H., and Barbar, T. X.: The influence of verbal or symbolic stimuli on salivary gland secretion. Ann. N. Y. Acad. Sci., 131:874, 1965.

Yurkstas, A. A.: The effect of missing teeth on masticatory performance and efficiency. J. Prosth. Dent., 4:120, 1954.

Yurkstas, A., Fridley, H. H., and Manly, R. S.: A functional evaluation of fixed and removable bridgework. J. Prosth. Dent., 1:570, 1951.

Yurkstas, A., and Manly, R. S.: Value of different test foods in estimating masticatory ability. J. Appl. Physiol., 3:45, 1950.

Zotterman, Y.: Olfaction and Taste. New York, The Macmillan Company, 1963.

Chapter Twenty-five

Influence of Nutrition on the Ecology and Cariogenicity of the Oral Microflora

W. J. Loesche, D.M.D., and R. J. Gibbons, Ph.D.

The nutrients available in the mouth, as well as the existing physical conditions, determine the kinds and quantity of bacteria which colonize the oral cavity, and also influence the organisms' metabolic and pathogenic potential. Despite its obvious importance, remarkably little detailed information is available concerning the in vivo nutritional requirements of the resident oral flora, or of the nutrients available in the mouth. Consequently, one must rely on the in vitro nutritional requirements of isolated bacteria as a guide and assume that similar requirements exist and are fulfilled in the oral cavity. This, of necessity, is artifactitious, for in the laboratory organisms are studied in pure culture using conditions of nutrient excess, though it is generally recognized that in vivo many nutrients are limiting. Further difficulties arise in attempts to determine available nutrients in vivo since substances present in the diet and oral fluids are, in some instances, modified by members of the oral flora. In addition, a variety of nutrients, particularly those in the vitamin or growth factor class, are synthesized de novo by one bacterium and may be available to fulfill the needs of neighboring organisms. Thus, even if the exact chemical composition of the oral secretions and the diet of an individual were known, it would still not be possible to state with exactness the nutrients available to any given microorganism residing in one of the microcosms of the mouth. Bearing these difficulties and deficiencies in mind, let us pursue what can be surmised concerning the nutrition of the oral flora.

GENERAL BACTERIAL NUTRITION

In order to discuss the nutrition of the indigenous oral flora, it is perhaps best to review some basic knowledge of the nutrition of microorganisms in general. In common with all other forms of life, microbes require an exogenous supply of chemical compounds for the maintenance of life. These compounds or nutrients provide the organisms with a source of energy, necessary for biosynthetic reactions, as well as with building blocks for the synthesis of cell proteins, carbohydrates, nucleic acids and lipids. Whereas only slight variation in nutrient requirements are observed in mammalian species, bacteria exhibit great nutrient diversity.

ENERGY SOURCE

All cells derive energy from the oxidation or transfer of electrons from a compound of high electrochemical potential to one of lower electrochemical potential. This energy is released in a stepwise fashion permitting its partial capture in the form of high energy chemical bonds present in compounds typified by adenosine triphosphate (ATP). These high energy compounds act as a currency of exchange, and when their hydrolysis is coupled to a biosynthetic reaction, their energy is transferred to enable the synthetic or energy-requiring reaction to occur. The chemical nature of the energy source used by various bacteria has been used by microbial nutritionists to subclassify the "chemotrophic" bacteria, i.e., those bacteria which gain energy from chemical oxidations. Organisms which obtain energy by oxidizing inorganic compounds (i.e., hydrogen sulfide, ammonia and hydrogen) are termed "lithotrophic," while "organotrophic" bacteria derive their energy exclusively from the oxidation of organic compounds. Most if not all human oral bacteria fit into this latter category.

Bacteria may also be grouped according to the nature of the final electron acceptor used in the process. Bacteria which can use only oxygen as the final electron acceptor are considered "strict aerobes." Organisms which can grow either in the presence or absence of oxygen are termed "facultative." They may be subdivided into "true" and "indifferent" facultative types. A "true" facultative organism, when grown aerobically, uses oxygen as the terminal electron acceptor, but when grown in the absence of oxygen (anaerobically) uses other compounds to accept the electrons. In the latter instance, an organic compound is frequently used for this purpose and the process is called "fermentation." An "indifferent" facultative bacterium carries out a fermentative metabolism at all times and is, as the name implies, indifferent to oxygen. Organisms which grow only in the absence of oxygen are "strict anaerobes" and exhibit various fermentative pathways. Another group of bacteria, though partially inhibited by atmospheric levels of oxygen, grow optimally at reduced oxygen tensions and are termed microaerophils." Facultative anaerobic and microaerophilic bacteria comprise the majority of the species normally indigenous to the mouth.

BIOSYNTHESIS

Besides displaying diversity in regard to their energy source and terminal electron acceptor, bacteria vary widely in the building blocks they require for biosynthetic purposes. Thus, an organism such as *Escherichia coli*, which commonly inhabits the intestinal canal of man and animals, possesses great biosynthetic capabilities and is able to synthesize all its cell constituents, using glucose as a carbon and energy source and an ammonium salt as a source of nitrogen. Other bacteria are more

fastidious and have reduced biosynthetic capabilities. For example, lactobacilli require several or, in some cases, all amino acids to be preformed and provided in a culture medium. In addition, several purines, pyrimidines, vitamins and growth factors are also required for their growth. Almost all oral bacteria are considered to be fastidious and, in fact, precise nutritional requirements are known for only a few species.

Chemically defined culture media, which contain ingredients whose chemical nature is known, have been developed for certain strains of lactobacilli, streptococci, peptostreptococci and fusobacteria. Such media tend to be costly and laborious to prepare and do not usually support optimal growth. Thus, though chemically defined media are the most informative from a nutritional point of view, they are not widely, or routinely, used. Rather, one uses complex, ill-defined natural media. These generally contain a peptone or a partially hydrolyzed protein which provides amino acids and peptides. Aqueous extracts or infusions of meat, liver, yeast, or soybeans are included to provide vitamins, growth factors, carbohydrates and minerals. Depending upon the species one wishes to cultivate, enrichments such as whole blood or ascitic fluid may also be added. Use of these ill-defined natural media permits cultivation of most of the bacterial species present in the mouth, but it is obvious that they shed little light on the exact nutrients required by a given bacterium.

COMPOSITION OF THE ORAL MICROBIOTA

Before we can apply these basic nutritional concepts to the oral flora we must have some knowledge of the component members. The mouth has access to microbial contamination from many sources. These contaminating organisms may persist for several days or weeks and then disappear. Consequently, suitable criteria must be established to eliminate these organisms from our consideration. Rumen microbiologists confronted with similar problems have defined parameters for inclusion of species in the indigenous rumen microbial flora. Simply restated, they are: (1) any species that is "always" isolated in "high" numbers from cattle in different parts of the globe, and (2) whose biochemical characteristics of indigenous species are compatible with the in vivo physiological environment (Bryant, 1959). Numerical dominance is assumed to reflect ecologic importance.

Application of the above principles to the human oral flora has led to a beginning understanding of the identity of the resident oral flora. Bacteria which are always isolated in high numbers, i.e., approximating 1 per cent or more of the cultivable flora, are considered as indigenous. These are the organisms which find the conditions of the mouth suitable for growth and can successfully compete for nutrients, thereby achieving numerical superiority. Many of the indigenous species are closely related to overt pathogens (i.e., oral streptococci, diphtheroids, staphylococci, *Neisseria* and spirochetes). Apparently, the evolutionary changes which permitted their successful colonization of the oral cavity included a loss of pathogenic potential. Within the framework of host parasite interactions, the indigenous species occupy neither the antibiotic nor the symbiotic extremes but rather an intermediate position aptly termed "amphibiosis" by Rosebury (1962). They are capable of initiating chronic infections, which in the mouth would include dental caries and periodontal disease.

The indigenous human oral flora is unique to the mouth of man. Organisms such as the coliforms which are characteristic of feces are not routinely isolated from the mouth. In fact, in vitro evidence suggests that *Escherichia coli* is inhibited by acid end products formed by certain oral bacteria. This is not true, however, of the oral flora of small laboratory rodents such as the hamster and rat which are commonly employed for dental research. These animals by virtue of their copraphagic habits regularly yield bacteria characteristic of feces from their mouths, and their fecal and oral flora have many species in common. For example, caries-potentiating bacteria inhabit the intestinal canal of caries-active hamsters, and feces from these animals has been used to transmit dental caries to other hamsters (Keyes, 1960).

Species frequently, though not always, present in moderate numbers (i.e., less than 1 per

cent of the total cultivable flora) are considered as members of the supplementary flora. The lactobacilli are prime examples of supplemental organisms, being regularly isolated from caries-active individuals, but being absent from individuals who are caries free. In general, organisms of the supplementary flora are thought to be dependent upon indigenous organisms either for nutrients or for creation of a suitable physical environment for their colonization. Bacteria sporadically isolated from the mouth are dismissed as transients and are assumed to be of no ecological importance. As might be suspected, a large array of organisms which would fall into this category has been isolated from the mouth.

Only chemo-organotrophic bacteria are indigenous to the oral cavity of man (Table 25-1). Obligate aerobes are rare and, with the exception of *Nocardia* species, they do not seem to be present in numbers which warrant their inclusion in the indigenous oral flora. Rather, the flora is facultative to anaerobic in character, and virtually all members will grow anaerobically. This point is important to grasp. Aerobic culture techniques permit the growth of aerobic and facultative organisms only. When simultaneous aerobic and anaerobic counts are performed, the anaerobic values are at least twofold higher. Any study of the indigenous flora of man which ignores the anaerobic species present is likely to grossly misinterpret the actual composition of the flora. A classic example of such a misinterpretation has been the significance of *E. coli* in the large intestine of man. Aerobic culture techniques revealed *E. coli* to be among the most numerous organisms in feces. However, when feces are cultured anaerobically, members of the genus *Bacteroides* were found to outnumber *E. coli* by 100 to 1 (Rosebury, 1962).

The anaerobic nature of the oral flora may appear paradoxical, as the mouth has much apparent access to the atmosphere. However, the penetration of oxygen into the gingival sulci, periodontal pockets, crypts of the tongue, saliva and interproximal areas is limited by diffusion, and that which does diffuse in is scavenged by the facultative organisms present, thereby creating a state of anaerobiosis. Conceivably, the first members colonizing the tooth surface and creating plaque could

TABLE 25-1.   GENERAL NUTRITIONAL REQUIREMENTS OF BACTERIAL SPECIES
INDIGENOUS TO THE MOUTH OF MAN

Group	Energy Source	Relation to Oxygen	Unusual Growth Requirements
1. GRAM POSITIVE COCCI			
Streptococcus species	Carbohydrates	Microaerophilic	
Staphylococcus species	{ Carbohydrates { Amino acids	Facultative	
Peptostreptococcus species	{ Carbohydrates { Nonvolatile acids	Anaerobic	
2. GRAM NEGATIVE COCCI			
Neisseria species	Carbohydrates	Facultative	
Veillonella species	{ Lactate { Nonvolatile acids	Anaerobic	
3. GRAM POSITIVE RODS			
Diphtheroids	Carbohydrates	{ Facultative { Microaerophilic { Anaerobic	
Lactobacilli	Carbohydrates	Microaerophilic	
Bacterionema matruchotii	Carbohydrates	Facultative	
Leptotrichia buccalis	Carbohydrates	Anaerobic	
Nocardia species	Carbohydrates	Aerobic	
4. GRAM NEGATIVE RODS			
Bacteroides melaninogenicus	{ Amino acids { Carbohydrates	Anaerobic	{ Hemin { Vitamin K
Bacteroides oralis	Carbohydrates	Anaerobic	Hemin
Vibrio sputorum	Unknown	Microaerophilic	Stimulated by NO_3^-
Spirillum sputigenum	Carbohydrates	Anaerobic	
Fusobacterium nucleatum	{ Amino acids { Carbohydrates	Anaerobic	
5. SPIROCHETES			
Treponema microdentium	Carbohydrates	Anaerobic	{ Isobutyric acid { Spermine { Eh of −185 to −220 mv
Treponema dentium	Amino acids	Anaerobic	Ascitic fluid
Borrelia vincentii	Unknown	Anaerobic	

be aerobic, but as the plaque thickens, the oxygen supply to the tooth surface diminishes and more anaerobic organisms become established.

CONDITIONS EXISTING IN THE MOUTH

The oral cavity presents some obvious advantages for microbial growth. The mouth is warm, moist and, as emphasized above, anaerobic. Nutrients would seem to be derived from several sources, as will be discussed subsequently. The pH range encountered in various sites is usually between 6.0 and 7.8, which is well suited for growth of most oral bacteria. In addition, the flow of saliva and the gingival crevice fluid help to maintain the pH within physiologic limits.

WARMTH

The oral bacteria are dependent upon their environment for the heat energy necessary for optimal enzymatic activity. Oral microorganisms are "mesophilic," growing best in the temperature range of 25 to 40° C. Mouth temperatures are optimal for growth of the oral flora and also discriminate against organisms which do not grow well at these temperatures, such as "psychophilic" organisms which grow at low temperatures (i.e., 25° C.

or lower), and "thermophilic" bacteria which flourish at elevated temperatures (i.e., 45° C. or higher). Both types of organisms are continuously introduced into the mouth since they are common contaminants of refrigerated and frozen products as well as plant produce. Their failure to become established is generally attributed to the temperature prevalent in the oral cavity.

MOISTURE

The importance of water as a nutrient for the oral flora should not be neglected. Bacteria have a water content of approximately 80 per cent and are primarily aquatic forms. The mouth is provident in this regard, for water is present in the diet and in the saliva and gingival crevice fluid which bathe the oral tissues. These bathing fluids are also involved in the transport of nutrients derived from food debris from one site to another and in the removal of bacteria and debris from dental stuctures, resulting in their subsequent elimination from the mouth via swallowing. For bacteria to survive in significant numbers they must find loci which have access to oral fluids, yet which are protected from washoff. The topography of the teeth and gingiva, especially the occlusal fissures and interproximal areas, provides such sites. The carious lesion and the periodontal pocket are microbial precipitated extensions of these niches.

ANAEROBIC CONDITIONS

The anaerobic nature of the indigenous flora has been emphasized previously. Whereas oxygen is the most essential nutrient for the mammalian host, it has less nutritional significance for the oral flora under normal conditions. In fact, oxygen is toxic for oral spirochetes, bacteroides and other oral anaerobes. The mechanism of this toxicity is not understood but it is believed to be related to the ease with which sulfhydryl groups of enzymes are oxidized to the inactive disulfide form. Also, certain enzymes important for anaerobic growth require a reducing environment which cannot be achieved in the presence of oxygen.

RATE OF GROWTH

Despite these seemingly ideal conditions and the astronomic number of 4×10^{10} cultivable organisms which has been obtained per gram of dental plaque and gingival debris, it seems clear that the bacteria inhabiting the oral cavity are not growing rapidly. It is possible to make approximate calculations to support this view, although it should be pointed out that direct experimental evidence is lacking. Saliva, which is the vehicle that removes bacteria from the oral cavity, contains approximately 10^8 cultivable organisms per ml. Thus, if an individual secretes and swallows a liter of saliva per day, approximately 10^{11} cultivable organisms will be removed from the oral cavity. This would represent approximately 2.5 gm. wet weight of oral bacterial debris. The data of Krasse (1954) and Gibbons et al. (1964a) suggest that approximately two-thirds of this material is derived from the tongue, with the remaining one-third, or about 830 mg., of debris lost per day from plaque, materia alba, the gingival sulcus and the cheek surfaces combined. If one estimates the total mass of bacterial debris present in these latter sites at any given time to be in the vicinity of 200 to 300 mg., then their indigenous microorganisms must average only three to four divisions per day in order to account for the 830 mg. of bacteria that are eliminated daily by swallowing. Evidently growth is limited in these sites by a lack of essential nutrients or, to some extent, by the presence of inhibitory substances. This over-all slow rate of growth has important implications in the ecology of oral bacteria and partially explains why organisms with markedly different potential growth rates are able to coexist in a balanced state. For example, gingival crevice debris contains both streptococci of the *Streptococcus mitis* type and spirochetes of the *Treponema microdentium* type. When these organisms are grown in pure culture in the laboratory under conditions thought to be ideal, the streptococci are observed to divide every 25 to 30 minutes, whereas the spirochetes require 6 to 8 hours. In the oral cavity these organisms exist in a relatively constant ratio to each other and consequently their rate of growth in vivo must be identical. Thus, it would seem that the oral spirochetes are growing near their potential growth rate in the mouth, whereas the oral streptococci are growing far below their potential rate.

SOURCES OF NUTRIENTS

DIET

Foodstuffs are the ultimate source of all bacterial nutrients. The oral bacteria may gain access to them during their ingestion and mastication or later after chemical modification of the foodstuffs by the body, as nutrients in saliva and tissue fluids. Foodstuffs which are not rapidly cleared from the oral cavity, i.e., adhesive or fibrous foods, are of considerable nutritive value for the oral flora. Sticky carbohydrate foods are implicated in the carious process, as their retention in the mouth allows maximum acid production by the plaque flora. Fibrous proteinaceous foods tend to be retained where interproximal contact points are poor or where periodontal pockets exist. The proteolytic and amino acid fermenting organisms of the gingival crevice area would seem to benefit in this case.

Dietary Carbohydrates. Carbohydrates comprise 50 to 60 per cent of the total calories in the average diet in the United States. Starch contributes the majority of these calories, but in recent years sucrose or sugar has increased in the diet so that it now comprises about 25 per cent of all the carbohydrates consumed. Glucose, lactose and glycogen are present in the diet but are not so plentiful. With dietary carbohydrates so abundant, it is not surprising that most oral bacteria, and in particular those that reside in dental plaque, ferment carbohydrates. Indeed, most of the oral streptococci actually require large quantities of carbohydrate for growth.

The freely soluble mono- and disaccharides (i.e., glucose , fructose, sucrose, lactose and maltose) are rapidly fermented by most oral species in vitro, acidifying carbohydrate broth to a pH between 4.5 and 6.0. One would suppose that if these carbohydrates were present in the diet they would be readily utilized by the oral flora. This appears to be the case, for in vivo studies have demonstrated that the

plaque pH can be lowered in a matter of minutes by applying a glucose or sucrose rinse, or by ingesting carbohydrate (Jenkins, 1960). Soluble carbohydrate-containing foods give lower pH values in toothbrush-inaccessible areas than do undissolved foods, indicating their more rapid utilization by the flora. As little as 1 to 2 gm. of dietary carbohydrate has been found to cause a significant drop in pH on the tooth surfaces. These and many other studies indicate that the soluble carbohydrates in vivo can be quickly metabolized by the plaque flora forming acid. The cariogenicity of dietary carbohydrate depends on its being made available to the plaque flora during ingestion, for when sucrose was fed to caries-susceptible animals by stomach tube to bypass the oral cavity, no caries resulted (Kite et al., 1950).

The less soluble carbohydrates present in the diet consist principally of starch and, to a lesser extent, glycogen. Before these can be metabolized by the oral flora, they must be degraded to their mono- or disaccharide constituents. The necessity for this additional step retards their immediate availability to the flora. However, as these polymers of limited solubility are more apt to become entrapped in crevices or between teeth, their comparatively slow breakdown can provide the flora with a source of carbohydrate over long periods of time. Indeed, Lanke (1957) found that the levels of carbohydrate in saliva remained elevated for 27 to 38 minutes following the ingestion of starchy cereal foods. The enzymes necessary for the initial hydrolysis of starch and glycogen are derived from two sources. Saliva is rich in alpha amylase, which is mixed with these polysaccharides during mastication. In addition, certain members of the oral flora, i.e., strains of diphtheroids, *B. melaninogenicus* and fusobacteria, excrete extracellular enzymes capable of hydrolyzing these polymers. As this initial phase of digestion occurs extracellularly, the soluble carbohydrate residues which are produced are made available to other members of the oral flora.

Dietary Protein. Dietary protein must be hydrolyzed to its component amino acids and small peptides before it becomes metabolically available to the oral flora. Certain members of the oral flora possess proteolytic enzymes which can attack gelatin, collagen and muco-proteins. In vivo, very little protein would be digested during transit through the mouth. However, in those situations in which fibrous foods are retained between the teeth, significant microbial digestion of proteins probably occurs. Several investigators have correlated an increased microbial proteolytic activity in saliva with periodontal disease. This reflects the sequence of food retention predisposing to protein hydrolysis which releases nutrients permitting microbial proliferation and resulting soft tissue pathology.

Other Dietary Nutrients. Approximately 30 per cent of the calories in an average United States diet are derived from lipids. However, little information exists relating these substances to the metabolism of the oral flora. Low levels of lipase and other esterases have been found in canulated saliva, leukocytes and a few oral bacteria (Chauncey, 1961). The possibility exists, therefore, that these nutrients may be made available to the oral flora. Individual examples exist in which certain members of the flora are stimulated or inhibited by a particular long chain fatty acid. Thus, oleic acid stimulates *Corynebacterium acnes* and possibly certain of the oral spirochetes, whereas it inhibits certain cariogenic hamster streptococci. The ecological significance of these interactions for the oral cavity are not understood, and clearly more information is required in this area.

Although the diet contains fat and water soluble vitamins as well as minerals, there are no data linking these nutrients to the microbial ecology of the mouth. Since most oral species require certain vitamins for growth, it is likely that their needs are at least partially satisfied by the dietary intake of these substances.

SALIVA

Whole saliva, which bathes the microorganisms living in the mouth, is composed of secretions from the parotid, submaxillary, sublingual and several lesser glands, and contains both cellular and soluble elements. Its cellular constituents would include leukocytes, sloughed epithelial cells, and dislodged microorganisms, while its soluble elements include proteins and glycoproteins, inorganic salts, low molecular weight compounds such as amino acids or urea, the products of host and bacterial cell metabolism, and soluble or partial breakdown products derived from the diet. Saliva is one of the major sources of nutrients for the oral microflora, as well as a vehicle for nutrient distribution.

Mucin. The major proteinaceous constituent of saliva is mucin. This glycoprotein is composed of one-third polysaccharide and two-thirds protein and is largely responsible for the viscous nature of saliva. Since oral bacteria are able to hydrolyze mucin to some extent, it is possible that it is a source of nutrients for them. The polysaccharide component would be of special interest, since saliva contains very low levels of free carbohydrate (less than 0.5 mg./100 ml.). With the recent demonstration that oral bacteria produce neurimidase, an enzyme which can attack the sialic acid moiety of mucin, it is possible that this glycoprotein could serve as a source of carbohydrate. Besides its possible nutritive value, mucin may function in a protective role for oral bacteria. This substance has long been known to enhance the pathogenicity of bacteria, presumably by coating the organisms, thereby rendering them less susceptible to phagocytosis. Mucin coating appears to be analogous to a bacterial capsule and can markedly increase the survival of bacteria under unfavorable conditions. Thus, in the mouth, bacteria coated with mucin could be expected to better survive the prolonged periods when growth is not possible, and also the wide extremes in pH such as one encounters in plaque.

Amino Acids and Vitamins. Saliva contains low levels of free amino acids and the vitamins that are required by oral bacteria. For example, the total quantities of free amino acids approximate 5 mg./100 ml., and the concentrations of riboflavin, thiamine and pyridoxine are 5, 0.7 and 60 μg./100 ml., respectively. These concentrations are well below those which pure cultures of bacteria require to attain maximal growth in vitro and, in fact, saliva itself is a poor culture medium for most microorganisms. However, since organisms in the mouth are growing slowly and are dividing only a few times each day, these quantities of salivary nutrients would seem to be of great significance in their nutrition.

Minerals. The minerals in saliva probably influence the oral flora in several ways. In terms of nutrition, saliva abounds in phosphate, bicarbonate, sodium, potassium, calcium and magnesium, which are required for growth and metabolism of all microorganisms. In addition, it probably provides them with cations required in trace quantities for their metabolism, such as manganese, cobalt, copper, iron and molybdenum.

Salivary minerals are also important contributors to the ionic strength of the oral environment. While the osmotic tolerance of most bacteria is great because of their rigid cell wall, it is nevertheless finite, and for growth to occur electrolytes are necessary. The ionic strength of saliva is only 25 per cent that of blood, which is suitable for most oral bacteria. However, it is too low for host phagocytes and epithelial cells and thus is responsible for their lysis and release of metabolites in the mouth.

Buffers. The relatively large concentrations of phosphate and bicarbonate in saliva are of special importance because of their buffering action. These anions, together with salivary proteins, result in a reasonably efficient buffering system. The continuous flow of saliva thus serves to neutralize acid end products and to maintain the microcosms in the mouth at an acidity suitable for growth. It is interesting to note that both the flow of saliva and its bicarbonate content are increased during periods of food ingestion. Local acid production in plaque is greatest during meals, and thus the increased buffering action of saliva at this time is probably of great significance in the neutralization and elimination of acids which could initiate dental caries.

MICROBIAL INTERACTIONS

The microbes themselves are an important source of nutrients. They may produce compounds which are required nutrients for other coinhabiting species. Also, the organic material of dead cells may be reutilized by the viable members of the flora. The in vivo microbial interactions are difficult to dissect out and study. More often than not several organisms are involved and, in addition, precise anaerobic conditions may be required. A simple but illustrative interaction involves lactate and members of the genus *Veillonella*. *Veillonella* are unable to ferment carbohydrates but actively ferment lactate and certain other organic acids (Rogosa, 1964). The lactic acid produced by members of the oral flora probably serves as the energy source for *Veillonella* in the mouth, and indeed the proportion of *Veillonella* increases slightly in carious plaque, presumably as a consequence of its greater lactic acid content. The dependence of *Veillonella* upon lactate producers is strikingly demonstrated by germ-free studies. *Veillonella* strains will not establish and colonize germ-free mice unless they are simultaneously introduced with lactate-producing organisms (Gibbons et al., 1964b). Thus, it would appear that the *Veillonella* species in the mouth derive their energetic needs at least from a by-product of microbial fermentation.

The situation differs somewhat with *Bacteroides melaninogenicus*. Many strains of this organism require as trace nutrients hemin and naphthoquinones of the vitamin K series. Sporadic bleeding occurring in the gingival crevice area probably provides the hemin, while neighboring bacteria produce the vitamin K which is required. Evidence for the microbial synthesis of vitamin K may be deduced from the fact that saliva as it leaves the salivary glands contains negligible quantities of this vitamin, whereas whole saliva, which has been exposed to oral bacteria, contains 15 μg. of vitamin K per liter. In addition, pure cultures of a variety of oral organisms synthesize naphthoquinones in vitro.

The microbial interactions which permit the in vitro cultivation of the small oral spirochete *Treponema microdentium* are illustrative of the importance of the anaerobic environment in vivo (Socransky et al., 1964). Conventional culture media failed to support growth of this organism in pure culture, but the organism grew readily in association with two other oral bacteria. However, no combination of lysates or culture filtrates of these supporting organisms would substitute for living cells until it was realized that these organisms were also lowering the oxidation-reduction potential (Eh) of the medium to a range suitable for spirochete growth. When the medium was poised at the critical Eh required for initiation of spirochete growth, filtrates could replace the living organisms. Subsequent investigation revealed that the essential nutrients produced by the supporting organisms could be replaced by isobutyric acid and the polyamines, spermine and spermidine. Thus, it seems clear from the examples cited that nutritional interactions among members of the oral flora are of great importance in the ecology of the oral cavity.

MISCELLANEOUS SOURCES OF NUTRIENTS

Small volumes of fluid enter the oral cavity through the gingival crevice. Brill and Bjorn (1959) have suggested that the physiologic function of this fluid is to remove bacteria and debris from the gingival crevice area. This fluid contains small amounts of proteins and its Na/K ratio suggests that it is an exudation from an underlying inflammation. As such it would provide small amounts of peptides, amino acids and other nutrients to the gingival flora. These could be important in the metabolism of the amino acid-fermenting species *Bacteroides melaninogenicus* and *Fusobacterium nucleatum*. The inflammatory nature of this fluid suggests that its volume may increase in periodontal disease and thus serves as a nutrient source for the microbial overgrowth which characterizes this syndrome.

Desquamated epithelial and leukocytic cells contribute to the nutrient pools of saliva and gingival crevice fluid. These host cells lack rigid cell walls and therefore are susceptible to alterations in osmotic conditions. They lyse in a hypotonic environment such as saliva, releasing metabolites and enzymes. Any glycogen released by the epithelial cells may be metabolically available to the oral flora. This is postulated to occur in the vagina where the glycogen released from desquamated vaginal epithelial cells is used as a source of carbohydrate for the vaginal lactobacilli. No evidence is available suggesting the magnitude or significance of this phenomenon in the mouth.

Bleeding of the gingival epithelium provides hemin, which is a required growth factor for *Bacteroides melaninogenicus* and some *B. oralis* strains. Conceivably, this may be a contributing factor in the localization of these organisms in the gingival crevice area.

A variety of experiments have demonstrated that low molecular weight compounds and ions can pass through the tooth. Thus, if a microbial nutrient is absent in dental plaque but present in the pulp fluids, it conceivably could migrate from the pulp to the tooth surface. The in vivo significance of this phenomenon for the nutrition of the plaque flora has not been determined.

NUTRITION OF MICROCOSMS IN DIFFERENT MICROENVIRONMENTS WITHIN THE ORAL CAVITY

The discussion thus far has related to the nutrient sources available to the oral flora as a whole. In the mouth all nutrients are not available equally to all resident bacteria. The nutrients present in gingival fluid would not seem to contribute significantly to the nutrition of bacteria present in the crypts of the tongue. Food retained in occlusal fissures does not influence the metabolism of bacteria deep in a periodontal pocket. Thus, it is more physiological to consider the nutrition of the oral flora in terms of the microenvironments present in the mouth. The flora of dental plaque, the gingival crevice area and the tongue differ somewhat from site to site (Table 25-2). The bacteria present in saliva represent the washoff from these sites and, as might be expected, most resemble the tongue flora in composition. The changes from plaque flora to gingival crevice flora are gradual, with many species common to both areas. The existence of these distinct microbial niches argues strongly that the nutrient supply and physical conditions differ from site to site within the mouth. We shall now consider each microcosm separately and attempt to understand how its microbial composition may reflect the nature of the available nutrients.

DENTAL PLAQUE

The surfaces of the teeth rapidly collect dental plaque, which is comprised principally

TABLE 25-2.   THE MOST NUMEROUS CULTIVABLE ORGANISMS INDIGENOUS TO VARIOUS SITES IN THE ORAL CAVITY OF MAN

	Dental Plaque[a]	Gingival Crevice[b]	Tongue[c]	Saliva[d]
Total anaerobic count	4.6×10^{10}/gm. wet wt.	3.5×10^{10}/gm. wet wt.	?	1.1×10^{8}/ml.
Total aerobic count	2.5×10^{10}/gm. wet wt.	2.0×10^{10}/gm. wet wt.	?	4.0×10^{7}/ml.

Approximate Distribution of Species as % of Total Anaerobic Count

1. GRAM POSITIVE COCCI				
Streptococcus mitis	28.0[a]	19.0[b]	30.0[c]	6.0[d, a]
Streptococcus salivarius	< 1.0	< 1.0	8.2	10.0
Enterococci	< 1.0	8.0	< 1.0	< 1.0
Micrococcus species	< 1.0	< 1.0	3.0	< 1.0
Peptostreptococcus species	13.0	7.0	4.2	?
2. GRAM NEGATIVE COCCI				
Neisseria species	3.5	< 1.0	3.0	2.0
Veillonella species	6.0	10.0	16.0	15.0
3. GRAM POSITIVE RODS				
Diphtheroid species ⎫	42.0	35.5	20.0	?
Nocardia species ⎬				
Actinomyces species ⎭				
Lactobacilli	< 0.01	< 0.1	< 0.1	< 0.1
4. GRAM NEGATIVE RODS				
Bacteroides melaninogenicus	< 0.5	5.0	< 0.5	< 0.5
Bacteroides oralis	5.0	5.0	4.0	?
Vibrio sputorum	1.0	5.0	< 0.5	?
Fusobacterium nucleatum	4.0	3.0	1.0	< 1.0
5. SPIROCHETES				
Treponema species	< 0.1	1.0	< 0.1	?
6. OTHERS			8.0	63.0

Data taken from:

a Gibbons et al., 1964b.
b Gibbons et al., 1963.
c Gordon and Gibbons, 1966, unpublished data.
d Richardson and Jones, 1958.

of microorganisms. Microscopic counts of over 10^{11} bacteria per gram wet weight of plaque and viable counts of over 10^{10} organisms per gram wet weight have been reported for this material (Table 25-2). The microorganisms present have direct access to the nutrients provided by saliva and also to those present in the diet, which probably accounts for the magnitude of their accumulation in this site. The plaque flora contains nearly equal numbers of facultative, aerobic and anaerobic bacteria. Carbohydrates seem to be especially important in their metabolism, as most species ferment a wide range of carbohydrates in vitro to terminal acidities of pH 4.5 to 5.5. Such acidities have also been observed in situ following the ingestion of soluble carbohydrates. The microbial acid end products found in vivo in plaque after carbohydrate ingestion

initially consist principally of lactic acid. However, appreciable quantities of acetic, propionic, butyric and isobutyric acids later appear, accompanied by a decrease in the lactic acid concentration. This suggests that the initial rapid fermentation of carbohydrates is mediated principally by plaque streptococci and diphtheroids. Other plaque bacteria such as the peptostreptococci, *Neisseria* and *Bacteroides* species carry out a slower carbohydrate fermentation which contributes to the delayed appearance of the volatile fatty acids. The *Veillonella* species, by converting lactic acid to acetic acid and carbon dioxide, are partly responsible for the decrease in lactic acid concentration.

Effect of Carbohydrates on Plaque Bacteria. As the ingestion of dietary carbohydrates is intermittent and unpredictable, conditions in the dental plaque could be imagined to range from feast to famine, depending upon the length of time since the last meal. A more stable condition regarding carbohydrate availability would obviously be more desirable for many plaque microorganisms, and in particular the streptococci, which have an absolute requirement for carbohydrate for their growth and energy metabolism. Thus, it is not surprising that many plaque organisms can convert excess dietary carbohydrates into extracellular slimes and intracellular glycogen-like, polysaccharides which may function as a carbohydrate storage mechanism. These polymers can be catabolized during fasting periods, thus expanding the time during which carbohydrates are available to the plaque flora.

Comparatively little is known regarding the nature or functions of extracellular capsular polysaccharides formed by plaque bacteria. However, considerable information is available concerning the intracellular glycogen-like reserve polysaccharides (Gibbons, 1964). Almost all carbohydrate-fermenting species present in plaque, with the exception of certain lactobacillus strains, form copious amounts of intracellular polysaccharide of the glycogen type when grown in vitro on culture media containing excess glucose, fructose, sucrose, maltose or lactose. In addition, the bacteria present in dental plaque in situ have been observed to contain accumulations of these polysaccharides. The quantity of organisms capable of forming these glycogen-like polysaccharides in plaque appears to be dependent on the level of dietary carbohydrate. Van Houte (1964) found that the proportions of polysaccharide-storing organisms decreased markedly in individuals changing to a diet essentially free of carbohydrates. Upon return to a normal carbohydrate-containing diet, the proportions of polysaccharide-storing organisms increased to starting values. No changes were observed in the bacterial composition of plaque during the dietary regimen, as judged from the examination of gram stained smears. This suggests that the presence or absence of dietary carbohydrate effected a change in the metabolic potential of the plaque bacteria.

Besides the quantity of carbohydrate present, the ability to form polysaccharide may also be dependent upon the nature of the carbohydrate ingested. Guatemalan children on a dietary regimen which included candy, cake and other sweets had almost three times as many polysaccharide-forming bacteria in their plaque as did comparable aged Guatemalan village children who ingested a diet high in unrefined starches but low in refined sugars (Loesche and Henry, unpublished data).

Caries, Carbohydrates and Plaque Bacteria. Polysaccharide production by the plaque flora may be an important, but not an exclusive, determinant in dental decay. Since Miller's classic experiments, acid production by plaque bacteria has been implicated in the carious process. Intracellular polysaccharide formation provides a mechanism by which the plaque bacteria can effectively capture soluble carbohydrates in a short time period, converting the mono- and disaccharides to a reserve energy compound which can be catabolized to acids during periods of substrate deprivation. Such a phenomenon would explain the persistent acidity found in plaque after a glucose rinse. Bacterial accumulation on the tooth surface is maximal during sleep. This probably results from the decreased nocturnal flow of saliva which fails to wash off the accumulated flora. Accordingly, the acid produced by the flora from stored carbohydrates remains in situ on the tooth surfaces for long periods of time, and pH levels adequate for decalcification of enamel may be achieved.

Correlations between polysaccharide production and dental caries have been found in two widely divergent cultures. Plaque removed from a group of caries-active individuals resident in Boston harbored 54 per cent polysaccharide-positive organisms, whereas a caries-inactive group had 29 per cent of its plaque flora as strong polysaccharide producers. Guatemalan children with low or no caries experience had only one-third as many polysaccharide-positive organisms as did a high caries group within the same village.

Polysaccharide production alone does not explain adequately the relationship between caries, carbohydrates and the plaque microflora. Dietary sucrose is far more cariogenic than glucose in experimental animal caries yet, theoretically, the flora should be able to form polysaccharide equally well from both substrates. Recently Krasse (1965) has demonstrated an effect of dietary soluble carbohydrates on the introduction and establishment of cariogenic streptococci in the mouths of hamsters. On a sucrose diet the streptococci persisted in the oral cavity and gave rise to caries. However, on a glucose diet these same bacteria failed to persist and no caries developed in the animals. Dietary sucrose, but not glucose, apparently provided a mechanism by which the cariogenic strains were selected. In our laboratory the rat and hamster cariogenic streptococci as well as human cariogenic strains have been found to form copius amounts of capsule material when cultured in sucrose but not in glucose broth. This capsular slime permits these bacteria to adhere to the walls of glass tubes and flasks. If this phenomenon occurs in vivo, then sucrose might permit the cariogenic bacteria to adhere to the tooth surface, thereby minimizing washoff from saliva. This working hypothesis, of course, needs experimental verification.

THE GINGIVAL CREVICE

Organisms residing in the gingival crevice area, or under pathologic conditions, in the periodontal pocket do not have so much exposure to salivary nutrients as does the plaque flora. However, they would appear to have greater access to nutrients derived from dead epithelial cells and they also have direct access to the nutrients present in gingival crevice fluid. The availability of dietary nutrients would seem limited, except in areas where food is retained between the teeth. At these sites of food retention microbial growth is more abundant and periodontal breakdown is increased. For example, in a periodontally involved mouth one can readily remove by scaling from 50 to over 200 mg. of gingival debris. However, in a healthy mouth only 10 to 20 mg. of gingival debris can usually be recovered. This microbial overgrowth undoubtedly is a reflection of increased nutrient availability in sites of food retention.

The gingival crevice flora is comprised of slightly more obligately anaerobic organisms than facultative types, i.e., 56 versus 44 per cent (Table 25-2). Perhaps because these organisms have less access to the carbohydrates present in the diet and consequently must rely on proteinaceous substrates derived from gingival crevice fluid and tissue transudates, one tends to find more amino acid-fermenting bacteria localized in this site. For example, *Bacteroides melaninogenicus* and *Treponema dentium*, which do not require carbohydrates for growth and are able to actively ferment amino acids, are found in high numbers only in the gingival crevice area. *Fusobacterium nucleatum*, an organism which preferentially ferments certain amino acids over glucose, is more numerous in this site than elsewhere in the oral cavity. The somewhat specific localization of these three amino acid-fermenting bacteria in the gingival crevice area may be of significance in periodontal disease, for all three species have been suggested as playing a possible etiologic role. It is interesting to note that *B. melaninogenicus* and *T. microdentium* both have unusual growth factor requirements which are fulfilled in the gingival crevice area through microbial interactions, as previously described.

THE TONGUE

The crypts and fissures of the tongue provide niches for microbial colonization. Because of the relatively large surface area involved, these organisms would appear to have great access to the nutrients present in saliva. Microorganisms on the tongue also have direct access to the nutrients provided by the dying and sloughed epithelial cells upon which they are growing. The exposure of tongue organisms to dietary nutrients would appear to be transient at best. However, some nutrients are probably derived from food debris which has become entrapped in the crevices of the teeth which, when subject to partial decomposition, may be transported to the tongue flora by saliva. The contribution of nutrients present in gingival crevice fluid would appear to be nil.

Because of the large surface areas exposed, the organisms on the tongue have greater contact with molecular oxygen than organisms present in other sites in the mouth and, in fact, two-thirds of the cultivable organisms are facultative types. Because aerobic metabolic processes are more efficient than the anaerobic, it seems probable that the bacteria present on the surfaces of the tongue are growing at a more rapid rate than those present on the teeth or in the gingival crevice areas. As a result of the differences in nutrients available, the flora of the tongue differs from that of plaque and gingival debris in several respects (Table 25-2). There are greater proportions of facultative streptococci present, and some of the strictly anaerobic oral species such as *B. melaninogenicus* and the oral spirochetes are not present in appreciable proportions. Subtle differences in the types of streptococci present are also found. For example, *Streptococcus salivarius* comprises a large proportion of the facultative streptococci on the tongue and in saliva, emphasizing the point that the organisms found in saliva are derived for the most part from the tongue. The precise reasons for these subtle differences in the composition of the oral flora from site to site within the mouth are not known, but it may be assumed that differences in the availability of various nutrients are mainly responsible.

SUMMARY

The foregoing discussion has attempted to demonstrate the existence of microbial ecologic niches within the oral cavity which are a consequence of differences in the available nutrients. Thus, the bacteria which reside in the gingival crevice areas and on the surfaces of teeth differ from each other in some respects and are, in turn, different from the bacteria which reside on the tongue and in saliva. Diet composition can affect their metabolic and pathogenic potential, as evidenced from the fact that diets high in refined sugars result in greater capsule and intracellular polysaccharide formation by the plaque bacteria, and also in acid production which leads to the formation of dental caries.

It is unfortunate that so few nutritional data are available concerning the oral flora from in vivo experiments, but rather deductions have had to be drawn from in vitro observations on pure cultures.

While animal experiments in the past have generally ignored the significance of the indigenous microorganisms, future studies relating nutrition to dental disease will have to take into account the effect of diet upon the dynamic oral microflora.

REFERENCES

Brill, N., and Bjorn, H.: Passage of tissue fluid into human gingival pockets. Acta Odont. Scand., *17*:11, 1959.

Bryant, M. P.: Bacterial species of the rumen. Bact. Rev., *23*:125, 1959.

Chauncey, H. H.: Salivary enzymes. J. Amer. Dent. Assn., *63*:360, 1961.

Gibbons, R. J.: The bacteriology of dental caries. J. Dent. Res., *43*:1021, 1964.

Gibbons, R. J., Kapsimalis, B., and Socransky, S. S.: The source of salivary bacteria. Arch. Oral Biol., *9*:101, 1964a.

Gibbons, R. J., Socransky, S. S., and Kapsimalis, B.: Establishment of human indigenous bacteria in germ-free mice. J. Bacteriol., *88*:1316, 1964b.

Gibbons, R. J., Socransky, S. S., de Araujo, W. C., and Van Houte, J.: Studies of the predominant cultivable microbiota of dental plaque. Arch. Oral Biol., *9*:365, 1964c.

Gibbons, R. J., Socransky, S. S., Sawyer, S., Kapsimalis, B., and Macdonald, J. B.: The microbiota of the gingival crevice area of man. II. The predominant cultivable organisms. Arch. Oral Biol., *8*:281, 1963.

Jenkins, G. N.: The Physiology of the Mouth. Oxford, England, Blackwell Scientific Publications, 1960.

Keyes, P.: Infectious and transmissable nature of experimental dental caries. Arch. Oral Biol., *1*:304, 1960.

Kite, O. W., Shaw, J. H., and Sognnaes, R. F.: An influence on dental caries incidence produced in rats by tube feeding. J. Dent. Res., *29*:668, 1950.

Krasse, B.: The proportional distribution of *Streptococcus salivarius* and other streptococci in various parts of the mouth. Odont. Rev., *5*:203, 1954.

Krasse, B.: The effect of caries-inducing streptococci in hamsters fed diets with sucrose or glucose. Arch. Oral Biol., *10*:223, 1965.

Lanke, L. S.: Influence on salivary sugar of certain properties of foodstuffs and individual oral conditions. Acta Odont. Scand., *15*: suppl. 23, 1957.

Miller, W. D.: The Microorganisms of the Human Mouth. Philadelphia, S. S. White Co., 1890.

Rogosa, M.: The genus *Veillonella*. I. General cultural, ecological, and biochemical considerations. J. Bacteriol., *87*:162, 1964.

Rosebury, T.: Microorganisms Indigenous to Man. New York, McGraw-Hill Book Co., 1962.

Socransky, S. S., Loesche, W. J., Hubersak, C., and Macdonald, J. B.: Dependency of *Treponema microdentium* on other oral organisms for isobutyrate, polyamines, and a controlled oxidation-reduction potential. J. Bacteriol., *88*:200, 1964.

Van Houte, J.: Relationship between carbohydrate intake and polysaccharide storing microorganisms in dental plaque. Arch. Oral Biol., *9*:91, 1964.

Chapter Twenty-six

The Cariogenicity of Different Foodstuffs

Basil G. Bibby, Ph.D., D.M.D.

The idea that specific foods have particular importance in causing dental caries can be traced to the earliest writings on the subject. Aristotle named figs as being of particular importance and the Arabian Mesu pointed to dates as the cause of dental decay. Interestingly enough, both commented on the sweetness of these foods. On this basis, one could say that the sweetness of foods and dental caries have been interrelated from earliest times. Although, as will be pointed out later, this is probably an oversimplification, it is interesting to note that from the time that dentistry became a specialty in its own right, the association between caries and sweet foods has been noted repeatedly. Pierre Fauchard (1746), the founder of the dental profession as we know it today, wrote that "all sugary food contributes not a little to the destruction of the teeth" and that "those who like 'les sucrenes' and use them frequently rarely have good teeth." When the leadership of dentistry moved to Great Britain, Thomas Berdmore (1768), the outstanding dental scientist of his time and dentist to King George III, wrote, "I am creditably informed that in the Low Countries, where sugar, tea, coffee and sweetmeats are used in excess, the people at an early age are remarkable for the badness of their teeth." He added advice which is as good as any we can give today: "Eat of them but seldom, and always wash the teeth after them." At that time, trade with the West Indies was expanding, and as sugar was coming into more general use, teeth were suffering accordingly.

As medicine and science advanced in Europe, the mechanisms which accounted for other disease processes were brought forward to explain dental decay. Nutritional, systemic and developmental factors were suggested as the causation of various diseases, and dental caries was included among them. Under the influence of such leaders as Bell (1830) these theories gained wide acceptance, but their dominance was weakened by improved clinical observations and the growth of the experimental method in scientific studies. Based partly on his dental study of men in the battlefields of the Napoleonic Wars, Parmly (1820) decided that caries "arises from uncleanliness of the mouth in interstices and irregularities of the teeth. . . . It has been a grand mistake to consider it of internal origin." To prevent caries he recommended the use of the toothbrush and silk for cleaning between the teeth.

William Robertson (1845) was the leader in the experimental approach. From his careful experiments he concluded caries was caused by "acids formed from lodgment of food in certain places" and that "prevalence of caries is due to preparation of food which renders it more likely to be retained about the crevices of the teeth." America's first significant contribution to caries study came from Westcott (1843), who decided that all vegetable and mineral acids attack the teeth and that such acids are formed in the mouth when vegetable substances are acted upon by "ferments." Italy's contribution came from Mantegazza (1864) who, at the time Pasteur was showing the importance of bacteria in fermentation processes, showed that teeth could be decalcified by being placed in a mixture of saliva and sugar. He suggested that the formation of lactic and acetic acid were responsible for this decalcification. Prior to Miller's (1890) studies of oral bacteria,

318

Magitot (1878), Milles and Underwood (1882), Leber and Rottenstein (1867), and others, concluded that fermentation processes brought about by oral fungi produced acids which cause tooth destruction. Miller's great contribution was that beginning from existing ideas he was able, by using refined bacteriological procedures, not only to show what types of mouth organisms produced acid, but to show that the acids they formed were capable of destroying the teeth, and further that the salts of these acids could in turn be recovered from the carious lesion. Thus, he was able to turn what was good theory into sound fact and put beyond reasonable doubt the explanation that teeth could in fact be destroyed by the action of mouth bacteria on carbohydrate foodstuffs.

Miller went further than this. He set out to determine which foods were most conducive to caries. He showed that tooth destruction was not produced by the action of mouth bacteria on protein materials, and that common foodstuffs, such as flour, potatoes and bread, produced more acid on fermentation in saliva than sugar. From this finding, he decided that potatoes and bread were more cariogenic than sugar. While this conclusion was not really justified, Miller must be given fullest credit for beginning a series of studies which, in a desultory fashion have continued to the present day without, as yet, giving the final answer to what types of carbohydrate foods are the most important in the causation of dental decay.

Since it would be a pointless exercise to attempt to evaluate the relative importance of different foods, unless one had a fairly definite opinion as to what part they play in caries causation, it is appropriate to restate, in the light of present knowledge, why the ideas of the early writers, which Miller consolidated into the chemico-parasitic theory of dental caries, have received almost complete acceptance by modern students of the subject.

THE CHEMICO-PARASITIC THEORY OF CARIES CAUSATION

LOCATION

The fact that caries occurs almost exclusively in those areas on the tooth surfaces where the self-cleansing mechanisms of the mouth are least effective is the starting point in the association between food and caries. The points at which food remnants persist for the longest period of time are those at which caries occurrence is highest. Although the frequency of occurrence in various areas of the teeth has been determined by many reliable investigators, satisfactory quantitation of the amount of food persisting in specific locations in the mouth has not been achieved, and as yet mathematical correlations between food adhesion and caries attack are not possible. By incorporating dyes into candies, MacGregor (1958) was able to show quite clearly that they were held in the specific locations at which caries occurs most frequently.

FOOD

The observation of the early investigators that there was more caries in the mouths of people indulging excessively in sweet foods has stood the test of time. While, as must be expected in any biological system, apparent exceptions can be found, it is quite obvious that no other correlation with food usage is nearly so consistent. Several studies show that wartime restrictions on sugar consumption were associated with dramatic reductions in the activity of dental decay. Likewise, studies such as those of Jay (1947) and of Becks et al. (1944) in which patients were placed on restricted carbohydrate diets, demonstrate that the controlled reduction of carbohydrate intake will reduce dental decay. The most dramatic demonstration of the importance of added sugar in the causation of dental caries was made in a study in which patients used a variety of sugar supplements over periods of one or more years. Clear-cut evidence was produced of a striking relationship between the addition of sugar to the diets and the amount of dental decay which occurred during the experimental periods. There were differences in the cariogenicity according to the nature of the sugar supplements. A point that cannot be stressed too much is that the frequency of eating the sugar-containing foods was an important factor influencing the amount of caries produced (Gustafsson et al., 1954).

MECHANISM

If, as is generally accepted, caries begins with the destruction of the predominantly inorganic enamel, then the agents necessary for the initiation of caries must be those which will destroy this most durable of all body tissues. The century-old demonstration that there is more acid in the areas on the teeth where caries occurs and that these are identical with the areas in which food persists suggests that acids formed from foods are the destructive mechanism. Since Miller (1890) and numbers before and after him have demonstrated that the action of the salivary bacteria on carbohydrates will decalcify enamel, and since as yet no one has been able to show comparable enamel destruction by other biological agents which are likely to be present in the mouth, it is easy to understand why acid formation from carbohydrates is seldom questioned as the activating cause of caries.

That acid is indeed the destructive agent of enamel is verified by recent work using advanced techniques of microradiography or polarizing microscopy. These demonstrate that lesions in every way identical with those occurring in natural caries in the mouth can be produced by the carefully controlled use of acids on the tooth surfaces. Moreover, it appears that it is only when conditions of exposure to acid are similar to those operating on the tooth surface in the mouth that experimental lesions identifiable with natural caries are produced; comparable destruction of enamel has not been produced by agents other than acid. An additional reason for believing that acid is the activating agent in caries is that agents such as fluorine, which make the teeth more resistant to decalcification, also produce clinical caries.

BACTERIA

The role of bacteria becomes important in explaining the origin of acids in the mouth. The conclusions of Miller et al., (1940) that a variety of mouth bacteria can produce sufficiently high concentrations of acid to decalcify the teeth has been adequately supported by subsequent dental investigators and is in keeping with general bacteriologic knowledge of the fermentative capacities of microorganisms. That bacterial action on carbohydrates is an essential part of the mechanism of caries has been indicated by demonstration of inhibition of caries in animals and perhaps in man by the use of agents which have no effect on bacteria other than inhibiting their ability to form acid from carbohydrates. Less definitive but more dramatic evidence of the essentiality of bacteria in caries has been provided by the ability of antibiotics such as penicillin to reduce caries in animals and by experiments using germ-free or gnotobiotic animals, in which diets which will produce dental decay in animals with normal bacterial population fail to do so in the animals in which no bacteria are present.

EXPERIMENTAL CARIES

Further proof of the dominant role of carbohydrates in caries causation has been provided by animal experimentation. This has shown that experimental caries is not produced in animals unless there is a fermentable carbohydrate in the diet and, in general, the higher its content the more caries that results. That actual contact between the carbohydrate and the teeth is necessary for the production of dental decay has been demonstrated by feeding caries-producing diets to animals by means of a stomach tube. When contact between the carbohydrate and the oral environment was avoided by this means, there was no dental decay, but the usual amount resulted when the same diet was eaten in the usual way. In man, the importance of direct contact between carbohydrate and tooth surface has been demonstrated by experiments in which carbohydrates were held in contact with the tooth in a retaining device. A very rapid decalcification of the enamel resulted.

In view of the above considerations and much kindred evidence which has not been presented, it becomes impossible to doubt that carbohydrates play a dominant part in tooth destruction. Thus, even though other destructive mechanisms may some day be shown to contribute to the initiation of caries, we must, if we are concerned with taking advantage of existing knowledge and using it today, direct our attention to seeing in what ways we can modify the carbohydrate factor operating on the tooth surface to make it less destructive than we now know it to be. Since it is impracticable and undesirable to eliminate carbohydrate foodstuffs from modern civilized diets, the most realistic approach to reducing their harmful effects on the teeth would seem to rest in substituting less destructive carbohydrate-containing foods for those which are most destructive. In this way, something in the nature of practical advice can be given to those who want to make the effort to avoid dental decay. Unfortunately, all the necessary information is not yet available. Hereafter the state of knowledge on this subject will be reviewed.

FREQUENCY

The influence of the frequency of eating on caries has been observed by many investigators since earliest times and deserves the greatet emphasis. It will be mentioned again later.

COMPARISONS OF THE CARIO-GENICITY OF DIFFERENT FOODSTUFFS

Although efforts to differentiate between the cariogenicity of different foodstuffs antedate Miller (1890), he seems to have offered the first quantitative assessments of their destructiveness to the teeth, which he based on the amount of acid they formed on fermentation in saliva. Since his time other approaches have been made in comparing cariogenicity. These include measurements of the acid present in expectorated saliva, of acid formation in plaque, of the pH of cavities, of food retention in the mouth—alone and in combination with fermentation tests—of comparative enamel decalcification and, finally, of caries production in animals.

ACID FORMATION IN INCUBATED SALIVA

Miller (1890) and others of the earlier investigators used comparisons of acid formation in

TABLE 26-1.   ACID FORMATION FROM FOODS FERMENTED IN SALIVA

	Descending Order of Acid Production
* Miller, 1890	Cornstarch, rice starch, potato starch, bread, glucose, cane sugar
* Pickerill, 1912	Pastry, white bread, brown bread, chocolate biscuit, apple, potato, bread, butter and jam, orange, cane sugar, rice
* Walkhoff, 1919	Bran, bread, cane sugar, wheat starch
* Belding, 1948	Rice, potato, corn bread, starch, macaroni, sucrose, glucose
† Möse, 1949	Potatoes, saccharose, wheat bread, maltose, honey
‡ Beck and Bibby, 1961	Lentils, green peas, oatmeal, wheat germ, rice, rye flour, macaroni, corn meal, yellow beans, soy flour, farina, graham flour, white flour, Minute rice, tapioca, cornstarch, potato starch, sucrose, glucose

* Titrated acidity.
† pH after 6 and 24 hours.
‡ Titrated acidity after periodic neutralization.

saliva as a means of rating the caries-producing capacities of different foods. In general, the indications of cariogenicity obtained by this procedure did not accord with clinical experience and it has not been used in recent years. Table 26-1 summarizes the conclusions reached by various investigators who have used this procedure. The finding that bread and vegetables produced more acid than sugar can be explained by the failure of the early investigators to take into account the considerable buffering capacity of the less refined foodstuffs. As a result, while a low pH which would inhibit acid formation was rapidly reached in sugar fermentations, this did not occur so soon in foods with some inherent buffering capacity, and acid formation continued for a much longer time before the inhibiting pH was reached, with consequent greater final acid formation.

To give a truer picture of the total acid-producing potential of foodstuffs when fermented by saliva we made titrations and neutralizations of fermenting food-saliva mixtures at six hourly and three more widely spaced intervals over a 48-hour period. This measurement gives information both on short-term acid formation and on what probably occurs

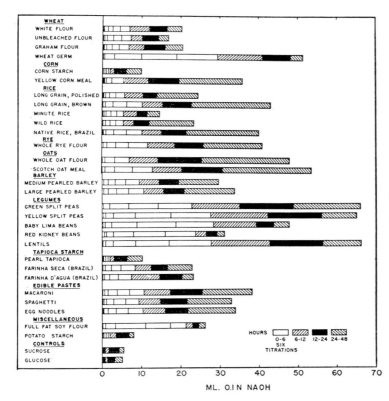

FIGURE 26-1. *Comparison of acid formation from foods incubated in saliva and neutralized at periodic intervals.*

when large accumulations of food are retained on the teeth, making it possible for acid production to continue over fairly long periods of time. Figure 26-1 shows that many common foodstuffs have a much greater total acid-forming capacity than sugar and suggests that the polysaccharides of foods contribute to giving them a high total acid-forming capacity. Whether they are important in caries production is probably determined by the readiness and speed with which they ferment and the nature of the sugar or other substance with which they are combined when eaten. It seems certain that at most the amount of acid produced from foods on fermentation is only one of the factors which determine their cariogenicity.

As an incidental finding in the studies on enamel decalcification, which will be discussed later, it was noted that in addition to the lack of parallelism between acid formation and the amount of enamel dissolved by fermenting foods, there was also a lack of a consistent relationship between their titratable acidities

and the pH's. This seems to indicate that more than one type of acid or different spectra of acids are formed during the fermentation of foods and, therefore, the amount of enamel destroyed in the mouth need not be proportionate to the amount of acid formed on the teeth. That different proportions or combinations of acids are formed is confirmed by the finding that different proportions of volatile and nonvolatile acids result from the fermentation of different common cereals with oral streptococci (Steinkraus et al., 1965).

Another aspect of food fermentation is worth noting, namely, that when acid foods such as carbonated beverages, fruit juice or honey are mixed with saliva they will not ferment because they are already at a level of acidity which inhibits the enzyme action of oral bacteria. Before active fermentation can begin, these foods must be buffered to near neutrality by the saliva. On the tooth surface this mixing with saliva would mean dilution and removal from the tooth surface before there is significant acid formation. Therefore, it seems likely that any decalcification produced by these foods will be caused mainly by their direct action upon the teeth rather than as the result of their fermentation products. This probably accounts for the low cariogenicity of fruits, juices and carbonated beverages.

ACID CONTENT OF EXPECTORATED SALIVA

The acid content of saliva after eating foods has been measured by making direct titrations of the saliva with alkali, by determining the pH of the expectorated saliva or by measuring its content of lactic acid. Table 26-2 summarizes the principal findings obtained by these procedures and shows that there is little agreement between them. This is attributable to the different measuring procedures used, uncontrollable variables relating to methods of using foods and collecting saliva, and patient-to-patient variations. Pickerill's (1912) finding that acid foods give rise to a more alkaline saliva is interesting because of the increased significance now being given to the capacity of the saliva to buffer acids on the tooth before they initiate decalcification of the enamel. In a sense, the findings of Oster et al. (1953) concerning the use of carbonated beverages confirm those of Pickerill.

TABLE 26-2. ACID IN EXPECTORATED SALIVA

	Descending Order of Acid Present
* Pickerill, 1912	Lemon, apple, orange, water, brown bread, cake, dry bread, carrot (boiled), bread and butter
† Neuwirth and Klosterman, 1940	Dextrose and sucrose, starch
‡ Haggard and Greenberg, 1949	Fruit juice, cola beverage, caramel, crackers, mixed meal, ice cream
† Ericsson, 1951	Potato, bread, sugar
‡ Oster et al., 1953	Water control, carbonated beverage
† Ludwig and Bibby, 1954	Fig cookie, toffee, white bread butter and jam, ice cream, potatoes (French fried), apple, cola beverage, orange juice, carrot

* Titrated acidity or alkalinity.
† Lactic acid in saliva.
‡ pH of saliva.

TABLE 26-3. ACID FORMATION IN PLAQUE

	Descending Order of Acid Formed
* Miller et al., 1940	lactose
† Lennon and Sullivan, 1955	Honey, treacle, brown sugar, white sugar, molasses
‡ Ludwig and Bibby, 1957a	Fig cookie, bread and jam, apple, Sucrose, glucose, maltose, starch, cola beverage, toffee, milk chocolate, potatoes (French fried), ice cream
Kleinberg and Jenkins, 1959	20% Glucose, glucose sweet, white bread, apple, cooked rice, 1% glucose

* Lactic acid production in plaque material.
† pH after 30 minutes in salivary concentrate.
‡ pH after 30 minutes.

ACID FORMATION IN PLAQUE

Stephan's (1948) demonstration of rapid acid formation in plaques and the return to neutral pH in a period of 30 to 60 minutes has stimulated considerable study of this phenomenon. Unfortunately, most of this work has used laboratory chemicals and little of it

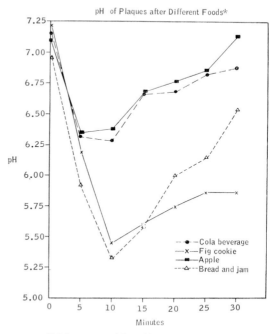

pH of Plaques after Different Foods*

- ● - Cola beverage
- × - Fig cookie
- ■ - Apple
- △ - Bread and jam

*All figures mean of five subjects

FIGURE 26-2. *pH curves given by four foodstuffs in in vivo plaques.*

has been directed toward comparing foodstuffs of the sort that are likely to come into contact with the plaque during ordinary meals. The principal findings on acid production by common foods are recorded in Table 26-3. The only published plaque pH curves for ordinary foodstuffs (Fig. 26-2) show that foods which adhere to the teeth produce a lower pH than those, such as beverages or fruit, which are rapidly eliminated. This is in keeping with Kleinberg and Jenkins' (1959) finding of lower pH's when larger quantities of food are placed on plaques. On the basis of the limited information available, the effects of various foods on plaque pH would seem likely to relate to the initiation of caries on the enamel surface. Unfortunately, the possibility of making direct correlations is complicated by the influence of the saliva on the plaque pH and a wide range in the biochemical activities of different plaques (Luoma, 1965).

ACID FOR-
MATION IN CAVITIES

The absence of salivary flow and the persistence of foodstuffs provide different condi-

tions in cavities from those operating during the initial breakthrough of the enamel surface. Therefore, foods which initiate caries may be less important in expanding existing cavities. The only available comparison of the effects of foodstuffs on the pH of cavities is given in Figure 26-3. This shows that a different sequence of events takes place in the initially acid cavity from that which takes place in the near neutral plaque on the enamel surface. These preliminary comparisons do not justify drawing any conclusions but they do show that more information is needed regarding the action of foods in established cavities.

AMOUNTS OF FOOD
RETAINED IN THE MOUTH

If food retention on the tooth surface is an important part of caries initiation, it seems logical that foods which are retained on the tooth surface in the greatest amounts would have the greatest potentialities for caries production. This hypothesis has led to a number of comparisons of food retention, most of

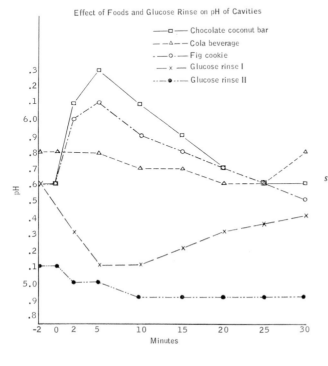

FIGURE 26-3. *pH curves given by several foodstuffs in open cavities.*

TABLE 26-4.   RETENTION OF FOOD IN THE
MOUTH

	Descending Order of Amounts of Food Retained
Volker and Pinkerton, 1947	Sweet sugar lump, chewing gum, sugar solution
Bibby et al., 1951	Fig cookie, chocolate ice cream, pastry, caramel, cracker, bread, potato, apple, soda beverage, orange juice, carrot
Lundqvist, 1952	Caramel, chocolate, cookie, pastry, bread, ice cream, potatoes, bread and butter, apple, fruit juice, lemonade, carrot
Lanke, 1957	Chewing gum, rye bread, biscuits, white bread, sweets, chocolate cream, macaroni, potato

which have measured food retention by making carbohydrate determinations, generally as glucose, in the expectorated saliva or in mouth washings. All but the most recent work on this subject has been fully reviewed by Lanke (1957). Selected results of some of the studies which offered comparative ratings are given in Table 26-4. Although the techniques of testing have varied in respect to the quantities of test food consumed, and the times and methods of recovery from the mouth, there is some similarity in the findings. In general, sticky foods, such as dried fruits or caramels, show greatest retention, and starchy foods, such as breads and potatoes, occupy an intermediate position, with fruit juices, fresh fruit and beverages being lowest. In general, this rating agrees fairly closely with clinical observation and the subjective experience of anyone who has made observations in his own

mouth. It is of some interest to note that some of the foods in the Lundqvist (1952) comparison were tested in the Vipeholm caries study (Gustafsson et al., 1954) and, in most instances, the amount of new decay occurring in the human users of the particular foodstuffs paralleled Lundqvist's laboratory tests on carbohydrate clearance from the mouth (Fig. 26-4). We found parallelism with our laboratory grading of foodstuffs in one animal study, but not in a second test with other foods. On the basis of the available evidence, it seems safe to conclude that the persistence of foodstuffs in the mouth bears some relationship to their caries-producing potential, but of itself this is not the principal determinant of cariogenicity.

In an attempt to supplement the findings on retention of foods and to determine a better indication of their cariogenicity, we combined them with a measurement of the acid formed when a unit amount of each food was fermented by salivary bacteria (Bibby et al., 1951). Because of possible deficiencies in our measurements of acid formation, only a small range of differences was found in acid production, and when these figures were multiplied by the much more widely spread retention figures, no changes were produced in the order of the "decalcification potentials" of the tested foods.

We later made comparisons of the retention of the same foodstuffs in different persons' mouths and found that, while the amounts retained were higher for some persons than others, there was a consistent parallelism in the position of the foods from one mouth to another (Ludwig and Bibby, 1957). The de-

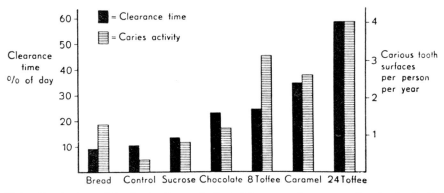

FIGURE 26-4. *Comparison of food retention as indicated by sugar clearance and caries production in experimental subjects. (From Lundqvist; reproduced by permission of the author.)*

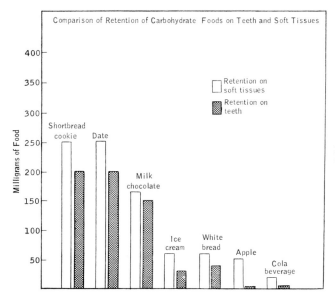

FIGURE 26-5. *Comparison of amount of food retention on teeth and mucous surfaces of the mouth.*

termination of how much of the foodstuff was actually retained on the teeth and how much on the mucous surfaces of the mouth was made with several foods. The findings, which are given in Figure 26-5, show that there is a considerable retention of carbohydrate on the mucous surfaces. Whether this has significance in contributing to caries or maintaining a source of acid over a long period is not known.

Interest in the use of the retention figure as an index of cariogenicity of foodstuffs has been active enough that efforts have been made to quantitate adhesiveness outside the mouth and thus avoid the complex and difficult experimental procedures which must be followed for intraoral quantitation. Toward this end Caldwell (1959) used an adhesion-measuring device and Beck and Bibby (1961) a centrifugal method. These gave some parallelism to intraoral figures, but the procedure of Beck and Bibby indicated that, depending upon concentration and other factors, some foods gave several levels of adhesiveness.

ENAMEL DECALCI-
FICATION BY FOODSTUFFS

Since the initial stage of caries is primarily a process of enamel decalcification, measuring the amount of decalcification produced by foods when incubated with saliva would seem to be a logical method for comparing their cariogenicity. That foods differ in this respect was clearly shown by Osborn et al. (1937) 30 years ago in tests on whole teeth incubated in saliva-food mixtures. Recently, several other methods, none of which is completely satisfactory, have been used to compare a number of carbohydrate foodstuffs. The results of the studies, which have produced reasonably quantitative results, are given in Table 26-5. The most consistent finding is that the less refined cereals and sugars produce less decalcification than the more refined forms. There is also evidence in Andlaw's (1960) results that the milk content of foods reduces the amount of decalcification produced. In Soni and Bibby's (1961) series it can be seen that the addition of phosphate reduces enamel decalcification by fermenting flour. In the work of both Soni and Bibby and Andlaw the amount of acid formed did not parallel the amount of enamel decalcified, which means either that the foodstuffs contained factors which protected the enamel against decalcification or that some of the acids formed were of types which had less capacity than others to dissolve enamel (Table 26-6).

Jenkins et al. (1959) have shown that the

TABLE 26-5.   ENAMEL DECALCIFICATION BY FOODS

	Descending Order of Enamel Decalcification
* Miller, 1890	Potato, bread, sugar
* Pickerill, 1912	Bread, apple, sugar
* Osborn et al., 1937	Flour (60-70% extn), fine mealie, wheat mealie, sucrose, flour (90% extn), whole mealie, crude sugar juice
† Jenkins et al., 1959b	White flour, brown flour
† Jenkins et al., 1959a	Honey, sucrose, brown sugar, golden syrup, cane juice, treacle
§ Andlaw, 1960	Farina, white flour, sucrose, chocolate cream, enriched flour, granulated sugar, polished rice, brown rice, potato starch, corn flakes, wheat flakes, white bread, cornstarch, corn meal, golden syrup, whole wheat, graham flour, wheat germ, All-Bran, maple syrup, milk chocolate, caramel, wheat bran
‡ Soni and Bibby, 1961	Wheat flour, honey, cracker, white bread, caramel, ice cream, corn flour, Shredded Wheat, shortbread, corn flakes, rye flour, wheat flour with germ oil, wheat flour with CaHPO$_4$, wheat germ
§ Thanik, 1962	White flour, graham flour, white rice, chocolate cream, corn flakes, yellow corn, wheat flakes, brown rice, wheat germ, 40% Bran, golden syrup, All-Bran, sugar, molasses, milk chocolate, bran, caramel, corn starch, maple syrup
¶ Khanna, 1964	Oat flakes, whole barley, 1st clear wheat flour, barley pearls, corn flour, corn bran, oat groats, whole corn, corn grits, corn meal, 2nd clear flour, whole oats, corn germ, barley hulls, oat hulls, North Star oats, whole wheat, wheat germ, wheat shorts, wheat bran

* Tooth decalcification.
† Chemical analysis.
‡ Polarized light.
§ Gravimetric method.
¶ Radioactive enamel.

TABLE 26-6.   ACID FORMATION AND ENAMEL DECALCIFICATION BY CERTAIN FOODS

Food	Acid Formed 0.05 NaOH	Enamel Dissolution (mg.)
Whole wheat bread	11.0	0.2
White bread	8.4	0.4
Corn flakes cereal	4.3	0.5
All-Bran cereal	16.2	0.1
Chocolate coconut bar	11.1	1.1
Milk chocolate	13.2	0.1
White flour	4.9	1.0
Graham flour	5.1	0.2

quite marked differences between the decalcifying effects of cereals from different states, which cannot be accounted for in the above ways.

In spite of some lack of agreement in the findings of different investigators, it seems clear that the amount of acid formed from the foodstuff is not solely responsible for the amount of enamel decalcification produced by the foods and that this is modified, in part at least, by the presence in foodstuffs of factors which protect the teeth against the action of acid.

CARIES IN ANIMALS

Only a small fraction of impressive numbers of experiments which have been carried out with animal caries has been directed toward trying to indicate the relative cariogenicity of the foodstuffs commonly used by man. Most of the work has been more concerned with establishing mechanisms of caries and has used pure laboratory compounds such as lactose or starch. Some of the findings in animal experiments which seem to be pertinent in comparison of the cariogenicity of foods demonstrate that the whole grain portions of cereals seem to be less conducive to caries than the refined types (Constant et al., 1952; Madsen and Edmonds, 1962), that fat (Shaw, 1949) and milk-containing (Shaw et al., 1959) diets are less cariogenic than those having no milk, and that caries decreased with an increase in the fluidity of the diet (Harris and Stephan, 1953). The effects on caries of phosphate and other

calcium or phytate content of foods may protect the enamel, and Thanik (1962) has offered evidence that the calcium and phosphorus content of foods are important in determining their effect on enamel solubility; however, with some foods, such as bran, other factors also play a part. Khanna (1964) found additions to animal diets is discussed in another chapter (Chapter Eight).

COMPARISONS IN MAN

Although a great many clinical impressions have been recorded and some individual observations in family groups have been made, the Vipeholm study in Sweden seems to have been the only well-organized comprehensive study in which comparisons of the cariogenicity of different foodstuffs have been made in man (Gustafsson et al., 1954). A summary of the results is given in Fig. 4-5. It was shown that less than half the quantity of sugar taken in bread with the meals produces more caries than more than twice the amount of sugar used mainly in liquid form. Although the fluidity of the high sugar content probably accounts for the difference, it is not impossible that it could be related to some dissimilarity in the products of the bread fermentation. It appeared that the sugar taken in chocolate was less cariogenic than that taken in the caramels and toffee. Subsequent experiments suggest this may be the result of the presence of enzyme inhibitors or enamel protective factors in the chocolate. The outstanding difference between the experimental groups related to the frequency of eating. It is apparent that dramatic increases in caries took place in groups which used sugar between meals. This occurred even when there was no over-all difference in the total sugar intake.

CORRELATIONS

It is difficult to draw clear-cut conclusions from the various experimental attempts which have been made to indicate the comparative caries-producing capacities of foods. The most consistent finding which revealed itself in almost all the testing methods is that sugar in liquid form is less conducive to caries than when it is used in a solid or semiliquid state. The importance in caries causation of the quantity of carbohydrate which is indicated in animal experiments is in keeping with the increased depression of pH and longer duration of acid found in human plaques treated with a higher sugar concentration, but the Vipeholm study seems to show that the frequency of eating is at least of equal importance. Some correlations between the findings on acid production in saliva, adhesiveness of foodstuffs to the teeth and their enamel decalcifying effects can be found with clinical caries, but these relationships are not consistent and therefore make it seem likely that none of these factors is of itself the dominant determinant of the caries attack. The evidence of the presence of natural enamel "protective" factors in less refined sugar and cereal products and in milk, which appeared in the decalcification studies, accords with the reductions these substances have been shown to produce in experimental animal caries; however, as yet there is no real evidence that these play a part in human caries.

Although it seems that patterns of correlation may be beginning to emerge and that these can be expected to grow clearer as more foods are studied and methods are further refined, it is not possible at this stage of experimentation to draw conclusions that would indicate beyond reasonable doubt which foods have the highest and which the lowest cariogenicity. Until this is possible, it is wise to make only tentative interpretation of experimental findings into clinical caries control. On the other hand, the differences that have appeared in the various properties of the foodstuffs which have been studied have encouraged the belief that when they are more fully understood it should be possible to take positive steps to reduce the cariogenicity of the more destructive of our common foods. It does not seem unrealistic to expect that in the future it will be possible to make noncariogenic confections and candies which could be substituted in our diets for the more destructive forms in which sucrose and other carbohydrates are now offered.

FREQUENCY OF EATING

On the basis of the evidence appearing above and a small number of clinical studies which cannot be reviewed at this time, it seems clear that the outstanding difference between cariogenic and noncariogenic diets is not in their sugar or carbohydrate content but

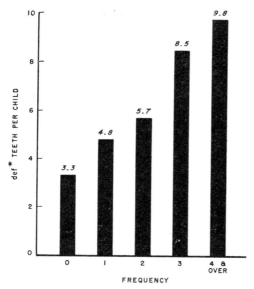

FIGURE 26-6. *Effect of between-meal eating on caries activity in children. (From Weiss and Trithart; reproduced by permission of the authors.)*

in the frequency of their use. This conclusion can be drawn from the Vipeholm study. It is also supported very nicely in a study by Weiss and Trithart (1960) which showed that the frequency of the use of in-between-meal snacks bears a direct relationship to the amount of caries in preschool children (Fig. 26-6).

In spite of the danger of negating the importance of what has been presented in this chapter, the writer states without any hesitation that the most important change that can be made in a dietary to make it less conducive to dental decay is to reduce the frequency of eating sweet foods. All other procedures which can be followed or recommended in the area of dental nutrition are of secondary importance to this one all-important step.

REFERENCES

Andlaw, R. J.: The relationship between acid production and enamel decalcification in salivary fermentations and carbohydrate foodstuffs. J. Dent. Res., 39:1200, 1960.

Becks, A., Jensen, A. L., and Millarr, B.: Rampant dental caries; Prevention and prognosis, a five year clinical survey. J. Dent. Res., 23:210, 1944.

Beck, D. J., and Bibby, B. G.: Whole saliva, salivary sediment, and supernatant as fermenting agents for foods. J. Dent. Res., 40:479, 1961.

Beck, D. J., and Bibby, B. G.: Acid production during

the fermentation of starches by saliva. J. Dent. Res., 40:486, 1961.

Belding, P. H.: Carbohydrates as acid producers. Dent. Items Interest, 70:297, 1948

Bell, T.: Anatomy, Physiology, and Diseases of the Teeth. Philadelphia, Carey and Lea, 1831.

Berdmore, T.: A treatise on the disorders and deformities of the teeth and gums. London, B. White, 1768.

Bibby, B. G., Goldberg, H. J. V., and Chen, E.: Evaluation of caries-producing potentialities of various foodstuffs. J. Amer. Dent. Assn., 42:491, 1951.

Caldwell, R. C.: Method of measuring the adhesion of foodstuffs to tooth surfaces. J. Dent. Res., 38:188, 1959.

Constant, M. A., Phillips, P. H., and Elvehjem, C. A.: Dental caries in the cotton rat. J. Nutr., 46:271, 1952.

Ericsson, Y.: On the salivary amylase and its significance in the caries process. Acta Odont. Scand., 9:89, 1951.

Fauchard, P.: Le chirurgien deutiste, au traité des dents. Paris, J. Mariette, 1728 (2nd ed., 1746).

Gustafsson, B. E., Quensel, C. E., Lanke, L. S., Lundqvist, C., Grahnen, H., Bonow, B. E., and Krasse, B.: Vipeholm dental caries study. The effect of different levels of carbohydrate intake on caries activity in 436 individuals observed for five years. Acta Odont. Scand., 11:232, 1954.

Haggard, H. W., and Greenberg, L. A.: The concentration of sugar in the saliva after ingestion of carbohydrate substances. Dent. Survey, 25:1788, 1949.

Harris, M. R., and Stephan, R. M.: Effect of mixing water in the diet on the development of carious lesions in rats. J. Dent. Res., 32:653, 1953. (Abstract.)

Jay, P.: The reduction of oral lactobacillus counts by the periodic restriction of carbohydrate. Am. J. Orth. & Oral Surg., 33:162, 1947.

Jenkins, G. N., Forster, M. G., and Speirs, R. L.: Influence of the refinements of carbohydrates on their cariogenicity. In vitro studies on crude and refined sugars and animal experiments. Brit. Dent. J., 106:362, 1959a.

Jenkins, G. N., Forster, M. G., Speirs, R., and Kleinberg, I.: Influence of the refinement of carbohydrates on their cariogenicity. In vitro experiments on white and brown flour. Brit. Dent. J., 106:195, 1959b.

Khanna, S. L.: Enamel Decalcification by Carbohydrate Foods from Different Geographic Areas. Thesis, University of Rochester, 1964.

Kleinberg, I., and Jenkins, G. N.: Further studies on the effect of carbohydrates substrates on plaque pH in vivo. J. Dent. Res., 38:704, 1959. (Abstract.)

Lanke, L. S.: Influence on salivary sugar of certain properties of foodstuffs and individual oral conditions. Acta Odont. Scand., 15(Suppl., 23):3, 1957.

Leber, T., and Rottenstein, J. B.: Untersuchungen über die Caries der Zähne. Berlin, August Hirschwald, 1867.

Lennon, D. F., and Sullivan, H. R.: Relative rates of acid production from "refined" and "natural sugars." Dent. J. Australia, 27:67, 1955.

Ludwig, T. G., and Bibby, B. G.: Acid formation in

the mouth after ingestion of different carbohydrate foods. J. Dent. Res., *33*:671, 1954. (Abstract.)

Ludwig, T. G., and Bibby, B. G.: Acid production and different carbohydrate foods in plaque and saliva. J. Dent. Res., *36*:56, 1957a.

Ludwig, T. G., and Bibby, B. G.: Further observations upon the caries producing potentialities of various foodstuffs. J. Dent. Res., *36*:61, 1957b.

Lundqvist, C.: Oral sugar clearance. Its influence on dental caries activity. Odontologish Rev., *3*(Suppl. 1):5, 1952.

Luoma, H.: Personal communication, 1965.

MacGregor, A. B.: A new method of demonstrating the retention of food stuffs in the mouth, with special reference to different forms of sweets. Proc. Roy. Soc. Med., *51*:41, 1958.

Madsen, K. O., and Edmonds, E. J.: Effect of rice hulls and other seed hulls on dental caries production in the cotton rat. J. Dent. Res., *41*:405, 1962.

Magitot, E.: Treatise on dental caries. Experimental and therapeutic investigations. Transl. by T. H. Chandler. Boston, Houghton, Osgood and Co., 1878.

Montegazza, T.: An experimental inquiry into the action of sugar and of certain acids upon the teeth. Brit. J. Dent. Sci., 7:49, 1864. (Transl.)

Miller, B. F., Muntz, J. A., and Bradel, S.: Decomposition of carbohydrate substrates by dental plaque material. J. Dent. Res., *19*:473, 1940.

Miller, W. D.: The Microorganisms of the Human Mouth. Philadelphia, S. S. White Dental Mfg. Co., 1890.

Milles, W. J., and Underwood, A.: An investigation into the effects of organisms upon the teeth and alveolar portions of the jaws. Amer. J. Dent. Sci., *15*(3rd Ser.):546, 1881-82; Brit. Dent. J., *3*:11, 1882.

Möse, J.: Versuche zur Ätiologie und Prophylaxe der Zahnkaries. Ztschr. Stomatol., *46*:1, 1949.

Neuwirth, I., and Klosterman, J. A.: Demonstration of rapid production of lactic acid in oral cavity. Proc. Soc. Exp. Biol. Med., *45*:464, 1940.

Osborn, T. W. B., Noriskin, J. N., and Staz, J.: A comparison of crude and refined sugar and cereals in their ability to produce *in vitro* decalcification of teeth. J. Dent. Res., *16*:165, 1937.

Oster, R. H., Prout, L. M., Pollack, B. R., and Shipley, E. R.: Salivary buffering capacity measured in situ in response to the acid stimulus found in some common beverages. J. Dent. Res., *32*:676, 1953. (Abstract.)

Parmly, L. S.: Lectures on the natural history and management of the teeth; the cause of their decay; the art of preventing its accession; and various operations, never hitherto suggested for the preservation of such teeth as it is too frequently considered necessary to extract. New York, 1820.

Pickerill, H. P.: The Prevention of Dental Caries and Oral Sepsis. London, Bailliere, Tindall & Cox, 1912.

Robertson, W.: A Practical Treatise on the Human Teeth, Showing the Causes of Their Destruction, and the Means of Their Preservation. London, Hayward and Moore; Birmingham, John Churchill, Princes St., Soho, 1841-42, 1845.

Shaw, J. H.: Carious lesions in cotton rat molars. J. Nutr., *38*:275, 1949.

Shaw, J. H., Ensfield, B. J., and Wollman, D. H.: Studies on the relation of dairy products to dental caries in caries-susceptible rats. J. Nutr., *67*:253, 1959.

Soni, N. N., and Bibby, B. G.: Enamel decalcification in food-saliva mixtures. J. Dent. Res., *40*:185, 1961.

Steinkraus, K., Gilmour, M., and Bibby, B. G.: *In* I.A.D.R. Proceedings, 1965.

Stephan, R. M.: Relative importance of polysaccharides, disaccharides and monosaccharides in the production of caries. J. Amer. Dent. Assn., *37*:530, 1948.

Thanik, D. D.: Effect of Calcium and Phosphorus of Foods on Enamel Decalcification. Thesis, University of Rochester, 1962.

Volker, J. F., and Pinkerton, D. M.: Factors influencing oral glucose clearance. J. Dent. Res., *26*:225, 1947.

Walkhoff, O.: Biologische Studien über das Wesen der Zahnkaries. Deutsch. Zhlkd. H. 42 (heft), 1919.

Weiss, R. L., and Trithart, A. H.: Between-meal eating habits and dental caries experience in preschool children. Amer. J. Public Health, *50*:1097, 1960.

Westcott, A.: Dissertation on dental caries. Amer. J. Dent. Sci., *4*(1st Ser.):31, 1843-44.

Chapter Twenty-seven

Fluoride and Caries Control—Mechanism of Action

FINN BRUDEVOLD, D.D.S., AND HAROLD G. McCANN, M.S.

The idea of utilizing fluoride in the control of dental caries originated from the observation that mottled teeth were conspicuously resistant to caries (Dean, 1938) and the discovery that mottling was associated with the use of drinking water containing fluoride in excess of 2 ppm (Dean and Elvove, 1937). A series of epidemiological studies subsequently established that protection against caries increased markedly with an increase of fluoride in the drinking water to a level of about 1 ppm and that further increase of fluoride concentrations provided only a slight additional protection (Dean et al., 1942). These findings are summarized in Figure 27-1. Mottling was not a problem in midwestern communities (temperate climate) when the level of fluoride was 1 ppm or less. In hot climates the optimal level of fluoride in the water, i.e., the level which affords maximal protection against caries without causing mottling, tends to be less than 1 ppm because of increased water intake (Galagen and Lamson, 1953).

Following these early observations, numerous communities initiated water fluoridation, and long-range, carefully conducted studies have demonstrated that 1 ppm of fluoride in the drinking water provides the same dental effects whether fluoride occurs naturally or is added by artificial means. Water fluoridation has been found to be safe and economical and to bring about a reduction in caries of up to 70 per cent (Ast and Fitzgerald, 1962).

Table 27-1 shows the development of water

FIGURE 27-1. *Relation between the amount of dental caries (permanent teeth) observed in 7257 selected 12 to 14 year old white school children of 21 cities of 4 states and the fluoride (F) content of public water supply. (Data from Dean et al., 1942.)*

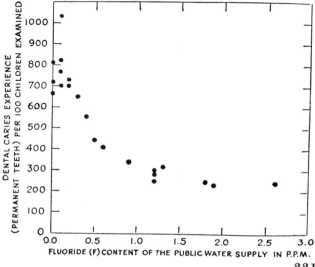

331

TABLE 27-1.   FLUORIDATION IN THE
UNITED STATES, 1945-1965*

Year	Number of Communities	Population
1945	6	231,920
1950	100	1,578,578
1955	1347	26,278,820
1960	2111	41,179,694
1965	2858	57,895,448

* Fluoridation census, 1965.

fluoridation in the United States from 1945 to 1965. Of the entire population of nearly 190 million, about 58 million were using controlled fluoridated water by September, 1965. Including the 7.7 million who have naturally fluoridated drinking water, 65.6 million, or 43.4 per cent of the population using public water supplies, consumed water containing 0.7 or more ppm of fluoride. Other countries are also increasingly adopting water fluoridation. Water fluoridation plants were known to be in operation in 41 countries in 1963. In Ireland all major water supplies have been fluoridated; in Canada, 22 per cent of the population used fluoridated water by June, 1965. At least a dozen countries have experimental fluoridation programs.

Since water fluoridation is suitable only in locations where communal water supplies are available, other vehicles for administering fluoride, including table salt and milk, have been studied, particularly in Switzerland. Fluoride pills or drops are used extensively in some low fluoride areas.

This chapter will review the voluminous literature on fluoride ingestion as related to caries in man. Current views concerning the mode of action of fluoride in inhibiting caries will also be discussed.

WATER FLUORIDATION AND DENTAL CARIES

Basically, the resistance to caries resulting from drinking fluoridated water is related to the amounts of fluoride which are deposited in the enamel surface. Since fluoride deposition is dependent on the period of time that fluoridated water is consumed, this discussion will deal with: (a) life-long, continuous fluo-

ride exposure, (b) prenatal versus postnatal fluoride exposure, and (c) pre-eruptive compared to posteruptive fluoride exposure.

FLUORIDE EXPOSURE THROUGHOUT LIFE

The low prevalence of dental caries in the permanent teeth of children who have used fluoridated water since birth or infancy is shown in Table 27-2. The data represent findings in several thousand 12 to 14 or 13 to 14 year old children from different cities in the United States and Canada. The number of decayed, missing and filled teeth (DMFT) per child ranges from 3.2 to 4.7 in the cities with controlled fluoridation, compared to 7.5 and 12.8 DMFT per child in the two control cities with fluoride-deficient water listed in the table. The reduction in caries as a result of fluoridation has been calculated to range from 50 to 70 per cent. The loss of permanent teeth per child, given in the last column of Table 27-2, is also significantly reduced in the fluoride-exposed children. It is worthy of note that the children in Aurora, which has a naturally fluoridated water supply (1.2 ppm. of fluoride), have a caries experience and tooth loss similar to those from cities with controlled fluoridation.

One important aspect of the inhibition of caries by water-borne fluoride is that the pro-

TABLE 27-2.   COMPARISON OF DECAYED, MISSING AND FILLED TEETH (DMFT) AND MISSING TEETH PER CHILD FOR SPECIFIED AGE GROUPS FROM COMMUNITIES WITH WATER SUPPLIES CONTAINING 1.2 PPM NATURAL FLUORIDE, 1.0 PPM CONTROLLED FLUORIDE, AND ONLY TRACES OF FLUORIDE*

	Fluoride in Water (ppm)	Age (yrs.)	DMFT per Child	Missing Teeth per Child
Aurora	1.2†	12–14	3.2	0.15
Grand Rapids	1.0	12–14	4.3	0.29
Evanston	1.0	12–14	4.7	0.06
Brantford	1.0	12–14	3.2	0.22
Newburgh	1.0	13–14	3.7	0.10
Kingston	0.1	13–14	12.8	0.92
Sarnia	0.0	12–14	7.5	0.75

* Data from Ast and Fitzgerald, 1962.
† Natural fluoride.

TABLE 27-3. PERCENTAGE REDUCTION IN CARIES IN DIFFERENT TOOTH TYPES—CONTINUOUS EXPOSURE TO 1.3 TO 2.2 PPM OF WATER-BORNE FLUORIDE*

| Tooth Type | Percentage Reduction | |
	Maxillary	Mandibular
Central incisor	85.1	92.6
Lateral incisor	84.5	100.0
Cuspid	80.7	100.0
First premolar	75.2	56.2
Second premolar	64.1	72.6
First molar	51.4	34.7
Second molar	54.3	33.5

* Data from Klein, 1948.

tective effect is greater in anterior than posterior teeth. For example, Dean et al. (1942) found that the children who lived in communities with 0.6 to 2.6 ppm of fluoride in the water supply had 90 to 95 per cent less caries experience in the four maxillary incisors than children from cities with fluoride-deficient water, while the corresponding decrease for the entire dentition was 50 to 70 per cent. The caries reduction in different types of teeth in children who had used fluoridated water all their lives is given in Table 27-3 (Klein, 1948). The decrease is greatest in the mandibular anterior teeth and least in the mandibular first and second molars. Klein suggested that the depressant effect of fluoride on caries may be of approximately equal potential for all the teeth and that the variation in per cent reduction of caries may be due to gradients in caries susceptibility. It has also been postulated, and recently verified, that the anterior teeth come into more intimate contact with water during drinking than posterior teeth and that they thus acquire more fluoride and, hence, greater caries resistance.

This concept of surface fluoride action led to the important development of topical fluoride treatments.

Although the caries-reducing effect of fluoridated water is less in deciduous than in permanent teeth, it is still significant. Studies from communities using water with controlled fluoride have demonstrated a decrease in caries experience of approximately 40 per cent in six to eight year old children. Again the effect is similar to that observed from naturally fluoridated water.

As a result of the reduction of caries in areas with fluoridated water, there is diminished premature loss of deciduous teeth and of first permanent molars among young children. Early loss of teeth is particularly detrimental, not only because of impairment of proper chewing at a critical age but because of a possible disturbance of the alignment of the permanent teeth. Five years of fluoridation in one city increased by 20 per cent the number of five and six year old children who had all their posterior deciduous teeth present and free of caries (Ast et al., 1951). Two studies have shown a decrease in malocclusion of approximately 20 per cent as a result of fluoridation (Hill et al., 1959; Ast and Fitzgerald, 1962). The marked improvement in the occlusion of 13 to 14 year old children who were born after fluoridation began is shown in Table 27-4. Among these children, 93 per cent had all first molars present, and 35 per cent had normal occlusion, compared to 65 and 13 per cent, respectively, in the neighboring control city with fluoride-deficient water.

The question is often asked whether adults benefit from water fluoridation. Weaver (1944) suggested that fluoride only postpones the onset of caries, but his study, which has been

TABLE 27-4. AVERAGE NUMBER OF MISSING TEETH AND PERCENTAGE OF CHILDREN WITH NORMAL OCCLUSION IN COMMUNITIES WITH AND WITHOUT FLUORIDATED WATER*

| | Age (years) | No. Children | Missing Teeth per Child | % Children with | |
				All First Molars Present	Normal Occlusion
Newburgh†	13–14	160	0.1	93	35
Kingston‡	13–14	142	0.9	65	13

* Data from Ast et al., 1962.
† Fifteen years of fluoridation.
‡ Fluoride deficient.

TABLE 27-5. MEAN DECAYED, MISSING OR FILLED TOOTH SURFACES (DMFS) FOR 1193 ADULT NATIVE RESIDENTS OF AURORA AND ROCKFORD, ILLINOIS*

| Age (yrs.) | Fluoride ppm in Water Supply | |
	Aurora (1.2)	Rockford (0.1)
20–29	19.8	45.8
30–39	26.8	49.1
40–49	32.8	49.8
50–59	32.5	49.1

* Data from Englander et al., 1964.

widely quoted, was concerned with an atypical group (mothers attending a maternity and child welfare center) rather than with a cross section of the population. Other well-documented studies have demonstrated beyond doubt that the low caries prevalence observed in children living in areas with fluoridated water persists during adult life. Table 27-5 shows the number of decayed tooth surfaces (DMFS) in comparable adult populations living in two cities with 1.2 and 0.1 ppm of fluoride in the water supply and ranging in age from 20 to 59 years (Englander et al., 1964). Those exposed to optimum fluoride had a mean caries score of 27 DMFS compared to 48 DMFS for those in the area with fluoride-deficient water. The difference between the two populations was most marked in the 20 to 29 year olds, but it was highly significant also in the oldest age group, 50 to 59.

It is worthy of note that tooth loss in adults is materially decreased as a result of water fluoridation. For example, among 40 to 44 year olds, Russell and Elvove (1951) found an average of 3 and 12 missing teeth, respectively, in persons who had used water containing 2.6 ppm and only traces of fluoride. It is evident from these and other data that fluoride decreases the prevalence of caries in all age groups and that it also inhibits the growth of carious lesions and thus prolongs the life span of the teeth.

A recent study of young adults showed an indication of decreased prevalence of periodontal pockets as a result of drinking fluoridated water (Englander et al., 1962). Although this finding needs to be confirmed it is in harmony with observations by Leone et al.

(1960), suggesting that use of water-borne fluoride may counteract osteoporosis. No difference was found in the status of oral hygiene or calculus formation in the groups drinking fluoridated and fluoride-deficient water. A beneficial effect on the alveolar bone by the ingested fluoride is therefore not unlikely.

PRENATAL VERSUS POST-NATAL FLUORIDE EXPOSURE

The belief that fluoride ingestion during pregnancy will benefit the dental health of the child is based on the knowledge that calcification of the deciduous teeth is initiated in utero and that fluoride will pass through the placenta to the fetus. In addition, the clinical observation that children born in the same year that fluoridation of water was initiated had greater caries experience than children of the same age group in the same community, born after the water supply had been fluoridated for one or more years, has been interpreted as evidence that prenatal exposure to fluoride is beneficial (Blayney, 1964). However, a large percentage of the children born in the same years that fluoridation started might have had their first fluoride exposure several months before or after birth, making any interpretation of pre- and postnatal effects doubtful.

It should be noted that the only study reported in the literature which specifically included children whose prenatal and postnatal exposure to fluoride was known showed no dental benefits from prenatal fluoride (Carlos et al., 1962). No difference in caries experience was observed in this study in deciduous teeth of 6 year old children who had used fluoridated water anywhere from birth up to three months of age and children of the same age who had been exposed to fluoride both pre- and postnatally. Relative to these findings, mention should be made of a recent study which demonstrated that the mineralization of deciduous teeth is not so advanced at birth as is generally believed (Reiss, 1961). The findings in this study, given in Table 27-6, showed that the amount of mineral present in the crowns of different deciduous tooth types at birth ranged from only 5 to 36 per cent of that of completely formed crowns. The

TABLE 27-6. EXTENT OF MINERALIZATION OF DECIDUOUS TEETH AT BIRTH*

	No.	Percentage of Crown Calcified at Birth
Upper central incisors	69	36
Other incisors	106	31
Cuspid	50	6
First molar	50	17
Second molar	50	5

*Data from Reiss, 1961.

TABLE 27-7. EFFECT OF WATER CONTAINING 3 PPM FLUORIDE USED FOR TWO YEARS ON DMF OF TEETH INITIALLY ERUPTED AND CARIES-FREE*

	Years Erupted at Start of Fluoride Exposure	Percentage New DMF	
		Fluoride Exposed	Control
Second molar	0	16	47
Second bicuspid	1–2	7	10
First molar	4–6	35	38

*Data from Klein, 1946.

deciduous molars which represent the greatest caries risk are mineralized only in the tips of the cusps in the prenatal period.

Although available evidence favors the view that prenatal fluoride exposure is unimportant from a dental point of view, additional studies carefully designed to distinguish between groups exposed to fluoride before and after birth are needed.

PRE-ERUPTIVE COMPARED TO POSTERUPTIVE FLUORIDE EXPOSURE

That appreciable resistance to caries can be derived from pre-eruptive use of fluoridated water may be deduced from the observation that children who consumed fluoride-containing water (8 ppm) during the first years of life and then, for a period of 12 years, used a water supply which contained only traces of fluoride, had less than half the DMF teeth of children who were born and raised on fluoride-deficient water (Dean et al., 1941). Deatherage (1943) in a study of military service men from Illinois was able to confirm that use of fluoridated water during only the first eight years of life resulted in a substantial reduction in adult caries. His findings also demonstrated an additional, but lesser, reduction in caries in persons who lived continuously in fluoridated areas beyond the age of eight years. From these findings it was concluded that a significant decrease in dental caries is derived from use of fluoridated water, not only in the pre-eruptive period but, to a lesser extent, also from such use after the teeth have erupted.

In regard to the posteruptive effect of fluoride Klein (1946) concluded that, among the teeth present in the mouth at the beginning of the exposure to fluoride, those most recently erupted were those most protected against caries attack. The data in Table 27-7, taken from Klein's study, were obtained from children who had been drinking water containing 3 ppm of fluoride for two years from the age of 10 to 12. The second molars, which were erupting when the fluoride exposure began, show marked reduction in caries, but the first molars which had been erupted for four to six years before being exposed to fluoride were not benefited. These findings were confirmed more recently by Backer-Dirks (1961), who found greater caries reduction after combined pre- and posteruptive fluoride exposure, and greater reduction in teeth which were erupting at the time fluoridation was introduced than in teeth which had been erupted for one or two years. He also reported greatest posteruptive fluoride effect in bucco-lingual surfaces, lesser effects in proximal surfaces, least effects in occlusal fissures, suggesting that posteruptive increase in resistance to caries is related to the accessibility of the enamel surface to external fluoride.

No significant inhibition of caries has been observed in the deciduous dentition as a result of posteruptive use of fluoridated water (Russell and White, 1961). Only children under two years of age at the time fluoridation began showed any decrease in carious deciduous teeth, but the permanent teeth benefited in all age groups studied. Evidently, fluoridated water must be used during the first year of life in order to bring about a reduction in carious deciduous teeth. It is possible that a study concerned with DMF surfaces, rather than DMF teeth, will reveal a posteruptive fluoride effect, but such data are not as yet available in the case of deciduous teeth.

FLUORIDE SUPPLEMENTATION BY VEHICLES OTHER THAN COMMUNAL WATER

Since the discovery that 1 ppm of fluoride in the water supply represents the optimal level in regard to caries inhibition, water has been considered the most suitable vehicle for fluoride supplementation. The fluoridation of a water supply is a duplication of nature's way of providing necessary fluoride. It is inexpensive, the fluoride will reach all the people in the community, and toxic effects are excluded by the self-limiting consumption of water. However, since only a portion of the world population has communal water supplies, other means of supplying fluoride have been sought, the most important ones being supplementation through table salt or milk or by tablets or drops. In controlled water fluoridation the question of dosage need not be answered since the individual variation in water consumption involves no hazards, and the amount of fluoride ingested is known to be optimal to the population at large. In contrast, successful fluoride supplementation by other means must be based on knowledge of what constitutes the optimal dosage of fluoride in different age groups, and at what time fluoride supplementation should be initiated.

According to available evidence, the critical period for development of mottled enamel is from birth to about six years of age. The enamel of all anterior teeth and the first molars and most of the enamel of the premolars calcifies during this period. Therefore, since mottled enamel can form only in the early years, the dosage is not critical for older children and adults but should be carefully regulated from birth up to about six years of age. Attempts to determine the optimal dosage have been made by estimating the amounts of fluoride ingested from water containing 1 ppm fluoride. Unfortunately, there is considerable uncertainty with regard to water consumption by children. We are concerned here with consumption of tap water and not with the water present in milk, carbonated beverages and solid foods, which usu-

TABLE 27-8. CALCULATED WATER CONSUMPTION OF CHILDREN[*]

Age (yrs.)	Body Weight (kg.)	Water Consumption in ml.	
		Range	Mean
1–3	8–16	390–556	470
4–6	13–24	520–745	620
7–9	16–35	650–930	790
10–12	25–54	810–1165	990

[*] Data from McClure, 1943.

ally contains minimal amounts of fluoride. Such a distinction has not always been made. Indeed, accurate quantitative knowledge based on measurements is greatly lacking in this very complex field of water consumption. McClure (1943) calculated the water consumption by children from available data of caloric requirements and body weight. His estimates of average daily water consumption in different age groups are given in Table 27-8; these have been used for determining fluoride requirements. According to these data, the daily supplement to children living in areas with fluoride-deficient water should be in the order of 0.5 mg. for one- to three year old children, increasing to 1.0 mg. in the age group of 10 to 12 years.

Data obtained from records of food and fluid intake of children show that the consumption of tap water is less and varies more than is suggested by McClure. Neuman (1957), in a study of New York City children under six years of age, found that the total daily fluid requirement averaged aproximately 1200 ml. for the one year olds and 1500 for the five year age group. Of the fluids consumed, only about 25 per cent was tap water, while over 70 per cent was dairy milk. There were marked variations in fluid intake, particularly in regard to water.

In a recent investigation which included 797 children residing in four states in the United States (Florida, Georgia, New Mexico and Michigan), the water intake also proved to be consistently less than one-half the total daily fluid intake in all age groups (Walker et al., 1963). The total fluid intake increased with age, but this increase was represented only in part by tap water. The data on water intake obtained by different investigators agree well and suggest that 0.3 to 0.4 mg. of fluoride is provided per day to children up to six years of age from fluoridated water.

The Council on Dental Therapeutics of the American Dental Association (1958) has suggested use of bottled water containing 1 ppm of fluoride as most feasible for fluoride supplementation during the first two years of life. For children between two and three years of age, a dosage of 1 mg. every other day is recommended, and for children older than three years, 1 mg. each day. In a recent editorial in a leading pediatric journal, a daily dosage of 0.5 mg. of fluoride was advocated for children up to the age of three years, and twice that level for children who had passed their third birthday (Schlesinger, 1963).

NATURAL FLUORIDES IN FOODS

Food-borne fluorides are as assimilable as water-borne fluorides but constitute a very small part of the total daily fluoride intake. Foodstuffs such as meat, eggs, vegetables, cereals and fruits have an average fluoride content of 0.02 to 0.70 ppm. Milk has a very negligible concentration of fluoride (about 0.1 ppm) and, like other body fluids, the level is virtually unaffected by the extent of fluoride ingestion by the cow. Fish may be relatively high in fluoride. For example, canned salmon, sardines and mackerel have been reported to contain 7 to 12 ppm. If tea, which has 75 to 100 ppm fluoride, were to become more popular as a beverage, it could conceivably contribute considerable fluoride to the dietary, for each cup contains approximately 0.12 mg. In areas in which water supplies contain negligible amounts of fluoride, approximately 0.2 to 0.6 mg. of fluoride is consumed daily from food (Armstrong, 1942; McClure, 1949).

FLUORIDE PILLS AND LOZENGES

The caries-inhibiting effect of fluoride pills and lozenges has been studied extensively in young schoolchildren. In most of these studies 0.5 to 1.0 mg. of fluoride per day was given beginning in the first or second grades. All investigators observed a reduction in caries in the range of 20 to 40 per cent after two or more years of fluoride ingestion.

It is worthy of note that Bibby et al. (1955) found 30 per cent inhibition of caries after only one year when the fluoride (1 mg. per day) was administered in lozenges which were dissolved in the mouth before swallowing, while no effect was obtained after the same period when fluoride was given in pills. These findings suggest that the local effect of supplementary fluoride on the tooth surface is of importance and that a method of administration which permits maximal contact of the fluoride with the teeth before swallowing may be most beneficial. Some support for this view may be derived from the fact that a greater inhibition in caries generally has been reported in studies when the children were instructed to dissolve the fluoride pills in the mouth before swallowing. However, controlled studies designed to test this concept are warranted.

Investigators agree that it is advantageous to begin taking pills at the earliest possible age. Thus, a study in Poland showed 26 per cent reduction in caries of the deciduous teeth of kindergarten children who were given fluoride for two years from the age of three, but there was no effect in the deciduous teeth of those children who started taking pills at the age of four or five years. In the latter group, the permanent teeth which erupted during the study did show caries inhibition. In another European study caries inhibition of 27, 24 and 20 per cent were observed in schoolchildren who began fluoride supplementation in first, second and third grades, respectively. These findings agree well with the observation made in the previously mentioned studies concerned with water fluoridation, viz., that exposure to fluoride during and soon after tooth eruption will provide protection against caries.

It is interesting that the studies in which fluoride was given to the children only on school days (about 200 days a year) have shown results comparable to those in which fluoride was provided throughout the year. Controlled studies are needed which are designed to determine which method of administration—every day or only a limited number of days—is preferable.

The effectiveness of continuous fluoride

administration in the pre-eruptive as well as the posteruptive period is demonstrated in the study of Arnold et al. (1960). Nearly two-thirds of the children in this study began to take fluoride tablets before their third birthday, and almost all before their sixth birthday. The number of years of ingestion of fluoride ranged from 1 to 15. No nonfluoride exposed controls were included, but comparison of the caries incidence with that found in Grand Rapids before and after fluoridation in similar age groups suggested that the fluoride pills produced a caries inhibition similar to that obtained from water fluoridation.

No detrimental effects from the supplemental fluoride have been reported in any of the studies cited. The dosage employed, from 0.5 to 1.0 mg. per day, must therefore be considered safe, provided, of course, that the drinking water contains only traces of fluoride. The study of Arnold et al. (1960) is of particular interest in regard to dosage because fluoride was taken by most of the children during and following the period of formation of the permanent teeth. Infants up to two years of age were given fluoride by incorporating 1 ppm of fluoride in all water they consumed. This was accomplished within the family by adding 1 mg. of fluoride (1 pill) to a quart of tap water. Children between two and three years swallowed 1 pill containing 1 mg. of fluoride every other day, and all children older than three years took 1 pill every day. On this regimen, four of 32 children appeared to have dental fluorosis, three instances of which were questionable, and one very mild. None of these cases was esthetically objectionable. The percentage of fluorosis in this study parallels that observed in Grand Rapids after water fluoridation (Arnold et al., 1962).

Considering all the evidence, it may be concluded that supplementation of fluoride intake by pills, drops or lozenges will cause a significant decrease in dental caries and that organized programs of administration in schools, beginning in kindergarten, are advisable in communities which are unable to fluoridate the water supply. The use of pills in households is also advisable but, unfortunately, experience has shown that even in highly educated groups, only about one-half the families continue giving their children

pills for the necessary number of years (Arnold et al., 1960).

FLUORIDATION OF KITCHEN SALT

Fluoridated kitchen salt containing 90 mg. of fluoride per kg. of salt has been available in certain parts of Switzerland since 1955. According to two progress reports, use of this salt will cause some reduction in caries of the permanent teeth of children, but the results are inferior to those obtained from water fluoridation. Table 27-9 lists the percentage reduction obtained in nine year old children of caries in different types of tooth surfaces. Included, also, are the corresponding data obtained after five and a half years of water fluoridation in Holland. Although these findings may not be strictly comparable, they do demonstrate the greater effectiveness of water fluoridation over salt fluoridation and suggest that the amounts of fluoride incorporated in the salt are below the optimal level. With a concentration of 90 mg. of fluoride per kg. of salt, an intake of 11 gm. of salt will provide 1 mg. of fluoride. The mean individual salt consumption in families with children in Switzerland varies between 3.6 and 6.8 gm. per day (Marthaler, 1961). It is now believed that the dosage of 90 mg. fluoride per kg. of salt should be at least doubled and that a level of 200 or 250 mg. per kg. may be required to produce the inhibition of caries achieved by water fluoridation.

Fluoridated kitchen salt can be manufactured and made available to the total population at low cost. This method of fluorida-

TABLE 27-9.   CARIES REDUCTION IN PERMANENT TOOTH SURFACES OF NINE YEAR OLD CHILDREN WHO USED FLUORIDATED KITCHEN SALT OR FLUORIDATED WATER FOR 5.5 YEARS

	Percentage Reduction in Carious Surfaces		
	Proximal	*Bucco-labial*	*Occlusal*
Fluoridated salt*	41	32	21
Fluoridated water†	66	66	28

* Data from Marthaler and Schenardt, 1962.
† Data from Backer-Dirks et al., 1961.

tion, therefore, seems suitable in areas where fluoridation of the water supply is impractical. However, the optimal concentration of fluoride in the salt must be determined in long-range clinical studies before it can be considered seriously. It should be noted that children use virtually no salt during the first two years of life and that deciduous teeth would receive little or no fluoride from fluoridated salt during their formation. Since the permanent dentition is calcified for the most part after the age of two, theoretically it could benefit as much from fluoridation of table salt as of water.

FLUORIDATION OF MILK

Milk has been suggested as a vehicle for fluoride because it is the food used most universally by infants and children during the period of tooth formation. To date only one study, involving a small group of children (65 experimental and 64 controls), has been reported in the literature (Rusoff et al., 1962). The children were served one pint of homogenized milk fortified with 1 mg. of fluoride as sodium fluoride each day at lunch, beginning at the age of six, for a period of three years. There was a significant reduction in caries in the teeth which erupted during the study, the results being comparable to those which have been obtained from water fluoridation. Eighteen months after cessation of the fluoride ingestion, a significant difference in caries rate was still retained in favor of the treated group, suggesting a prolonged fluoride effect.

Although the caries inhibition obtained with fluoridated milk shows that the fluoride added to the milk is available for uptake by the teeth, it is of interest that fluoride in milk is absorbed at a slower rate than fluoride ingested in water (Ericsson, 1958). This may be because it combines with milk calcium to form the relatively insoluble calcium fluoride or because it complexes with the organic matter, principally casein (Ericsson, 1958). A portion (about 25 per cent) of the 0.1 ppm fluoride normally occurring in cow's milk is also bound to casein.

As in the case of salt fluoridation, well-controlled, long-term clinical experiments must precede serious consideration of milk as a vehicle for fluoride.

MECHANISM OF ACTION OF FLUORIDE IN CARIES INHIBITION

The fact that exposure to both water-borne fluoride during childhood only and to topically applied fluoride will cause caries reduction demonstrates that the anticaries action of fluoride is related to its accumulation in the enamel. Since dissolution of mineral is a predominant phase of the carious lesion, and the dissolution is brought about by bacterial metabolites, fluoride could inhibit caries by increasing the resistance of the enamel to demineralization, or by exerting antibacterial effects. In the following discussion the deposition of ingested fluoride in enamel will be considered first, followed by the possible role of the antibacterial and solubility effects of enamel fluoride.

CHEMISTRY OF ENAMEL AND ITS REACTION WITH FLUORIDE

Enamel mineral consists primarily of relatively impure hydroxyapatite or carbonate apatite containing, on a dry basis, about 36 per cent calcium and 17.5 per cent phosphate, with a calcium to phosphorus ratio of approximately 2.08. X-ray and electron microscope studies show that the enamel crystals have an apatite structure and measure on the average 3000 to 5000 Å in length and 500 to 1200 Å in width (Frank et al., 1960). Gas adsorption measurements indicate a surface area of 20 to 30 square meters per gram compared to about 100 for dentin and 150 or more for bone. The unit cell, or smallest repeating unit in the crystals, is $Ca_{10}(PO_4)_6(OH)_2$. All these ions may be partially replaced by other ionic species, particularly on the surface. The sodium may replace surface calcium (Neuman and Neuman, 1958); carbonate may replace phosphate in both interior and surface positions (Trautz, 1960). As a result of such substitutions, a great number of different elements are present in tooth mineral in small concentrations, which vary according to the composition of food and drinking water. Mean values for some of these minor constituents, as compiled from the literature, are 2.5 per cent carbon dioxide, 0.77 per cent sodium, 0.42 per cent magnesium and 0.25 per cent chloride.

Of major importance is the substitution of fluoride for the hydroxyl group. Since the fluoride ion is only slightly smaller than hydroxyl ions, this ionic exchange can take place without otherwise disturbing the lattice structure (Neuman et al., 1950). The reaction may be expressed by the equation:

$$Ca_{10}(PO_4)_6(OH)_2 + 2F^- \longrightarrow$$
$$Ca_{10}(PO_4)_6F_2 + 2\,OH^-$$

Fluoride may also be acquired by dissolution of calcium and phosphorus and reprecipitation with extraneous fluoride, or fluoride derived from dissolved material, as fluorapatite. This process appears to be of considerable importance in the incipient carious lesion, as will be discussed later. When the hydroxyls are completely replaced by fluoride, fluorapatite is formed (3.8 per cent F). Complete saturation of enamel with fluoride does not occur in vivo. The acquisition of fluoride is confined almost entirely to the surface of the crystals and does not involve the body of the crystals to any significant extent. As a result, relatively low concentrations of fluoride are capable of saturating the crystal surface and thus give the entire crystal chemical properties approaching those of fluorapatite.

UPTAKE OF FLUORIDE

Since some fluoride is always present in water and food, even in low fluoride areas, it is always found in teeth and bones. In the early studies in this field, fluoride determinations were made on the total enamel and total dentin of teeth obtained from various communities with different levels of fluoride in the drinking water. A remarkable correlation between the mean fluoride levels in the different tooth groups and the concentrations of fluoride in the drinking waters was observed (McClure and Likins, 1951). As the fluoride level in the drinking water increased from 0.1 ppm to 1.0 ppm and then to 7.6 ppm, the mean fluoride value in the enamel rose from

90 ppm to 135 ppm and to 660 ppm. Corresponding values in the dentin were 200 ppm, 360 ppm and 1300 ppm fluoride. The higher levels in dentin reflect the smaller crystal size, since most of the uptake is on the surface of the apatite crystals.

It was difficult to relate this slight increase in the fluoride content of the enamel (about 50 ppm) with the very significant reduction in dental caries observed with increase in the water fluoride from 0.1 to 1.0 ppm. However, this difficulty was resolved when it was found that the concentration of fluoride in surface enamel is several times greater than in the enamel as a whole and that water-borne fluoride markedly increases the deposition of fluoride in the outermost enamel layers (Brudevold et al., 1956; Isaac et al., 1958). This process begins while the enamel is calcifying and continues through the pre- and posteruptive life of the tooth. There are three stages in the deposition of fluoride. First, small amounts, reflecting the low levels of fluoride in tissue fluids, are incorporated into the enamel crystals while they are being formed. After the enamel has been laid down, deposition continues in the surface enamel. Diffusion of fluoride from the surface inward is restricted to the outer portion, since the body of the enamel becomes inaccessible as a result of high mineralization. After eruption, the surface enamel acquires fluoride from water, food, supplementary fluoride and saliva.

The uptake of fluoride is known to depend on the amount of fluoride ingested and on the time of fluoride exposure. This is summarized in Figure 27-2, which shows the rate of fluoride uptake by the surface enamel (Brudevold et al., 1960). The data were obtained from analyses of a great number of teeth from patients of different ages from areas with 0.1, 1.0, 3.0 and 5.0 ppm fluoride in the water supply. The rate of uptake in surface enamel is very rapid during the first years of its existence and then tapers off to a level closely related to the fluoride concentrations in the drinking water. The higher the level in water, the greater is the initial rate of deposition. Hence, in areas with fluoridated water protection against caries is acquired at the critical time of tooth eruption.

The marked increase in the deposition of

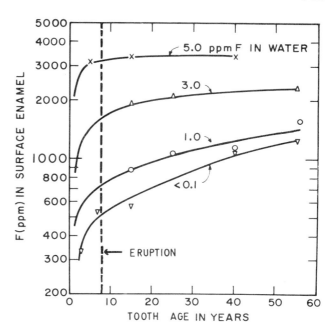

FIGURE 27-2. *Parts per million of fluoride in ash of surface enamel of teeth from areas with different levels of fluoride in the water supply. The dotted vertical line represents the time of eruption. (From Brudevold et al., 1960.)*

fluoride in surface enamel with increased levels of fluoride in the drinking water is shown in Table 27-10 (Isaac et al., 1958). As the level of fluoride in the water supply increases from 0.1 ppm to 1.0, the fluoride concentration in the surface enamel increases from about 500 to about 900 ppm. At 5.0 ppm of fluoride in the water the enamel surface concentration increases to over 3000 ppm. The relatively slight increase, from 500 to 900 ppm, must account for the caries inhibition brought about by water fluoridation. That enamel acquires caries resistance by such a small fluoride increase is due to its crystalline structure. As previously mentioned, fluoride deposition in the fully formed enamel crystal is confined almost entirely to the surface and does not involve the body of the crystal. Since the surface area of enamel crystals is small and only partly accessible, minute quantities of fluoride will fill available surface positions. These strategically placed surface ions, although few in number, affect the properties of the entire crystal. The reaction between the crystal surface and the low concentrations of fluoride in body fluids, food and drinking water progressively changes the composition of the surface from hydroxyapatite to fluorapatite.

As already stated, this reaction is limited

TABLE 27-10. PARTS PER MILLION OF FLUORIDE IN SUCCESSIVE LAYERS OF ASHED ENAMEL OF PERSONS UNDER 20 YEARS OF AGE AND LIVING IN COMMUNITIES WITH 0.1, 1.0 OR 5.0 PPM FLUORIDE IN THE WATER SUPPLY*

Enamel Layer	0.1 ppm Fluoride (Buffalo, N.Y.)	1.0 ppm Fluoride (McKinney, Texas)	5.0 ppm Fluoride (Post, Texas)
1	499	889	3370
2	162	363	1710
3	108	255	1124
4	63	171	926
5	76	133	811
6	42	129	570

* Data from Isaac et al., 1958.

almost entirely to the crystals located in the surface of intact enamel. However, fluoride does penetrate the partly demineralized enamel of intial carious lesions. Owing to porosity, fluoride diffuses freely into these lesions and reacts with the residual apatite crystals. Considering the increased flow of ions, the finding in radioactive studies that initial carious lesions may take up ten times more fluoride than adjacent sound enamel is not surprising. Nor is it surprising that fluoridated

TABLE 27-11. CARBON DIOXIDE AND FLUORIDE (PER CENT) IN ENAMEL LAYERS OF TEETH
FROM AREAS WITH DIFFERENT LEVELS OF FLUORIDE IN THE DRINKING WATER*

Fluoride ppm in Water Supply	Layers					
	First		Second		Third	
	Fluoride	CO_2	Fluoride	CO_2	Fluoride	CO_2
0.1	.050	1.66	.006	1.88	.011	2.04
1.0	.066	1.55	.023	1.71	.012	1.97
1.6	.097	1.58	.036	1.70	.026	1.78
2.5	.156	1.30	.087	2.27	.057	2.76
5.0	.337	1.79	.171	1.90	.113	2.08

* Data from McCann, Rasmussen and Brudevold, 1964.

water retards the growth of carious lesions (Backer-Dirks, 1961).

SECONDARY CHANGES IN ENAMEL

The possibility has been suggested that enamel, like bone, may have secondary changes in composition associated with the deposition of fluoride and that such changes may affect the physical properties of the mineral. Studies by Zipkin et al. (1960) showed lower carbonate and citrate and higher magnesium in the bones of persons residing in areas with fluoridated water. These findings, however, apparently do not apply to enamel. Thus, even the teeth of rats given water containing 100 ppm fluoride had a carbonate content identical to the teeth of those given no fluoride, although the bones of the fluoride-treated animals contained considerably less carbonate (McCann and Bullock, 1957).

Findings on human teeth are shown in Table 27-11. It is apparent that the carbonate level is not related to fluoride ingestion. Although the carbonate is always lower and the fluoride always higher in the outermost layer, and the carbonate concentration increases while the fluoride decreases in the underlying layers, they are independent phenomena. The carbonate remains relatively constant regardless of the amount of fluoride uptake. Although carbonate may be a factor in enamel solubility, caries resistance cannot be explained by a fluoride-carbonate relationship.

Magnesium was present in increased concentrations in the enamel of rats given very high concentrations of fluoride (100 ppm) in the drinking water (McCann and Bullock, 1957). However, no such relationship has

been observed in human teeth in which the fluoride exposure is, of course, much lower. Comparison of human enamel from patients in areas which have water with high and low amounts of fluoride has shown no differences in magnesium content (Ockerse, 1943). Recent observations also suggest that there is no relationship between deposition of fluoride and magnesium in human enamel (Table 27-12).

Although no data are available in regard to a possible relationship between fluoride and citrate in enamel, the low concentrations present (0.1 per cent) and the observation that the distribution of citrate closely follows that of the organic material tend to preclude any association. Since the deposition of fluoride in enamel is independent of the deposi-

TABLE 27-12. MAGNESIUM IN OUTER
ENAMEL LAYER*

(Teeth from areas with different levels of fluoride in the water)

Fluoride in Water (ppm)	Fluoride in Enamel Layer (per cent)	Magnesium in Enamel Layer (per cent)
1	0.027	0.196
	0.045	0.203
	0.066	0.226
	0.484†	0.294
1.5	0.095	0.190
	0.096	0.183
	0.100	0.282
	0.250	0.180
2.5	0.113	0.262
	0.140	0.270
	0.156	0.266
	0.250	0.185

* Data from McCann and Brudevold, 1965.
† Probably treated with topical fluoride.

tion of carbonate, magnesium and citrate, the inhibition of caries observed in areas with fluoridated water is attributable to the fluoride directly rather than to secondary changes in enamel composition.

INHIBITION
OF BACTERIAL
ACID PRODUCTION

According to Lilienthal and Martin (1956), much larger concentrations of fluoride ions are required to inhibit acid production in plaque (30 to 40 ppm) or to stop bacterial growth (250 ppm) than are likely to be present in the oral environment. Salivary fluoride rarely exceeds 0.1 ppm, and the concentration is similar in persons living in areas with only traces and up to 3 ppm of fluoride in the water supply (McClure, 1941). In contrast to saliva, appreciable amounts of fluoride may accumulate in dental plaque. A recent study showed mean concentrations of 6 ppm (wet weight) and maximal concentrations of 60 ppm (Hardwick, 1961). Although it is not known in what form this fluoride is present, there are good reasons for believing that only a minor fraction is ionized. Most of the fluoride is likely to be combined with calcium phosphates, which accumulate rapidly in plaque material (Schroeder, 1963). Jenkins (1959) has proposed that acid formed in the plaque could release fluoride from plaque on the enamel surface, thus providing the concentrations needed to inhibit additional production of acid. He observed that the same concentration of fluoride caused increasing inhibition with decrease in pH, such small concentrations as 6 ppm being effective at pH 5. However, there is no evidence that fluoride is dissolved in such high concentrations at these pH's. On the contrary, fluoride is taken up by enamel mineral extremely fast at acid pH, leaving only traces in solution. A recent study by Helstrom (1960) has shown that the rate of clearance of fluoride from the mouth is rapid and that the concentrations required to inhibit salivary acid production are present for only a few minutes after a mouth rinse with 0.1 per cent fluoride solution. All this evidence makes it unlikely that the inhibiting action of fluoride on acid

production is responsible for its caries-reducing effect.

EFFECT ON DISSOLUTION
OF TOOTH MINERAL

Numerous studies have shown that deposition of fluoride in the enamel will retard acid dissolution of the enamel mineral. This ability of fluoride to slow down the dissolution process is commonly believed to be its mode of action in inhibiting caries. This is a logical concept in view of the fact that the demineralizing acids in carious lesions are produced intermittently and are capable of decalcifying tooth mineral for only a short period before being neutralized by saliva. Because of the rapid and reversible change in pH in carious lesions, it would be reasonable to expect that retarded decalcification would slow down or inhibit the carious process. However, a great number of ionic species, some of which are present normally in the enamel, have the same or greater capacity than fluoride to retard acid dissolution (Manly and Bibby, 1949); and yet, among these solubility-reducing agents, only fluoride has been shown to reduce caries. In this category, ineffective agents against caries include stannous tin, lead, zinc and zirconium salts. These salts react with the enamel to form insoluble phosphates or hydroxides, which act as barriers against ionic diffusion and slow down the dissolution process in test tube experiments. They do not, however, reduce caries when applied topically, according to several clinical studies. It is apparent, therefore, that caries will develop regardless of the presence in enamel of diffusion barriers in the form of extraneous precipitates and that fluoride must counteract dissolution of tooth mineral by a mechanism which is unique and different from the one of interfering with ionic diffusion. Recent studies suggest that this is indeed the case.

It has already been mentioned that fluoride is present in enamel as fluorapatite. It is incorporated in the apatitic crystal lattice as an integral part and does not occur as a heterogeneous precipitate. Fluorapatite is known to be less soluble in acid than hydroxyapatite, but its solubility is directly dependent on the concentration of fluoride in the solution. As

TABLE 27-13. CALCIUM PHOSPHATE
COMPOUNDS FOUND IN CALCULUS

	Formula	Molar Ca/P Ratio
Brushite	$CaHPO_4 \cdot 2H_2O$	1:00
Octacalcium phosphate	$Ca_8H_2(PO_4)_6$	1:33
Whitlockite	$Ca_3(PO_4)_2$	1:50
(Always contains Mg substituted for some Ca—up to 10 atom per cent)		
Hydroxyapatite	$Ca_{10}(PO_4)_6(OH)_2$	1:67

previously mentioned, surface enamel from a 1 ppm fluoride area contains about 900 ppm fluoride compared to 500 ppm in surface enamel from a low fluoride area. Although most of this fluoride is located at the surface of the apatite crystals, it does not surface saturate the crystals with fluorapatite and, according to theoretical calculations, the solubility should be reduced only by approximately 10 per cent.

Other components of the enamel mineral which increase its solubility, including carbonate and magnesium, could obscure such a minor effect. Actual findings from solubility studies on powdered enamel originating from high and low fluoride areas tend to substantiate the view that solubility differences are relatively minor. A relationship between concentrations of fluoride in layer samples and rate of acid dissolution was consistent in regard to surface-subsurface enamel from the same tooth group but did not always apply when different tooth groups were compared (Isaac et al., 1958).

More pertinent is a study on the rate of dissolution of phosphate from normal surfaces of deciduous intact teeth of children from geographic areas with 0.25, 1.4 and 2.0 ppm fluoride in the drinking water (Jenkins et al., 1962). While less phosphate leached away from the teeth of children from areas with fluoridated water, the difference was significant only in those drinking water with 2 ppm. The same trend was found in permanent teeth, but the difference was not significant even though permanent teeth receive much greater protection from fluoride than do deciduous teeth. It seems unlikely in view of all this evidence that reduced solubility as measured in vitro is the main cause of caries reduction. Some

additional factor responsible for the unique action of fluoride must be found.

In the carious process, loss of mineral is not simply a case of dissolution. It is known that certain components are preferably dissolved in the carious lesion, e.g., carbonate, magnesium, sodium (Johansen, 1963). Others, including fluoride, zinc, lead, aluminum and iron, accumulate in the mineral residue. In addition, since saliva is supersaturated in respect to enamel (Vogel, et al., 1965), mineralization can occur in the mouth at normal pH, as is evidenced by the formation of calculus. Partly demineralized enamel may also remineralize. This is suggested by a microscopic study of replicas of affected enamel taken at intervals over a prolonged period of time (Mannerberg, 1960), and by a polarizing microscopy and microradiograph study which showed that saliva can deposit mineral in enamel (Johansson, 1965). Considering the relatively small and intermittent production of acid in the dental plaque, deviation from equilibrium in carious lesions must be slight, and an alternating process of demineralization and reprecipitation may be assumed to take place. The crucial role of fluoride in this environment appears to be that of favoring reprecipitation of apatite.

Under various conditions and particularly in the acid pH range, a number of calcium phosphate compounds may form in addition to apatite. In Table 27-13 these compounds are listed along with their chemical formulas and calcium to phosphate ratios. The lower the calcium-phosphate ratio, the less basic and, in general, the more soluble the compound. All these materials are commonly present in dental calculus and one of them, whitlockite, has been observed in carious lesions (von Vahl, et al., 1964). From what is known of the formulation of the salts listed in Table 27-13, it is likely that all of them, and not only whitlockite, will precipitate in the acid carious environment. In the presence of sufficient amounts of fluoride, however, only apatite can form. Thus, Newesely (1960) observed that as little as 10 to 100 μg. of fluoride per liter prevented the formation of octacalcium phosphate. Grøn and Messer (1964) showed that brushite and whitlockite are converted to apatite by small amounts of fluoride at pH levels as low as 2.5. In addition to the ability of fluoride to induce formation of apatite under conditions which in the absence of fluoride would favor formation of more acid and more soluble compounds, other and unique effects of fluoride on the calcium phosphate system have also been observed.

Increase in recrystallization of apatite in the presence of fluoride was demonstrated in radioactive studies by McCann and Fath (1958). Brudevold et. al. (1961), showed that as little as 0.2 ppm of fluoride significantly increased the rate of precipitation of hydroxyapatite from supersaturated solutions of calcium and phosphate in the pH range of 6.2 to 7.4. Zipkin et al. (1962) and Eanes et al. (1965) found an increase in apatite crystal size in human and rat bone by the in vivo action of fluoride. McCann (1966) has observed that small amounts of fluoride maintain the apatite structure of the crystal surface from neutral pH to a pH below 6. Even with very limited concentration of fluoride, the apatite form is maintained. Hydroxyl ions rather than fluoride ions predominate in the crystal structure, but fluoride favors the formation of the apatite lattice. In the absence of fluoride and at acid pH the surface structure of apatite crystals readily changes to other lattice forms such as octacalcium phosphate.

All these effects are different facets of one basic mechanism, involving the great affinity of fluoride for calcium and phosphate, and the fact that these ions can combine only in the spacial arrangement of the apatite lattice. The presence of relatively low concentrations of fluoride in a crystal maintains this stable structure even at relatively low pH.

The effect of fluoride in combatting dental caries is thus seen to be a much more complex and specific mechanism than simple solubility reduction. It is the result of the unique relationship of fluoride to the dental hard tissues which no other substance can replace.

REFERENCES

Armstrong, W. D., and Knowlton, M.: Fluorine derived from food. J. Dent. Res., 21:326, 1942.

Arnold, F. A., Jr., Likins, R. C., Russell, S. L., and Scott, D. B.: Fifteenth year of the Grand Rapids fluoridation study. J. Amer. Dent. Assn., 65:780, 1962.

Arnold, F. A., Jr., McClure, F. J., and White, C. L.: Sodium fluoride for children. Dent. Prog., 1:12, 1960.

Ast, D. B., Allaway, N., and Draker, H. L.: Prevalence of malocclusion, related to dental caries and lost first molars, in a fluoridated city, and a fluoride deficient city. Amer. J. Ortho., 48:106, 1962.

Ast, D. B., Finn, S. B., and Chase, H. C.: Newburgh-Kingston caries fluorine study. III. Further analyses of dental findings including the permanent and deciduous dentitions after four years of water fluoridation. J. Amer. Dent. Assn., 42:189, 1951.

Ast, D. B., and Fitzgerald, B.: Effectiveness of water fluoridation. J. Amer. Dent. Assn., 65:581, 1962.

Backer-Dirks, O.: Bias in clinical testing of caries-preventive dentifrices. In Caries Symposium, Zurich. Berne and Stuttgart, Hans Huber, 1961, pp. 37-47.

Backer-Dirks, O., Houwink, B., and Kwant, G. W.: Some special features of the caries preventive effect of water fluoridation. Arch. Oral Biol., 4:187, 1961.

Bibby, B. G., Wilkins, E., and Witol, E.: A preliminary study of the effects of fluoride lozenges and pills on dental caries. Oral Surg., 8:213, 1955.

Blayney, J. R.: Evanston dental caries study. XXIV. Prenatal fluorides—value of water-borne fluorides during pregnancy. J. Amer. Dent. Assn., 69:291, 1964.

Brudevold, F., Amdur, B. H., and Messer, A.: Factors involved in remineralization of carious lesions. Arch. Oral Biol. (Spec. Suppl.), 6:304, 1961.

Brudevold, F., Gardner, D. E., and Smith, F. A.: The distribution of fluoride in human enamel. J. Dent. Res., 35:420, 1956.

Brudevold, F., Savory, A., Gardner, D. E., Spinelli. M., and Speirs, R.: A study of acidulated fluoride solutions. I. In vitro effects on enamel. Arch. Oral Biol., 8:167, 1963.

Brudevold, F., Steadman, L. T., and Smith, F. A.: Inorganic and organic components of tooth structure. Ann. N. Y. Acad. Sci., 85:110, 1960.

Carlos, J. P., Gittelsohn, A. M., and Haddon, W.: Caries in deciduous teeth in relation to maternal ingestion of fluoride. Pub. Health Rep., 77:658, 1962.

Council on Dental Therapeutics: Prescribing supplement of dietary fluoride. J. Amer. Dent. Assn., 56:589, 1958.

Dean, H. T.: Endemic fluorosis and its relation to dental caries. Pub. Health Rep., 53:1443, 1938.

Dean, H. T., Arnold, F. A., Jr., and Elvove, E.: Domestic water and dental caries. V. Additional studies of the relation of fluoride domestic waters to dental caries experience in 4425 white children, aged 12 to 14 years, of 13 cities in 4 states. Pub. Health Rep., 57:1155, 1942.

Dean, H. T., and Elvove, E.: Further studies on the minimal threshold of chronic endemic dental fluorosis. Pub. Health Rep., 52:1249, 1937.

Dean, H. T., Jay, P., Arnold, F. A., Jr., and Elvove, E.: Domestic water and dental caries. I. A dental caries study, including L. acidophilus estimations of a population severely affected by mottled enamel and which for the past 12 years has used a fluoride-free water. Pub. Health Rep., 56:365, 1941.

Deatherage, C. F.: Fluoride domestic waters and dental

caries experience in 2026 white Illinois selective service men. J. Dent. Res., 22:129, 1943.

Eanes, E. D., Zipkin, I., Harper, R. A., and Posner, A. S.: Small-angle x-ray diffraction analysis of the effect of fluoride on human bone apatite. Arch. Oral Biol., 10:161, 1965.

Englander, H. R., DePalma, R., and Kesel, R. G.: The Aurora-Rockford, Illinois study. Effects of water having a naturally occurring fluoride on dental health of young adults. J. Amer. Dent. Assn., 65:614, 1962.

Englander, H. R., Reuss, R. C., and Kesel, R. G.: Roentgenographic and clinical evaluation of dental caries in adults who consume fluoridated versus fluoride-deficient water. J. Amer. Dent. Assn., 68:14, 1964.

Fricsson, Y.: The state of fluorine in milk and its absorption and retention when administered in milk. Acta Odont. Scand., 16:1, 1958.

Fluoridation Census. Washington, D. C., U. S. Department of Health, Education, and Welfare, U. S. Public Health Service, 1965.

Frank, R., Sognnaes, R. F., and Stern, R.: Dental aspects of mineralization with special reference in the metastructural pattern of enamel. In Sognnaes, R. F. (ed.), Calcification in Biological Systems. Washington, D. C., American Association for the Advancement of Science, 1960, pp. 163-202.

Galagen, D. J., and Lamson, G. G., Jr.: Climate and endemic fluorosis. Pub. Health Rep., 68:497, 1953.

Grøn, P., and Messer, A. C.: The effect of fluoride and magnesium on the hydrolysis of dicalcium-phosphate. J. Dent. Res., 43:866, 1964. Abstract.

Hardwick, J. L.: The fluoride content of plaque: Its possible anti-enzymatic action. In Caries Symposium, Zurich. Berne and Stuttgart, Hans Huber, 1961, pp. 112-119.

Helstrom, I.: Fluorine retention following sodium fluoride mouth washing. Acta Odont. Scand., 18:263, 1960.

Hill, I. N., Blayney, J. R., and Wolf, W.: Evanston dental caries study. XIX. Prevalence of malocclusion of children in a fluoridated and control area. J. Dent. Res., 38:782, 1959.

Isaac, S., Brudevold, F., Smith, F. A., and Gardner, D. E.: The relation of fluoride in the drinking water to the distribution of fluoride in the enamel. J. Dent. Res., 37:318, 1958.

Jenkins, G. N.: The effect of pH on the fluoride inhibition of salivary acid production. Arch. Oral Biol., 1:33, 1959.

Jenkins, G. N., Armstrong, P. A., and Speirs, R. L.: Laboratory investigations on the relation of fluorine to dental caries on Tyneside. Proc. Roy. Soc. Med., 45:517, 1962.

Johansen, E.: Chapter in Mechanism of Hard Tissue Destruction. Washington, D. C., American Association for the Advancement of Science, 1963.

Johansson, B.: Remineralization of slightly decalcified enamel studied by polarized light microscopy. J. Dent. Res., 44:64, 1965.

Klein, H.: Dental caries (DMF) experience in relocated children exposed to water containing fluorine. II. J. Amer. Dent. Assn., 33:1136, 1946.

Klein, H.: Dental effects of community waters accidentally fluoridated for nineteen years. II. Differences in the extent of caries reduction among the different types of permanent teeth. Pub. Health Rep., 63:563, 1948.

Leone, N. C., Stevenson, C. A., Besse, B., Hawes, L. E., and Dawber, T. R.: The effect of the absorption of F. II. A radiological investigation of 546 human residents of an area in which the drinking water contained only a minute trace of fluoride. A.M.A. Arch. Indust. Health, 21:326, 1960.

Lilienthal, B., and Martin, N. D.: Investigations of the anti-enzymatic action of fluoride at the enamel surface. J. Dent. Res., 35:189, 1956.

Manly, R. S., and Bibby, B. G.: Substances capable of decreasing the solubility of tooth enamel. J. Dent. Res., 28:160, 1949.

Mannerberg, F.: Appearance of tooth surface, as observed in shadowed replicas in various age groups, in long-time studies, after tooth-brushing in cases of erosion and after exposure to citrus fruit juice. Odont. Revy, 11 (Suppl. 6): 1960.

Marthaler, T. M.: Zur Fraze des Fluorvollsalzes: erste klinische Resultate. Schweiz. Msch. Zahn., 71:671, 1961.

Marthaler, T. M., and Schenardt, C.: Inhibition of caries in children after 5½ years of fluoridated table salt. Helv. Odont. Acta, 6:1, 1962.

McCann, H. G.: Solubility relationships between fluorapatite and hydroxyapatite. (To be published.)

McCann, H. G., and Brudevold, F.: In Environmental Variables in Oral Disease. Washington, D.C., American Association for the Advancement of Science, 1966, pp. 103-128.

McCann, H. G., and Bullock, F. A.: The effect of fluoride ingestion on the composition and solubility of mineralized tissues of the rat. J. Dent. Res., 36: 391, 1957.

McCann, H. G., and Fath, E. H.: Phosphate exchange in hydroxylapatite, enamel, dentin and bone. II. Effects of fluoride on the exchange. J. Biol. Chem., 231:869, 1958.

McCann, H. G., Rasmussen, S., and Brudevold, F.: Carbonate and fluoride relationships in human dentin. J. Dent. Res., 43:866, 1964. Abstract No. 289.

McClure, F. J.: Domestic water and dental caries. Fluorine in human saliva. Amer. J. Dis. Child., 62: 512, 1941.

McClure, F. J.: Ingestion of fluoride and dental caries. Quantitative relations based on food and water requirements of children 1 to 12 years old. Amer. J. Dis. Child., 66:362, 1943.

McClure, F. J.: Fluorine in foods: survey of recent data. Pub. Health Rep., 64:1061, 1949.

McClure, F. J., and Likins, R. C.: Fluorine in human teeth studied in relation to fluorine in the drinking water. J. Dent. Res., 30:172, 1951.

Neuman, H. H.: The milk and water intakes of small children. A survey of drinking habits. A.M.A. Arch. Pediat., 74:456, 1957.

Neuman, W. F., and Neuman, M. W.: The Chemical Dynamics of Bone Mineral. Chicago, University of Chicago Press, 1958.

Neuman, W. F., Neuman, M. W., Main, E. R., O'Leary, J., and Smith, F. A.: The surface chemistry of bone. II. Fluoride deposition. J. Biol. Chem., *187*:655, 1950.

Newesely, H.: Darstellung von Oktacalciumphosphat (Tetracalciumhydrogentrisphosphat) durch homogene Kristallisation. Mh Chem. (Monatshefte), *91*: 1020, 1960.

Ockerse, T.: The chemical composition of enamel and dentin in high and low caries areas in South Africa. J. Dent. Res., *22*:441, 1943.

Reiss, L. Z.: Strontium 90 absorption by deciduous teeth. Science, *134*:1669, 1961.

Rusoff, L. L., Konikoff, B. S., Frye, J. B., Johnston, J. E., and Frye, W. W.: Fluoride addition to milk and its effect on dental caries in school children. Amer. J. Clin. Nutr., *11*:94, 1962.

Russell, A. L., and Elvove, E.: Domestic water and dental caries. VI. A study of the fluoride-dental caries relationship in an adult population. Pub. Health Rep., *66*:1389, 1951.

Russell, A. L., and White, C. L.: Dental caries in Maryland children after seven years of fluoridation. Pub. Health Rep., *76*:1087, 1961.

Schlesinger, E. R.: Dental caries and the pediatrician. Amer. J. Dis. Child., *105*:1, 1963.

Schroeder, H. E.: Inorganic content and histology of early dental calculus in man. Helv. Odont. Acta, *7*:17, 1963.

Trautz, O. R.: Crystallographic studies of calcium carbonate phosphate. Ann. N. Y. Acad. Sci., *85*:145, 1960.

von Vahl, J., Hohlung, H. J., and Frank, R. M.: Elektronenstrahlbeugung an Rhomboedrisch aussehenden Mineralbildungen in Kariosem Dentin. Arch. Oral Biol., *9*:305, 1964.

Vogel, J. J., Naujoks, R., and Brudevold, F.: The effective concentrations of calcium and inorganic orthophosphate in salivary secretions. Arch. Oral Biol., *10*:523, 1965.

Walker, J. S., Margolis, F. J., Teate, H. L., Weil, M. L., and Wilson, H. L.: Water intake of normal children. Science, *140*:890, 1963.

Weaver, R.: Fluorine and dental caries. Further investigations on Tyneside and in Sunderland. Brit. Dent. J., *77*:185, 1944.

Zipkin, I., McClure, F. J., and Lee, W. A.: Relation of the fluoride content of human bone to its chemical composition. Arch. Oral Biol., *2*:190, 1960.

Zipkin, I., Posner, A. S., and Eanes, E. C.: The effect of fluoride on the x-ray diffraction pattern of the apatite of human bone. Biochim. Biophys. Acta, *59*:255, 1962.

Section Five

Applied Nutrition
in Dental Practice

Chapter Twenty-eight

Interviewing and Counseling the Patient on Normal Diet and Meal Planning

Charlotte M. Young, Ph.D.

NUTRITION COUNSELING

Nutrition counseling may be thought of in terms of who, what, when, how and why. Let us turn first to the *why*.

WHY

Nutrition and nutrition counseling are an integral part of preventive and therapeutic dentistry. Preventive dentistry is inadequate unless definitive dietary advice and nutrition counseling are included. Since the nutrition, diet and food habits of the patient may well be affecting his dental status, the nutrition interview (with the appraisal so obtained) is a valuable diagnostic tool, giving a possible clue to the basis of the dental difficulties. As will be seen in subsequent chapters, manipulation of food habits and food intake also plays a role in therapy. Furthermore, patients are most appreciative of the opportunity to discuss the problems they encounter in eating during the course of certain dental procedures and situations. Most patients are grateful for helpful suggestions given during a session of nutrition counseling. The research minded dentist may be interested in knowing that many research contributions could be made by practicing dentists if they systematically gathered data on dietary habits, food and nutrient intake and various conditioning factors and related them to the dental and periodontal

states encountered. Such contributions are needed and give real satisfaction to those who combine the stimulation of research with an active practice.

WHAT

There are several different kinds of nutrition interviews. All are concerned, first, with the patient's present food intake and food habits. This information is basic whether the interview is primarily for diagnosis or for obtaining background information on which to build therapeutic procedures. The closer a therapeutic regimen can be to the patient's present eating pattern and circumstances, the more likely it is to be followed.

Nutrition interviews differ essentially in degree of complexity and quantitativeness. The simplest, which serves the purpose in many instances, is the *food habit inquiry*, essentially qualitative in nature. Simply put, what are the patient's usual eating habits? What kinds of food are usually eaten? What is the pattern of eating? This information may form a basis for diagnosis or for planning and instructing the patient in some special diet.

A second type of nutrition interview is the *dietary history* of the Burke (1947) type, which is directed at getting a more quantitative estimate of nutrient intake over a period of time. It is a time-consuming procedure and requires a high degree of skill. Except in special long-

351

term research studies it is probably not the method dentists would find most useful.

A third type of nutrition inquiry is what might be called the *searching interview,* in which one is concerned primarily with the "reason behind the action." This type of interview is useful when one is concerned with knowing more about the patient and his living habits in general. The food inquiries are largely a means of getting the patient to talk. Since food is intimately concerned with many aspects of an individual's life, insights into his personality and general living situation come out in a discussion that is centered on his food habits.

WHO

Good dietary interviewing requires skill, time and some background knowledge of what goes into forming food habits, their significance and the factors which affect them. If a dietary interview or nutrition counseling is worth doing at all, it should be worth doing well. The counselor should realize that food intake and eating habits are among the most complex facets of human behavior and that when one attempts to influence them he is not dealing with physical nourishment alone. Instead, he is using an intellectual approach to the highly involved behavior pattern of food acceptance, which is a composite of biochemical, physiological, psychological, sociological, cultural and educational factors. This is why it is one of the most difficult aspects of human behavior to change and why the interviewer needs to learn as much as possible about the person being interviewed.

People cling to their customary food habits, especially when disquieting events are taking place in their lives; food to them has far more significance than mere physical nourishment. This realization may cause the dentist to question whether the advantage to be gained by the patient as a result of the special diet or dietary alteration is worth the disruption of his security in his present patterns. However, once the decision has been made in favor of such disruption, every effort should be made to fit the therapy to the patient and his immediate circumstances insofar as is consistent with the therapeutic goals. The dentist should

not delude himself that giving a printed diet list to an unprepared patient is a form of therapy. It is a mistake to think he can pull a therapeutically appropriate printed diet list from a file, tell the patient to follow it and feel that he has discharged his responsibility. Actually, poor diet therapy may be worse than none at all, for the patient may be switched from a "not too bad" regimen to a hodgepodge resulting from attempts by the patient to fit what he only partially understands to what he usually does and wants to do. If a therapeutic diet is worth prescribing, it is worth presenting in such a manner and adapting to the patient in such a fashion that there is some reasonable hope the patient may be able to follow it (Young, 1960).

Obtaining adequate dietary information or giving nutritional guidance effectively depends on good rapport between the patient and the interviewer. Therefore, the interviewer needs to be someone who recognizes and understands the needs and attitudes of the patient. The basic attitudes and needs encountered in patients include: fear, expressed as anxiety, insecurity, or various other means used to cover up fear; the need to protect or preserve his self-respect, often expressed as a need to defy authority; and the need for dependency and love.

The expressions of fear or anxiety by patients are many and diverse. They include impatience with the interviewing, restlessness and physical uneasiness, self-consciousness, extreme talkativeness or extreme reticence, the "know-it-all" attitude, contemptuousness, or depression and discouragement and a "what's-the-use" attitude. Unwillingness to change the pattern of diet or of living, either by statement or by act, often represents fear of leaving the secure, the known or the accepted. As indicated earlier, this may be a sound reason for not changing the patient's life routine any more than absolutely necessary. Apparent indifference on the part of the patient may be a basic defense mechanism to cover up fear or it may be an attempt to preserve self-respect. In either case, the therapist may accomplish most by ignoring the indifference and proceeding with instruction with even greater care, interest and reassurance than if the patient appeared vitally interested. Fear is also seen in apprehensiveness on the part of the patient: "What is all this going to involve?" Because of the basic fear and anxiety often present in patients, scare and threat tactics should usually be avoided.

The patient's need to preserve his self-respect is often expressed as a "superior" attitude, defiance, hostility or anger. The hostility may be toward the therapist, the patient's family, even his friends or the world in general. Anger and irritability for no apparent reason may be directed against the dentist or anyone in his office. Often the anger is best ignored; however, in some cases, it is helpful to encourage the patient to talk about the reasons for his anger. Frequently, patients show a need to establish their prestige and position with the therapist. Sometimes a "superior" attitude is the result of fear of the loss of status, prestige, social acceptance, or the feeling of personal failure and inadequacy by the patient. Such a reaction is assoicated with the need for self-respect which has been threatened. Defiance and aggressiveness toward those who try to tell the patient what to do or to change his patterns may represent his need to maintain his own integrity or an immature need to defy authority. Resentment, rebellion against restriction, shame at failure to follow instructions—all may be associated with a need to maintain self-respect. At the other extreme is the patient's great exhilaration over any accomplishments he achieves in therapy, which increases his self-respect.

The patient's need for dependency is manifested in many ways: his apparent enjoyment in coming to the dentist, his faithfulness in keeping appointments, his various means of vying for the therapist's attention, his desire for special attention and instruction, his gratitude toward the therapist which may be expressed in many ways, and his relief of symptoms (Young, 1957).

Each patient is an individual and has his own attitudes and needs which must be considered. Some knowledge of what the attitudes may mean will help in understanding and dealing more effectively with the patient.

An effective nutrition counselor should have the ability to put the patient at ease by a warm, friendly manner backed by absolute sincerity and honesty. Counseling is a matter of building the right climate so that the patient will feel free to talk and be comfortable

enough to be able to make maximum use of his intellect in absorbing what he is being told. Empathy, the ability to project oneself into the role of the patient with sensitivity to his needs and feelings but without overidentification, is a great asset to any therapist. A willingness to listen as well as tell and a patient, unhurried, unjudging, relaxed manner with the patient also help, as do flexibility, insight and practicality.

Who the counselor is in terms of his training is not so important as that he or she be the right *kind* of person, be he or she a nutritionist, dietitian, dental hygienist, office helper or the dentist himself. Ideally, one could wish the dentist himself could take on the responsibility for nutrition counseling, for surely the words of no one else would have more prestige or status with the patient. Anything he tells the patient has significance over and above what anyone else in his office may say. At the very least, it is his responsibility to give the counselor and her function prestige in the eyes of the patient; to that end he should make provision for receiving periodic reports from either the counselor or the patient, but preferably both. To be effective, nutrition counseling needs to be correlated with the rest of the dental therapy.

WHEN

Good nutrition counseling takes time. One of the greatest weaknesses in many interviews is insufficient time to do a satisfactory job. It takes time for a patient to feel at ease with the interviewer. Even in a qualitative or more superficial food habit inquiry, little of real value can be accomplished in the 10 or 15 minutes often allotted. It takes at least a half hour in an initial interview for the patient to be at ease and to give or accept useful information. In shorter periods one usually gets only very superficial or "expected" answers. Sufficient time must be provided either through a longer initial interview or, preferably, by follow-up contacts of shorter duration. Often the patient cannot absorb all the information given to him during his first visit, and later many questions arise in his mind as he tries to carry out instructions. It is important that these questions be dealt with at a subsequent visit and not disregarded. Careful consideration of small practical points may make the difference in whether instructions will be followed, and the willingness of the interviewer to help on these serves to impress the patient with the importance of his food habits to his dental welfare. Follow-up visits also impress on the mind of the patient the association between dental health and good food habits (Young, 1961).

Quantitative or "searching" types of dietary interviews take longer and seldom can be done adequately even by an experienced interviewer in less than an hour. Fortunately, as indicated earlier, this is not the type of interview which is likely to be used often by the dentist.

Two or three additional points may be made concerning the *when* of the dietary interview or nutrition counseling. The interview or counseling should not be done at the end of an exhausting session with the dentist when the patient is weary, apprehensive or distracted. If the patient seems too tired or for some reason is unable to concentrate on the matter at hand, it is probably better for the interviewer just to try to become acquainted with the patient and to make a new appointment for a more auspicious time. Also, there is a limit to the number of good diet inquiries and instructions that an interviewer can handle in one day, especially if they are of a similar nature. Hence, it is often well to combine the interviewer's job with some other function in the office so that only part of her day is spent in interviewing and she comes to each interview fresh and enthusiastic.

WHERE

The right kind of interviewer with adequate time available is more important than *where* the interview is conducted. However, some privacy is desirable, no matter how small the enclosure, because food habits are very personal and some degree of privacy implies respect for their personal character. Also freedom from distractions aids in obtaining the recall needed for dietary histories or in obtaining the careful reflection and concentration required for instruction. This is difficult amid the bustle, confusion and distraction found in an unenclosed corner of a busy office.

HOW

There are no standard or uniform procedures which can or should be laid down as applicable to all patients in the hands of all therapists. Techniques should be adjusted to the needs of the individual patient at the time instruction is being given and to the skills of the particular therapist or nutritionist.

It may be useful to talk not only with the patient but with the person responsible for the planning, purchase, preparation and service of the patient's food. It is far better to interview children separately from their parents, though information from both may be useful. Certainly, if changes are to be made in the food available to children, or other members of the family, or in its preparation, the person responsible for the food service, presumably the homemaker, needs to be informed and her help solicited. However, in the interest of accomplishing possible changes in food habits and in not leading to emotional conflicts centered on the food intake of a family member, the homemaker should be cautioned against too aggressive an attempt to influence by word of mouth what the patient eats (Chapter Eighteen). Instructions to the patient and admonitions concerning food intake might better be left to the therapist. The homemaker's function is best limited to seeing that the needed food is available to the patient.

In the initial interview the interviewer should try to understand the patient's personality, the general circumstances of his life (in particular those which may have influenced his present dental condition and his eating habits), his reaction to whatever is the dental difficulty which brought him to the dentist, what motivations may be meaningful to him to get him to follow any new food patterns or habits, and what special resistances need to be overcome. One must be concerned not only with what the patient eats but with *why* the patient has been selecting and eating his past and present foods. The latter inquiry takes into consideration not only the present complaint and illnesses but also past nutritional history, the family history, food habits, personal oral hygiene habits, systemic conditioning factors and social history.

The counselor tries to build a climate such that the patient feels relaxed and at ease and may talk more freely, remember better and receive instructions more easily. The interviewer wishes to know all he can about the patient: from his record, from any other therapists involved and most especially from the patient himself. This can be accomplished by helping the patient to talk freely and then "reading between the lines." If the dentist does the interviewing he probably has the advantage of an already established rapport as well as some knowledge of the background of the patient. Others may start by asking such general questions as the patient's height, weight, age, the composition of his family, what he does, where he lives, etc. Often the patient's extraneous comments are of far more help than answers to specific questions. Once the patient is relaxed and at ease, more specific questions related to food intake and food habits may be covered quite quickly: where and with whom he eats; when and under what circumstances; perhaps food preparation and storage facilities; possibly shopping facilities; money available for food; regularity of eating patterns; the usual eating pattern as to kind of food, amount, and variations; between-meal eating—what, how much, when, under what circumstances.

A simple form may be devised for recording all the various items into which inquiry is made as well as a form for a simple dietary history. There are certain advantages to having the patient record his diet for a period before the actual diet interview. Often the simplest possible form with headings of "when," "where," "what," and "how much" may be best, with instructions to the patient to record "everything that you put in your mouth and swallow." The record probably should be kept for at least seven consecutive days which, of course, includes a week end, when food habits are sometimes different.

EVALUATION OF DIETARY INFORMATION

There are many ways in which the dietary interview material may be summarized and evaluated. Which method is best depends upon the use to which the dentist or interviewer wishes to put the material obtained and, of course, also on the material itself or the type of inquiry. Good sense implies that one will not apply a quantitative calculation and evaluation to qualitative and nonquantitative information!

QUANTITATIVE CALCULATION

The most quantitative and carefully obtained diet history may be worthy of actual calculation by nutrient, using appropriate tables [e.g. Appendix I, or *Nutritive Value of Foods*, House and Garden Bulletin No. 72 (1960), or Agricultural Handbook No. 8 (Watt and Merrill, 1963)] after the diet has been summarized. In the summary all like foods are added together, e.g., all forms of milk, so that only one multiplication will need to be made for each food. Nutrients are then totaled for comparison with some evaluation tool such as the Recommended Dietary Allowances (Food and Nutrition Board, 1964), having in mind the bases for these allowances and that one is then evaluating only *dietary* status, not *nutritional* status. Much time may be saved if it is remembered that there is no point in taking the time to calculate nutrients in which there is no specific interest insofar as the dental condition is concerned.

QUALITATIVE EVALUATION

However, for the more usual qualitative type of dietary inquiry evaluation will probably be in terms of the 4 Food Groups discussed in Chapter Fourteen. If the diet under study is of a fairly typical United States pattern, the 4 Food Groups evaluation serves reasonably well (Chapter Twenty-nine). However, one should always bear in mind that many different combinations of foods will meet the Recommended Dietary Allowances. Hence, if the individual's food pattern is that of another culture or if it is restricted in the *kinds* of foods used, even though those used may be in fairly large quantities, one should not make the mistake of deciding that the diet is inadequate because it does not meet the 4 Food Groups pattern. If there is any question of the nutrient adequacy of such a diet, it may be well to calculate even this qualitative information in terms of nutrients and to compare the results to the Recommended Dietary Allowances or some other appropriate evaluation tool (Young and Musgrave, 1951).

Simplified Calculation. If one is interested primarily in either the carbohydrate, protein or fat content of the diet, a somewhat simplified method of calculation known as the Exchange System may be used. This is a system developed originally as a means of standardizing diabetic diets throughout the country by a joint committee of the American Diabetes Association, the American Dietetic Association and the Diabetes Section of the United States Public Health Service (Caso, 1950; Diabetes Guide Book for the Physician, 1956). This method will be described later. It is more useful as a means of calculating a diet prescription and for instructing an individual in this prescription than as a means of calculating diet histories made up of many mixed dishes.

RESPONSE TO NUTRITION COUNSELING

Before we turn to some comments on counseling with regard to the normal diet and meal planning, a word might be said as to what one may expect in terms of patient response to counseling, particularly when dietary changes are involved. The remarks of Dr. Hilde Bruch concerning her experiences working with diabetic children and their families over a period of 20 years have pertinence here. Dr. Bruch (1961) says that in her experience none of the children found restriction in the freedom of food choice easy, but that the children reacted in different ways, depending upon the emotional health and stability of their families. She could recognize roughly three patterns. The first and most desirable pattern was one of *rational acceptance* of the new situation with consistent, though not rigid, cooperation. There might be some confusion at first which did not interfere with

readiness to learn. Not infrequently the whole family adjusted their eating habits to make the diet of the patient not too difficult. In these families other situations also were handled in a spirit of mutuality and without resentment for the member who required special consideration. These patients, whose families had been rated as stable, congenial and capable of handling the difficulties of diabetes, did well in the long run.

The pattern of Dr. Bruch's second group was characterized by *excessive cooperation.* Instructions were carried out in an overperfectionistic manner. There was rigid adherence to the diet with much anxiety about the slightest deviation. The diet and other tasks were not seen in relation to their function but as a sort of magic that would bring about cure by obedient attention to all rituals. These patients often found it difficult to accept changes in their regimen as changes became necessary. Family relationships were smooth but tense, and there was usually an anxious and over-rigid mother.

The pattern of Dr. Bruch's third group was one fraught from the start with *difficulties in accepting* the disease and its *necessary restrictions.* In all cases, symptoms of emotional disorder and severe family discord had preceded the illness. These patients not only had difficulty in controlling their diabetes but in making an adjustment in any aspect of living.

Dr. Bruch's experience with diabetic patients has significance far beyond this one disease. It coincides almost exactly with what we have experienced with obese patients (Young et al., 1955). In fact, it seems generally applicable to all patients whose dietary patterns must be changed. The dentist or the counselor should be aware of these types of responses on the part of patients and not feel frustrated or unique in his experience when all patients do not follow his plans for them as he would like. Patients of the first group are ones who do rather well when they are well motivated. To be well motivated the patient needs to understand the reasons for the dietary changes in terms suitable to his age and education. As Dr. Bruch points out, too much information may be frightening if it is misunderstood, and too little may make the sophisticated feel uneasy. As we have emphasized repeatedly, instructions must be adjusted insofar as consistent with the therapeutic purposes to the patient's customary food habits and daily routine of living.

The "perfectionistic" rigid patient requires special attention and care. So often the individual who starts out in this fashion becomes so anxious that he abandons all diet instructions and, indeed, often drops from all care as a means of getting away from an anxiety-producing situation. Individuals of the third group who are openly uncooperative and emotionally upset probably will never carry through on dietary or other instructions. Actually, for most of these latter individuals, help with emotional problems would probably need to precede any cooperation in a change in dietary pattern.

COUNSELING IN NORMAL DIET AND MEAL PLANNING

As a simple guide to normal diet and meal planning the 4 Food Groups published in 1958 by the United States Department of Agriculture (Food for Fitness: A Daily Food Guide, 1958*) have proved very useful in instructing patients. Table 28-1 gives the food groups, number of servings, average size of serving and the principal nutrient contributions of each group of foods to the diet.

In using the Daily Food Guide the patient should be instructed to select the main part

* Copies of the guide may be purchased in quantity for a nominal sum from Superintendent of Documents, U. S. Government Printing Office, Washington 25, D.C.

TABLE 28-1.   FOOD FOR FITNESS—DAILY FOOD GUIDE

Food Groups	No. Servings	Size Serving (Adult)	Nutrient Contribution to Diet
MILK GROUP			
Milk (cheese and ice cream may replace part of milk)	3-4 servings—Children 4 or more servings—Teenagers 2 or more servings—Adults 4 or more servings—Pregnant women 6 or more servings—Nursing mothers	1 8 ounce glass of fluid milk or its equivalent in calcium 1½ cups cottage cheese, 1⅓ ounces Cheddar cheese, or 2 cups of ice cream contain the amount of calcium found in 1 glass of fluid milk	Calcium Riboflavin Protein Vitamin A (whole milk) Vitamin D (if fortified) Thiamine
VEGETABLE-FRUIT GROUP			
Dark green or deep yellow vegetables (broccoli, kale, carrots, squash, etc.)	1 serving at least every other day	½ cup	Vitamin A Iron in dark green vegetables
Citrus fruits, cantaloupe, strawberries, broccoli, tomatoes, raw cabbage, pepper or other foods rich in vitamin C	1 serving	½ cup or portion normally eaten, as 1 orange, ½ grapefruit, etc.	Ascorbic acid (vitamin C)
Other fruits and vegetables, including potatoes		½ cup or portion as ordinarily eaten, as 1 apple, 1 banana, or 1 potato	Varying amounts of several nutrients
MEAT GROUP			
Meat, fish, poultry, eggs, dried beans, dried peas, nuts	2 or more servings	2 or 3 ounces lean boneless cooked meat, fish or poultry 2 eggs, 1 cup cooked dried beans, dried peas or lentils 4 tablespoons peanut butter	Protein Iron B vitamins Vitamin A (liver, egg yolks)
BREAD-CEREAL GROUP			
Whole grain, enriched or restored breads and cereals	4 or more servings If no cereal, use 5 servings of bread	1 slice bread, 1 ounce ready-to-eat cereal, ½ to ¾ cup cooked cereal, macaroni, noodles, rice or spaghetti	B vitamins Iron Protein

of his diet from the broad 4 Food Groups. At least the minimum number of servings should be chosen from each of the 4 groups. The *quantities* of each food to be used, i.e., the *size of serving*, will vary with the age and physiological status of the patient (Chapter Fifteen). In general, servings for children will be smaller and for teen-agers extra large or increased in number. Women in the second and third trimester of pregnancy and nursing mothers also require more foods.

To round out the meals and to provide sufficient energy to meet requirements, additional foods may be chosen from the 4 Food Groups or from "other foods" not listed in the groups. "Other foods" include such items as butter, margarine, other fats, oils, sugars or unenriched refined grain products, which either may be added to foods during preparation or at the table or may be additional ingredients in mixed dishes or baked goods. The best index of sufficient energy is the support of normal growth in children and the maintenance of body weight at a level most favorable for health and well-being in the adult. For many people there is advantage to the inclusion of some rich source of animal protein at each meal, i.e., meat, fish, poultry, eggs or milk products.

The foods listed in this guide may be put together in many different combinations and distributed over the day in various fashions consistent with the individual's usual pattern of eating. The actual foods used from the four groups, the way in which they are put together, prepared and served will vary with cultural and other influences. The patient's previous habits are the best indication of how he may wish to use the food and should be followed unless there is a really important reason for modification.

The Daily Food Guide is one way of helping the normal individual to choose food wisely. By following it he is reasonably sure of getting the nutrients needed from a variety of everyday foods. One of the most important points to remember in the use of any guide is flexibility and the lack of rigidity. Also, it is well to remember that the Daily Food Guide is best adapted to the general food pattern in the United States. If one is dealing with markedly different cultural or subcultural groups, adaptations may be necessary.

INSTRUCTION FOR A CALCULATED DIET

In most instances, the dentist would have no need for the calculated diet in the instruction of his patients with regard to normal diet and meal planning. However, there may be exceptions when for some reason there is need to control one of the principal nutrients, for example, carbohydrate, which may be desirable in certain cases of dental caries or periodontal disease. In many instances, the control or elimination from the diet of the use of either sugar or products made with sugar will be sufficient to accomplish the purpose. When further carbohydrate control is desirable, calculation by means of the Exchange System developed by the joint committees indicated earlier may be useful.

THE EXCHANGE SYSTEM

In the Exchange System all foods are divided into six groups of exchanges with an assigned composition for each group (Table 28-2). A list of equivalents of essentially the

TABLE 28-2.   COMPOSITION OF FOOD EXCHANGES*

List	Group Exchanges	Measure	Gm.	C. (gm.)	P. (gm.)	F. (gm.)	Calories
1	Milk	½ pint	240	12	8	10	170
2A	Vegetables	as desired	..	..	..	..	..
2B	Vegetables	½ cup	100	7	2	..	36
3	Fruit	varies	..	10	..	..	40
4	Bread	varies	..	15	2	..	68
5	Meat	1 oz.	30	..	7	5	73
6	Fat	1 tsp.	5	..	..	5	45

* From Diabetes Guide Book for the Physician. 2nd ed. New York, American Diabetes Association, Inc., 1956.

same composition has been developed for each group as shown in List 1 through List 6. For example, in Table 28-2 we see that the first group listed is Milk Exchanges and that 1 cup of milk contains 12 gm. carbohydrate, 8 gm. protein, 10 gm. fat and 170 calories. In List 1 of Table 28-3 ½ cup (120 gm.) of evaporated milk, or ¼ cup (35 gm.) of powdered milk, or 1 cup (240 gm.) of buttermilk may be substituted for 1 cup (240 gm.) of whole milk and is the nutrient equivalent of 1 Milk Exchange. If any of the forms of milk used are fat free, two Fat Exchanges (see List 6) should be added.

Vegetable Exchanges have been divided into two categories, one of which, A, may be used as desired because the foods listed contain negligible carbohydrate and calories; the other, B, is limited to ½ cup or 100 gm., with an average value of 7 gm. carbohydrate, 2 gm. protein and 36 calories. The nutrient equiv-alents of one Vegetable Exchange may be found in List 2B.

SELECTION OF DIET PRESCRIPTION

In the use of the Exchange System, the diet prescription is based on the caloric needs of the patient. Caloric requirements, in turn, are influenced by present weight in relation to desirable weight, activity and body size.

To simplify the selection of a diet prescription the committees have developed nine sample meal plans of various caloric, nutrient and food exchange values (Table 28-4).

TABLE 28-3.   FOOD EXCHANGES*

List 1. Milk Exchanges
CARBOHYDRATES, 12 GM.; PROTEIN, 8 GM.; FAT, 10 GM.; CALORIES 170

	Meas.	Gm.
†Milk, whole	1 cup	240
Milk, evaporated	½ cup	120
†Milk, powdered	¼ cup	35
†Buttermilk	1 cup	240

List 2. Vegetable Exchanges
A. THESE VEGETABLES MAY BE USED AS DESIRED IN ORDINARY AMOUNTS—CARBOHYDRATES AND CALORIES NEGLIGIBLE

Asparagus	Greens	Lettuce
Broccoli	Beet	Mushrooms
Brussels sprouts	Chard	Okra
Cabbage	Collard	Peppers
Cauliflower	Dandelion	Radishes
Celery	Kale	Rhubarb
Chicory	Mustard	Sauerkraut
Cucumbers	Spinach	String beans,
Escarole	Turnip	young
Eggplant	Tomatoes	Summer squash

B. VEGETABLES: 1 SERVING EQUALS ½ CUP. EQUALS 100 GM. CARBOHYDRATES, 7 GM.; PROTEIN, 2 GM.; CALORIES, 36

Beets	Peas, green	Turnips
Carrots	Pumpkin	Winter squash
Onions	Rutabaga	

* The food exchange method is based on material in "Meat Planning with Exchange Lists," prepared by committees of the American Dietetic Association and the American Diabetes Association, Inc., in co-operation with the Chronic Disease Program, Public Health Service, Department of Health, Education, and Welfare.

† Add **2** Fat Exchanges if fat free.

TABLE 28-3—(*Continued*)

List 3. Fruit Exchanges
CARBOHYDRATES, 10 GM.; CALORIES, 40

	Meas.	Gm.
Apple	1 small (2″ diam.)	80
Applesauce	½ cup	100
Apricots, fresh	2 medium	100
Apricots, dried	4 halves	20
Banana	½ small	50
Berries: straw., rasp., black.	1 cup	150
Blueberries	⅔ cup	100
Cantaloupe	¼ (6″ diam.)	200
Cherries	10 large	75
Dates	2	15
Figs, fresh	2 large	50
Figs, dried	1 small	15
Grapefruit	½ small	125
Grapefruit Juice	½ cup	100
Grapes	12	75
Grape juice	¼ cup	60
Honeydew melon	⅛ (7″ diam.)	150
Mango	½ small	70
Orange	1 small	125
Orange juice	½ cup	100
Papaya	⅓ medium	100
Peach	1 medium	100
Pear	1 small	100
Pineapple	½ cup	80
Pineapple juice	⅓ cup	80
Plums	2 medium	100
Prunes, dried	2 medium	25
Raisins	2 tbsp.	15
Tangerine	1 large	100
Watermelon	1 cup	175

List 4. Bread Exchanges
CARBOHYDRATES, 15 GM.; PROTEIN, 2 GM.; CALORIES, 68

	Meas.	**Gm.**
Bread	1 slice	25
Biscuit, roll	1 (2″ diam.)	35
Muffin	1 (2″ diam.)	35
Cornbread	1 (1½″ cube)	**35**
Flour	2½ tbsp.	20
Cereal, cooked	½ cup	100
Cereal, dry (flake & puffed)	¾ cup	**20**

TABLE 28-3.—(*Continued*)

List 4. Bread Exchanges

CARBOHYDRATES, 15 GM.; PROTEIN, 2 GM.; CALORIES, 68

	Meas.	Gm.
Rice, grits, cooked	½ cup	100
Spaghetti, noodles, etc., cooked	½ cup	100
Crackers, graham (2½" sq.)	2	20
Oyster	20 (½ cup)	20
Saltines (2" sq.)	5	20
Soda (2½" sq.)	3	20
Round, thin (1½" diam.)	6-8	20
Vegetables		
Beans and peas, dried, cooked (Lima, navy, split peas, cowpeas, etc.)	½ cup	90
Baked beans, no pork	¼ cup	50
Corn	⅓ cup	80
Parsnips	⅔ cup	125
Potatoes, white, baked, boiled	1 (2" diam.)	100
Potatoes, white, mashed	½ cup	100
Potatoes, sweet or yams	¼ cup	50
Sponge cake, plain	1 (1½" cube)	25
Ice cream (omit 2 fat exchanges)	½ cup	70

List 5. Meat Exchanges

PROTEIN, 7 GM.; FAT, 5 GM.; CALORIES, 73

	Meas.	Gm.
Meat and poultry (med. fat) (beef, lamb, pork, liver, chicken, etc.)	1 oz.	30
Cold cuts (4½" sq. x ⅛" thick)	1 slice	45
Frankfurter	1 (8-9/lb.)	50
Fish: cod, mackerel, etc.	1 oz.	30
Salmon, tuna, crab	¼ cup	30
Oysters, shrimp, clams	5 small	45
Sardines	3 medium	30
Cheese, Cheddar, American	1 oz.	30
cottage	¼ cup	45
Egg	1	50
*Peanut butter	2 tbsp.	30

List 6. Fat Exchanges

FAT, 5 GM.; CALORIES, 45

	Meas.	Gm.
Butter or margarine	1 tsp.	5
Bacon, crisp	1 slice	10
Cream, light, 20%	2 tbsp.	30
Cream, heavy, 40%	1 tbsp.	15
Cream cheese	1 tbsp.	15
French dressing	1 tbsp.	15
Mayonnaise	1 tsp.	5
Oil or cooking fat	1 tsp.	5
Nuts	6 small	10
Olives	5 small	50
Avocado	⅛ (4" diam.)	25

* Limit use or adjust carbohydrates.

TABLE 28-4. AMERICAN DIETETIC ASSOCIATION SAMPLE MEAL PLANS

Diet	A. MEAL COMPOSITION			
	Carbohydrate (gm.)	Protein (gm.)	Fat (gm.)	Calories (cal.)
1	125	60	50	1200
2	150	70	70	1500
3	180	80	80	1800
4	220	90	100	2200
5*	180	80	80	1800
6*	250	100	130	2600
7*	370	140	165	3500
8	250	115	130	2600
9	300	120	145	3000

B. TOTAL DAY'S FOOD IN SAMPLE MEAL PLANS

Diet	Milk	Exchanges					
		Veg. A	Veg. B	Fruits	Bread	Meat	Fat
1	1 pt.	as desired	1	3	4	5	1
2	1 pt.	as desired	1	3	6	6	4
3	1 pt.	as desired	1	3	8	7	5
4	1 pt.	as desired	1	4	10	8	8
5*	1 qt.	as desired	1	3	6	5	3
6*	1 qt.	as desired	1	4	10	7	11
7*	1 qt.	as desired	1	6	17	10	15
8	1 pt.	as desired	1	4	12	10	12
9	1 pt.	as desired	1	4	15	10	15

* These diets contain more milk and are especially suitable for children.

In the selection of an appropriate prescription and sample meal plan for an adult, it is recommended that the therapist refer to a table of desirable weights and make an appraisal of the usual amount of activity in which the patient is involved. Table 28-5 is an index prepared by the committee as a guide to the caloric needs for adults. On the basis of caloric need, that one of the nine plans best suited to the patient is then selected.

The selection of a diet prescription for children is more difficult than for adults. Sufficient calories must be provided for proper growth and development and to meet energy needs. Three sample plans for children have been included among the nine diet or meal lists prepared by the committees (Table 28-4):

	Calories	
Meal Plan 5	1800	For children around 8 years of age
Meal Plan 6	2600	For children 10-12 years of age and girls 13-18 years of age
Meal Plan 7	3500	For boys 13-18 years

TABLE 28-5. GUIDE TO CALORIC NEEDS FOR ADULTS

| Weight and Activity Status | Below 5'6" | | Above 5'6" | |
	Calories	Diet No.	Calories	Diet No.
Desired weight				
Sedentary	1800	(3)*	2200	(4)
Moderately active	2600	(8)	3000	(9)
Very active	3000	(9)	3500	(7)
Overweight				
Sedentary	1200-1500	(1 or 2)	1500-1800	(2 or 3)
Moderately active	1800	(3)	2200	(4)
Very active	2200	(4)	2600	(8)

Column group header: Height (spanning Below 5'6" and Above 5'6")

* Numbers refer to Table 28-4, Section B.

Diets for children at different ages and with different caloric requirements can easily be developed by adding additional amounts of the food exchanges, especially the bread and meat exchanges, Lists 4 and 5 (approximately 70 calories per exchange) of Table 28-3.

A periodic check on the patient's weight is the best index of whether the prescribed diet is meeting or exceeding his caloric needs. On the basis of this observation, adjustments in the caloric level of the diet may be made (Table 28-5).

DEVELOPMENT OF A DIET PRESCRIPTION

The development by the committees of the nine sample diets or meal plans has been convenient. However, using the Group Exchanges (Table 28-2) it is easy for the dentist to calculate his own diet plan to fit the diet prescription he may wish to use. Let us take an example:

Usually the amounts of milk, vegetable and fruit exchanges basic to the intended meal pattern are planned first. Carbohydrate is then balanced by adding the carbohydrate contributed by these groups, subtracting it from the total carbohydrate required by the prescription and converting the residual carbohydrate to bread exchanges. In our example 24 + 7 + 30, or 61, subtracted from 90 gives a residual of 29 gm. of carbohydrate or approximately the equivalent of two bread exchanges (30 gm.). The meat exchanges are determined by adding the protein contributed by the milk, vegetable and bread exchanges and subtracting this from the total protein prescribed in the diet. Finally, the number of fat exchanges to be added is determined by subtracting the fat in the milk and meat exchanges from the total fat prescribed and dividing by five, the number of grams in each fat exchange.

Once the food exchanges to be included for the day have been determined, they may be arranged in a variety of patterns to suit the

Prescription:

Carbohydrate gm.	**90**	
Protein gm.	80	
Fat gm.	80	
Calories	1400	

List	Group	No. of Exchanges	Measure	C (gm.)	P (gm.)	F (gm.)
1	Milk	2	1 pint	24	16	20
2A	Vegetables	As desired	As desired	..	..	..
2B	Vegetables	1	½ cup	7	2	..
3	Fruit	3	Varies	30	..	..
4	Bread	2	Varies	30	4	..
5	Meat	8	8 oz.	..	56	40
6	Fat	4	4 tsp.	..	..	20
				91	78	80

patient and the menus changed from day to day by use of the six exchange lists (Table 28-3). For example, the following are two menus which might be evolved from the prescription which was converted into a meal plan:

Breakfast:

Orange juice—½ cup	Grapefruit—½ small
Fried eggs	Puffed Wheat—¾ cup
Eggs—2	Milk, whole—1 8 oz.
Butter—1 tsp.	cup
Toast—2 slices	Coffee, black
Butter—1 tsp.	
Coffee, black	

Lunch:

Tomato stuffed with	Ground beef patty—4
chicken salad	oz.
Tomato—1 whole	Buttered green peas—
Chicken—2 oz.	½ cup
Celery—as desired	Butter—1 tsp.
Lettuce—as desired	Cottage cheese on let-
Mayonnaise—2 tsp.	tuce—¼ cup
Milk, whole—1 8 oz. cup	Buttermilk—1 cup
Cantaloupe—¼	Applesauce—½ cup

Dinner:

Roast beef—4 oz.	Frankfurters—3
Asparagus—½ cup	Sauerkraut—as desired
Carrots—½ cup	Mashed potatoes—½
Milk, whole—1 cup	cup
Pear—1 small	Butter—1 tsp.
	Blueberries—⅔ cup
	Cream, 20%—¼ cup

Many other combinations could be made using the exchange lists.

SOME POINTS ON INSTRUCTING DIFFERENT AGE GROUPS

Always the method of instruction needs to be adjusted to the age, education and apparent comprehension and personality of the patient with whom one is dealing. Earlier in this chapter it was indicated that for children both the child and parent or person responsible for providing the child's food probably should be instructed but that, for best results, they should be instructed separately. The essential points of instruction should be put on paper so that they may be referred to later, but this does not take the place of going over the material very carefully with the patient. Too much should probably not be put on the paper or it will be confusing. If the patient is given the essentials during his first visit, it is

better to bring him back later for follow-up instruction and to ask questions after he has tried to follow the plan proposed for him. By then he realizes what he does not know or understand; he has run into practical problems he did not anticipate at the time of the original instruction. Also, he is probably better able to listen because he is less frightened, he knows you better and he has made some adjustments.

In an earlier chapter we have discussed some of the problems involved in trying to change food habits (Chapter Eighteen) and pointed out that the kind of appeal that may be effective in changing food habits will vary at different ages. The particular characteristics of teen-agers in relation to effective means of working with them were considered at some length and might well be reviewed at this point.

Two points might be stressed with regard to instructing adults. Beware of the adult who knows too well what he usually does in the way of eating unless that person is a homemaker or someone particularly trained to be concerned with food. Except for the latter two groups, too easy and too precise a recall of food intake makes one suspicious that he may be dealing with a food faddist. Most people are not particularly conscious of what they eat, and it takes some thought to recall what they have had. Another point to remember is that an individual with many food dislikes or with a rigid attitude with regard to food is apt to be one with emotional problems.

Older people have, in general, the same foods needs as younger adults except that their caloric needs are less. They may wish to eat smaller meals more frequently, and there should be no contraindication for this so long as the total food intake for the day represents an adequate diet and is not a series of coffees, teas, toasts, cakes and cookies. Older people often fall into the traps of the food faddists because of their desire to be as young, well and agile as previously. They need to see that food fads are not miraculous, that indeed an adequate normal diet of usual foodstuffs will serve their purposes better. A problem sometimes encountered with older people is lack of adequate facilities for food purchase, storage and preparation. It is well during the course of nutrition counseling to inquire into the facilities available. Lack of appetite may also be a problem because of a less keen sense of taste and of smell as well as diminished exercise. Use of somewhat more seasoning, the appreciation of the importance of an attractive meal in terms of color, flavor and texture, and the encouragement of exercise within the tolerance limits all may help to increase appetite. Problems incidental to chewing will be discussed in Chapter Thirty-four.

REFERENCES

Babcock, C. G.: Psychologically significant factors in the nutrition interview. J. Amer. Diet. Assn., *23*:8, 1947.

Bruch, H.: Social and emotional factors in diet changes. J. Amer. Dent. Assn., *63*:461, 1961.

Burke, B. S.: The dietary history as a tool in research. J. Amer. Diet. Assn., *23*:1041, 1947.

Caso, E. K.: Calculation of diabetic diets. J. Amer. Diet. Assn., *26*:575, 1950.

Diabetes Guide Book for the Physician. 2nd ed. New York, American Diabetes Association, Inc., 1956.

Food for Fitness: A Daily Food Guide. Washington, D. C., U. S. Department of Agriculture Leaflet No. 424, 1958.

Food and Nutrition Board: Recommended Dietary Allowances. Sixth Revised Edition. Washington, D. C., National Academy of Sciences–National Research Council, Publ. 1146, 1964.

Nutritive Value of Foods. Home and Garden Bull. No. 72, U. S. Department of Agriculture. Washington, D. C., U. S. Gov't. Printing Office, 1960.

Watt, B. K., and Merrill, A. L.: Composition of Foods. Agricultural Handbook No. 8, U. S. Department of Agriculture. Washington, D. C., U. S. Gov't. Printing Office, 1963.

Young, C. M.: Teaching the patient means reaching the patient. J. Amer. Diet. Assn., *33*:52, 1957.

Young, C. M.: Interviewing the patient. Amer. J. Clin. Nutr., *8*:523, 1960.

Young, C. M.: Nutrition counseling for the dental patient. J. Amer. Dent. Assn., *63*:469, 1961.

Young, C. M., Moore, N. S., Berresford, K. K., and Einset, B. M.: What can be done for the obese patient? A report of a study in an experimental clinic. Amer. Pract. & Digest Treat., *6*:685, 1955.

Young, C. M., and Musgrave, K.: Dietary study methods. II. Uses of dietary score cards. J. Amer. Diet. Assn., *27*:745, 1951.

Chapter Twenty-nine

A Step-by-Step Technique
for the Dietary Counseling of
Patients with High Caries Susceptibility

Abraham E. Nizel, D.M.D.

It is essential that dietary counseling be carried out in an objective, logical, realistic and above all personalized manner if it is to be an effective and enduring method for the control of dental caries. The purpose of this chapter is to suggest just such a methodology.

There are several questions that need to be answered in addition to describing the how-to-do-it procedure of dietary counseling for patients with rampant caries. "Are dentists qualified and expected to give this specialized service?" "Is the dentist's fundamental knowledge of food and nutrition adequate and sound enough so that his clinical application can be considered rational?" "How can patients be motivated to cooperate and to continue on this dietary regimen?"

ARE DENTISTS EXPECTED
TO GIVE DIETARY GUIDANCE?

An emphatic "yes" is the answer to this question on the responsibility of the dentist for providing dietary guidance. Of all the professional people in the healing arts, dentists are perhaps in one of the best positions to inform the public with respect to food and nutrition. They have regular contact with their patients over longer and sometimes more frequent periods than even the family physician. They, like the physician, develop a patient rapport and are recipients of a respect and confidence which makes for receptiveness

on the part of the patient with respect to advice on health matters. For these several reasons, recognized knowledgeable experts, educators and scientists in medicine and nutrition are strongly recommending and even urging dentists to counsel their patients on proper diets.

For example, Dr. Scrimshaw, Chairman and Professor of Nutrition and Food Science at Massachusetts Institute of Technology stated at a Conference on Nutrition Teaching in Dental Schools; "Giving dietary advice should be as much a part of their (dentists') practice of dentistry as restoring teeth" (Scrimshaw, 1966). He continued: "The plain fact is that few dental educators and fewer still practicing dentists have really acted as if dietary habits conducive to oral pathology, including dental caries, were their responsibilitiy. Furthermore, it is essential that the teaching of nutrition in dental schools create in the dentist an equally strong sense that he is practicing dentistry when he is giving advice on choice of foods to minimize oral difficulties as well as when he is working with his own hands in the oral cavity." Dr. Scrimshaw's sentiments were echoed at this and previous conferences by colleagues in allied professions and sciences. There is no question in their minds nor should there be in the dentist's that as a health advisor it is incumbent upon him to prescribe and advise on proper food and diet selection for ameliorating oral disease. There should be no doubt that dietary counseling knowledgeably performed is definitely within the dentist's province and responsibility.

WHAT IS OUR PRESENT KNOWLEDGE ON THE EFFECTS OF NUTRIENTS AND FOOD ON DENTAL CARIES?

This question has, in effect, already been answered in detail in several of the preceding chapters. However, in order to bring these threads of information together so that some concepts and the direction of applied practice will emerge, a brief review is given here.

NUTRITIONAL INFLUENCES ON SUSCEPTIBILITY OF THE TOOTH TO DECAY

A nutritional influence implies a systemic endogenous effect, which simply means that there is a transfer of nutrients to the developing unerupted tooth through the vascular circulatory channels in and around the tooth. This occurs during the periods of matrix formation and initial calcification.

The first scientific investigations showing a relationship between nutrients and dental health were the classic studies of Mellanby (1918) with vitamin D and of Wolbach and Howe (1925; 1933) with vitamins A and C. For example, it was reasoned that a deficiency of vitamin C could cause a decrease in protein utilization and affect adversely the integrity of a proteinaceous mesenchymal tissue like dentin so that it might be irregularly formed and hypoplastic. However, dentin affected this way has not been shown to be unusually caries susceptible.

There is some proof from animal studies that feeding high sucrose diets starting during pregnancy and lactation will predispose the molars of the offspring of these dams to caries (Sognnaes, 1948). There are also data to show that the molars in offspring of protein-deficient females will be smaller in size, the cusps underdeveloped, the eruption time delayed, and the teeth more susceptible to caries (Paynter and Grainger, 1956). These studies suggest that nutritional deficiencies induced early enough might influence the dental genetic pattern.

A diet which has a calcium-phosphorus ratio imbalance will produce an altered tooth size and shape as well as a higher than usual caries susceptibility (Paynter and Grainger, 1956). Several investigators have found that a high phosphorus–low calcium ratio diet fed during the period of tooth development and continued through the early posteruptive maturing and fully matured stages imparts a significant caries resistance (Sobel et al., 1960; Nizel and Harris, 1960). A similar observation was made when an ash supplement of natural foods was added to a cariogenic diet and fed to hamsters beginning at eight days of age. If the same mineral supplement is fed to littermates starting at 21 days of age, some of the caries-inhibiting properties are lost (Nizel, 1952). Evidently, there is some modifying influence by these nutrients on caries susceptibility if they are introduced during the active mineralizing period of tooth development.

Fluorides ingested at optimal levels are by far the most effective caries-inhibiting nutrient yet discovered for both experimental animals and man. This is particularly true if fluorides are available in either naturally or artificially fluoridated water supplies and ingested from the start of tooth development. For example, Bunting (1928) observed that children born and raised in Minook, Illinois, which has 2.5 ppm natural fluorides in its water supplies, had considerably less caries than children who moved to Minook after their teeth were completely matured and erupted. Since then, a well controlled and well designed long-term study comparing caries experience of children living in Newburgh, New York, where the water was artificially fluoridated with 1 ppm fluoride with that of Kingston, New York, children who drank untreated water, has confirmed the importance of early fluoride administration. For example, the six to nine year olds of Newburgh had 57 per cent less caries than those of Kingston; the 10 to 12 year olds, 52 per cent; the 13 to 14 year olds, 48 per cent; and the 16 year olds, 40 per cent (Ast, 1956).

LOCAL EFFECTS OF FOOD ON CARIES SUSCEPTIBILITY

There is increasing evidence which strengthens the validity of the chemico-parasitic theory of Miller (1890) in which he postulated that acid produced from bacterial enzymatic action on a carbohydrate food substrate could be the prime and initial attacking force on the tooth surface in the caries process. The findings that the presence of bacteria in the oral cavity is necessary (Orland, et al., 1954), that the local contact of food with the tooth is mandatory (Kite, et al., 1950) that the food must be a carbohydrate (Shaw, 1954), and that it must be retentive and must be eaten frequently (Gustafsson, et al., 1954) all strengthen the arguments of those who champion the concept that the etiology of dental caries is primarily a local oral mechanism. Furthermore, there is much evidence from dental surveys that condemns as cariogenic our modern sophisticated diets, which are high in sugar content, low in detergency and soft in consistency (Lossee, 1965). These diets not only affect the newly erupted teeth of children but the fully matured teeth of adults also, indicating again that a local enamel surface effect is the initial action for the carious lesions.

As was shown previously, fluorides act both locally and systemically. Furthermore, the outer surface of the enamel continues to increase its fluoride content with age and at all times is the site in the tooth at which the maximum fluoride is concentrated.

Phosphate compounds fed experimental animals as supplements to an otherwise cariogenic diet have proved to be effective in inhibiting caries development. The preferable phosphate compound is one that combines the readily hydrolyzed sodium or potassium cation with trimetaphosphate, a cyclical phosphate anion (Nizel et al., 1962). Other phosphates have some effect but not in the same order of magnitude as trimetaphosphate. One of the mechanisms suggested for its action is that there may be an iso-ionic exchange between the phosphate of tooth apatite and that of the oral environment, particularly the dental plaque. This inhibiting effect of phosphate on human caries has yet to be proved. When it is, restoring the phosphorus in staple foods such as breads and cereals to the level at which it is found in the whole cereal grain before its refinement would be a practical and worthwhile public health measure for caries prevention.

In conclusion, it seems that the local deleterious effect of some foods on the tooth can overpower any innate caries resistance built up during tooth development. However, it should be added that this does not in any way minimize the importance of proper systemic nutrition during tooth development for imparting some caries resistance. Retentive fermentable carbohydrates, usually as a component of dental plaque in local contact with the tooth, are instrumental in *initiating* dental caries. However, again it should be pointed out that the *progress* of the carious action can be mitigated either by the innate resistance of the tooth or by modifying the mineral nutrient influence of the oral environment with fluorides and possibly with phosphates or elements yet to be discovered. It also should be pointed out that if one period in the tooth developmental and maturation cycle (preeruptive, newly erupted or posteruptive) were to be selected as the most sensitive to nutrient change, it would be the newly erupted stage. This would mean that dietary guidance would be of maximal benefit to the one to three year old for preservation of the integrity of his deciduous teeth and to the 6 to 14 year old for preservation of his permanent teeth.

HOW ARE PATIENTS MOTIVATED TO COOPERATE?

Motivation to accept dietary guidance is perhaps the single most important factor in caries control and prevention. The patient must be given reasons meaningful to him if he is to change some of his usual food habits and even to deny himself what he considers pleasurable foods. The motivating factors can be social or economic or assurance that this is the most painless dental service.

The dentist can dwell on the social aspect by emphasizing the favorable esthetic results that will be realized from good nutrition. Every teen-ager needs to know that good food is the foundation of health and vigor and glowing good looks. The pretty smile and slim figure, clear skin, bright eyes, shiny hair, firm muscles, good posture and boundless energy are the benefits derived from a varied and balanced diet.

The economic aspect can also be made quite vivid by the bill rendered for restorative dental services. Dentistry can cost many dollars if extensive prosthesis and crown and bridge work become necessary. The dental results of neglect, abuse and overindulgence in sweets can be expensive.

Despite the many new and modern techniques and tools that are available to mitigate dental pain and discomfort, no one will deny that minimization of even this is desirable. Few people look forward to any health service, least of all dental. Prevention is the best and least painful treatment, and through dietary guidance this goal can be readily realized.

A fee commensurate with the amount of time spent in dietary counseling must be charged if the patient is to cooperate and the dentist to perform the service. This makes the service objective, and sets an understandable and appreciated value on the service. Based on previous experiences, there will be approximately a 50 per cent reduction in caries which, of course, will provide for a savings and serve as a meaningful return on the investment. Furthermore, if the dentist can utilize trained auxiliary personnel in helping to execute the service, this will increase its likelihood of being implemented.

Finally, maximal cooperation will be realized if the objectives of diet prescription are explained. The diet must be individualized to satisfy, whenever possible, the likes and dislikes as well as the daily routine of the patient.

TYPES OF DIET ANALYSES USED IN CARIES CONTROL

The questions of who is qualified to prescribe diets, how much time need be alloted for giving this advice and the specific locale in the dental office where this service should be performed has been very adequately covered in Chapter Twenty-eight. In this section we are going to detail the how-to-do-it procedure.

Before a diet can be prescribed, an analysis of what has been eaten must be made so that deficiencies can be determined and suggestions made for their correction. Essentially, there are three techniques of dietary analysis: a detailed quantitative determination of each nutrient in a food; an Exchange System which measures the approximate carbohydrate, protein and fat content of a food; and a simplified qualitative procedure that evaluates adequacy on the basis of amounts of foods found in the basic 4 Food Groups as compared to the Recommended Daily Allowances.

Detailed quantitative analysis of nutrient adequacy is determined by referring to a food composition table such as is found in Appendix I for the exact amounts of nutrients found in each food. The amounts are totaled and then compared with the Recommended Dietary Allowance. This procedure is recommended in some instances because it can serve as an exercise in discovering unsuspected nutrient values of foods. In a sense, it is dealing with the basic chemistry of foods, which has merit when variations in amounts of vitamins or minerals are critical. However, for control of dental caries, with the exception of fluorine, no critical level for any other nutrient has been established; therefore, this technique is not one that would be deemed practicable.

The qualitative *Exchange System* of foods used for calculating diabetic diets could be used for providing a controlled carbohydrate intake for caries control. For those clinicians who would like to adopt this technique for dietary analysis, a description of the procedure is presented in Chapter Twenty-eight.

The simplified *qualitative technique*, the third procedure for diet analysis, used in the Tufts plan (to be described in the next few pages) uses the basic 4 Food Groups in recommended amounts as a basis for determining dietary adequacy. This is a comparatively simple food guide which is not only valuable for rapidly checking the patient's food intake but also serves as an easily understood teaching device for the patient.

Table 29-1 used by the Michigan group and Table 29-2 used by the Oregon group are two examples of qualitative techniques for prescribing diets for caries control.

TABLE 29-1.　MICHIGAN PLAN FOR CARIES CONTROL*

This diet plan contains no sugar or foods prepared with sugar. It is low in starch also. *The first two weeks*, potatoes, and foods prepared with flour, such as bread, crackers, and thickened gravy, are not permitted.

The complete diet plan consists of three two-week parts, as follows:

PART I

Preliminary dietary period: *two weeks*. The diet contains approximately 100 gm. of carbohydrate (no matter what age group), with protein and calories adequate for the age and activity of the patient.

PART II

Foods from Part I; bread is added. Vegetables are increased to desirable amounts, including potato. None is to be prepared with sugar. All vegetables eaten must be fresh, frozen, dried or canned without sugar.

PART III

Continue Part II, adding as much sugar as is desirable at one meal during the day. This sugar is to be taken with the meal—not between meals.

If the lactobacillus count has not increased *during the two-week period* on Plan III, the diet is then unrestricted.

Patient cautioned to not overdo on sweets. Best to have a recheck every three to six months. If count is again high, patient is so advised and asked to return to the diet.

Advantages

　　a. Restricted diet often of shorter duration on this plan than on Oregon plan.
　　b. Usually more rapid lowering of the lactobacillus count.

Disadvantages

　　a. This plan is in the high cost bracket—high protein and high fat—and in the majority of the clinic patients' cases, it is often economically difficult or impossible.
　　b. In the younger children it is very difficult to achieve best cooperation because of drastic change, not only in elimination of sweets but also in restriction of breads, potatoes, etc.
　　c. When many children eat at the school lunchrooms or carry lunches, it is difficult to achieve adequate supervision or cooperation.
　　d. Less chance of altering taste for sweets and faulty food habits in a short-range program for diet regimen.

*Used by permission of the University of Michigan School of Dentistry.

TABLE 29-2.　OREGON PLAN FOR CARIES CONTROL*

PART I

Diet plan follows the basic 4 Food Groups pattern.

Calories, fat, carbohydrate and protein are calculated for each individual, considering age, height, bone structure, etc.

Diet is restricted in carbohydrates in that allowance is lowered at least one fourth to one third below recommended allowances, but is never lowered as drastically as in the Michigan Plan.

The protein allowance is calculated *at least* 10 per cent above the recommended allowances.

All free sugar and concentrated sweets are excluded—in other words, no sugar or foods prepared with sugar are allowed. This also means no honey, molasses, raw sugar or the like.

Length of time on the diet as planned is variable, depending on individual differences, and on each individual's cooperation.

PART II

Patients are kept on the diet as planned until *at least three consecutive sets of zero counts (or very low level—100 to 300) are received* (lactobacillus counts). *Then* and *only then* is the patient allowed to add sugar in desired amounts at *one meal* only—*none between meals*.

When several sets of counts show no increase, the patient is then allowed to go off the diet regimen. At that time the patient should be cautioned to keep concentrated sweets to a minimum and preferably at mealtimes.

It is well to recheck the patient in three to six months, by checking the oral cavity and taking additional saliva samples to determine the count. If the count is again high, patient should be so advised and asked to return to diet regimen.

Advantages

　　a. This plan is more economically feasible for the average family at the present time.
　　b. The plan sets forth a good procedure and background for future use in developing a better understanding of a well balanced diet. Following the "Basic 4" makes for better dietary habits for both general health and, specifically, dental health.
　　c. The Oregon Plan is a well balanced diet and adaptable to a long-range program.
　　d. Good food habits are more firmly established by a long-range program. "Breakovers" may occur, but to a lesser degree than in a short-range program.

Disadvantages

　　The lactobacillus count lowers more slowly.

* Used by permission of the University of Oregon Dental School.

BASIC CONCEPTS

ETIOLOGICAL CLASSIFICATIONS

It is a basic tenet in medicine and dentistry that the management of any problem is on a sounder basis if one understands and deals with the cause rather than the effect of the disease. Therefore, a classification of the patient with high caries susceptibility will serve as a useful tool in providing a rational treatment plan and will also make the prognosis easier.

A Juvenile Caries-susceptible Patient. In this category are classified those children who develop rampant caries in their deciduous dentition. There is a strong possibility that a genetic or systemic factor plays a major role in predisposing the patient to his high caries susceptibility. In these cases the dentist is faced not only with the local environmental dietary factors but also with an innate poor tooth quality. The challenge here is great.

A Patient with Caries Based on Psychiatric Influences. These patients are characterized by excessive and bizarre eating habits. They have cravings for particular foods such as sweets. For such patients to eat a pound of chocolates or an inordinate quantity of dried fruits like raisins is not unusual. They claim that they have a "sweet tooth" addiction and "can't shake the habit." If this inordinate desire is based on a deep-seated psychological or emotional disturbance, the prognosis is not encouraging unless the basic emotional problem can be resolved.

A Patient with Adolescent Caries. This is, by far, the more prevalent or more "ordinary" type. Local factors predominate here. With this type of patient, the why of the diet is influenced by environmental circumstances which can at least be modified or managed to a degree; therefore, the prognosis for the patient who falls within this classification is favorable.

DIETARY PRINCIPLES

According to the Tufts plan for dietary control of caries, the fundamental principle in prescribing a diet is to recommend a normal diet as a base line and then to modify it to suit the individual's needs, keeping in mind the therapeutic goal. A normal diet is one which provides the essentials of good nutrition, such as the 4 Food Groups, in the amounts recommended.

In general, dietary modifications are made (1) by altering the physical texture of food, or (2) by increasing or decreasing some nutrient, or (3) by eliminating a specific food. Applying these principles to the caries-susceptible patient, one might institute the following modifications: For changing texture, a sticky retentive diet would be changed to one that is detergent and easily cleared from the oral cavity. If individual nutrients should be emphasized, fluorides in particular and minerals in general as well as increased use of sources of protein are reasonable on the basis of present scientific evidence. The foods that definitely must be deleted are sticky, readily fermentable carbohydrates, which means eliminating foods that are made with sugar (cakes, candy, etc.) and those that are rich in natural simple sugars (dates, raisins, figs, etc.).

METHODOLOGY

First Visit

STEP 1. SELECTION OF PATIENTS

A valid argument might be made for giving dietary counseling to all patients just as we give tooth brushing instructions to all patients because both are important adjuncts in the control and prevention of the local factors responsible for dental caries. No doubt as the application of the art and science of nutrition becomes more widespread as an important tool in the practice of preventive dentistry, more and more patients will receive the benefits of dietary counseling. For the present, from a practical and realistic standpoint, the patient for whom this service is indispensable is the one with high caries susceptibility. He, above all, must receive the benefits of this service. In addition, children who are wearing bands and appliances that

TABLE 29-3.   THE DETERMINATION OF
RELATIVE CARIES EXPERIENCE*

Age	Rampant	High	Low	Re-sistant
2	3 +	0–2	0	0
3	6	2–5	0–1	0
4	9	3–8	0–2	0
5	9 + def 1 DMF	4–8	1–3	
7	10 + def 5 + DMF	7–9 def 1–4 DMF	1–6 def	0
9	10 + def 5 + DMF	6–9 def 3–4 DMF	0–5 def 0–2 DMF	0 0
11	8 + DMF	5–7 DMF	1–4 DMF	0
13	12 +	6–11	1–5	0
15	15 +	9–14	3–8	0–2
17	18 +	9–17	4–8	0–3
19	21–32	14–20	7–13	0–6
20–24	26–32	18–25	9–17	0–8
25–29	28–32	20–27	12–19	0–11
30–34	28–32	20–27	13–19	0–12
35–39	29–32	22–28	14–21	0–13
40–44	32	22–32	14–21	0–13

*Adapted from Grainger, R. M., and Nikiforuk, G.: Canad. Dent. Assn. J., *26*:531, 1960.

are a necessary part of orthodontic treatment might well be considered as good candidates for this service.

Some criteria for classifying a patient as having rampant caries are: (1) The relatively high caries experience according to the number of def or DMF teeth at a specific age (Table 29-3). (2) The rapid rate of new and recurrent caries. (3) The location of the lesions on surfaces usually considered caries immune, e.g., interproximal surfaces of lower anterior teeth or the labial surfaces of anterior and the lingual surfaces of posterior teeth. (4) The erosive, chalky white and diffuse areas of decalcification, particularly around an existing restoration. (5) The light tan or ivory color and soft mushy consistency of the decay.

STEP 2. INSTRUCTION OF PATIENT IN KEEPING OF FOOD DIARY

Actual food intake is, as a rule, not adequately ascertained by direct questioning about the usual eating patterns and meal practices. Important details may be omitted or may be inaccurate because remembering the kinds, amounts and preparation of foods

is difficult. In the case of a child, the parent will often comment on how well her child eats and how infrequently he partakes of sweets. This sort of information should be gratefully received but should be countered by the comment that specific amounts of each food must be recorded so that a meaningful diet evaluation can be made.

The preferred method for recording food intake is to use a specific diary form similar to the one shown in Table 29-4. On this diary are printed instructions for what to list and how to record the information about foods eaten at mealtime and between meals. It is kept for at least five consecutive days that include a week-end day or a holiday. Since there is usually a considerable difference between meals on those days and ordinary weekdays, both types of days must be taken into consideration in order to arrive at a true representation of a food intake pattern. Usually the responsibility for keeping the diary is assigned to the parent unless the patient is old enough to be relied on. It is also important to instruct the patient that no change or "improvement" be made in the customary meals during the period that the food diary is being kept; doing so will prevent a true and accurate evaluation of his usual food intake patterns.

The patient should record every type of food or beverage that is consumed during the day and evening, no matter where it is eaten —at home, at work, at school, at play, at the corner "sweet shoppe" or at snack time while watching television or movies or while reading, etc.

Instruct the patient to record, (1) all foods and mixed dishes, for each meal or between-meal snack, (2) the amounts in household measures (teaspoons, tablespoons, servings, cups), and (3) the type of preparation (fried, boiled, baked, broiled, raw).

Fruits, vegetables, bread, meat and cheese should be described according to their physical nature, as raw, refined, whole, chopped, hard or soft. The method of preparation, whether fresh, frozen, dried or canned, is useful information and should also be included.

The sequence of foods as eaten should be accurately recorded to help establish the detergency of the diet.

TABLE 29-4

FOURTH DAY

Food: Quantity: Prepared:

BREAKFAST:
None

10:00 A.M.

LUNCH:
2 glasses milk
2 bologna sandwiches
(4 slices white bread)
(4 slices bologna)

3:00 P.M.
1 coca-cola
1 egg & olive sandwich
(2 slices white bread)

DINNER:
2 servings roast turkey
1 serving mashed potato
1 serving boiled carrots,
olive, cranberry jelly
2 glasses milk

EXTRAS: Bedtime
1 peanut butter sandwich
(2 slices whole wheat bread)
1 glass milk
1 glass raspberry Tonic
1 glass orangeade Tonic
2 glasses grape Tonic
3 juices Lebrun-covered
jelly candies

FIFTH DAY

Food: Quantity: Prepared:

BREAKFAST:
None

10:00 A.M.
1 piece toast
1 glass grape juice
1 glass milk

LUNCH:
2 glasses milk
1½ turkey sandwiches
(3 slices bread) (½ tsp
2 teas butter   mayonnaise)

3:00 P.M.

DINNER:
2 servings pork chops
1 serving mashed potatoes
1 serving string beans
1 glass milk
1 black & white cookie

EXTRAS:
1 glass orangeade tonic
1 peanut butter sandwich
(2 slices whole wheat bread)
1 glass milk

FOOD INTAKE DIARY

Of: John Doe

FOR THE FIVE DAYS (include one WEEKEND DAY or a HOLIDAY)

INSTRUCTIONS

1. Record every type of food consumed, solid or liquid, at mealtime, between meals, at the soda fountain, while watching television. Record also, candies, Lifesavers, gum, cough drops or syrups.

2. For each meal, list the food, preparation, (fried, boiled, etc.) and amount in household measures (1 t., 1 T., 1 cup (8 oz.), 1 4 oz. glass, no. of pieces).

3. For fruits and vegetables, record whether raw, fresh, frozen or canned.

4. Record amount of sugar or sugar products and cream or milk added to cereal, beverages or other foods.

5. Record foods in the order in which they are eaten.

6. Particular information on extras is most important to us. Do not leave out the smallest detail.

Patients should also note the amount of sugar added to foods and beverages; the amount of fats and dressing added to vegetables and salads; and the amount of butter, margarine, jam or jelly spread on bread.

Any between-meal snacks such as cookies, cakes, chewing gum, lozenges, cough drops and cough syrup are as important to record as the main course of any meal.

Demonstration of Recording of Previous 24-Hour Intake. In order to make clear the kind of detail that is desired in this diary, it is best to demonstrate by actually recording for the patient his previous 24-hour food intake. The questioning usually begins with the meal or snack that the patient has eaten just prior to his office visit. For example, if the patient is seen at 1:30 P.M., the first meal to record is the noonday meal which he consumed perhaps an hour before.

The questioning may be along the following lines: What did you eat at lunch today? (The patient might answer, "A sandwich and milk." This is an incomplete answer, and the patient should be advised of the type of descriptive detail that is desired.) What was the filler for the sandwich? What kind of bread? rye? soft? white? toasted? Was there mayonnaise or butter spread on the bread? What did you have as a beverage? If milk, was it plain or flavored? Were the sandwich and milk consumed together or was the entire sandwich eaten first and the meal finished with milk? What did you eat for dessert? If cookies, how many? What kind?

Then working backwards in recording the food intake, we next ask questions like, What did you eat between luncheon and breakfast? (The patient might answer, "A cup of coffee and a piece of cake at 10:00 A.M.") Did you use sugar in the coffee? How much? Was it coffee cake, plain cake or frosted cake? Was the piece large, medium or small? Did you chew any gum or suck on any hard candies?

How about breakfast? (Just juice, toast and coffee," the patient answers.) Is this the sequence in which you ate the foods? What kind of juice and how much? How many cups of coffee and how many teaspoonfuls of sugar in each one?

The patient is next asked to recall what he ate at bedtime as well as during the pre-vious evening. A record of nibbling while watching television or snacks before retiring, particularly when not followed by toothbrushing, is important information that should not be overlooked.

It is sometimes difficult for the patient to recall immediately all the specific foods eaten at the previous evening's dinner. However, if a little time for concentration is allowed, the patient usually recalls the main course first and then the accompanying side dishes. It may take a little suggestive reminding: Did you have potatoes with your meal? Did you put butter on your potatoes? What kind of salad did you have? Was there dressing on it? What did you have for dessert? Was a beverage taken with the dessert?

The last eating period that is inquired about and which will complete a 24-hour intake for this hypothetical patient is the midday or after-school snack. Did you stop and eat something at the corner drugstore or soda fountain on the way home from school? What kind of snack did you find at home?

This demonstration of diary keeping will give the patient a much clearer idea of the detailed information that is expected of him. The 24-hour intake that has just been recorded can be used as the first day. For the next four days an attempt should be made to record each meal and snack as it is eaten. The patient should not trust his memory or his good intentions to record this information later. Proper keeping of the diary is one of the first indications of the type of cooperation that can be expected when a diet prescription is given. Actually, the diagnosis and treatment plan can be only as accurate and as useful as the information on which it is based.

Second and Third Visits

DETERMINING THE "WHY" OF THE DIET

The "why" of the diet, as the word implies, is that portion of the history that reveals the *reasons for food selection.* This part of the diagnostic work-up is important because the tailoring of the final recommendations and enlisting the *cooperation* of the patient, as well as the ultimate success of the entire counseling procedure, hinge on an understanding of the "why" of the diet. The reasons for food selection, in general, are elicited from a personal and family history which deal specifically with environmental or systemic or psychological factors.

PERSONAL AND FAMILY HISTORY

The environmental factors are those concerned with everyday living experiences. Pertinent information should be tactfully sought on the occupation of the family breadwinner, the patient's economic status, his cultural or ethnic or religious background, his daily routines, how often he eats during a day, and his general living conditions and home life.

The systemic considerations pertain to physical conditioning factors, such as allergies or spurt growth periods, which can influence significantly the food selection and nutritional requirements of the patient. For children under six, the nutritional history of the mother during pregnancy, whether the child was breast or bottle fed, and the circumstances surrounding the weaning of the child are factors that need to be considered. From a nutrition standpoint, the weaning period may have been a stressful experience. Not only could there have been psychological trauma, but appeasement and bribing with sweets might have been tried in order to overcome an unpleasant situation.

Psychologically, one must be alert to detect strained family relationships or domestic discord. Other factors like jealousy or sibling rivalry, adolescent rebellion, overprotective or completely disinterested parents as well as oversolicitous relatives should be determined. Patients with a craving for sweets might well be tagged as "carboholics" because they are psychologically in the same category as compulsive eaters, alcoholics or smokers.

CLINICAL EXAMINATION

The objective part of the examination consists of noting general appearance, facial expression, obesity or emaciation, and normal or abnormal skeletal development. Skin changes of dryness and erythema, evidence of paresthesia and peripheral neuropathy, or generalized muscular weakness are possible signs of impaired nutritional status. The lips and mouth may reflect nutritional anemia when there is a generalized pallor of the oral and lingual mucosa. Cheilosis, stomatitis, swollen interdental papillae and papillary atrophy of the tongue are all signs that may suggest nutritional inadequacy.

Fourth Visit

DIETARY HISTORY AND COUNSELING

In order to give the dietary counseling procedure the importance it deserves, this visit should be devoted *exclusively* to it. Furthermore, the parent or homemaker should accompany the child or adolescent because it is only with her complete understanding and cooperation that this procedure can be successful.

STEP 1. EXPLAIN THE PROCESS OF DENTAL DECAY

To justify the dentist's role in providing this nontechnical type of service, the process of dental decay should be explained to the patient.

It will emphasize the etiological role of food and nutrition in dental caries. The simplest and most easily understood description is one that uses a chemical reaction type explanation:

1. FOOD $+$ BACTERIA $\longrightarrow$ ORGANIC ACIDS
2. ACID $+$ TOOTH $\longrightarrow$ STARTS DECAY

Of the three variables, food, bacteria and tooth, it will be readily agreed by the patient that food is the easiest to manipulate and manage.

STEP 2. CIRCLE IN RED ALL FOODS SWEETENED WITH SUGAR

The patient is asked to circle in red all the foods, beverages or confections recorded in his five-day diary that contain sugar. Dried fruits like figs, dates, prunes, apricots and raisins should also be circled.

This procedure does two things: First it separates the good foods (the foundation foods) from the less nourishing (the "empty calorie," sweet foods). Secondly, it vividly points up any diet that is unbalanced and non-nutritive. The number of red-circled foods are always more than the patient suspected.

STEP 3. EVALUATE THE ADEQUACY OF THE FOUNDATION FOOD INTAKE

The Diet Evaluation Summary, Table 29-5, lists the 4 Food Groups in the left-hand column. Each of the first five columns across the top of the chart represents a single day and the sixth column is for the average per day value. The next three columns record the recommended amounts of standard servings for children, adolescents and adults. The actual average intake in column six is compared with the recommended allowance in the column next to it, and the difference between the two is recorded in the last column. If the

patient is eating as much or more than is recommended the symbol O.K. is placed in the last block. If he eats less than the amount recommended, a minus (—) quantity should be tallied.

A valuable aid in nutritional education is to allow and even encourage the patient to tally his own food intake and place the markings in the appropriate blocks of the food groupings. As the patient records the tally he will be learning the foods that belong to each of the 4 Food Groups and will undoubtedly begin to appreciate the difference between the types of foods he thought he was eating and the types he actually consumed.

The Tally Technique. We transpose the information from the patient's food diary to the chart by classifying the foods into one of the 4 Food Groups. Each average serving or portion (Appendix II defines the amount of food considered a serving) is credited as a single unit and is recorded as a chit mark (/) in the appropriate column. Five servings would be recorded as ////.

Half portions are recorded as ½; for example, two eggs are considered one serving. If the patient ate one egg, the tally is ½ in the meat column. When a leaf of lettuce

TABLE 29-5

FOOD GROUPS	1st Day	2nd Day	3rd Day	4th Day	5th Day	Ave. per Day	Child	Adol.	Adult	Difference
								Recommended Amounts		
MILK GROUP	/	//	///	/	//	2	3-4 serv.	4 or more serv.	2 serv.	child −2 adol. −2 adult ok
MEAT GROUP	//	/	0	//	/	/+	2 or more servings			−/
VEGETABLE-FRUIT GROUP Total No. Serv. (Including those rich in Vitamin C & Vitamin A)	//	/	///	////	0	2	4 or more servings (Including ----- ----- 1 serving --1 serving every other day----)			−2
BREAD-CEREAL GROUP Enriched or Whole grain	////	//// //	////	//// /	///	4	4 or more servings			ok

DIET EVALUATION SUMMARY

and a slice or two of tomato are included as part of a sandwich, this is tallied as a ½ portion in the vegetable-fruit column.

A specific example of how the tallying technique is applied can be demonstrated as follows:

A lunch composed of a tuna fish sandwich on enriched toasted white bread with butter, a side dish of cole slaw, a cup of milk and an apple would be tallied thus:

The tuna fish is entered as a single serving in the MEAT GROUP. Two slices of bread were used for the sandwich; therefore, two chit marks are entered in the BREAD GROUP. The cole slaw and the apple are recorded in the VEGETABLE-FRUIT GROUP as two chit marks, and the cup of milk as one chit mark in the MILK GROUP.

FOOD GROUP	1st DAY
MILK	/
MEAT	/
VEGETABLE-FRUIT	//
BREAD-CEREAL	// .

The average intake per day is calculated by dividing the sum of the units on the chart by five and recording the number in the average column. The average is then compared to the recommended amounts and the difference is recorded in the "Difference" column as O.K. or minus (—).

FOOD	1st Day	2nd Day	3rd Day	4th Day	5th Day	Av.	Rec. Amts. Ch.	Rec. Amts. Adol.	Rec. Amts. Ad.	Diff.
MILK	//	//	/	//	//	2	4	4	2	O.K.

STEP 4. DETERMINE FORM AND FREQUENCY OF SWEETS

After the sweets listed in the five-day diary have been classified according to their form and the number of times that the teeth have been exposed to them tallied, as shown in Table 29-6, a total for each category can be made and the grand total recorded. If this

figure is multiplied by 20 (the approximate number of minutes it takes for the pH of the dental plaque to regain its neutrality after exposure to sweets) the patient will be vividly impressed with the undesirable effects.

STEP 5. MAKE DIAGNOSIS

With information on what the patient eats and why he selects these foods, a provisional diagnosis can be made; for example:

1. Inadequate diet as a result of a lack of knowledge about fundamental foods coupled with lack of adequate funds.

2. Use of sweets as a reward food by parents or grandparents as countermeasure for an unpleasant situation (overprotective parent, sibling rivalry, frustration, tensions, etc.).

3. An existing allergy to wheat products and, therefore, excessive use of sweets as a source of carbohydrates.

STEPS TAKEN IN MANAGING THE "WHY" AND THE "WHAT" OF THE DIET—THE DIET PRESCRIPTION

Managing the "Why" of the Diet

A fundamental concept of the management of any disease is removal of the cause or factor which conditions it. From the list of some of the typical impressions just given, we can see that there may be secondary conditioning factors such as systemic disease (e.g., allergy), psychological problems or social problems that are the basis for the poor diet.

These must be dealt with first. If the secondary factors are ignored, the time spent in devising a new diet may have been wasted because the patient may find it impossible to cooperate until these problems have been brought under control. In practice, the dentist can only recognize that such conditions exist and refer the patient to the physician, psychiatrist or social worker (whomever is indicated) for

TABLE 29-6

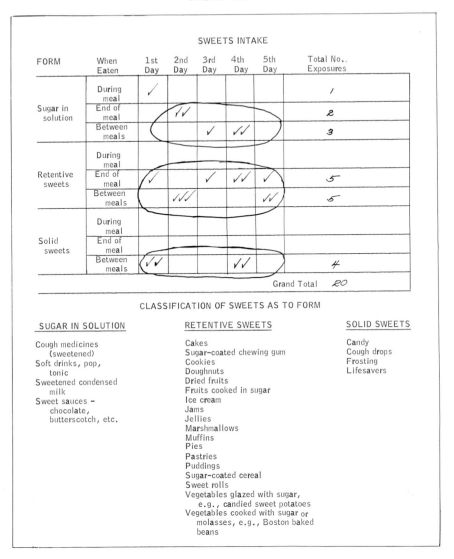

SWEETS INTAKE

FORM	When Eaten	1st Day	2nd Day	3rd Day	4th Day	5th Day	Total No.. Exposures
Sugar in solution	During meal	✓					*1*
	End of meal		✓✓				*2*
	Between meals			✓	✓✓		*3*
Retentive sweets	During meal						
	End of meal	✓		✓	✓✓	✓	*5*
	Between meals		✓✓✓			✓✓	*5*
Solid sweets	During meal						
	End of meal						
	Between meals	✓✓			✓✓		*4*

Grand Total   *20*

CLASSIFICATION OF SWEETS AS TO FORM

SUGAR IN SOLUTION	RETENTIVE SWEETS	SOLID SWEETS
Cough medicines (sweetened)	Cakes	Candy
Soft drinks, pop, tonic	Sugar-coated chewing gum	Cough drops
Sweetened condensed milk	Cookies	Frosting
Sweet sauces – chocolate, butterscotch, etc.	Doughnuts	Lifesavers
	Dried fruits	
	Fruits cooked in sugar	
	Ice cream	
	Jams	
	Jellies	
	Marshmallows	
	Muffins	
	Pies	
	Pastries	
	Puddings	
	Sugar-coated cereal	
	Sweet rolls	
	Vegetables glazed with sugar, e.g., candied sweet potatoes	
	Vegetables cooked with sugar or molasses, e.g., Boston baked beans	

actual management. Most patients, however, present superficial problems. A few typical problems with which the dentist should be able to cope and some suggested solutions are presented here:

Problem. One of the most common environmental factors that influences food selection is cost.

Solution. To cope with this problem there are several suggestions that can be made. Expensive cuts of meat are not more nutritious; they are more palatable with less preparation.

With the use of present-day meat tenderizers and condiments even the less expensive cuts can be made tasty.

Expensive protein foods may be stretched by making mixed dishes; for example, meat and vegetable casseroles, ragouts and curries. Leftover foods can be put to good use by concocting interesting recipes. Nonfat milk powder is considerably less expensive than fluid milk; its use is highly recommended. Margarine is nutrionally equal to butter and costs about one third as much. A good cookbook will give suggestions for making appetizing dishes from lower cost foods. Some suggested low cost menus are given in Table 29-7.

Problem. Citizens of foreign birth usually have a dual problem: they are often in a low income group and consequently are financially unable to purchase their traditional foods because they are more expensive in this country. Because of this they buy the most available and least expensive foods, which are usually breads, cereals and sweets.

Solution. People of every nationality have dishes and foods that are composed of each of the basic food groups. The dentist who has a knowledge of food habits of other nationalities can suggest the good points of the national diet. For example, cheese may be a more acceptable food from the milk group to Italian patients than fluid milk. Encouragement in the use of domestic cheeses whenever possible rather than the more expensive imported ones will help the food budget.

Problem. Adverse working conditions and hours as well as daily routine may limit daily menus to frequent snacks rather than a few complete meals.

Solution. Suggest better sleep habits so that the patient can arise in time for a more balanced breakfast. Although most people who have to carry their noonday meal from home use sandwiches, these can have a nutritious filling of meat, fish, cheese or egg. It is quite possible that salads and even some mixed dishes can be taken to work in plastic containers. Good meals will increase the quality and satiety value of the diet so that snack eating will be less necessary. If snacks are needed, they may be selected from those listed in Table 29-8.

Problem. The patient may use sweets to overcome psychological tension.

Solution. The pleasant taste of sweets is sometimes subconsciously used to overcome unpleasant situations and frustrations. The dentist, in a nondirective manner, can suggest that the patient or the parent consider the factors that might affect this excessive craving for sweets. For example, the parent may be asked how the patient gets along with friends

TABLE 29-7.   SAMPLE ECONOMICAL MEALS TO AID IN CARIES CONTROL

Breakfast

Scrambled egg	Shredded wheat with	Hot oatmeal, milk
Whole wheat toast and	milk, no sugar	Toast with margarine
margarine	Toast with margarine	Milk
Milk—children and	Milk	Coffee
adolescents	Coffee, unsweetened	Orange juice
Coffee, unsweetened—	Grapefruit, fresh	
adults		
Whole orange		

Lunch or Supper

Corn chowder	Egg salad sandwich with	Grilled cheese and
Crackers or biscuits with	lettuce	tomato sandwich
margarine	Milk	Dill pickles (if desired)
Milk	Tea, unsweetened	Milk
Tea, unsweetened—	Crackers and cheese	Pear
adults	Celery curls	
Apple		

Dinner

Braised pork chops and	Baked fish fillets	Italian spaghetti with
potatoes	Baked potato with sour	meat balls
Scalloped tomatoes	cream and chives	Tossed green salad
Raw cabbage salad	Green beans, margarine	Crusty bread, margarine
Rolls, margarine	Hard rolls, margarine	Orange-banana fruit cup
Sliced peaches with	Melon	Milk
cream	Milk	Tea or coffee,
Milk	Tea, unsweetened	unsweetened
Tea or coffee, unsweetened		

TABLE 29-8. SUGGESTED BETWEEN-MEAL SNACKS FOR A
PERSON WITH RAMPANT CARIES

FRUITS:	Raw: apple, fresh apricots, melons (watermelon, cantaloupe), cherries, grapefruit, grapes, oranges, peaches, pears, pineapple, tangerines. Fresh fruit juices (unsweetened).
VEGETABLES:	Raw: cabbage, cauliflower, carrots, celery, cucumbers, lettuce and other salad greens, radishes, tomatoes, turnips. Tomato juice.
MILK:	Unsweetened
CHEESE:	All kinds—cheddar, cottage or cream
NUTS:	All kinds
COMBINATIONS:	Apple slices with cheddar cheese Cream cheese balls rolled in chopped nuts Celery stuffed with cottage cheese and chives Baloney slices spread with cream cheese and cut in wedges Shrimp on toothpicks with tomato cocktail sauce made without sugar Deviled eggs Pear halves spread with cream cheese and sprinkled with chopped nuts

or members of the family. The parent may reveal that the child is having social difficulties. The dentist may then inquire whether this may have affected the eating pattern. In many cases the parent will quickly grasp the relationship between the child's social problem, his unhappiness or his temporary emotional disturbances and the eating of candy and other sweets. At this point, there is usually no difficulty in arriving at some practical recommendation.

Problems which emanate from such temporary situations are amenable to manipulation. There are others that are of a more profound and permanent nature. The dentist should never attempt to break through any serious level of resistance nor should he offer a diagnosis which may be strongly resisted by the patient. In these latter cases, professional psychiatric advice should always be sought.

Case histories describing superficial emotional problems are briefly recounted here.

A 15 year old boy who had recently lost his parents and had to move to a new city to live with relatives developed rather peculiar eating habits. Because he was lonely and had not made any new friends, he went to the movies four or five afternoons a week. While at the movies he ate four to six bananas, five candy bars, etc. This had been going on for several months before he was seen. His relatives were advised of the boy's need for companions his own age. It was suggested that he might join the local Y.M.C.A. or community center where supervised athletics and other social activities were available.

Twin girls about seven years of age, one of whom had rampant caries and the other a much lower caries incidence, were referred for consultation by the pediatrician. According to the pediatrician, the disturbing feature was that the child with the caries problem was the healthier of the two. The sickly child had required almost constant medical care from birth and would have been the one expected to be more caries susceptible. After interviewing the parent, it became evident that the healthy child did not receive even the average amount of maternal attention. Catering to the sick child had been so demanding that the mother just had no time for the other twin. To compensate for this deficiency the grandparents took over and showed their affection for and attention to the healthier child by giving her generous amounts of sweets. Obviously, this excess of sweets, a result of the misdirected good intentions of the grandparents, was the cause of the decay.

A 19 year old boy who had recently become more aware of the opposite sex was concerned about his "bad breath." To offset this, he indulged in chain eating of cough drops and hard candies (two or three packages a day). He was reassured that he could not develop "bad breath" if his teeth were restored and cleaned and if he would carry out proper oral hygiene at home. Toothbrushing instructions were given, and mouthwashes and a diet of protective foods with emphasis on detergency were advised.

Managing the "What" of the Diet

The specific dietary advice that is given the patient should be written out on some type of instruction sheet or prescription pad. It should be designated as a diet prescription, which will emphasize its importance, and the patient will respect it as he does any other medical or dental prescription. Furthermore, the patient will have tangible evidence of the service and will not need to depend on memory because he can always refer back to the written prescription whenever the question arises as to which foods he may include in his diet or must avoid.

STEP 1.
COMMEND THE PATIENT

It is important to commence a counseling procedure on a positive and constructive note. Patients do not take too well to being criticized or scolded at the very outset. Since the food evaluation chart will probably show that the recommended allowances were met in at least one or two food groups, a good starting point is to commend the patient on this. Urge him to continue this good practice.

STEP 2.
SUGGEST IMPROVEMENTS

Again referring back to the evaluation chart, it will be readily seen (and may even be commented on by the patient) that he is deficient in one or two food groups. To overcome this, positive recommendations for improving the amounts and adequacy of the deficiencies should be made. There is a wide variety of foods in each food group and it is rare that the food likes of the patient cannot be suited by judicious substitution. Variety is the keynote in selecting a balanced diet. In fact, variety in food consumption is the best practical means of providing a nutritious diet.

STEP 3.
DELETE FROM
THE DIET THE
SUGAR SWEETENED FOODS

By re-examining the sweets intake chart, the grand total of the number of exposures to sweets, the most prevalent type, and the fre-

quency with which they were eaten will be noted. Since the form and frequency of use of sweets are the two most pressing factors in caries production, it must be emphasized that there can be absolutely no compromise with respect to the deletion of retentive sweets from the diet. The patient will simply not realize any significant caries reduction unless this one rule is firmly followed. Using analogies of medical problems such as mild diabetes or obesity which are controlled only by diet restriction might bring home the point. Caries cannot be controlled without elimination of sticky sweets from the diet.

STEP 4.
SUGGEST
SNACK SUBSTITUTES

It is important to realize that if snacking is a habit of long standing, it is futile and unrealistic to expect immediate abandonment of between-meal nibbling. The alternative suggested here is selection of desirable snacks such as raw fruits, raw vegetables, Cheddar cheese, nuts, etc. (Table 29-8). However, if the patient is constantly reminded that increasing the total food intake at each meal will satisfy appetite and hunger, it is possible that eventually the number of between-meal snacks will be reduced.

STEP 5.
SUGGEST IMPROVE-
MENTS OF THE MENU

Starting with breakfast and the customary dishes and foods that are eaten, suggest slight modifications. The rule is to *improve the quality not the quantity* of the food so that cooperation and acceptance will be more likely. For example, if the patient is accustomed to eating doughnuts and coffee sweetened with sugar, suggest as a substitute coffee sweetened with Sucaryl (or better, unsweetened) and muffins or toast. Do not attempt to list a five-course breakfast and expect the patient to cooperate if for years he has been in the habit of eating only two food items for breakfast. *Gradual* improvement not *drastic* change is a more realistic goal.

The same procedure of suggesting gradual

TABLE 29-9.   GENERAL SUGGESTIONS FOR FOODS TO BE INCLUDED
AND AVOIDED AT EACH MEAL

	Include	*Avoid*
Breakfast		
Eggs	Any style (bacon or sausage if desired)	Syrup on French toast
	French toast with butter or margarine	
Bread (with butter) or Cereal	Whole grain or enriched toast or hard variety (dark rye, Ry-Krisp)	Dried cereals with sugar, syrups, jams, jellies, buns, pastry or sweet rolls
	Cereal, cooked, with butter and milk	
Fruit, citrus	Whole fruit, fresh, (orange, grapefruit, melon)	Adding sugar to fruit
	Frozen or unsweetened canned juice	
Milk	Plain or buttermilk	Chocolate or other flavored and sweetened milk
Beverage	Coffee, tea, water, as desired	Sugar
Lunch		
Soups, juice	Fresh or canned soups, vegetable juice or unsweetened fruit juice	
Sandwich	Toasted or dark bread—Meat, fish, fowl, egg or cheese fillings	Raisin or cinnamon bread, jam, jellies or honey filling
Salads	Any combination of fresh fruits or vegetables, raw carrot or celery sticks	
Milk	Plain or buttermilk	Any flavored milks or soda pop
Desserts	Raw apples, oranges, tangerines or other fruits in season	Cakes, cookies, pies, pastry, bananas, ice cream, raisins, figs, dates or other dried fruits
Water	As desired	
Dinner		
Soup, juice	Fresh or canned soups, vegetable juice or unsweetened fruit juice	
Meats, fish	Beef, lamb, pork, veal, fowl or fish (liver at least once a week)	
Vegetables	1 or 2 portions, especially green or yellow, and potatoes	Candied sweet potatoes or glazed vegetables
Salad	Any combination of fresh fruits or vegetables, raw carrot or celery sticks	
Desserts	Raw apples, oranges, tangerines or other fruits in season	Same as lunch
Milk	Plain or buttermilk	Any sweetened milk
Beverage	Coffee, tea, water, as desired	Soda pop

changes is used for lunch. If the patient's lunch has been a jelly sandwich and coffee, do not prescribe vegetable soup, tuna casserole, raw salad, bread and butter, milk and fruit. Merely suggest that, instead of jelly, he use tuna fish for a filler and include milk as a beverage or perhaps coffee or tea without sugar, plus some carrot strips for a final food. Let that be sufficient for the first change. In the weeks that follow, a gradual build-up of quantity and variety can be attained by using the Exchange List of foods given in Chapter Twenty-eight (Table 28-3). The ultimate goal is to have the patient include and avoid foods at each meal as suggested in Table 29-9. Sample diets, depending on the age, sex, size and activity of the patient, are shown in Tables 29-10, 11, 12 and 13.

TABLE 29-10.   A SAMPLE DIET FOR A 6 YEAR OLD CHILD

1. Recommended food intake for a day:

Milk group...................... 3 or more servings
Meat group...................... 2 or more servings
Vegetable-fruit group.............. 4 or more servings
Bread-cereal group................ 4 or more servings
(Additional foods from the above groups plus fats may be
used to meet individual's energy needs.)

2. Meal Plan

Breakfast:

Milk or meat group—1 serving
Bread-cereal group—2 servings
Vegetable-fruit group—1 serving

Lunch:

Meat group—1 serving
Bread-cereal group—2 servings
Milk group—1 serving
Vegetable-fruit group—1 serving

Dinner:

Meat group—1 serving
Vegetable-fruit group—2 servings
Bread-cereal group—1 serving
Milk group—1 serving

3. Menu

Breakfast:

Wheat flakes, milk
Toast, butter
Milk to drink
Whole orange

Lunch:

Egg salad sandwich—consisting of
whole wheat bread
hard cooked egg
lettuce
butter or mayonnaise
Milk
Apple and cheese strips

Dinner:

Hamburg patties
Scalloped potatoes
Buttered green beans
Bread, butter
Milk
Celery and carrot strips
Peach

TABLE 29-11.   A SAMPLE DIET FOR A 9 YEAR OLD CHILD

1. *Recommended food intake for a day:*

Milk group.......................... 3 to 4 cups
Meat group.......................... 2 servings or more
Vegetable-fruit group.............. 4 servings or more
Bread-cereal group................. 4 servings or more
(Additional servings of any of the above food groups plus
    fats may be eaten to meet energy needs.)

2. *Meal Plan:*

Breakfast:

Meat group—1 serving
Bread-cereal group—2 servings
Milk group—1 serving
Vegetable-fruit group—1 serving

Lunch:

Meat group—1 serving
Bread-cereal group—2 servings
Milk group—1 serving
Vegetable-fruit group—1 or 2 servings

Dinner:

Meat group—1 serving
Bread-cereal group—1 serving
Milk group—1 serving
Vegetable-fruit group—2 or more servings

Between meals:

Milk group
Vegetable-fruit group

3. *Menu*

Breakfasts:
Fried egg
Toasted English muffin
    with butter
Milk
Grapefruit (fresh)

Shredded wheat with milk
    and sliced fresh peach
Toast and butter
Milk
Orange juice

Lunches:
Macaroni and cheese
Grilled frankfurter
Milk
Roll and butter
Cabbage-pineapple salad

Egg salad sandwich with
    lettuce and mayonnaise
Dill pickles
Milk
Salted peanuts

Dinners:
Broiled liver
Mashed potato
Broccoli, butter
Tossed salad
Bread, butter
Fresh pineapple

Hamburg patties
Creamed potatoes
Carrot and celery strips
Corn on the cob
Bread, butter
Fruit gelatin mold
Milk

TABLE 29-12.   A SAMPLE DIET FOR A 13 YEAR OLD GIRL

1. Recommended food intake for a day:

Milk group...................... 4 or more servings
Meat group...................... 2 or more servings
Vegetable-fruit group.............. 4 or more servings
Bread-cereal group................ 4 or more servings
(Additional foods from the above groups plus fats may be
used to meet energy needs.)

2. Meal Plan

Breakfast:

Meat or milk group—1 serving
Bread-cereal group—2 servings
Vegetable-fruit group—1 serving

Lunch:

Meat group—1 serving
Bread-cereal group—2 servings or more
Vegetable-fruit group—1 serving or more
Milk group—1 serving or more

Dinner:

Meat group—1 serving
Vegetable-fruit group—2 servings
Bread-cereal group—1 serving
Milk group—1 serving

Bedtime:

Milk group—1 serving

3. Menu

Breakfast:

Scrambled eggs
Toast, butter
Milk
Grapefruit (fresh)

Lunch:

Tuna fish casserole
Cole slaw
Milk
Crackers and cheese
Pear (fresh)

Dinner:

Meat loaf
Buttered carrots
Tossed green salad
Crusty rolls
Milk
Strawberries and cream

Bedtime:

Milk
Popcorn

TABLE 29-13.   A SAMPLE DIET FOR A 15 YEAR OLD BOY

1. Recommended food intake for a day:

Milk group...................... 4 or more servings
Meat group...................... 2 or more servings
Vegetable-fruit group.............. 4 or more servings
Bread-cereal group................ 4 or more servings
(Use additional servings of the above food groups and fats
   to meet energy needs.)

2. Meal Plan

Breakfast:

Meat group—1 serving
Bread-cereal group—2 to 4 servings
Milk group—1 serving
Vegetable-fruit group—1 serving

Lunch:

Meat group—1 serving
Vegetable-fruit group—2 servings
Bread-cereal group—2 to 4 servings
Milk group—1 serving

Dinner:

Meat group—1 serving
Vegetable-fruit group—3 servings
Bread-cereal group—2 servings
Milk group—1 serving

Bedtime or snack:

Milk group—1 serving
Bread-cereal group—1 serving
Vegetable-fruit group—1 serving

3. Menu

Breakfasts:

Fried eggs, with bacon if desired	Scrambled eggs
Toast, butter	Whole wheat toast, butter
Milk	Milk
Orange	Grapefruit

Lunches:

Hamburgers	Corn beef sandwiches on rye
Cole slaw	bread (with lettuce)
Milk	Milk
Cheese and crackers	Cheese and crackers
Apple	Pear

Dinners:

Pot roast	Pork chops
Mashed potatoes with gravy	Rolls (hard)
Carrots	Baked potato
Tossed salad	Green beans
Rolls, butter	Carrot strips
Watermelon	Plums

Bedtime or snacks:

Popcorn	
Milk	Milk
	Apples

Subsequent Visits

A follow-up visit should be scheduled for four to six weeks later. The patient is asked to complete a five-day food diary in the same manner as he did the first one.

The food diary is re-evaluated and the results of this evaluation are compared to the original plan to note whether recommendations have been followed. Misinterpretation, misunderstandings and problems that have arisen during this period are discussed. Menu changes are recommended, if necessary.

Patients' misconceptions of food composition can sometimes be surprising. For instance, a patient with high caries susceptibility mentioned that she had been constipated and, rather than use a laxative, had been nibbling on "some natural dried fruits that do not contain refined sugar." Unfortunately, the patient had not realized that natural sugars of dried fruits are just as cariogenic as refined sugar and sugar products. She was advised to increase her roughage and bulk with cellulose foods and to eliminate the dried fruits.

Continuous follow-up is very helpful in maintaining patient cooperation. The personal interest that the clinician takes in his patients is rewarded in many ways.

CONCLUSION

Dietary counseling is an effective and necessary service for caries-susceptible patients. It must be an integral part of the total oral health service. The dietary recommendations for caries-susceptible patients should be written specifically as a prescription after due consideration of the reasons for food selection. The diet itself should be individualized, balanced and varied by using a selection of the foods found in the 4 Food Groups. Sticky sweet foods should definitely be condemned and detergent foods used as substitutes. It is interesting that the optimal nutritional status for dental health, as suggested here, is essentially no different from that for general good health.

REFERENCES

Ast, D. B., Smith, D. J., Wachs, V. B., and Cantwell, K. E.: Newburgh-Kingston caries-fluorine study. XIV. Combined clinical and roentgenographic dental findings after ten years of fluoride experience. J. Amer. Dent. Assn., 52:314, 1956.

Bunting, R. W., Crowley, M., Hard, D. B., and Keller, M.: Further studies on the relation of bacillus acidophilus to dental caries. Dent. Cosmos, 70:1002, 1928.

Gustafsson, B. E., Quensel, C. E., Lanke, L., Lundqvist, C., Grahnen, H., Bonow, B. E., and Krasse, B.: The Vipeholm dental caries study; effect of different levels of carbohydrate intake on caries activity in 436 individuals observed for five years. Acta Odont. Scand., 11:232, 1954.

Kite, O. W., Shaw, J. H., and Sognnaes, R. F.: Prevention of experimental tooth decay by tube feeding. J. Nutr., 42:89, 1950.

Lossee, F. L.: Enamel caries research: 1962-1964. J Amer. Dent. Assn., 70:1428, 1965.

Mellanby, M.: An experimental study of the influence of diet on teeth formation. Lancet, 2:767, 1918.

Miller, W. D.: The Microorganisms of the Human Mouth. Philadelphia, S. S. White Dent. Mfg. Co., 1890.

Nizel, A. E.: The Cariogenic Properties of Similar Foodstuffs Grown in High and Low Caries Areas: the Influence of Trace Elements. Thesis, Tufts University, School of Dental Medicine, May, 1952.

Nizel, A. E., and Harris, R. S.: Phosphate and dental caries. I. Effect of metaphosphoric acid in diet of weanling hamsters on dental caries development. J. Amer. Dent. Assn., 60:193, 1960.

Nizel, A. E., Baker, N. J., and Harris, R. S.: The effect of phosphate structure upon dental caries development in rats. Int. Assn. Dent. Res., 1962. (Abstract.)

Orland, F. J., Blayney, J. R., Harrison, R. W., Reyniers, J. A., Texler, P. C., Wagner, M., Gordon, H. A., and Luckey, T. D.: Use of germfree animal technic in the study of experimental dental caries. I. Basic observations on rats reared free of all microorganisms. J. Dent. Res., 33:147, 1954.

Paynter, K. J., and Grainger, R. M.: The relation of nutrition to the morphology and size of rat molar teeth. J. Canad. Dent. Assn., 22:519, 1956.

Scrimshaw, N.: Objectives of Conference on Nutrition Teaching in Dental Schools. J. Dent. Educ., 1966 (in press).

Shaw, J. H.: Effect of carbohydrate-free and carbohydrate-low diets on the incidence of dental caries in white rats. J. Nutr., 53:151, 1954.

Sobel, A. E., Shaw, J. H., Hanok, A., and Nobel, S.: Calcification. XXVI. Caries-susceptibility in relation to composition of teeth and diet. J. Dent. Res., 39:462, 1960.

Sognnaes, R. F.: Caries conducive effect of a purified diet when fed to rodents during development. J. Amer. Dent. Assn., 37:676, 1948.

Wolbach, S. B., and Howe, P. R.: The effect of the scorbutic state upon the production and maintenance of intercellular substances. Proc. Soc. Exp. Biol. Med., 22:400, 1925.

Wolbach, S. B., and Howe, P. R.: The incisor teeth of albino rats and guinea pigs in vitamin A deficiency and repair. Amer. J. Pathol., 9:275, 1933.

Chapter Thirty

The Rational Use of Vitamins in Clinical Practice[*]

WILLIAM J. DARBY, M.D., PH.D.

Supplementary vitamins may be prescribed as multivitamin preparations with or without added minerals of varying composition, as preparations of individual vitamins, as concentrates or supplements of vitamin-rich foods (yeast, liver extract, etc.), or in other forms. They may be taken orally or parenterally, frequently are available in a wide range of dosages, and the majority of preparations are obtainable over the drug counter as nonprescription items. Patients seem to like to "take vitamins" and often expect the physician and dentist to advise or prescribe them, especially for young children, pregnant women or older persons. The clinician himself frequently prescribes vitamins for his patients without seriously considering the need or benefit, often as an assumed precaution against faulty nutrition without making any critical attempt to determine whether the patient is well nourished. Many dentists regard all patients as being guilty of malnutrition without giving them the benefit of trial.

The present article will examine some of the facts bearing on the requirements for vitamins and the need for vitamin supplementation. It will not attempt to review in detail all the claims and counterclaims made relative to the subject. This has been done elsewhere (Culver, 1949; Goldsmith, 1956; Ruffin and Cayer, 1944).

* Adapted from an article, Recent Advances in Applied Nutrition, ed. by S. L. Halpern, which appeared in the Medical Clinics of North America, Vol. 48, No. 5, September, 1964.

GENERAL CONSIDERATIONS

In order to appreciate the proper use of supplementary nutrients in medicine and dentistry, several considerations must be understood: (1) The normal source of essential nutrients, including vitamins, is food or, in one or two instances, synthesis by the gastrointestinal flora (particularly in ruminants). (2) Except under unusual circumstances, a sufficient quantity of those nutrients required by man is provided by an appropriately varied diet such as is usually consumed in the United States. (3) Administration of supplementary essential nutrients is indicated when there is a readily recognizable deficiency state. (4) The nonspecific general tonic-like effect commonly attributed to the administration of supplements is merely a placebo effect (Ruffin and Cayer, 1944; Wolf, 1950; Wolff et al., 1946). (5) The daily need for nutrients may be increased by conditioning factors, and the existence of such factors may justify protective supplementation. (6) The indiscriminate and unnecessary prescription of vitamins or supplementary nutrients has no more justification than does unnecessary medication of any sort. (7) Nutrient or vitamin supplements, although differing widely in price, are not inexpensive. (8) Gross overdosage with certain nutrients, including some vitamins, may be harmful or even fatal (A.M.A. Council on Foods and Nutrition, 1959; Bauer and Freyberg, 1946; Donegan et al., 1949; Nieman and Obbink, 1954). (9) The administration of even harmless medication in lieu of the establishment of a definite diagnosis or the institution of a proper therapeutic regimen is poor medical and dental practice.

One of the therapeutic experiences most gratifying to the clinician is to witness the response of a deficiency disease to specific replacement therapy. Prevention of such a disease is, of course, more satisfying. Prevention is usually best accomplished by assuring a continuing good dietary intake through a varied food consumption pattern which includes daily an appropriate diversity of the basic 4 Food Groups: milk group, meat group, vegetable-fruit group and bread-cereals group.

This basic plan supplies the adult with one half to two thirds of needed calories and four fifths or more of the allowances of iron, thiamine, niacin and riboflavin as recommended by the Food and Nutrition Board. Fat spreads and oils, desserts and other foods not included in these groups are added to fulfill the Recommended Dietary Allowances (Food and Nutrition Board, 1964).

STORAGE OF VITAMINS

The intake by an individual of a particular nutrient may range over a one- to four- or fivefold variation without producing any effect on health. The capacity of the body to store nutrients is such that for ascorbic acid it requires 4 to 16 weeks of deprivation to lower the serum levels of the healthy, well nourished adult to zero concentration, 17 to 21 weeks for the appearance of minor clinical evidences of deficiency, and five to eight months for interference with wound healing to occur (Medical Research Council, 1953). The depletion time for thiamine or riboflavin seems to be of this same order of magnitude (Horwitt et al., 1956). By contrast, it requires six months to seven years or more to deplete a well nourished patient of vitamin B_{12} to a point of detectable hematologic relapse (macrocytosis, minimally detectable decrease in erythrocyte count) (Darby et al., 1958). To reduce the adult stores of vitamin A to a point of clinically demonstrable functional impairment, 10 to 20 months of depletion are required (Medical Research Council, 1949). Short periods of deprivation are, therefore, well tolerated by adults, although for infants the depletion time

for many nutrients is much shorter because of lower initial stores and the relatively higher requirements for growth.

In general, the water soluble vitamins are excreted in the urine and sweat, but fat soluble factors are not. Hence, depletion of the latter group usually requires a particularly long period of deprivation. The long depletion time of the water soluble vitamin B_{12} is, in part, a result of the enterohepatic circulation of this nutrient. We have demonstrated that vitamin B_{12} has a biological half-life of the order of 407 to 770 days and that only some 0.1 per cent of body stores is lost per day; hence, the long but variable depletion time (Bozian et al., 1963; Darby et al., 1958).

From these several considerations it is evident that a normally varied, easily obtained diet can be depended upon to meet adequately the needs of the usual person, and that on such a diet the stores are sufficient to tide the subject over ordinarily encountered periods of deprivation, bridging brief intervals of anorexia or of purposeful withholding of foods. Indeed, during brief or moderate periods of starvation the tissues that are catabolized provide a metabolic mixture of high nutrient value. Therefore, there is little reason to give supplements during starvation of a few days' duration.

QUALITY OF USUAL NUTRIENT INTAKE

The United States dietary is often deprecated as having been rendered nutrient-poor by undue artificiality as a result of circumstances of agricultural production, of food processing, of preparation or of storage. None of these deprecations is justified in fact. Modern methods of production, processing and storage assure good nutrient retention and a more dependable level than old-fashioned home procedures. Furthermore, they render readily available throughout the year quantities of foodstuffs and a variety previously impossible. They have eliminated the limited seasonal availability of most foods. In addition, many universally available foodstuffs have been enhanced in nutrient value by the deliberate, controlled addition of nutrients or concentrates to the foodstuffs during the process of preparation. Examples are the enrichment of flour, maize, rice and cereals with niacin, riboflavin, thiamine, iron and, sometimes, calcium; the controlled addition of vitamin D to milk; the iodinization of salt; the incorporation of vitamin A in margarine and other fats; the inclusion of iron in certain infant foods; the addition of milk solids to bread and cereals; the enhancement of the ascorbic acid content of fruit juices; and the fluoridation of water. These well-considered measures assure an increased supply of many nutrients in the United States diet.

The control of the quantity of nutrients added by means of standards and definitions of identity established by the Food and Drug Administration and the limitation of vehicles that may be employed in supplementation assure the consumer that only significant and appropriate quantities of nutrients are incorporated (Gunderson et al., 1963). The prevention of overproliferation of nutrients is especially important in that it prevents excessive intakes. For example, the Commissioner of Foods and Drugs has adopted a policy that limits the addition of fluorine compounds to foods to that amount incident to the fluoridation of public water supplies and to the tolerances permitted under the pesticide regulations. Any other preparation containing added fluoride, with the exception of dentifrices, is limited to sale by prescription only. The author considers undesirable the combination of fluoride with polyvitamin formulas.

DOES THE "AVERAGE PERSON" NEED VITAMIN SUPPLEMENTS?

Reflection on these points indicates the uselessness of routine supplementation of the ordinary diet with the multivitamin preparations or other nutrient supplements so commonly advocated and widely followed with or without the advice of the physician and dentist. In considering this question, the Council on Foods and Nutrition of the American Medical Association states: "Nutrition surveys in several areas of the United States have indicated that a variable fraction of certain segments of the population is not receiving sufficient varieties of foods to supply vitamins in amounts necessary to meet the

Recommended Dietary Allowances. Generalization of these findings as a basis for vitamin supplementation of healthy individuals is not rational. The methodology employed in these surveys and the standards used for interpretation have varied considerably. It is necessary for the clinician to evaluate each person individually. Correction of inadequacies should then be instituted, preferably by a proper diet, although supplementation with vitamins may be necessary until dietary adjustments are made and the body stores repleted. Avoidance of excessive or unecessary supplementation is, of course, desirable."

In considering whether supplementation is desirable or necessary, it is the responsibility of the dentist to review with the patient the nature of his diet and to evaluate the need for supplements. A reasonable dietary history and appraisal can be made from a few minutes of discussion with the patient (Goldsmith, 1959). Failure to make such an appraisal often leads to unnecessary expense to the patient for nutrient supplements.

The author is aware of no acceptable evidence that an intake of vitamins greater than that provided by the Recommended Dietary Allowances is beneficial unless the patient has become grossly depleted or has a clearly identifiable condition that results in a definitely higher than normal need. Excessive and unnecessary supplementation is to be avoided, just as one avoids any other type of unwarranted therapy. It is an unnecessary expense to the patient, promotes uncritical appraisal by the physician of the patient's illness and may be potentially harmful. The well established syndromes of hypervitaminosis A and D resulting from excessive intakes (Bauer and Freyberg, 1946; Donegan et al., 1949; Nieman and Obbnik, 1954) and the possibility of inducing a dependency, such as in the pyridoxine-dependent syndrome (Hunt et al., 1954) should further serve to make one avoid unnecessary supplementation and to advise patients against unnecessary self-medication with vitamins and nutrient concentrates.

PROTECTIVE SUPPLEMENTATION WITH VITAMINS

On the other hand, there is an indication for the protective use of supplementary nutrients (vitamins as well as some inorganic essentials) for the prevention of deficiency states in persons who are especially exposed to the possibility of a deficiency, such as some conditioning factor (e.g., chronic, excessive loss of blood; malabsorption; hyperthyroidism; and the like), or for meeting unusual physiologic needs not readily supplied by dietary adjustments (e.g., iodine or vitamin D supplementation during pregnancy, vitamin D supplementation for young infants) (A.M.A. Council on Foods and Nutrition, 1959; Darby, 1957; Goodhart, 1960).

Protective therapy may be needed because of adherence to an abnormally restricted diet, such as may occur in some psychiatric patients who grossly distort their dietary pattern, or, more rarely, in instances of prolonged rigidly limited therapeutic regimens. In modern diet therapy the latter are increasingly rare. The basic design of a therapeutic diet is only a modification of a normal, nutritionally adequate dietary in which there is control of quantity of nutrients, physical characteristics or feeding pattern, or removal of an undesirable substance. If the therapeutic diet is so planned as to assure adequate intake of essential nutrients, supplementation becomes totally unnecessary.

Again, it is the responsibility of the clinician to know the nature of the therapeutic regimen imposed and through discussions with the patient to ascertain the extent to which the prescribed regimen is followed; he can thereby decide whether supplementation is desirable.

During certain physiologic periods there is an increase in the requirement for some nutrients. During growth of any type, whether it be the growth of the infant or child or during pregnancy or lactation, the diet, to be adequate, must contain sufficient calories, protein and essential nutrients to meet the higher metabolic demands and the needs for building new tissues. The magnitude of these needs is reckoned in the Food and Nutrition Board's Recommended Dietary Allowances and will, in almost all instances, be met by proper dietary adjustment. The few special situations in which supplementation is advisable are stated by the Council on Foods and Nutrition of the

American Medical Association (1959) as follows:*

"Infants and Children. The daily diet of the artificially fed infant should be supplemented with vitamins C and D if the diet does not supply 30 mg. of vitamin C and 400 U.S.P. units of vitamin D. The diet should be brought up to these amounts, with care exercised that the intake of vitamin D is not excessive. The requirement of the breast-fed baby for vitamin D is not accurately known, but it is accepted practice to advocate 400 U.S.P. units of vitamin D supplement daily. Administration of vitamin D, and in artificially fed babies, of vitamin C should be started with the introduction of artificial feeding. Too often administration of vitamin C is delayed even into the second month. When administration of the vitamins is started, the amount of vitamin D is often too great and the amount of vitamin C too small. Maximum calcium and phosphorus retentions are obtained with 300 to 400 U.S.P. units of vitamin D daily. Not only are retentions no greater with large amounts, but the use of 1,800 U.S.P. units or more daily for several months decreases appetite and, as a consequence, reduces the total retentions of calcium and phosphorus and slows linear growth. Infants receiving unfortified skimmed milk formulas also require supplements of vitamin A (1,500 U.S.P. units daily).

"Healthy children fed adequate amounts of wholesome foods need no supplemental vitamins except vitamin D, which should be supplied throughout the growth period. An adequate intake of vitamin D-fortified homogenized milk or reconstituted evaporated milk (1½ to 2 pt. daily) provides the vitamin D required. The physician should determine the approximate amount of vitamin D supplied by foods before supplementing the diets. In certain instances, physicians (and dentists [W.J.D.]) may wish to supplement the diets of infants and children with preparations containing vitamins A, C, D, and certain B vitamins. The Council believes that such preparations containing the B complex are not needed for routine use but would be of value for children with special problems. It is important that the growing child be introduced to

* J.A.M.A., *169*:41, 1959. Used by permission.

a wide variety of wholesome foods, since food is the normal source of nutrients.

"*Adults.* Healthy adults receiving adequate diets have no need for supplementary vitamins except during pregnancy and lactation, when 400 U.S.P. units of vitamin D daily are required if the intake of vitamin D-fortified milk is low. In these periods of physiological stress, if any doubt exists as to the adequacy of the previous or present diet, supplementary vitamins in addition to vitamin D should be administered.

"Supplementary vitamins are useful during periods of emotional illness which result in bizarre food habits or greatly diminished food intake. The choice of vitamin preparations to be used to insure a desirable nutrient intake in such instances should be based upon the physician's evaluation of the patient's dietary pattern.

"When restricted or nutritionally inadequate diets are prescribed for pathological conditions, vitamin mixtures as supportive supplements are indicated. Examples of conditions in which such diets may be instituted include allergic states and chronic diseases of the gastrointestinal tract. Vitamin supplementation also is indicated when it is necessary to employ parenteral feeding. The character of the supplementation required will depend on the diet, the nutrients administered, and the period of time the regimen is maintained.

"In any prolonged illness associated with decreased food intake or in other situations in which an individual is unable or unwilling to eat an adequate diet, the dentist must decide whether supplementation is necessary. The extent of the illness or the nature of the dietary restriction should be evaluated to determine whether the level of vitamin supplementation should be equal to allowances under normal physiological conditions or in excess of them."

The use of supplementary vitamins as protective therapy in those instances just discussed should be at a level of about half the Recommended Dietary Allowances. In the supplementation of rigidly restricted therapeutic diets or when prolonged illness or other causes significantly reduce food intake, preparations containing one to one and a half times the Recommended Dietary Allowances are useful. For the treatment of frank deficiency states, preparations providing three to five times the Recommended Dietary Allowances are preferable.

NUTRIENTS RECOGNIZED AS REQUIRED FOR MAN

Recommended dietary allowances have not been established for pyridoxine, pantothenic acid, folic acid or vitamin B_{12}, or indeed for a number of other nutrients that are recognized as essential for man. At the present time, the following nutrients are recognized as essential for man by the Food and Drug Administration:

vitamin A	vitamin K
vitamin D	calcium
thiamine	iron
riboflavin	phosphorus
niacin	iodine
vitamin B_6	copper
vitamin B_{12}	magnesium
ascorbic acid	manganese
folic acid	zinc
pantothenic acid	sodium
vitamin E	potassium

Quantitative evidence pertaining to the human requirement of several of these nutrients was considered by the Food and Nutrition Board as inadequate to establish a recommended dietary allowance. The existing evidence on requirements is summarized in the Sixth Edition of Recommended Dietary Allowances (Food and Nutrition Board, 1964). As a guide for the dentist, the author suggests that reasonable allowances of these nutrients for good nutrition may be the following:

Pyridoxine: infants, 400 to 500 μg. per day; adults 2.0-3.0 mg. per day

Vitamin B_{12}: 1 to 5 μg. daily

Folic acid: 0.25 to 0.1 mg. per day

Pantothenic acid: 10 mg. per day

Tocopherol: 10 to 30 mg. daily; probably should be estimated with consideration of the "E:PUFA ratio," i.e., mg. d-α-tocopherol: grams of polyunsaturated fatty acid in the diet. It has been suggested that any diet with an E:PUFA ratio of less than 0.6 should be regarded as probably deficient in E (Harris and Embree, 1963).

Vitamin K: The level of need cannot be estimated at present because of uncertainty concerning the amount derived from intestinal flora.

VITAMIN-MINERAL MIXTURES

Many supplementary mixtures contain both minerals and vitamins. Concerning such combinations, the Council on Foods and Nutrition of the American Medical Association (1959) states that "although certain supplemental vitamin mixtures with calcium, iron, or with both minerals have proved useful, there is no good evidence to support the inclusion of the 12 or more mineral elements essential for man. Few of these minerals are likely to be lacking, even in restricted diets. When iron is needed as a dietary supplement, it should be given as such in most instances. Iron and calcium might be included as optional ingredients in certain supplemental vitamin mixtures, for example, for administration during pregnancy. A combination of calcium and vitamin D in stabilized form may be useful. Sodium, chlorine, and iodine are usually supplied by iodized table salt. Supplementation with copper is rarely needed, since it is usually adequately supplied to the diet."

The effectiveness of magnesium, zinc, cobalt and molybdenum in such mixtures is not supported by good evidence. Some recent evidence does indicate a probable role for zinc and the possibility of a deficiency of this element occurring in growing children (Prasad et al., 1963). Excessive intakes of cobalt induce polycythemia in animals and man. The role of molybdenum as a nutrient remains to be clarified, but animal studies indicate some interrelationship with copper metabolism. Molybdenum toxicity has long been recognized (Fairhall et al., 1945). Excessive intakes of all the inorganic nutrients are to be avoided.

MIXTURES OF HEMOPOIETIC FACTORS

Vitamin-mineral mixtures are especially likely to be encountered among the hematinic preparations. Preparations containing vitamin B_{12}, folic acid, iron, copper, cobalt, ascorbic acid or pyridoxine in various combinations have been put forward. Although some justification for use of certain of these combinations may be possible on the basis of knowledge of their metabolic effects on clinically occurring combinations of deficiencies, the use of such mixtures is seldom logical and more often than not serves as a substitute for a more definitive diagnosis. In some situations uncritical use of these mixtures may be directly injurious to the patient or may delay proper diagnosis and institution of appropriate medical care.

The status of combinations of vitamin B_{12} and folic acid is rather widely understood. Any preparation containing more than 0.1 mg. of folic acid in the daily recommended dose is a prescription item. This precaution is taken in order to reduce the possibility of masking pernicious anemia by the administration of a multivitamin preparation that might contain sufficient folic acid to produce a hemopoietic response without, at the same time, protecting against the neurologic effects of vitamin B_{12} deficiency. Similarly, at present the Food and Drug Administration requires that oral vitamin B_{12} preparations containing intrinsic factor or its concentrate must be labeled with a prescription legend, since patients taking these preparations should be subject to continuous medical review and supervision.

The occurrence of anemia in chronic scurvy is not uncommon and there is considerable evidence that vitamin C is important in the metabolism of folic acid. In addition, it is clear that vitamin C in large quantities favorably influences the absorption of iron from the gastrointestinal tract. Accordingly, one finds various combinations of vitamin C and other hematinics proposed for use. The author is unaware, however, of any evidence indicating that the addition of ascorbic acid has made possible the response of a patient with iron deficiency or folic acid deficiency to the specific factor who would not otherwise have responded to iron or folic acid, respectively. On the other hand, in some cases of sprue the evidence is convincing that some of the patients who are deficient in folic acid become secondarily deficient in vitamin B_{12} and, hence, that administration of both these agents is therapeutically valuable.

As previously noted, the administration of cobalt salts in sufficient quantity is followed by the development of polycythemia. Any hemopoietic effect in man of inorganic cobalt salts appears to be related to this phenomenon rather than to replacement of a deficiency.

Evidence for the efficacy of copper in the treatment of anemias in man is exceedingly meager; if copper deficiency does occur in the United States, it must be a rare phenomenon. A limited number of cases of an anemia attributed to pyridoxine deficiency have been reported (Harris et al., 1956). In neither instance, however, does there appear to be evidence that these conditions are sufficiently frequent to justify combining the factors for use with iron, folic acid or vitamin B_{12}.

MASSIVE VITAMIN THERAPY

From time to time suggestions have been put forth that various ill-understood conditions appear to respond to massive quantities of one or another vitamin. These suggestions have usually followed within a relatively short time the recognition of the clinical usefulness of a vitamin in the treatment of a deficiency disease. Subsequently, such claims have almost invariably been found to be overstated and have been discarded. Massive doses of vitamins do have pharmacologic action (Molitor and Emmerson, 1948) and when the claims for therapeutic benefits from large doses have been substantiated, it would appear that they have been the result of the pharmacologic rather than the physiologic effects of the substances. On the other hand, the pharmacologic effects of excessive doses may at times be undesirable, particularly those of vitamins D and A, and, less conclusively, folic acid, K and perhaps niacin and ascorbic acid.

Prolonged ingestion of vitamin A in excess of 50,000 U.S.P. units daily can lead to toxicity; therefore, a prescription is required for vitamin A preparations containing greater than this amount as the daily recommended dose (A.M.A. Council on Foods and Nutrition, 1959). Hypervitaminosis A in children may be manifest as anorexia, weight loss, irritability, fretfulness, pruritus, seborrheic lesions of the skin, and fissures at the angles of the mouth,

followed by hepatomegaly, alopecia, hydrocephaly, pain in the long bones, hyperostosis and tender swellings in the extremities, and cortical thickness. Symptoms in the adult are similar but milder; there is less joint and bone pain. Exophthalmos and menstrual alterations as well as changes in pigmentation of the skin have been reported, and increased intracranial pressure has been noted. In acute vitamin A poisoning, a hemorrhagic tendency has been described. Patients receiving repeated daily dosage with more than 25,000 U.S.P. units of vitamin A should be followed carefully for evidence of any toxic symptoms and should have periodic measurements of the serum vitamin A levels. Indeed, it is difficult at present to identify a reason for administering such large dosages.

Hypervitaminosis D has occurred not infrequently in young children and in adults treated with massive quantities of this vitamin. Again, repeated dosing with quantities of 50,000 U.S.P. units of vitamin D may lead to intoxication, and preparations containing these quantities require a prescription.

Early symptoms of intoxication with vitamin D include anorexia, nausea, headache, polyuria, nocturia and diarrhea, weakness, fatigue, pallor and lassitude, renal damage with metastatic calcification, and depression (A.M.A. Council on Foods and Nutrition, 1959; Bauer and Freyberg, 1946; Donegan et al., 1949). Adults have been reported to develop a normocytic hypochromic anemia with azotemia. An increase in the serum calcium level to above 11 mg. is an indication for stopping vitamin D supplements.

Large quantities of folic acid administered to experimental animals have resulted in acute renal damage. No such lesions have been reported in man, but the dosage required to produce them in animals is such that it would appear prudent to avoid repeated daily injection of more than 20 mg. of pteroylglutamic acid in young infants. Ascorbic acid taken orally in half-gram or larger amounts may have a mild diuretic effect. This does not appear to be associated with any significant lesion. The "flushing" or vasodilating effect of nicotinic acid (not of the amide) is well known and protects against the administration of excessive quantities of this vitamin without specific indication.

TABLE 30-1.   USEFUL LEVELS FOR BIOCHEMICAL SCREENING OF
DEFICIENCY*

Ascorbic acid	Plasma level	< 0.1	mg./100 ml.
Vitamin A	Plasma level	<10	mcg./100 ml.
Carotene	Plasma level	<20	mcg./100 ml.
Riboflavin	Erythrocyte content	<10	mcg./100 ml. of red blood cells
Thiamine	Urinary excretion	<10	mcg./6 hrs.
		<27	mcg./gm. creatinine
Riboflavin	Urinary excretion	<10	mcg./6 hrs.
		<27	mcg./gm. creatinine
Nicotinic acid	Urinary excretion of N'methylnicotinamide	< 0.2	mg./6 hrs.
		< 0.5	mg./gm. creatinine

* Based on Interdepartmental Committee on Nutrition for National Defense standards classified as "deficient."

THERAPEUTIC USE OF VITAMINS

In the treatment of classical vitamin deficiency states the physician and dentist should critically appraise the manifestations of disease in the patient and the contribution of any conditioning factor. Conditioning factors should be corrected by appropriate measures and specific vitamin therapy instituted to replace such deficiencies as do exist. As was noted previously, one should use preparations containing three to five times the Recommended Dietary Allowances for such replacement. This evokes prompt clinical responses insofar as the portion of the patient's syndrome that is properly attributed to vitamin deficiencies is concerned. When the patient fails to respond to an adequate quantity of a specific nutrient, one should suspect an error in diagnosis.

There is no advantage in continuing a therapeutic dosage level of vitamin supplements for longer than about two weeks. The dentist and dietitian should make a vigorous effort to correct the dietary habits of a patient. In many cases, during the period of dietary re-education it is useful to continue supplementation for longer than the initial two-week period. When this is done, the continuing dosage should be at the lower levels of one half to one and a half times the Recommended Dietary Allowances. After recovery of the patient and establishment of good dietary habits there is no need for further supplementation unless the deficiency was caused by some uncorrectable conditioning factor.

Fortunately, one is not entirely dependent upon clinical findings or dietary history in assessing a vitamin deficiency as a cause for disease. Biochemical measures are useful and are increasingly widely available (Pearson, 1962). Some common examples are listed in Table 30-1. Other informative studies include serum content of vitamin B_{12} and vitamin B_{12} absorption studies; serum level of folic acid and white cell folic acid content; vitamin C tolerance and white cell platelet level of ascorbic acid; xanthurinic acid excretion after a tryptophan load test (pyridoxine); FIGLU (formiminoglutamic acid) before and after folic acid; erythrocyte transketolase for thiamine; serum tocopherols; prothrombin as a measure of vitamin K; Ca:P ratio in serum; alkaline phosphatase and even direct measurement of vitamin D in blood; hydroxyphenyl derivatives after a tyrosine load test. When doubt exists, biochemical determinations should be employed in order to sharpen diagnosis just as one depends on laboratory studies to aid in defining more clearly the etiologic factors in infectious diseases.

SUMMARY

Criteria for assessing the nutritional adequacy of a diet have been indicated. The over-all quality of the United States diet is considered to be good. Broad-scale supplementation with vitamin preparations is neither advisable nor necessary.

Supplementation is desirable as a protective measure for the prevention of deficiency states in individuals who are especially vulnerable because of a conditioning factor, who must meet some unusual physiologic need, or who are subjected to prolonged abnormally restricted dietaries. In such instances, supplementation at one half to one and a half times the level of the Recommended Dietary Allowances suffices. In therapy of overt nutritional disease, three to five times these levels for periods of two weeks or so and correction of the conditioning factor and dietary pattern usually suffice to bring the patient back to repletion.

In all instances, it is the responsibility of the dentist to know the nutritional background of the patient and the nature of any dietary therapy and to assess the need for supplementation. Dietary and biochemical guidelines are suggested.

Reference is made to the hypervitaminoses and the dangers of unlimited supplementation, as well as to useful biochemical screening procedures.

REFERENCES

A.M.A. Council on Foods and Nutrition: Vitamin preparations as dietary supplements and as therapeutic agents. J.A.M.A., *169*:41, 1959.

Bauer, J. M., and Freyberg, R. H.: Vitamin D intoxication with metastatic calcification. J.A.M.A., *130*:1208, 1946.

Bozian, R. C., Ferguson, J. L., Heyssel, R. M., Meneely, G. R., and Darby, W. J.: Evidence concerning the human requirement for vitamin B_{12}. Am. J. Clin. Nutr., *12*:117, 1963.

Culver, P. J: Vitamin supplementation in health and disease. New Eng. J. Med., *241*:970, 1011, 1050, 1949.

Darby, W. J.: Some considerations pertaining to the proper use of supplementary vitamins. J. Chron. Dis., *6*:178, 1957.

Darby, W. J., et al.: The development of vitamin B_{12} deficiency by untreated patients with pernicious anemia. Am. J. Clin. Nutr., *6*:513, 1958.

Donegan, C. K., Messer, A. L., and Orgainin, E. S.: Vitamin D intoxication due to Ertron: report of two cases. Ann. Int. Med., *30*:529, 1949.

Fairhall, L. T., Dunn, R. C., Sharpless, N. E., and Pritchard, E. A.: The Toxicity of Molybdenum. Public Health Bulletin No. 293, U.S.P.H.S., 1945.

Food and Nutrition Board: Recommended Dietary Allowances. Sixth revised edition. Publication 1146. Washington, D.C., National Academy of Sciences–National Research Council, 1964.

Goldsmith, G. A.: Current status of the clinical use of antibiotic-vitamin combinations. New Engl. J. Med., *254*:165, 1956.

Goldsmith, G. A.: Nutritional Diagnosis. Springfield, Ill., Charles C Thomas, 1959, p. 164.

Goodhart, R. S.: Rational use of vitamins in the practice of medicine. Postgrad. Med., *27*:663, 1960.

Gunderson, F. L., Gunderson, H. W., and Ferguson, E. R., Jr.: Food Standards and Definition in the United States. New York, Academic Press, 1963, p. 269.

Harris, J. W., Whittington, R. M., Weisman, R., Jr., and Horrigan, D. L.: Pyridoxine responsive anemia in the human adult. Proc. Soc. Exp. Biol. Med., *91*:427, 1956.

Harris, P. L., and Embree, N. D.: Quantitative consideration of the effect of polyunsaturated fatty acid content of the diet upon requirements for vitamin E. Am. J. Clin. Nutr., *13*:385, 1963.

Horwitt, M. K., Harvey, C. C., Rothwell, W. S., Cutler, J. L., and Haffron, D.: Tryptophane-niacin relationships in man. J. Nutr., *60*:43, Suppl. I, 1956.

Hunt, A. D., Jr., Stokes, J., Jr., McCrory, W. W., and Stroud, H. H.: Pyridoxin dependency: report of a case of intractable convulsions in an infant controlled by pyridoxin. Pediatrics, *13*:140, 1954.

Interdepartmental Committee on Nutrition for National Defense: Manual for Nutrition Surveys. 2nd ed. Washington, D.C., U.S. Govt. Printing Office, 1963.

Medical Research Council: Vitamin A Requirement of Human Adults. Special Report Series No. 264, London, 1949.

Medical Research Council: Vitamin C Requirement of Human Adults. Special Report Series No. 280, London, 1953, p. 179.

Molitor, H., and Emerson, G. A.: Vitamins as pharmacologic agents. Vitamins Hormones, *6*:69, 1948.

Nieman, C., and Obbink, H. J. K.: The biochemistry and pathology of hypervitaminosis A. Vitamins Hormones, *12*:69, 1954.

Pearson, W. N.: Biochemical appraisal of the vitamin nutritional status in man. J.A.M.A., *180*:49, 1962.

Prasad, A. S., Miale, A., Jr., Farid, Z., Sandstead, H. H., Schulert, H. R., and Darby, W. J.: Biochemical studies on dwarfism, hypogonadism, and anemia. Arch. Int. Med., *111*:407, 1963.

Ruffin, J. M., and Cayer, D.: The effect of vitamin supplements on normal persons. J.A.M.A., *126*:823, 1944.

Wolf, S.: Effects of suggestion and conditioning on the action of chemical agents in human subjects—the pharmacology of placebos. J. Clin. Invest., *29*:100, 1950.

Wolff, H. G., et al.: The use of placebos in therapy. New York J. Med., *46*:1718, 1946.

Chapter Thirty-one

Nutritional Requirements
of the Surgical Patient

ROBERT S. HARRIS, PH.D.

INTRODUCTION

Without question, good nutrition is important to the patient who is about to undergo oral surgery. If well nourished, he will probably experience less discomfort, have fewer complications, feel better and recover more rapidly.

Although it is true that the classical deficiency diseases are relatively rare in this country, it is equally true that mild subclinical deficiencies which produce biochemical lesions and/or modify body function are relatively common. Nutritional deficiencies may develop rapidly. Vitamin deficiencies sufficient to reduce resistance to infection and to delay wound healing may develop in a previously healthy person whose dietary intake has been markedly altered because of a chronic or an acute disorder. Nutritional deficiencies may develop in a person consuming an adequate diet. It is not what or how much he eats, but what he absorbs and utilizes that counts. A "conditioned" malnutrition (see Chapter Twenty-one) may develop in a subject as a result of vomiting, diarrhea, pain, intestinal obstruction, persistent blood loss, difficulties in the digestion or metabolism of foods, antimetabolites, etc.

If there is sufficient time, the oral surgeon should correct any nutritional deficiencies in his patient in preparation for surgery, and see to it that he receives adequate nutrition following surgery.

Surgical patients often require unusual amounts of some nutrients, while the needs for other nutrients remain unchanged. It is desirable to discuss some of the stress factors responsible for these alterations in nutrient requirements and to comment on some of the nutrient factors which are critical to rapid recovery from surgery. A discussion of the actual management of surgical patients follows in the next chapter.

METABOLIC RESPONSE TO INJURY

Surgery causes a metabolic stress which provokes a reduction in blood pressure as a result of the liberation from the tissues of substances such as histamine, 5-hydroxytryptamine and bradykinin. These substances cause an increase in tissue permeability (Florey, 1961). Failure of oxygen transport (anoxemia) is perhaps the major factor in early response to injury (Guyton and Crowell, 1961). This metabolic ebb period is characterized by electrolyte disturbance, hyperglycemia and other changes (Levenson et al., 1961). It is usually followed by a metabolic flow period in which cell activity is increased and tissue damage is repaired. During these first few days there is an increase in the urinary excretion of nitrogen (urea) and sulfur, a significant loss in creatine and phosphorus, yet little change in the creatinine or sodium excretion.

It is not always possible to preserve nitrogen balance by prescribing high intakes of protein and energy-rich diets, for nitrogen losses may be extraordinary (Cuthbertson, 1960). Tissue injury stimulates the hypothalamic-pituitary-adrenal chain mechanism. It is remarkable that the loss of body protein far exceeds the amount of tissue damaged. This is because the protein loss is more general than local, and is due only in part to hemorrhage in the trauma area and to disuse (Cuthbertson, 1932). Following serious injury, more than a kilogram of protein may be lost in a week's time.

The rise in body temperature (up to 4°F.) is generally noted as a sequel to moderate or serious injury and seems to be related to the endogenous deamination of amino acids and the excretion of urea (Cuthbertson, 1932). The non-nitrogenous residue is used as a source of energy. Traumatic fever provokes an increase in metabolism which enhances the healing process by accelerating the repair processes.

An effort should be made to stimulate the patient's appetite during the repair process. Protein-rich diets become increasingly important as the amount of injured or lost tissue increases. It is also important that these diets be adequate in carbohydrate content so as to ensure the most efficient use of the dietary amino acids for tissue protein synthesis. Provision should be made to meet an increased requirement for potassium, riboflavin and ascorbic acid.

Abbott et al. (1957) contend that much of the nitrogen deficit and loss of body weight during the first five to ten days following surgery are the result of semistarvation diets taken by patients who are unable, or are not permitted, to ingest adequate food. On the other hand, Wilkinson (1961) has demonstrated that severe surgical injury also causes a loss in body tissues.

THE HEALING OF WOUNDS

The way in which a patient is managed during the lag period of wound healing (first three to five days) may influence the physiology of fibroblastic fusion of wound surfaces. Nutritional considerations are involved in wound healing in relation to the distribution of body electrolytes, the hydration of tissues and the supply of amino acids, ascorbic acid, vitamin K, and other nutrients. Edema or dehydration may lengthen the lag phase. Dehydration or overhydration may change the wound surface and disturb the chemistry of the wound-space contents. Patients who do not respond to high protein diets should receive transfusions of whole blood or plasma if they appear to be malnourished or show lowered serum protein concentration or a lowered blood volume.

ASCORBIC ACID AND WOUND HEALING

Ascorbic acid is required for the formation of collagen and its precursors. (See Chapter Thirteen.) The development of tensile strength in the healing wound can be independent of fibrogenesis (Peacock, 1962). Ascorbic acid would be of little or no importance to the healing process if it operated only by affecting the production of fibrous tissue, and if there were not an interrelation between the collagen level of a wound and its tensile strength.

Catchpole (1957) has observed alterations in the connective tissue and the plasma of scorbutic guinea pigs which suggest an alteration in the chemistry of the ground substance, the continuous matrix which surrounds the cells, and the fibrillar elements of connective tissue. The ground substance appears to break down into simpler compounds which may adversely affect the organization of collagen. This means that ascorbic acid deficiency impairs wound healing by reducing the rate of formation of collagen and by disorganizing these fibers.

Old guinea pigs retain less ascorbic acid in their tissues than young ones (Abt and von Schuching, 1963). This decreased retention may be a factor in the poor healing of wounds commonly observed among older people.

Rinehart et al. (1938) contend that several common surgical conditions are associated with low ascorbic acid levels in the blood. Parenteral or oral administration of 100 to 300 mg. of ascorbic acid daily should remove the possibility of impaired wound healing as a result of ascorbic acid deficiency.

PROTEIN DEFICIENCY AND WOUND HEALING

Thompson et al. (1938) demonstrated that hypoproteinemia causes a delay in wound healing and is a major cause of wound dehiscence. Hypoproteinemic dogs showed a marked delay in fibroblastic proliferation and a subsequent delay in wound healing.

ANTIBODY PRODUCTION

The antigen-antibody relationship has been studied extensively in animals and human beings. Studies in human subjects indicate that antibody formation is inhibited by deficiencies of pantothenic acid (Cuthbertson, 1932; Hodges et al., 1962a), pyridoxine (Hodges et al., 1962b), and protein (Budiansky and DaSilva, 1957; Hodges et al., 1962c; Olarte et al., 1956; Wohl et al., 1949). Studies in animals have revealed that deficiencies of vitamin A, ascorbic acid, several of the B vitamins, amino acids and protein may limit the production of antibodies (Axelrod et al., 1947, 1958; Green and Anker, 1954).

NUTRITION AND RESISTANCE TO INFECTION

The nutritional status of a patient can influence his resistance to infection. (See Chapter Twenty.) Many people in the United States have been made more sensitive to infection because of malnutrition. This group includes the teen-agers, the elderly, the food faddists, and those who have depressed appetites and are debilitated as a result of chronic disease. Infections commonly cause loss of appetite and prompt a change from nutritionally rich protein foods to nutritionally inferior carbohydrate foods.

Infections upset the amino acid balance and increase the excretion of nitrogen, reduce blood vitamin A levels (Jacobs et al., 1954; Shank et al., 1944), precipitate thiamine deficiency (Smith and Woodruff, 1961), precipitate folic acid deficiency, increase the vitamin B_{12} requirement (Campbell and Pruitt, 1952), and increase the urinary excretion of ascorbic acid (Nassau and Scherzer, 1924).

Infections and infestations provoke anemia even when the iron intake appears to be adequate (Roche and Perez-Gimenez, 1959). Almost any type of chronic infection will promote the development of anemia by shortening the life span of the red cells and interfering with erythrocyte production (Wintrobe et al., 1947).

Plasma iron concentrations have been reported to be low following infections (Cartwright et al., 1946; Scrimshaw et al., 1953) and surgery (Feldthusen et al., 1953; Baird et al., 1957). Recently, Baird and Podmore (1963) confirmed this observation on rabbits and human beings and concluded that it is the result of an impaired release from iron depots into the plasma.

Tuberculosis increases the excretion of calcium and phosphorus (Johnston, 1953), and diarrheas induced by infections cause serious losses in sodium, chloride, phosphorus and potassium, occasionally causing death (Calloway and Spector, 1954).

There is a systemic relation of nutrient intake and resistance to infection. Depletion of vitamin reserves is accompanied by derangements in the activity of many cellular enzymes, and it is evident that the lowered resistance to injury and infection has a complex explanation. Trauma may act as a conditioning factor to increase the susceptibility of tissues (e.g., mucous membranes and skin) to infection.

NUTRIENTS OF SPECIAL IMPORTANCE TO THE SURGICAL PATIENT

WATER

The maintenance of water balance between the intracellular and extracellular fluids of the body is of extreme importance to the surgical patient. Water retention is largely dependent on the amount of body sodium, and sodium excretion is dependent on body water. Water is obtained from ingested foods and liquids and from metabolism of foodstuffs. Water is lost from the body in the urine, feces, breath, sweat and perspiration. The surgical patient with normal kidneys should receive at least 1500 cc., and preferably 2500 cc., of water daily. Of course, excessive fluid losses will increase the water requirements.

FATS

Since fats yield twice as much energy per gram as carbohydrates and protein, their use in intravenous preparations has received considerable attention. Several commercial preparations are available. No more than 50 gm. of fat should be given during a 24-hour period because hyperpyrexia increases with fat dosage. It is unfortunate, also, that a patient cannot be given intravenous fat preparations daily for more than about 10 days because of toxic effects. When eventually the pyrexia problem of intravenous fat feeding has been solved, it will be possible to provide patients with 2800 calories per day and to make protein hydrolysate therapy practical and thus permit complete intravenous therapy for the surgical patient.

CARBOHYDRATES

Carbohydrates are the simplest and safest means of providing the postoperative patient with his calorie requirements. In acute situations this energy food may be given intravenously in a 5 per cent aqueous solution. The amount which may be given intravenously is limited by the efficiency of its utilization. Three liters of 5 per cent glucose will provide only 600 calories in 24 hours. Fifty grams of protein as hydrolysate add only 200 calories. Although the average patient requires about 2500 calories, a minimum of 2800 calories per day is required before protein can be used efficiently for tissue repair rather than as a source of calories (Calloway and Spector, 1954). Concentrations of glucose above 5 per cent may be used for only a few days because they may provoke thrombosis and obliterate veins.

Whenever possible, carbohydrates should be ingested either in aqueous solutions or in foods or food mixtures.

PROTEIN

Probably the most critical nutrient to the surgical patient is protein. Protein supplies the amino acids required for the growth and repair of tissues, for the transport of lipids, and for the maintenance of blood volume, serum protein and total circulating blood cell mass. Amino acids are required also for the formation of antibodies, antigens, enzymes and hormones. While a daily intake of 60 to 80 gm. of protein is adequate for the needs of an active and healthy adult consuming sufficient calories, even as much as 120 gm. of protein may not be adequate during the first stages of recovery from surgery.

Infection, trauma and immobilization accelerate the breakdown of body proteins and decrease the utilization of food protein. If the total calorie intake is not adequate, the body will deaminize the amino acids in food protein to provide energy for the body, thus increasing still further the need for protein. Unlike carbohydrates and fats, proteins are not stored in the body in appreciable amounts. Failure to ingest protein for several days may cause a serious reduction in the protein content of body cells, serum and intercellular fluid, and may thus interfere with the maintenance and repair of the tissues.

It is important to comment here on the nutritional quality of food proteins. The proteins in human tissues are composed of more than 20 amino acids. Some of these amino acids can be synthesized in the body from other amino acids and, therefore, are called "nonessential"; that is, it is not essential that they be provided in the diet. Those amino acids which cannot be synthesized by the body tissues in sufficient amounts to meet the needs for tissue repair or growth are called "essential"; they must be supplied in the daily diet. Food proteins contain various mixtures of these essential and nonessential amino acids. Food proteins from vegetable sources (cereals, legumes, vegetables, fruits) contain approximately 30 per cent essential and 70 per cent nonessential amino acids. Food proteins from animal sources (meat, fish, milk, eggs) contain about 55 per cent essential and 45 per cent nonessential amino acids. Foods from animal sources are generally more nutritious because they contain higher proportions of the essential amino acids.

The quality of vegetable proteins can be improved by judiciously combining several of them in the same meal. The proteins of corn and beans are complementary proteins in this way. Several years ago we demonstrated on ourselves that when corn and beans were eaten together in four daily meals during one week they were 25 per cent better as a source of protein than when corn was alternated with beans in four meals a day during a week (Malaspinas and Harris, 1953). Complementary foods must be eaten in the same meal to provide the maximum possible nutritional value.

Thus, while vegetable proteins can be improved by eating them together, it requires knowledge, intelligence and persistence to do it most efficiently. No combination of plant proteins has been found to be equivalent to the better animal proteins. It may be concluded that whenever economically possible, the proteins supplied in the diet of surgical patients should be from animal sources (eggs, milk, meat, fish).

Protein deficiency is best treated by prescribing a nutritious diet rich in quality protein. Sometimes this approach is not practical for patients with gastrointestinal problems, infections or blood dyscrasias. These difficult patients may require the tube feeding of food, protein hydrolysates or amino acid mixtures. Nutrients given by tube are not so well absorbed and assimilated and may produce diarrhea. Protein hydrolysates can be given intravenously, but utilization is poor and only about 50 gm. can be administered during 24 hours. It should be remembered that when the daily calorie intake is inadequate, a considerable portion of the amino acids will be deaminated and used to meet energy requirements. As a result, there is insufficient amino acid for tissue repair and maintenance.

Clinically, the earliest sign of protein deficiency is loss in body weight; pathologically, the earliest signs are reduced blood volume and reduced total circulating red blood cell mass. Reductions in hematocrit, hemoglobin and serum protein values become evident somewhat later.

Edema develops when the stores of body protein are depleted and the serum protein level is lowered. The development of edema is accelerated when excess water or sodium is ingested. The trauma induced by surgery and by injections causes serum proteins to escape through capillary vessels to produce localized edema.

Protein deficiency markedly affects bone healing. Rhoads (1953) observed that animals with normal serum protein showed good callus formation within five weeks after the division of a bone, whereas hypoproteinemic animals showed little evidence of callus formation after 11 weeks. Possibly the hypoproteinemia interfered with the fibroblastic repair that normally precedes callus formation. Possibly, also, it is related to the hypocalcemia which is known to be caused by hypoproteinemia.

VITAMINS

Thiamine. Thiamine is required for the metabolism of carbohydrates. Carbohydrate metabolism may be seriously altered long before the appearance of the clinical symptoms of beriberi (malaise, irritability, fatigue, loss of appetite, neuritic pain, etc.). A patient may develop thiamine deficiency during five days of intravenous glucose feeding (Sydenstricker, 1941).

Riboflavin. Riboflavin is essential for metabolic oxidation processes, including protein metabolism and tissue repair. Deficiency of it has been shown to delay wound healing.

Pantothenic Acid. Pantothenic acid is required for carbohydrate, fat and protein metabolism and is especially involved in adrenal cortex function and antibody production.

Folic Acid. Folic acid is a hemopoietic and a leukopoietic agent. It serves to restore and to maintain the red cell mass and aids the development of leukocytes which defend the body against infection.

Ascorbic Acid. Ascorbic acid is required for the formation of intercellular substances (collagen, reticulum, dentin, cartilage, bone matrix) and, therefore, plays a role in the formation and repair of teeth and bones, the healing of wounds, the maturation of red cells, the utilization of iron and the maintenance of blood hemoglobin levels.

Vitamin K. Vitamin K is an antihemorrhagic factor which is occasionally effective in the prevention of postoperative bleeding, edema and ecchymosis (Morgan and Christensen, 1963).

SUMMARY AND CONCLUSIONS

Excellent nutrition is important to the presurgical and postsurgical patient because it helps to promote wound healing, increases resistance to infections and speeds convalescence.

Oral surgery patients often are unable to chew foods because of poor dentition, swelling and pain; this encourages faulty food selection and inanition, which lead to malnutrition and undernutrition.

The nutritional requirements of patients recovering from oral surgery may be abnormally high, in some instances as much as five to ten times the normal requirement.

Diets rich in quality protein, ascorbic acid and vitamin B complex are especially desirable. Foods such as liver, meat, eggs, milk and citrus fruits, as well as whole grain or enriched cereals, are recommended.

A variety of diets and diet mixtures are available which will provide adequate nutrition to surgical patients. Some are solids, while others are gruels, purées or clear liquids. They may be taken by mouth, by nasal tube, by stomach tube or by intravenous injection.

Patients should progress from liquid to solid diets as rapidly as possible because nutrients are better utilized when given by mouth, toxic effects are less likely, and because the patient is happier when eating normally.

Because the nutritional effects of surgery are variable and unpredictable, it is advisable to prescribe multivitamin-mineral capsules or tablets to oral surgery patients until recovery is complete.

REFERENCES

Abbott, W. E., Krieger, H., Holden, W. D., Bradshaw, J., and Levey, S.: Effect on body weight and nitrogen balance in surgical patients. Metabolism, 6:691, 1957.

Abt, A. F., and von Schuching, S.: Aging as a factor in wound healing. L-ascorbic-1-C_{14} acid catabolism and tissue retention following wounding in young and old guinea pigs. Arch. Surg., 86:627, 1963.

Axelrod, A. E.: The role of nutritional factors in the antibody responses of the anamnestic process. Amer. J. Clin. Nutr., 6:119, 1958.

Axelrod, A. E., Carter, B. B., McCoy, R. H., and Geisinger, R.: Circulating antibodies in vitamin-deficiency states. 1. Pyridoxine, riboflavin, and pantothenic acid deficiencies. Proc. Soc. Exp. Biol. Med., 66:137, 1947.

Baird, I. McL., and Podmore, D. A.: Studies in iron metabolism after surgical operation. Clin. Sci., 25:323, 1963.

Baird, 1. McL., Podmore, D. A., and Wilson, G. M.: Changes in iron metabolism following gastrectomy and other surgical operations. Clin. Sci., 16:463, 1957.

Budiansky, E., and DaSilva, N. N.: Formação de anticorpos na distrofia pluricarencial hidropigênica. O. Hospital (Rio de Janeiro), 52:251, 1957.

Calloway, D. H., and Spector, H.: Nitrogen balance as related to caloric and protein intake in active young men. Amer. J. Clin. Nutr., 2:405, 1954.

Campbell, R. E., and Pruitt, F. W.: Vitamin B_{12} in the treatment of viral hepatitis. A preliminary report. Amer. J. Med. Sci., 224:252, 1952.

Cartwright, G. E., Lauritsen, M. A., Jones, P. J., Merrill, I. M., and Wintrobe, M. M.: The anemia of infection. I. Hypoferremia, hypercupremia, and alterations in porphyrin metabolism in patients. J. Clin. Invest., 25:65, 1946.

Catchpole, H. R.: Healing of Wounds. New York, McGraw-Hill Book Co., 1957, p. 34.

Cuthbertson, D. P.: Observations on disturbance of metabolism produced by injury to limbs. Quart. J. Med., 25 (I.N.S.): 233, 1932.

Cuthbertson, D. P.: Parenteral fluid therapy in relation to the metabolic response to injury. Surg. Gynec. Obstet., 107:105, 1960.

Feldthusen, U., Larsen, V., and Lassen, N. A.: Serum iron and operative stress. Acta Med. Scand., 147:311, 1953.

Florey, H.: Exchange of substances between the blood and tissues. Nature, 192:908, 1961.

Green, H., and Anker, H. S.: On the synthesis of antibody protein. Biochim. Biophys. Acta, 13:365, 1954.

Guyton, A. C., and Crowell, J. W.: Dynamics of the heart in shock. Fed. Proc., 20 (No. 2, Part III): 51, 1961.

Hodges, R. E., Bean, W. B., Ohlson, M. A., and Bleiler, R. E., Factors affecting human antibody response. III. Immunologic responses of men deficient in pantothenic acid. Amer. J. Clin. Nutr., 11:85-94, 1962a.

Hodges, R. E., Bean, W. B., Ohlson, M. A., and Bleiler, R. E.: Factors affecting human antibody response. IV. Pyridoxine deficiency. Amer. J. Clin. Nutr., 11:180, 1962b.

Hodges, R. E., Bean, W. B., Ohlson, M. A., and Bleiler, R. E.: Factors affecting human antibody response. V. Combined deficiencies of pantothenic acid and pyridoxine. Amer. J. Clin. Nutr., 11:187, 1962c.

Jacobs, A. L., Leitner, Z. A., Moore, T., and Sharman, I. M.: Vitamin A in rheumatic fever. Amer. J. Clin. Nutr., 2:155, 1954.

Johnston, J. A.: Nutritional Studies in Adolescent Girls and Their Relation to Tuberculosis. Springfield, Ill., Charles C Thomas, 1953.

Levenson, S. M., Einheber, A., and Malm, O. J.: Nutritional and metabolic aspects of shock. Fed. Proc., *20* (No. 2, Part III): 99, 1961.

Malaspinas, A. A., and Harris, R. S.: Evaluation of concurrent feeding of complementary food proteins to man. Food Technol., *21* (Suppl. 7), 1953.

Morgan, D. H., and Christensen, R. W.: A clinical re-evaluation of the effectiveness of vitamin K_1 in oral surgery. Amer. J. Orthoped., *5*:202, 1963.

Nassau, E., and Scherzer, M.: Scurvy and infection. Klin. Wchnschr., *3*:314, 1924.

Olarte, J., Cravioto, J., and Campos, B.: Inmunidad en el niño desnutrido. Bol. Med. Hosp. Infant. Mex., *13*:467, 1956.

Peacock, E. E.: Some aspects of fibrogenesis during the healing of primary and secondary wounds. Surg. Gynec. Obstet., *115*:408, 1962.

Rhoads, J. E.: Supranormal dietary requirements of acutely ill patients. J. Amer. Diet. Assn., *29*:897, 1953.

Rinehart, J. F., Greenberg, L. D., Olney, M., and Choy, F.: Metabolism of vitamin C in rheumatic fever. Arch. Intern. Med., *61*:537, 1938.

Roche, M., and Perez-Gimenez, M. E.: Intestinal loss and reabsorption of iron in hookworm infection. J. Lab. Clin. Med., *54*:49, 1959.

Scrimshaw, N. S., Morales, J. O., Salazar, B. A., and Loomis, C. P.: Health aspects of the community development project, rural area, Turrialba, Costa Rica, 1948-1951. Amer. J. Trop. Med., *2*:583, 1953.

Shank, R. E., Coburn, A. F., Moore, L. V., and Hoagland, C. L.: The level of vitamin A and carotene in the plasma of rheumatic subjects. J. Clin. Invest., *23*:289, 1944.

Smith, D. A., and Woodruff, M. F. A.: Deficiency diseases in Japanese prison camps. Med. Res. Coun. (London) Spec. Rept. Series, *274*:63, 1961.

Sydenstricker, V. P.: Importance of vitamin therapy in preparation and post-operative care of surgical patients. South. Surg., *10*:592, 1941.

Thompson, W. P., Ravdin, I. S., and Frank, I. L.: Effect of hypoproteinemia on wound disruption. Arch. Surg., *36*:500, 1938.

Wilkinson, A. W.: Starvation and operation. The Lancet, *ii*:783, 1961.

Wintrobe, M. M., Greenberg, G. R., Humphreys, S. R., Ashenbrucker, H., Worth, W., and Kramer, R.: The anemia of infection. III. The uptake of radioactive iron in iron-deficient and in pyridoxine-deficient pigs before and after acute inflammation. J. Clin. Invest., *26*:103, 1947.

Wohl, M. G., Reinhold, J. C., and Rose, S. B.: Antibody response in patients with hypoproteinemia with special reference to the effect of supplementation with protein or protein hydrolysate. Arch. Intern. Med., *83*:402, 1949.

Chapter Thirty-two

Nutritional Management of Patients with Acute Oral Infection or Oral Surgical Problems

DAVID WEISBERGER, D.M.D., M.D.

Discussions of some basic concepts that interrelate nutrition with infection (Chapter Twenty) and nutrition with wound healing (Chapter Thirteen) and detail the nutritional requirements of the surgical patient (Chapter Thirty-one) are found elsewhere in this book. The pragmatic application of these concepts to the clinical problems that an oral surgeon must cope with is the purpose of this chapter. Here we will discuss the nutrient needs of the surgical patient in terms of specific foods, meal plans and suggested menus. Further, we will focus on important considerations that will help to make comfortable, acceptable and satisfying the feeding and eating experience of the patient with acute oral problems.

DIETARY CONSIDERATIONS IN ACUTE ORAL INFECTIONS

There are some acute problems occurring in the oral cavity which require special consideration with regard to the nutritional management of the patient. Many of these problems arise because of the painful mucosa associated with acute infections, and the pain is increased when food is taken into the mouth. Fortunately, most of these acute oral infections are of relatively short duration, lasting only from seven to ten days. In this short period, profound malnutrition does not occur. The main problem in such instances is to enable the patient to ingest food in order to satisfy his hunger and to maintain a degree of nutrition sufficient to assist in the healing process.

Since many of these acute infections of the oral cavity occur in infancy or early childhood, much assistance from the parents is needed. The first problem is to relieve the pain when food is ingested. For this purpose a topical anesthetic solution is selected. The most satisfactory is Dyclonine 0.5 per cent; a solution that has been used on hundreds of patients without side reactions. It is used about five minutes before feeding and is applied either in the form of a spray directed to the oral cavity or by means of a piece of gauze soaked in the solution and then held in contact with the various areas of the oral cavity. The anesthesia is effective for approximately one half hour and may be repeated as often as necessary. During the effective period a liquid diet consisting of fruit juice, refined cereal, creamed soups and desserts can be fed. After a few days, as healing progresses, a soft solid diet can be substituted. The soft solid diet may include minced meats or strained baby foods, eggs (except fried or hard boiled), and mashed vegetables as well as any liquids. When it is desirable to increase vitamin intake, this can be done by the addition of a vitamin solution such as Vi-Penta to the liquids.

DIET PRIOR TO ORAL SURGERY

Whenever possible, surgery should be delayed until the nutritional status of the patient has been made optimal.

To assure an adequate tissue nutrient

reserve, the surgeon should assess the adequacy of the patient's diet, especially in protein, ascorbic acid, riboflavin and vitamin A. The hemopoietic nutrients such as iron, B_{12} and folic acid deserve special consideration because blood loss from surgery may affect their level. It is true that in the great majority of instances the levels of these hemopoietic nutrients are not seriously affected by routine minor oral surgical procedures. Nevertheless, if the slightest advantage can be gained by improving the diet to include adequate amounts of these nutrients and those which will promote wound healing, the attempt should be made. Some consideration should also be given to the need for vitamin K supplementation,

especially if there is a previous history of postoperative bleeding after tooth extraction and if the prothrombin time is below normal.

The method of applying these principles to daily practice is to start dietary guidance at least a week prior to surgery. The advice should emphasize the importance of eating complete protein foods, such as eggs, milk, meat, fish or poultry. Meat, especially the glandular types like liver, kidney and heart, as well as fish, poultry and eggs not only provide protein but are also rich in iron and B complex vitamins. Other food sources of these nutrients are green leafy vegetables, whole or enriched cereals and dried fruits. Liberal use of milk and cheese provides calcium, phos-

---

TABLE 32-1. HIGH PROTEIN–HIGH CALORIE–HIGH VITAMIN AND MINERAL DIET*

Food for the Day

Milk: 4 cups, to be used as beverage, on cereal, in cooking
Meat, fish, poultry or liver: 6 or more ounces (two generous servings)
Eggs: 2
Cheese: as desired; may replace meat or milk, or be eaten in addition
Butter, margarine, oil: 2 tablespoons or more; also cream, bacon, cream cheese
Fruit: 2 or more servings, one of which should be citrus
Vegetables: 2 servings; one dark green leafy or at least one orange every other day
Potato: 1 or more servings
Cereal: 1 or more servings
Bread: 4 or more slices
Crackers: as desired
Desserts: 2—custard, gelatin, tapioca pudding, rice pudding, ice cream sherbet, plain cake and cookies

Meal Plan	*Sample Menu*
Breakfast	
Fruit	½ cup orange juice
Cereal	1 serving oatmeal, milk or cream
Eggs	1 or 2 eggs, any style
Bread, butter	2 slices toast, butter, jelly
Beverage	Coffee
Dinner	
4 oz. or more meat	2 good-sized slices roast beef *or* 2 chops
Potato or substitute	Baked potato, butter
Vegetables (2)	Buttered carrots
	Tossed salad with French dressing
Dessert	Tapioca pudding
Beverage	Milk
Lunch or supper	
Substantial soup	Cream of asparagus or chicken rice soup
Meat, fish, cheese, or egg in sandwich or main dish	Hamburger on roll *or* macaroni baked with cheese
Vegetables	Lettuce, tomato with mayonnaise
Fruit	Canned peaches
Beverage	Milk
Evening snack	
	Milk or eggnog and sandwich *or* crackers with cheese or peanut butter *or* milk and cake or cookies

Sugar and other sweets may be eaten in addition to basic foods listed above, but should not replace them.

* This diet will supply at least 120 gm. protein and 2500 or more calories.

phorus and vitamin D for bone repair. Patients are prone to omit foods rich in ascorbic acid, which is so important in providing proper intercellular material and collagen for wound healing. For this reason a pint of orange juice (two 8 oz. cups—equal to 200 mg. of ascorbic acid or about three times the recommended amount) should be taken daily for at least a week preceding oral surgery. If there is any idiosyncrasy to orange juice, there are other combinations of fruit juices on the market which are fortified with vitamin C. Some contain acerola cherry juice, which is an exceedingly rich source of vitamin C. Other fruit juices such as cranberry juice have a pure ascorbic acid added. If there is no difficulty in eating raw fruits and vegetables like grapefruit, tomatoes, strawberries, cantaloupe and green peppers, they can be recommended as excellent sources of ascorbic acid.

A sample high protein–high calorie–high vitamin and mineral diet is given in Table 32-1.

DIET AFTER ORAL SURGERY

One of the first questions the patient will ask after the removal of his teeth or any other oral surgical procedure is, "What can I eat?" A vague answer should be avoided. It is best to give a specific list of foods or beverages.

If the patient has had one of the general anesthetics like ether which causes a disturbance of fat metabolism, it is best to avoid foods high in fat content for the first 12 hours or so after surgery. Post nauseum, the patient should be given as many clear fluids as he can tolerate. In addition to water, beverages like cola drinks, ginger ale, apple juice and orange juice as well as clear broths or bouillon, clear tea or black coffee with sugar are usually easily tolerated. Assuring an adequate fluid intake is most important.

In cases of oral surgery done under local anesthesia, in addition to the above mentioned beverages, sherbets, junkets, custards, gelatin and ice cream may be advised. If the patient is hungry, gruel or cereal topped with sugar and milk as well as milk, egg or egg-

nogs, or some type of strained chicken, pea or vegetable soup could be suggested.

Whichever of these dietary suggestions are adopted should be written out with the other routine postoperative instructions.

On the first or second postoperative day, when the patient is questioned about his comfort, further instructions may be given concerning the diet. The patient may continue to use liquid foods to prevent wound irritation if he has had some extensive surgical procedure; however, this diet can be made more adequate by using foods from the 4 Food Groups. For example, he may have the following foods in each group:

Milk group: Use milk in all forms. For extra nourishment nonfat milk solids may be added to regular milk. Add 3 tbsp. of milk solids to each 8 oz. glass of whole milk. Soft, plain ice creams are also soothing. Milk shakes and malted milks are recommended. Cream may be added to milk if desired.

Meat group: Eggs in the form of strained eggnogs may be used in a liquid diet. They may also be used in soft baked custards. Strained baby meats may be diluted with milk or bouillon to make soups.

Vegetable-fruit group: Citrus juices such as orange and grapefruit are highly recommended, as well as tomato juice and other fruit and vegetable juices. Puréed fruits and vegetables, such as baby foods, may be used. Strained fruits may be used with milk as a drink, and strained vegetables may be cooked with milk and butter and other seasonings as soup.

Bread-cereal group: Strained gruels.

Included in Table 32-2 is a full liquid diet that furnishes 2000 calories and 80 grams of protein per day.

From the liquid diet the patient can progress to a soft diet if and when his mouth and

TABLE 32-2. FULL LIQUID DIET

Basic foods to include daily		
Milk	1 qt.	As beverage, cocoa, with cereal, in soups; dry skim milk may be added.
Eggs	2 or 3	Soft custard, eggnog
Vegetables	2–4 tbsp.	Strained, with meat, in cream soup, strained juice
Fruit	1 pt.	Strained juice
Cereals	2 servings	Gruel
Fats	As tolerated	Cream; butter in soups, cereals
Desserts	2 or 3 servings	Custard, junket, gelatin, ice cream, sherbet
Soups	1 or 2 servings	Clear broth, beef juice, cream soup
Miscellaneous	As desired	Tea, coffee, ginger ale, sugar

Sample meal plan

Breakfast	*Dinner*	*Lunch or supper*
Fruit juice, strained	Strained cream soup	Beef broth with strained vegetables
Gruel with butter, milk	Custard	Eggnog
Coffee, cream, sugar	Milk, hot beverage	Sherbet, hot beverage

Between-meal snacks
Fruit juices, milk drinks, allowed desserts, etc.

---

alveolar ridges allow. Then he may be allowed to have:

Milk group: Soft cheeses such as cottage cheese and cream cheese, in addition to milk and ice cream.

Meat group: Tender or ground meats, chicken, fish without bones, and well cooked legumes.

Vegetable-fruit group: Cooked or canned fruits and ripe bananas; vegetables without skins or seeds, mashed potatoes.

Bread-cereal group: Soft bread, macaroni, noodles, rice, spaghetti, cooked breakfast cereals with milk, and read-to-eat flaked and puffed wheat and rice.

Even without being specifically told, the patient will gradually go to the firmer foods as the soreness in his mouth disappears. The important point to stress is that adequate diets can easily be eaten if sufficient amounts of foods from the 4 Food Groups are eaten each day.

NUTRITION FOR PATIENTS WITH FRACTURED JAWS

In the nutritional management of patients whose jaws are immobilized because of a fracture, any type of solid food can be ingested if it is first processed in a Waring Blendor or other similar apparatus.

All these patients have sufficient space in the posterior portions of the mouth to allow this type of food to pass into the area of the mouth from which it can be swallowed. Many of these patients have, as well, spaces created by previous extractions. Frequently, teeth have been lost in the accident which produced the jaw fracture. If it is believed that the food selected by the patient may be low in vitamin content, liquid vitamin supplements may be added.

It is important to instruct the patient to cleanse the mouth thoroughly after meals. This can be done to some degree of satisfaction by simply rinsing the mouth with water.

MANAGEMENT OF FEEDING PROBLEMS IN PATIENTS WITH ORAL CANCER

In patients who have had extensive mutilating surgery of the mouth for the treatment of carcinoma, feeding becomes a serious problem. It is very difficult for the patient to take nourishment and, if food is taken by mouth, it is difficult to maintain cleanliness of the oral tissues. It is, of course, imperative to maintain good nutrition in these patients, and the use of a naso-esophageal tube is recommended.

Various commercial nutrient powders, liquids, and liquid vitamin preparations which are formulated with measured amounts of all nutrients required may be used for tube feeding. Cruder mixtures such as milk thickened with cooked farina or rice meal may also be used. Although there are several commercial products designed for use on tube feeding, the patient should ingest a vitamin-mineral tablet or capsule daily to ensure optimum intake of these critical dietary essentials, since they may not survive storage in prepared food mixtures.

The mixture for nasogastric tube feeding should contain adequate amounts of quality protein, carbohydrates, fat, minerals and vitamins to meet all therapeutic requirements. Samples of tube feedings are given in Table 32-3. A smooth thin polyethylene tube is well

tolerated and can be left in place without change for as long as four months without clogging. Small, frequent feedings should be given, and the stomach should be aspirated periodically to remove trapped air. In most instances, these patients are hospitalized for approximately ten days. Tube feedings are continued until the patient can take food into the mouth and swallow it. When this stage is reached a soft solid diet is prescribed.

In patients who have had cancericidal doses of radiation for the treatment of oral carcinoma, we are faced with a problem of painful mouth, commencing about ten days after radiation therapy has been started. At this time the oral intake of food is painful, and difficulty in swallowing ensues. The diet should be soft solid in character. Emphasis in

TABLE 32-3.   FOUR STANDARD TUBE FEEDINGS

1. *Waring Blendor*		2. *Low Sodium*		3. *High Calorie*		4. *Gevral*	
Milk	1 qt.	Lonalac	1¼ c.	Milk	1½ qt.	Milk	1 qt.
Milk, dry skim	½ c.	Protinal	2 oz.	Milk, dry skim	1 c.	Milk, dry skim	2 c.
Cream	½ c.	Cream	½ c.	Cream	1 c.	Gevral	2 oz.
Eggs	2	Eggs	2	Eggs	4	Sugar	½ c.
Baby meat	2 cans	Sugar	½ c.	Sugar	¾ c.	Water	1 qt.
Baby veg.	2 cans	Water	1¾ qt.	Water	100 cc.		
Baby fruit	2 cans	* Orange juice	1 c.	* Orange juice	1 c.		
Sugar	⅓ c.						
Water	200 cc.						
*Orange juice	1 c.						
Approx. volume 2000 cc.		2000 cc.		2000 cc.		2000 cc.	
Carbohydrate 210		205		320		240	
Protein	95	90		120		155	
Fat	85	85		130		45	
Cal. 2000 (1 cal./cc.)		1925 (1 cal./cc.)		3000 (1½ cal./cc.)		2000 (1 cal./cc.)	
Na (approx.) 1600 mg.		210 mg.		1700 mg.		1800 mg.	

Remarks

Similar to normal diet; calories can be increased by addition of cream, dry skim milk, etc. See #3. * Not added to formula.	Dextri-Maltose #2 or Dexin could also be used as source of carbohydrate.	Dextri-Maltose #2 or Dexin could also be used as source of carbohydrate.	High protein, low fat; vitamins, minerals added. Fluid can be increased if necessary because of high protein content.

NEW MILK-FREE FORMULAS FOR TUBE FEEDING†

(For patients who cannot tolerate milk or milk products)

Ingredient	Amount	Volume (cc.)	Calories	Protein (gm.)	Carbohydrate (gm.)	Fat (gm.)
Baby meat	3 jars	300	270	40		12
Baby vegetables	3 jars	300	80	2	17	
Eggs, raw	4	150	300	24.5	1	22
Gevral powder	75 gm.		250	45	18	
Sugar of Dexin	100		400		100	
Water	400 gm.	400				
Totals		1150	1300	112	136	34

† Adapted from McDowell, F., and Metzmeier, C.: New York J. Med.: *61*:2061, 1961.

TABLE 32-4.   FOODS ALLOWED ON A SOFT SOLID DIET

Fruit	Any cooked fruits, finely chopped; raw bananas
Milk	Milk, cocoa, eggnog
Eggs	Soft-cooked, poached, scrambled, creamed, omelet
Meat	Minced veal, lamb, beef, liver, fowl; finely chopped fish
Cheese	Cottage, cream, any melted cheese sauce
Bread	Refined, enriched, white, light rye, light graham (remove all crusts)
Cereals	Refined, enriched; also noodles, macaroni, spaghetti, rice
Vegetables	Cooked and finely chopped carrots, asparagus tips, beets, strained peas, squash, beans, spinach; potatoes may be baked, mashed, boiled, creamed
Soups	Broth, bouillon, strained vegetable soups, strained cream soups
Beverages	Carbonated beverages, tea, coffee
Desserts	Plain cakes, custard, gelatin, simple puddings without nuts, plain ice cream or sherbet, fruit whips

the soft solid diet should be placed on appearance and consistency of food. For instance, fried foods are considered inadvisable because this cooking process frequently produces sharp, nontender particles. All foods can be seasoned according to the patient's personal preference (Table 32-4).

Pain may be controlled with the application of Dyclonine before food is taken. Following the subsidence of the acute radiation reaction, the mouth is usually dry because of the effect of radiation on the major and minor salivary glands. The patients are instructed to eat some solid foods along with frequent sips of liquids to assist in swallowing. It has been found that many of the patients treated by radiation have a serious loss of appetite and should be instructed in the use of vitamin supplements. When it is practical, foods should be processed in a blender and the food made appetizing by the use of condiments. Every attempt should be made to satisfy the particular tastes of the patient.

Another problem arising in patients who have had surgery for cancer of the oral cavity is the loss of palatal structure, which then allows food taken into the mouth to pass up into the nasal cavity. It is very helpful if a temporary denture, with or without teeth, is constructed and inserted into the mouth to cover the defect. Such a procedure boosts the morale of the patient as well as helping to maintain adequate nutrition.

CONCLUSION

It can most certainly be said that consideration for the nutritional needs of the patients with surgical or acute oral problems is as necessary a part of the total surgical management as the anesthetic, x-rays, etc. Furthermore the morale and emotional well-being as well as the oral and general physical well-being may be served well. Diets can be made adequate and balanced in all the necessary nutrients that promote wound healing and reduce infection even though the texture of the diet is liquid or soft or the method of administration is by naso-esophageal tube.

Chapter Thirty-three

The Role of Diet and Nutrition in the Management of Gingival and Periodontal Disease

ABRAHAM E. NIZEL, D.M.D.

Food, depending on its consistency, can exert locally both a beneficial and detrimental effect on gingival and periodontal tissues. It can contribute significantly to partially preventing or delaying the onset of disease; on the other hand, it can serve as a nidus for calculus formation and consequent inflammation. The proper nutrients in the proper amounts which are made available through the digestion and metabolism of food to all body tissues, including those of the gingiva and periodontium, can systemically increase the resistance of these tissues to disease. A deficiency of these nutrients can accelerate and increase the intensity of an inflammatory process that has been initiated by a local challenge or irritant. A nutritional deficiency, for all practical purposes, is never the prime etiological factor of periodontal disease. However, it can be an important contributory factor. These axioms are based on an accumulation of data from a number of different laboratories using different experimental sets of circumstances and designs in an attempt to define the specific role of local and systemic influences of food and their nutrients.

Unfortunately, in the process of the rapid growth and development of the science of nutrition, some unsubstantiated claims and distortions have arisen. This has made thoughtful and discriminatory investigators and clinicians look with jaundiced eye on diet and nutrition as a meaningful aid in the management of some diseases, particularly gingival and periodontal disease. However, in their efforts not to be labeled as fellow-travelers of the nutrition faddist, some clinicians have done a complete about turn and have assumed almost an intransigent negative attitude toward diet and nutrition as an aid in the management of disease.

A second situation which has contributed to this antinutrition bias has been the confusion over the meaning of the terms diet and nutrition. This is regrettable because a consequence resulting from the physical effects of the food of the diet will not necessarily be the same as that resulting from the chemical effects of the nutrients derived from food. Oftentimes both have been considered to be involved when, in reality, only one should have been. Diet and nutrition have been used carelessly and inaccurately as synonymous terms when actually they are quite different: Nutrition is a bodily process dealing with the ingestion, digestion, transport, utilization and excretion of food. Diet, in contradistinction, refers to a regimen of food which supports nutrition. In short, diet deals with food; nutrition deals with nutrients derived from food. Since food has a physical consistency and a chemical nature, whereas nutrients are solely chemicals, the differences in their effects can be important. Food exerts primarily a local oral environmental effect. On the other hand, nutrients have an endogenous systemic effect on the oral tissues.

The object of this section is to present our

408

knowledge in the relation of food and nutrition to gingival and periodontal disease as it has developed from epidemiological surveys, clinical trials and animal experiments. The differentiation between local effects of food and the systemic effects of nutrients will be emphasized. To translate our present knowledge on food and nutrition and periodontal disease into practical terms, dietotherapy for the acute and chronic diseases that affect gingival and periodontal tissues will be shown.

LOCAL EFFECTS OF FOOD

The physical consistency of the natural, processed and prepared foods can by its retentive or detergent character influence the gingiva and periodontium positively or negatively. If the food is soft and sticky it will adhere to and accumulate on the teeth, forming plaque and serving as a nidus for calculus formation, the prime gingival and periodontal irritant. If, on the other hand, the food is firm and detergent it will clean, stimulate, massage and strengthen these tissues. In general, firm foods require chewing, which provides the basic physiologic need of all tissues—use. Disuse produces atrophy, degeneration and lowered resistance to infection; normal use promotes tissue growth and health. On the other hand, excessively coarse granular diets have been shown to produce extensive periodontal disease, either because of overuse or frank injury to the supporting tissues (Auskaps et al., 1957; Cohen, 1960; Person, 1961; Stahl et al., 1958).

GINGIVAL CIRCULATION

Foods of firm consistency will increase the number, distribution and tone of the capillaries (Burwasser and Hill, 1939; Pelzer, 1940). The circulation of the gingivae will be improved because the hard foods will mechanically improve and increase the interchange of nutrients between the blood and the tissues. In short, this will improve the metabolism and vitality of the gingivae.

EPITHELIUM

The degree of keratinization of the stratified squamous epithelium which affords protection against trauma or other injurious agents is affected by the frictional qualities of the diet. Nature reacts to mechanical irritation by increasing the thickness of the hornified layer of epithelium. Without this protection, chemical and bacterial irritants can make inroads into the gingival tissue and produce inflammation (O'Rourke, 1947; Weinmann, 1940).

PERIODONTAL MEMBRANE

Chewing, by its mechanical action, produces a compression and expansion of the periodontal spaces around the teeth which, in turn, stimulates the removal of waste products through the venous system and lymphatics and the entry of nutrients into the periodontium via the arterial system. Brekhus et al. (1941) have shown that firm foods promote formation of a dense fibrous suspensory structure in the periodontal membrane by increasing circulation and fibroblastic activity. Coolidge (1937) found that the width of the periodontal membrane directly related to the intensity of masticating function.

ALVEOLAR BONE

The maintenance of proper balance between bone resorption and new bone formation is materially aided by hard foods. Disturbances of this balance by the inadequate function induced by soft foods will produce atrophic changes and lower the threshold of bone activity.

GINGIVAL DEPOSITS

Calculus and plaque contribute the most important local etiologic agents in periodontal disease, ranging from gingivitis to severe periodontitis (Keyes and Likins, 1946; King and Glover 1945; Egelberg, 1965b). Calculus, by virtue of its mechanical hardness, is thought to act as a mechanical traumatic agent. Similarly, plaque and impacted food impinge on the gingiva and act as mechanical irritants. Therefore, the prevention and the elimination of these irritants is of prime importance in the management of periodontal disease.

BACTERIA

The consistency of the diet can influence both the kinds and amounts of microorganisms found in the gingival crevices and pockets. Animals fed a hard diet showed minimal amounts of mixed flora in the gingival pockets; whereas, when the same animals were changed to a soft diet, the oral flora was increased considerably and was dominated by fusiform bacilli and spirochetes. Gingival inflammation was noted in the latter group and not in the former. Furthermore, the flow of tissue fluids was decreased and there was a concomitant decrease in the amount of antimicrobial agent, gamma globulin, when firm hard foods which require vigorous mastication were not eaten (Krasse and Brill, 1960).

CARBOHYDRATE DIETS

Since carbohydrates, particularly granular sucrose mixed in a soft diet, have been found to be retentive around rat molars, it was concluded by Shaw and Griffiths (1961) that this nutrient could be an important etiological agent in periodontal disease—primarily because of its physical rather than chemical nature. This conclusion was strengthened by the lack of adherent masses of food debris on teeth when carbohydrate was replaced by lard. Others have also implicated carbohydrates (Klingsberg and Butcher, 1959; Mitchell and Johnson, 1956). In contrast is the finding of Carlsson and Egelberg (1965a), who reported that the addition of sucrose to a basic protein-fat diet of hard or soft consistency did not affect the formation of the plaque or the development of gingival inflammation. As an explanation for the difference in their findings compared to most other investigators, Carlsson and Egelberg hypothesized that this may be the result of differences in bacterial flora between their experimental animals

(dogs) and those of the other (rats). Furthermore, rats eat four times more frequently (20 versus 5) than dogs. However, in a recent study on the effect of frequent and large quantities of sucrose on early plaque formation in man, Carlsson and Egelberg (1965b) concluded that the plaque may be the result of the production of extracellular polysaccharides of plaque bacteria rather than acid precipitation of salivary mucoids.

TUBE FEEDING

The mandatory presence of food to serve as a substrate for calculus formation is not proved according to Egelberg (1965b). He found that there was abundant growth of plaque bacteria even without direct contact of food with the teeth. In this instance, it was postulated that the substrate for these bacteria might be transudates and exudates of epithelial tissue as well as conditions which favor the growth of compatible bacteria. Obviously, there are several postulates here that need further intensive investigation before they can be deemed acceptable.

CONCLUSION AND SUMMARY

It can be said that the rate of calculus formation is directly related to the stickiness and retentiveness of the food in the diet. Foods that are rough and fibrous rub against the teeth and can actually mechanically cleanse them, perhaps even better than a toothbrush. They are, in a manner of speaking, a natural toothbrush.

Of the different types of nutrients that have been indicated as cariogenic, sucrose appears to be the one about which the evidence seems to be most convincing. Furthermore, because carbohydrate-rich diets do not provide prolonged satiety and can lead to frequent snack eating, it seems logical to assume that there will be an ever-present accumulation of food debris around the teeth when diets are rich in sweets.

All this simply means that choosing a balanced diet of a wide variety of foods of firm detergent consistency will prevent food accumulation, slow down the rate of calculus formation, and might well contribute significantly to the prevention of gingival and periodontal disease.

NUTRITIONAL SYSTEMIC EFFECTS

Like all labile tissues, the periodontal tissues are affected by stresses of everyday living, of which diet and nutrition are one. Specific nutrients have been associated with maintaining the integrity of the epidermal, mesodermal and calcified tissues of the periodontium; e.g., vitamins A, B complex and C might affect epithelial and collagenous tissues, and calcium, phosphorus and vitamin D might affect mineralized tissue. These are undeniable facts. The confusion and debate that arises are whether these nutrients are primary or secondary in action. In other words, does lack of them precipitate disease or contribute to it? The evidence seems to point to a modifying action rather than an initiating one.

PROTEIN

Protein deficiency in laboratory animals has definitely been shown to make the periodontal tissues more susceptible to inflammatory and degenerative disease, particularly in the presence of a local irritant, be it a foreign body, soft food or a wound. (See Oral Relevance section of Chapter Six.) However, in man the evidence that periodontal disease has greater prevalence among individuals whose diets are low in protein is not convincing (Bradford, 1959), even though Cheraskin and Ringsdorf (1964) have reported that they noted a reduction in gingivitis in a group of humans after four days of protein supplementation. In fact, for the most part, it appears that no correlation exists. There certainly is no evidence that periodontal disease can be initiated by a deficiency in this nutrient alone.

VITAMIN C

The role of vitamin C in periodontal disease seems to follow a pattern similar to that of protein with respect to not being a prime etiological factor. Originally, Boyle et al. (1937) had suggested that diffuse alveolar atrophy was produced in animals by a vitamin C-deficient diet. However, Glickman (1948) has shown that local irritants like food debris must be present to initiate gingival inflammation and that the vitamin C deficiency will then accentuate the destruction of the periodontal membrane and alveolar bone. Waerhaug (1958) has also shown that vitamin C deficiency is a contributory rather than a causative factor in gingivitis and periodontitis. There have been several studies in which the effects of vitamin C supplements on gingival score, sulcus depth and clinical tooth mobility have been measured. The conclusion reached by these investigators is that vitamin C is beneficial (El-Ashery et al., 1964). Certainly, more objective research needs to be done to confirm this important conclusion. Clinical observations like these are suspect unless the differences between health and disease are great enough and clear enough so that one does not have to rely on small statistical differences to separate fact from fancy.

However, because vitamin C is necessary for normal metabolism of endothelium and connective tissue, its absence will cause altered capillary permeability and ready gingival bleeding. A deficiency will also prevent repair of the periodontium if some destruction has already taken place. For further details see Chapter Thirteen.

VITAMIN B COMPLEX

Nicotinic acid deficiency has not been shown convincingly to be the prime causative agent in either gingivitis or acute ulcerative necrotizing gingivitis. However, in dogs there has been a correlation between deficiencies of the B complex vitamins and nonspecific gingival and periodontal lesions (King, 1943). Here, too, the local irritating factor is considered the precipitating mechanism and the systemic nutritional deficiency a contributing one.

PRINCIPLES OF THERAPEUTIC DIETS

Since food factors can contribute both locally and systemically to the etiology and the protraction of disease in general and gingival and periodontal disease in particular, it is logical and even essential to consider methods of implementing a modification of the normal pattern of food intake to cope with these problems.

In general, dietary modifications are made with respect to (a) frequency of eating, (b) quantitative increase, decrease or elimination of one or more nutrients or food or (c) qualitative alteration of the physical consistency of food. In medicine, for instance, the calorie restriction diet used in obesity and the sodium restricted diet used in congestive heart failure are examples of diets in which some element of nutrition is decreased. An example of increase in a nutrient is the use of protein in convalescence following surgery. An example of elimination of a food is demonstrated in the management of allergies when specific foods have to be completely withdrawn from the diet because of patient sensitivity.

Diets can also be modified qualitatively by altering the physical characteristics or the texture of the diet—liquid, soft, bland, high residue, etc. In medicine this is an important consideration, for instance, in the management of certain gastrointestinal disorders such as ulcers.

There are certain general principles that should be incorporated in the prescription and formulation of therapeutic diets. These are:

1. The 4 Food Groups provide all the essentials of good nutrition; therefore, they should serve as a fundamental framework on which modifications are built.

2. The likes and dislikes of the patient as well as the personal habits and environmental conditions, such as economic, cultural, religious and physiological factors which influence food selection, should be considered.

3. A simple explanation of the specific disease process should be given to the patient and the homemaker of the family so that they understand the objective of the therapeutic diet; thus, they will be motivated to cooperate and to make the necessary changes in the household routine.

4. Special diets should be considered as just that—special. They are needed as a rule to overcome a special or abnormal situation and should be used only as long as this condition prevails. Once normal status has been achieved, the special diet should be discontinued and the patient should resume his normal diet regimen.

DIETARY CONSIDERATIONS IN ACUTE GINGIVAL DISEASE

Dietary management for this condition, as for other oral problems, is simply an adjunctive procedure and serves as a supportive measure to ensure more rapid restoration of tissues to their normal state. Diet by itself cures nothing; local treatment is always the primary consideration.

The previous 24-hour food intake and an evaluation of its relative adequacy on the basis of the recommended allowances of the 4 Food Groups are the starting point. Questions that relate to the length of time this dietary pattern has been followed and to the environmental factors that have influenced the selection of these foods are necessary in order to prescribe a realistic therapeutic regimen with which the patient can cooperate. For example, have there recently been personal situations that caused tensions? Are there budgetary problems that influence food selection? What is the patient's daily occupation and routine? Are there any current medical problems?

In addition, observations should be made of the general physical appearance, weight, state of malaise and lassitude, presence of fever, as well as degree of oral sensitivity and masticatory impairment. The oral examination should include not only the degree of gingival tissue alteration, but the color and topographical changes of the tongue, as well as the color of the oral mucosa; the fissures and lesions of the lips should be carefully examined for signs of anemia or suspected vitamin deficiencies.

There are no simple practical biochemical or bacteriological laboratory proceedings that need to be carried out for re-enforcement of the diagnosis with the exception of a serial determination of hemoglobin levels for suspected iron deficiency anemia. As a rule, simple clinical observation and a complete history will suffice.

Since this is an acute situation, the signs and symptoms will be severe, the onset will be sudden, but the response will be equally dramatic and rapid. Therefore, the special diet need be followed for only a few days, and then resumption of the normal maintenance dietary practices should be encouraged.

As a rule, the foods most often found lacking in the average diet are protein foods and a source of vitamin C.

The following advice for the patient is typical of what should be included in prescribing and formulating a special diet.

Frequency of Eating Periods. Use the scientific nibble approach to fulfilling your food requirements. In other words, instead of trying to eat three large meals eat six to eight small ones. Only one or two foods need to be eaten each time, but select a large enough variety of foods so that the eating experience will be pleasant and will satisfy the recommended daily allowance of two servings of milk, two of meat, four of fruits and vegetables and four of bread and cereals.

Quantitative Considerations. Use additional portions of the milk and meat groups because they are the best sources of protein. These include milk, cottage cheese, ice cream, eggs, tuna fish, salmon, boiled chicken, chopped meat, chopped liver.

Pay particular attention to ingesting more that the usual amounts of vitamin C-rich foods. If they cannot be eaten in their usual form, use them as juices. In addition to fruits like oranges, grapefruit, strawberries, cantaloupes, guava, mangoes and papayas, use vegetables such as cooked turnip greens, green peppers, tomato juice and baked potatoes.

Qualitative (Textural) Considerations. Plan your menu so that the general consistency of the diet is liquid and/or soft, omitting the spicy or sharp-tasting foods during the acute period. Choose foods that suit your individual taste and economic status. For variety select some of the following food exchanges in each of the food groups:

Milk group: Use milk in all forms. For extra nourishment nonfat milk solids may be added to regular milk. Add 2 tbsp. of milk solids to each 8 oz. glass of whole milk. Soft, plain ice creams are also soothing. Milk shakes and malted milks are recommended. Cream may be added to milk if desired.

Meat group: Eggs in the form of strained eggnogs may be used in a liquid diet. They may also be used in soft or baked custards. Strained baby meats may be diluted with milk or bouillon to make soups.

Vegetable-fruit group: Citrus juices such as orange and grapefruit are highly recommended, as well as tomato juice and other fruit and vegetable juices. Puréed fruits and vegetables, such as baby foods, may be used. Strained fruits may be used with milk as a drink, and strained vegetables may be cooked with milk and butter and other seasonings as soup.

Bread-cereal group: Strained gruels.

A sample menu of a full liquefied diet using the above suggested foods might be as follows:

Breakfast
> Pineapple-grapefruit juice
> Strained oatmeal gruel
> Coffee, with cream and sugar if desired

Midmorning
> Orange eggnog

Noon
> Strained cream of tomato soup
> Custard with cream
> Tea or coffee

Midafternoon
> Chocolate milk shake

Evening
> Strained pea soup
> Spanish cream (or any custard pudding)
> Tea or coffee

Bedtime
> Cocoa

The obvious purpose of this diet is to satisfy the normal requirement for all nutrients in a form which requires no chewing, just swallowing. Hopefully, this diet will be used only for a day or so. The patient should then change to a soft type diet consisting of the following variety of foods in the 4 Food Groups. An individualized meal plan can be readily made up from this assortment of foods.

Milk group: Soft cheeses such as cottage cheese and cream cheese, in addition to milk and ice cream.

Meat group: Tender or ground meats, chicken, fish without bones, and cooked legumes.

Vegetable-fruit group: Cooked or canned fruits and ripe bananas; vegetables without skins or seeds, mashed potatoes.

Bread-cereal group: Soft bread, macaroni, noodles, rice, spaghetti, cooked breakfast cereals with milk, and ready-to-eat flaked and puffed cereals, such as wheat or corn flakes and Puffed Wheat and Rice.

Even without being specifically told, the patient will gradually change to normal foods of regular texture as the soreness of his mouth lessens. The important point to stress is that a diet can be made adequate if sufficient amounts of foods from the 4 Food Groups are eaten each day regardless of their consistency.

It should be added that a patient with acute necrotizing ulcerative gingivitis will often be febrile. This does *not* preclude prescribing a full diet. The free administration of food does not, as once believed, raise the temperature of the febrile patient. In fact, the basal metabolism is usually increased significantly during a fever, which means that nutrient requirements will be correspondingly increased. An underfed febrile patient will be thrown into a negative nitrogen balance, which will interfere with the normal defense mechanisms of the body and delay healing.

Vitamin Supplementation. On the basis of increased nutrient requirements during the period of stress and because the patient has probably already used up some of his vitamin reserve, vitamin supplementation is desirable and indicated. Furthermore, because of his dental impairment, the patient probably has been unable for days or weeks to eat the usual fresh fruits and vegetables, the natural source of his vitamin intake. He may have bizarre eating habits or be on a restricted diet because of a systemic condition. In short, he may have eaten an unbalanced diet over a prolonged period.

Most deficiency states involve multiple factors; therefore, the type of vitamin preparation to supply is a multivitamin. Since fat soluble vitamins A and D are usually stored well and are not readily depleted, it is not necessary to include these in the vitamin therapy for an acute condition.

The multivitamin preparation most often indicated is a combination of the B complex and C. The B complex vitamins in most commercial preparations are usually thiamine, riboflavin and niacin, with or without pyridoxine and calcium pantothenate. The therapeutically effective dosage that is recommended is three to five times the Recommended Dietary Allowances. If the dentist will remember that the recommended allowance for riboflavin or thiamine under normal circumstances is about 2 mg., any multivitamin preparation can be considered therapeutic if it contains 6 to 10 mg. of each of these vitamins, plus the other water soluble vitamins in correspondingly increased therapeutic doses. Different brand names of these vitamin preparations of therapeutic dosage are given in Appendix III.

As soon as adequate amounts of foods from the 4 Food Groups can be ingested, vitamin supplementation can be suspended. In fact, the type of diet suggested as an adjunct in the management of chronic periodontitis should be adopted as soon as feasible.

DIETARY CONSIDERATIONS IN CHRONIC PERIODONTAL DISEASE

An integral part of the diagnosis and therefore total management of the patient with chronic periodontal disease should be diet and nutritional considerations.

Since systemic nutritional problems have been overstressed but infrequently found to be associated with chronic periodontal problems, there has risen an antipathy to this correlation, However, it is not valid to disregard completely the possibility of such a relationship. It is the purpose here merely to provide knowledge essential to making a nutritional diagnosis that might in some instances help the clinician to discover a state of malnutrition in a patient with chronic periodontal disease.

DIAGNOSTIC STEPS

Medical History. As the patient walks

into the operatory to seat himself in the chair, the patient's stature, gait, vigor and general health can be noted. While the dentist engages the patient in conversation he will gain an impression of the patient's emotional attitude, alertness and intelligence. The clinician should be alerted to the possibility of a nutritional problem if the complaints of the local oral problems such as "bleeding gums," "bad taste," or "itchy gums" are accompanied by systemic complaints of "tiredness," "no appetite," "upset stomach," "losing weight rapidly," etc.

Some symptoms that suggest malnutrition are apathy, photophobia, visual disturbances at night, sore lips, sore tongue and ulcerated angles of the mouth, digestive disturbances, diarrhea, dyspnea, edema and paresthesia. It should be emphasized that these symptoms, of course, may occur in other conditions that are not even concerned with nutritional deficiencies.

The presence of secondary conditioning factors that can lead to a nutritional deficiency, even in the presence of an adequate diet, need to be ruled out. Questions that relate to the presence of impaired ingestion as well as increased requirements for nutrients, such as are brought on by a loss of appetite due to infection, loss of teeth, food allergy, special diets, alcoholism or nausea of pregnancy, are proper and should be answered insofar as possible. Other conditioning factors in addition to psychological problems which need to be ascertained during the history taking are diseases of the gastrointestinal tract such as diarrhea (in which there is excessive loss of water soluble vitamins) and achlorhydria (which produces improper absorption of calcium, phosphorus, iron and ascorbic acid). The presence of uncontrolled diabetes (which interferes with the utilization of glucose and causes diuresis), cirrhosis of the liver and thyroid disease are important clues to the possibility of a malnutrition syndrome. Also, the physiological stress periods such as the growth spurt, pregnancy and lactation, which increase nutritional requirements, should be noted.

Diet History. The next step is to acquire information about the patient's daily diet. This includes not only the kinds but also the amounts of food eaten. Of particular interest in the patient with periodontal disease are the consistency and the detergency of the diet. By consistency we mean the degree of density or firmness of the food, which indicates its stimulatory effect. Detergency is the ability of a food to clear itself from the mouth and also to cleanse the teeth. Detergent foods are usually fibrous and, therefore, are of firm consistency, but not all firm foods are detergent. This differentiation should be kept clearly in mind when making an evaluation of the patient's diet. A method for obtaining dietary information is to provide the patient with a food diary form which has sufficient space to record each meal and between-meal snack for at least five days. One of the five should be a week-end day or holiday to obtain a representative dietary pattern. It is important to emphasize to patients that they should not conceal any information nor should they try to anticipate what the dentist may deem desirable and make changes in their customary dietary intake during the period of diary keeping.

Method for Evaluating Adequacy of Food Intake. The evaluation of the five-day food diary consists of using the simple quantitative and qualitative standards such as "Food for Fitness—A Daily Food Guide,"* which is based on the 4 Food Groups. If there is a particular indication or desire for details about nutrient intake, tables of food compositions can be used (Appendix I). However, in the majority of cases, the dentist will find the simple qualitative procedure adequate.

A total food intake chart such as shown in Table 14-5 may be used, in which foods are classified into the four groups. (See Chapter Fourteen.)

The procedure for evaluating the diet is to credit in the appropriate column with a chit mark each serving of food eaten. As a guide to what constitutes a single serving of a particular food and in which of the 4 Food Groups the food is classified, consult Appendix II. Mixed dishes are also found in this table. For example, a meal consisting of a serving of macaroni and cheese, a slice of bread and butter, 1 glass of milk and a few carrot strips would be tallied as follows: Macaroni and cheese is classified as 1 serving of the bread-cereal group and 1/2 serving of the milk group. Therefore, a single chit mark goes in the

* United States Department of Agriculture, Leaflet No. 424.

bread-cereal group and ½ credit in the milk. One slice of bread, one cup of milk and carrot strips are classified as single chit marks in the bread-cereal group, milk group and vegetable-fruit group, respectively.

When all the transposing of the individual foods is completed, the total number of servings for each group is added and divided by five to arrive at an average, which is recorded in the average per day column. This daily average is subtracted from the recommended amounts and the difference is recorded in the difference column in terms of servings.

If a person does not meet the recommended amounts of the standard, it does not necessarily mean that a nutritional deficiency exists. It merely suggests that the patient is consuming less than the amounts of nutrients which are considered desirable for maintaining good nutrition in healthy persons in the United States. These amounts represent goals to be striven for rather than requirements. By maintaining these standards the person is assured of not being nutritionally deficient so long as there are no secondary conditioning factors to interfere with the utilization of the food.

Evaluating the Consistency of the Diet. Of particular importance in the dietary management of a patient with chronic periodontal disease is a consideration of the physical nature of the food. In Table 33-1 is a chart that may be used for this purpose. Each of the 4 Food Groups can be subdivided into two basic physical forms from a periodontal standpoint, the stimulatory and the nonstimulatory. The nonstimulatory types would be the liquid and soft, chopped, processed and well cooked foods, whereas the stimulatory types would be the hard, solid, raw, slightly cooked and dry foods. As the foods from the diet are being credited onto the evaluation chart, the marks should be inserted in the proper subdivision. For example, if the patient ate cream cheese, this chit mark should be next to the "Liquid, soft" subdivision of the milk group; on the other hand, if the patient ate American processed or Swiss cheese, the chit mark would be in the "Hard" subdivision of the milk group. In the other food groups whole portions of fish or chicken, meats that are baked, broiled or fried are considered hard; vegetables that are slightly cooked, like broccoli, carrots, and green beans, as well as potato skins, may be considered of firm consistency; crusty and toasted breads as well as dry cereals are also considered firm enough to produce some stimulatory effect on the periodontium.

If emphasis on lack of firm consistency is desired, this can be achieved by circling with a red pencil the chit marks in the subdivisions,

TABLE 33-1. TOTAL FOOD INTAKE CHART WITH SPECIAL CONSIDERATION TO THE PHYSICAL FORM OF THE FOOD

FOOD GROUPS	PHYSICAL FORM	1st Day	2nd Day	3rd Day	4th Day	5th Day	Average per Day	Recommended Intake	Diff.
MILK	Liquid, Soft	/	//		/	//	1½	2 Serv.	-½
	Hard		/		/				
MEAT	Soft, Chopped	/				/	2	2+ Serv.	0
	Solid		//	//	//	/			
VEGE-TABLE-FRUIT	Juice, processed			/		//	1+	4+ Serv.	-3
	Raw, Firm, Partly Cooked		/		//				
BREAD-CEREAL	Soft, Cooked	HH	HH	//	///	HH //	5	4+ Serv.	+1
	Dry, Crusty Toasted		//		//				

"Liquid," "Soft," "Chopped," "Processed" and "Cooked."

Noting the Detergency of the Diet. The detergency or oral clearance of the diet is favorable if the dessert is a raw fruit such as apples, apricots, cherries, grapefruit, grapes, melons, oranges, pears, peaches, pineapple and tangerines or a raw vegetable like carrots, cauliflower, celery, cucumber, peppers, radishes or combinations of these. If fluids like coffee, tea, milk, fruit and vegetable juices are used as a dessert alone, this is also favorable from an oral clearance standpoint. However, the combination of a bread-cereal group food with a fluid is undesirable because it will leave food debris. The favorable detergent foods may be circled with blue pencil on the patient's diary to impress the patient with his desirable food selections.

Evaluating Food Habits. Other factors that need to be considered in a dietary history are environmental conditions that influence food selection and food habits. Food selection is influenced by such factors as restrictions for medical reasons, economic conditions, and cultural and religious beliefs. Some typical questions that may help elicit information about food selection are as follows: Are you eating a special type of prescribed diet which restricts certain foods? Is your diet restricted by any cultural or religious beliefs? How often do you eat in restaurants? Do you carry lunches from home? Do you have to prepare your own meals or does someone else do it for you? What are your cooking facilities?

Food habits are concerned with attitudes the patient has toward food. Does he feel that mealtime is a pleasurable and relaxed period of the day, or is eating considered a necessary chore? Does he omit meals? Does he take time to eat breakfast? Does he eat his meals at regular times during the day?

Inspection. The dentist through inspection alone can observe the presence of gross physical signs that might lead him to suspect a nutritional inadequacy.

It is easy, for example, to see whether a patient is fat or thin. If need be, the clinician can appraise the muscular and skeletal development and estimate the subcutaneous fat layer by the skin fold thickness test. It should be between 10 and 25 mμ in the upper posterior aspect of the arm (midpoint between the tip

of acromium and the tip of the olecranon with the elbow in 90° flexion). The general appearance of apathy and pallor can be noted. The facial skin may show fine, scaly, greasy desquamation in the nasolabial folds, around the ears, and at the acanthi of the eyes. The general skin may be dry and scaly, pebbled and pigmented or purpuric with petechiae. The palms of the hands may be excessively red or pale, and the nails may be brittle, ridged or spoon-shaped. The eyes may have circumcorneal or conjunctival injection. The lips may have a cheilosis accompanied by ulcerations at the angles of the mouth. The tongue may show papillary atrophy or hypertrophy and variations in color may range from pale pink to scarlet red to a magenta. The gingiva may show marginal redness and swelling. Numerous other physical signs which may be associated with malnutrition have been described in Chapter Nineteen.

Laboratory Aids. A few laboratory tests have been found useful in determining the nutritional status of population groups; however, they are not always applicable to the individual.

These four steps—medical history, dietary history and evaluation, general clinical inspection, and laboratory data—make up the necessary procedure for a nutritional diagnosis. From a practical standpoint, it will be found that laboratory procedures are seldom used, but the diet evaluation, medical history and clinical inspection are invariably considered.

DIETARY MANAGEMENT

If the nutritional diagnosis suggests that a systemic condition exists which is interfering with food utilization, the patient should be referred to the family physician for management.

However, if the nutritional diagnosis suggests that there is simply an inadequate food intake, it is the dentist's responsibility to advise the patient how to acquire an adequate diet. It should be made clear that diet for periodontal health is really not much different from that for general health. The emphasis is placed on food texture. Firm and detergent foods serve as adjuncts in carrying out proper oral physiotherapy, an indispensable factor for success in periodontal therapy.

Like all preventive, treatment or maintenance regimens, the dietary one should be based on the diagnostic findings—specifically, what is the patient eating and why is he eating it? The "what" of the diet is made evident in the dietary evaluation and history. The "why" of the diet which is concerned with reasons for food selection is based on the medical background, general inspection, and information from personal and family histories.

It is always important to keep in mind the motivating factors that determine the food habits and eating patterns when prescribing a diet. If the diet is oriented with an appreciation for this important point, then the diet will be individualized, assuring a more cooperative patient and a better chance of a successful result. In short, radical changes should be avoided. Only subtle ones that the patient almost prescribes for himself should be instituted. This is a realistic approach and one that can be rewarding.

STEP 1. IMPROVE
TOTAL DAILY INTAKE

To demonstrate to the patient which food groups have been deficient in his diet, show him the results of his diet evaluation which is based on a comparison of his actual intake with the standard recommended amounts. Emphasis should be placed on those food groups which have been shown to be deficient, and these foods should be increased to at least the recommended amounts. The two food groups most often found deficient are the milk group and the vegetable-fruit group. For example, if the patient is eating only one vegetable a day, say potatoes, then he should be advised that he ought to include at least one citrus fruit and one green or yellow vegetable, as well as some other raw fruit or vegetable each day.

STEP 2. IMPROVE THE
BALANCE OF EACH MEAL

After the total intake has been improved, foods from each of the 4 Food Groups should be advised for each of the three meals when possible. This provides a balanced diet because all the essential nutrients (carbohydrates, fats, proteins, vitamins, minerals and water) are provided in adequate amounts at each meal. The areas of weakness in most meal planning are usually the lack of a protein food for breakfast, the absence of a vegetable for lunch and inadequate sources of calcium for dinner. There may be some environmental factors, such as lack of facilities for preparing breakfast or lack of time for eating an adequate meal, that must be taken into consideration when prescribing meals.

STEP 3. SUGGEST AN
INDIVIDUALIZED MENU

Menu planning is simply the interpretation of food groups into dishes and foods that are found in each of the food groups. Prescribe the dishes that the patient likes as long as it does not interfere with the main objective. To improve the patient's menus, suggest that a *variety* of different foods from the food groups be used. For example, if the vegetable for dinner one day is broccoli, the next day it could be corn and on the third day it could be cauliflower, and on the fourth day it could be carrots. Here again such environmental factors as cost must be taken into consideration. Menus can be suggested that are high in nutritional value but low in cost.

STEP 4. INCLUDE
FOODS OF FIRM CONSIST-
ENCY AND DETERGENCY

As a rule, the evaluation will show the physical consistency of the diet to be soft. To improve it, one might suggest at least one food of firm consistency for each meal. Each of the food groups has firm as well as soft foods. For example, in the milk group there is hard cheese; in the meat group there are broiled steaks, chops and chicken fried on the bone; in the vegetable-fruit group, besides raw fruits and vegetables, corn on the cob and skins of baked potatoes can be suggested; in the bread group, crusty bread, Melba toast and Ry-Krisp are excellent for providing a firm type of food.

From the food diary the type of dessert

should be noted as far as its detergency is concerned. Substituting a raw fruit or vegetable for desserts like pie and coffee is usually all that is necessary. The following detergent fruits can be suggested: raw apples, apricots, cherries, grapefruit, grapes, melons, oranges, pears, peaches, pineapple and tangerines. Raw vegetables like carrots, cauliflower, celery, cucumbers, tossed salads and cole slaw should be used liberally, during the latter part of the meal if possible. Another suggestion might be the use of carrot strips or celery as snacks between meals.

The following are sample menus that incorporate all the suggestions in Steps 1 to 4; they include foods typical of three different nationalities:

Italian

Breakfast
>　Egg, soft boiled
>　Italian bread
>　Coffee
>　Orange

Lunch
>　Lasagne, cheese
>　Salad greens
>　Crusty bread, cheese
>　Fresh peach
>　Milk

Dinner
>　Antipasto, with cheese
>　Scallopini
>　Italian bread
>　Watermelon
>　Tea or coffee

Jewish

Breakfast
>　Cottage cheese
>　**Bagel**
>　**Coffee**
>　Orange

Lunch
>　Salmon salad
>　Pumpernickel bread
>　Carrot, celery curls
>　Fresh peach with sour cream
>　Milk

Dinner
>　Broiled chicken
>　Broccoli
>　Baked potato
>　**Tomato-cucumber salad**

>　Crusty bread
>　Watermelon
>　Tea or coffee

American

Breakfast
>　Fried eggs and bacon
>　Toast and butter
>　Coffee
>　Orange slices

Lunch
>　Tuna fish salad sandwich
>　Wedge of lettuce
>　Milk
>　Pineapple-strawberry fruit cup

Dinner
>　Baked pork chop
>　Baked potato
>　Broccoli
>　Tossed salad
>　Crusty bread, butter
>　Milk
>　Raw apple

CONCLUSIONS AND SUMMARY

The role of diet in the management of gingival and periodontal disease is primarily one of prevention or maintenance. Only when there is clear evidence that acute systemic, nutritional problems exist can it be said that food has truly a curative effect. In light of our present limited knowledge about the etiology of gingival and periodontal disease, it appears that local factors are the prime consideration. The benefits derived from food can best be explained on the basis of its local oral physiotherapeutic action. Furthermore, the potential for good is great enough that judicious use of dietary guidance should be seriously considered as routine an office procedure as instructing the patient on such other oral physiotherapy measures as toothbrushing and interdental massage.

REFERENCES

Auskaps, A. M., Gupta, O. P., and Shaw, J. H.: Periodontal disease in the rice rat. III. Survey of dietary influences. J. Nutr., *63*:325, 1957.

Boyle, P. E., Bessey, O. A., and Wolbach, S. B.: Experimental production of diffuse alveolar **bone** atrophy type of periodontal disease by diets deficient in ascorbic acid (vitamin C). J. Amer. Dent. Assn., *24*:1768, 1937.

Bradford, E. W.: Food and the periodontal diseases. Proc. Nutr. Soc., *18*:75, 1959.

Brekhus, P. J., Armstrong, W. D., and Simon, W. J.: Stimulation of muscles of mastication. J. Dent. Res., *20*:87, 1941.

Burwasser, P., and Hill, T. J.: The effect of hard and soft diets on the gingival tissues of dogs. J. Dent. Res., *18*:398, 1939.

Carlsson, J., and Egelberg, J.: Local effect of diet on plaque formation and development of gingivitis in dogs. II. Effect of high carbohydrate versus high protein-fat diets. Odont. Rev., *16*:42, 1965a.

Carlsson, J., and Egelberg, J.: Effect of diet on early plaque formation in man. Odont. Rev., *16*:112, 1965b.

Cheraskin, E., and Ringsdorf, W. M.: Periodontal pathosis in man: IX. Effect of combined versus protein supplementation upon gingival state. J. Dent. Med., *19*:82, 1964.

Cohen, B.: Comparative studies in periodontal disease. Proc. Roy. Soc. Med., *53*:275, 1960.

Coolidge, E. D.: The thickness of the human periodontal membrane. J. Amer. Dent. Assn., *24*:1260, 1937.

Egelberg, J.: Local effect of diet on plaque formation and development of gingivitis in dogs. I. Effect of hard and soft diets. Odont. Rev., *16*:31, 1965a.

Egelberg, J.: Local effect of diet on plaque formation and development of gingivitis in dogs. III. Effect of frequency of meals and tube feeding. Odont. Rev., *16*:50, 1965b.

El-Ashery, G. M., Ringsdorf, W. M., and Cheraskin, E.: Local and systemic influences in periodontal disease: IV. Effect of prophylaxis and natural versus synthetic vitamin C upon clinical tooth mobility. Inter. J. Vit. Res., *34*:202, 1964.

Glickman, I.: Acute vitamin C deficiency and periodontal disease. I. The periodontal tissues of the guinea pig in acute vitamin C deficiency. J. Dent. Res., *27*:9, 1948.

Keyes, P. H., and Likins, R. C.: Plaque formation, periodontal disease, and dental caries in syrian hamsters. J. Dent. Res., *25*:166, 1946.

King, J. D.: Nutritional and other factors in trench mouth with special reference to the nicotinic acid component of the vitamin B. complex. Brit. Dent. J., *74*:113, 1943.

King, J. D., and Glover, N. E.: The relative effects of dietary constituents and other factors upon calculus formation and gingival disease in the ferret. J. Path. Bact., *57*:353, 1945.

Klingsberg, J., and Butcher, E. O.: Aging, diet and periodontal lesions in the hamster. J. Dent. Res., *38*:421, 1959.

Krasse, B., and Brill, N.: Effect of consistency of diet on bacteria in gingival pockets in dogs. Odont. Rev., *11*:152, 1960.

Mitchell, D. F., and Johnson, M.: The nature of the gingival plaque in hamsters—production, prevention and removal. J. Dent. Res., *35*:651. 1956.

O'Rourke, J. T.: The relation of the physical character of the diet to the health of the periodontal tissues. Amer. J. Orth. & Oral Surg., *33*:687, 1947.

Pelzer, R.: A study of the local oral effects of diet on the periodontal tissues and the gingival capillary structure. J. Amer. Dent. Assn., *27*:13, 1940.

Person, P.: Diet consistency and periodontal disease in old albino rats. J. Periodont., *32*:308, 1961.

Shaw, J. H., and Griffiths, D.: Relation of protein, carbohydrate and fat intake to the periodontal syndrome. J. Dent. Res., *40*:614, 1961.

Stahl, S. S., Miller, C. S., and Goldsmith, E. D.: Effect of various diets on the periodontal structure of hamsters J. Periodont., *29*:7, 1958.

Waerhaug, J.: Effect of C avitaminosis on the supporting structure of the teeth. J. Periodont., *29*:87, 1958.

Weinmann, J.: Keratinization of the human oral mucosa. J. Dent. Res., *19*:57, 1940.

Chapter Thirty-four

Food and Nutrition for the New Denture Wearer, Particularly the Geriatric Patient

ABRAHAM E. NIZEL, D.M.D.

The dentist may have to consider himself a failure in the matter of patient acceptance and toleration of what is otherwise considered a technically successful prosthesis. Why this contradiction? The answer is simply that the denture can only be as successful as the health of the tissue upon which it rests. This problem of poor tissue toleration and chronic sore mouth under dentures is accentuated and hastened in the geriatric patient by endocrine changes and nutritional aberrations.

The factors that contribute to the food and nutritional needs of the elderly denture patient are really the same as those of any other geriatric patient with the additional consideration that the texture of the food might need to be modified. What follows, then, is really a discussion of geriatric nutrition and an expansion of what was covered under the subsection Old Age in Chapter Twenty. In addition, some specific dietary advice is suggested to make easier the difficult adjustment period that patients with new dentures are likely to experience.

AGING

Some of the signs and symptoms of physiological aging are drying of tissues, retardation of cellular activities, degeneration of cells, decline of digestive juices, lower metabolic rate, decreased tissue elasticity and impaired taste perceptibility. Through molecular biology we have begun to understand the mechanism of aging. Theoretically, the process of aging is simply the death of cells (Watkins, 1965). The explanation of this cell death is that the synthesizing sites of ribonucleic acid (RNA), namely DNA, are damaged as aging takes place. Since RNA is necessary for the enzymatic activity of the cells, any interference with the production of RNA will interfere with cell life. Thus, defective RNA produces less enzymes which, in turn, promotes cell death, hastens aging and manifests itself clinically in the mouth as dry, atrophied, and friable, oral mucosa. If this theory be valid, no doubt future research will provide answers on how to control and manipulate DNA-RNA molecules so that they will function continuously at an optimum to insure enzyme synthesis and thus at least slow down the aging process. For now, however, we must concern ourselves with pragmatic and available biological tools such as food and nutrition education.

CONSIDERATIONS IN GERIATRIC NUTRITION

Historically, the importance of nutrition for promoting the health of the geriatric patient can be traced back to the year 430 B.C. when Hippocrates, the father of medicine, wrote this aphorism, "Old men have little warmth and they need little food which produces warmth; too much only extinguishes the warmth they have." Later, Galen, Hippocrates' disciple, advised weak old men to take small amounts of food three times a day. In

421

the eighteenth century Luigi Cornaro, a Venetian nobleman, advocated frugal diets and attributed his own longevity to them. He lived to the age of 98 (Gordon, 1964).

However, the nutritional scientist of today is concerned not only with extending the number of years of man's life, but also with extending the prime of his life. Sherman (1930) accomplished the lengthening of life and increased the vigor of rodents by supplementing their diets with individual nutrients, calcium, protein, vitamins A and B complex and a combination of all four of these nutrients in the form of whole milk. The latter supplement he found to be most effective in improving even an adequate diet and thus making it an optimal one. Other investigators have demonstrated that the life span of the rats could be increased and development of diseases retarded by restricting calorie intake to a minimum (Saxton, 1945). *Optimal nutrition evidently not only adds years to life but also gives life to years,* at least on the basis of these animal experiments. As the specialty of geriatrics matures, this will, no doubt, prove to be a truism for man, too.

Many of the common ills of older people— fatigue, loss of appetite, indigestion, constipation, sleeplessness, overweight—often can be traced to poor food habits and a poor nutritional state. It follows, then, that the obvious answer to the problem is to educate the individual patient on what is best for him from a nutritional standpoint.

Sound nutrition for the elderly is fundamentally not different from that for the mature adult. However, certain characteristics inherent in the aging process add unique facets to geriatric nutrition:

1. State of dentition.
2. Diminished sensitivity of taste and smell.
3. Decrease in physical activity and diminished appetite which lower caloric needs.
4. Decrease in over-all food intake, which means that the proportion of essential nutrients in the food should be high.
5. Physiological factors such as changes in the digestive processes, which tend to decrease absorption.
6. Need for adequate protein to keep up nitrogen balance.
7. Psychological factors like loneliness, imaginary grievances, apathy and long-standing, ingrained, faulty eating habits.

THE FOOD REQUIREMENTS OF THE OLDER PERSON

Each period of the life cycle, childhood, adolescence, adulthood and old age, quantitatively requires different food intakes. However small these quantitative differences may appear, they must be met to overcome the peculiar stress of the specific period.

Calories. Because of a decrease in basal energy and in physical activity, the older person differs from the adult of young and middle age in his caloric requirements, which are 10 to 20 per cent lower. Therefore, to prevent overweight, the older person must reduce his caloric intake. (Since thiamine requirements are intimately associated with energy needs, this means that slightly less than the usual amount of this B vitamin is adequate for the older person.) We should recognize that in older people a disproportionate share of the total energy is provided by the readily available and relatively low cost foods, bread and cereals. Furthermore, it is significant that sweets and desserts remain important in the dietary pattern and very often furnish as much as 20 per cent of the day's calories. Lyons and Trulson (1956) did an interesting survey on the food practices of people over 65 years of age living at home. They found that these patients ate enough bread but not nearly enough of other foods, especially green and yellow vegetables and milk.

Protein. Ohlson, et al. (1952) have shown in a survey that the older age group does not meet its protein needs, which may be 10 per cent above the usual allowance of 1 gm. per kg. of body weight. Evidently the lay person has the notion that only meat provides protein; and since meat is expensive, adequate protein intake is considered to be beyond the means of the senior citizen. This is erroneous. Chopped meat and chopped liver are relatively inexpensive food items and are equally as nutritious as prime steaks. In fact, they are really more desirable for the denture-wearing patient because they are easier to chew. Milk, another excellent source of protein, which can be purchased in the low cost dry form and reconstituted by adding water, should be highly recommended. Those who do not care to drink milk as a beverage can be encouraged to use it on cereals, in soups and in puddings. At least 25 per cent of the total calorie intake should be protein.

Fats. Hypothetically, saturated fats and atherosclerosis are still considered to have a possible cause and effect relationship. Since the unsaturated vegetable oils have some beneficial action on reducing serum cholesterol levels and possible consequent atherosclerosis, emphasis should be on selection and use of this type of fat by the older person. Avoidance of fried foods, and even of sauces and gravies, makes good sense. A suitable diet for the elderly should contain enough fat to provide about 25 per cent of the calories.

Calcium. Calcium intake is often low and may contribute to development of senile osteoporosis. As mentioned previously, many elderly folk who will not drink plain milk will dilute it with water, use it in tea or coffee or add it to puddings. Other good sources of calcium that the older person can readily manage are cheeses and ice cream. Even though adaptation of the body systems to low calcium intake is possible, it has been found that in women from 40 to 60 years of age calcium is lost from the body unless they average 1 gm. a day intake (Ohlson et al., 1952). This need can almost be met by the intake of a pint of milk or its equivalent in cheese.

Iron. Iron deficiency anemia, found in 12 per cent of women and 5 per cent of men over the age of 65, can be prevented or corrected by adequate intake of liver, red meats and green vegetables like brussel sprouts, cabbage, peas, beans and lentils.

Ascorbic Acid. Another nutrient which has been found to be low in the diet of older people is ascorbic acid. Since ascorbic acid is found mainly in fresh fruits and vegetables, some planning is needed to include a rich source of vitamin C daily. Many canned juices are now being fortified with additional amounts of vitamin C. An older person may prefer tomato juice with his dinner to orange juice at breakfast.

Vitamin A. It is important that vitamin A needs be stressed and this can be done very nicely by suggesting dark green and deep yellow fruits and vegetables which are not only nutritionally good but add color, interest and zest to a meal. Potatoes and other tuber vegetables are good foods so long as other vegetables are not excluded from the menu.

Water and Bulk. Adequate water and bulk intake are vital both to promote proper urinary excretion and to avoid constipation. A total daily intake of about two quarts of liquid is recommended. For older folk, drinking water in smaller amounts at frequent intervals is advised. Dried fruits and their juices, like prunes, apricots or figs, are excellent foods for prevention of constipation, one of the older folks' most common complaints.

Salt. To prevent the edema associated with congestive heart failure and hypertension, excessive use of salt and salty foods should be avoided. In the matter of salt and taste perception, Taylor and Doku (1963) have shown that these are appreciably lowered in patients whose palates are fully covered. For example, in young people 23 to 44 years old 87 per cent could detect 0.4 per cent concentration of salt. In contrast, in the 60 to 88 year old group only 33.3 per cent of the group with palates uncovered and 14 per cent of those whose palates were covered could perceive the taste of salt at these low concentrations.

THE TECHNIQUE FOR NUTRITIONAL GUIDANCE OF GERIATRIC DENTURE PATIENTS

As mentioned earlier, nutrition education can be meaningful to the patient only if it is individualized and personalized to meet his nutritional needs and his peculiar idiosyncrasies and habits. If this is not done, nutrition remains a theoretical abstract concept rather than an objective, meaningful art and science. Nutrition must be applied as a definitive practical office procedure because it will motivate the dentist to perform this service and the patient to accept it.

The steps that are advocated are: (1) determination through history of food habits, the "why" of the patient's diet, (2) evaluation of the adequacy of his diet by comparison of the actual intake with the recommended allowances, and (3) provision of an individualized diet prescription with due consideration to the "why" and "what" of his dietary practices, as well as to the chewing limitations of his oral prosthesis.

THE "WHY" OF THE DIET

Food habits of any age group but particularly of the elderly person are not easily altered because they are longstanding and have a set pattern. Clinging to well known and familiar foods provides him with a feeling of security, individuality and independence. However, this must not deter the clinician from attempting to modify the patient's food habits if he knows that this is necessary to achieve a state of better health. Nevertheless, the dentist must appreciate that if he does not consider the social, psychological, economic and other environmental factors that influence the patient's food selection, his dietary advice will simply fall on deaf ears. In short, consideration of the ecology of each individual patient is significant.

Although there is no sure way to motivate people to alter their food habits, a sensible approach is to help the patient to help himself through better understanding of the reasons why he accepts or rejects foods. One must try to help the patient to analyze his own food habits objectively, to recognize the origin of any prejudices, and then point up the advantages of change (Beeuwkes, 1960).

These are some of the questions that one might ask the patient to elicit the "why" of his diet. Do you live alone? Do you prepare your own meals? What are the cooking facilities? Does your income preclude purchasing nutritionally important foods? Do you have any medical problems like high blood pressure, diabetes or gastric ulcers that influence food selection? Which foods do you like or dislike? Are you having problems chewing your food?

EVALUATING THE ADEQUACY OF THE DIET

Either by a 24-hour recall technique or by the patient's keeping a food diary for five consecutive days, including a week-end day, the general food intake pattern can be ascertained. The points in the food diary that should be noted are the number of eating periods, whether each of the meals is balanced (i.e., consists of foods from each of the 4 Food Groups), whether the kinds and amounts of food ingested compare favorably with the recommended amounts of two, two, four and four servings of the milk, meat, fruit-vegetable and bread-cereal groups, respectively.

To quantitate the adequacy the following simple procedure is followed:

1. Mark on a diet evaluation sheet (Table 29-5) in the proper block each average serving or portion of food listed for each meal by a single chit mark. One credit is given for an 8 oz. glass of milk, 4 oz. glass of fruit or vegetable juice, ½ cup of fruit or vegetable or 1 medium size raw fruit or vegetable, 3 oz. of lean meat, fish or poultry, 2 eggs, 1 slice of bread, ¾ cup of cooked cereal, 1 oz. of ready-to-eat cereal.

2. Compare with the Recommended Dietary Allowances to determine whether the patient is eating less, as much or more than the recommendation.

THE DIET PRESCRIPTION

On the basis of the information acquired from the history of food habits and the actual food intake, the clinician should now be able to prescribe realistic modifications and improvements of the patient's diet.

1. Commend the patient on the good foods that he has selected.

2. Suggest improvements in the food groups in which he is inadequate. It is important to make gradual rather than drastic changes. "Dietary improvements should be based on the patient's present habits, substitutions not denials, evolution not revolution" (Harris, 1961).

3. These general suggestions should now be made more interesting by prescribing a variety of foods and menus based on the 4 Food Groups that will suit the individual's needs, tastes and ability to masticate the food.

DIET SUGGESTIONS FOR THE EARLY PERIOD FOLLOWING INSERTION OF A DENTURE

In the case of the patient who is a new denture wearer, the ability to manage the physical consistency of the food is an important consideration. The process of eating food actually involves three steps: biting or incising, chewing or pulverizing and, finally, swallowing. The incising or biting of food is actually a grasping and tearing action and involves opening the mouth wide, which might cause a dislodgment of the denture by the action of overtensed muscle attachments. When the leverage force of the incising action is exerted in the anterior segment, the only equal and opposite force to prevent dislodgment is the post dam compression of the soft palate. In short, the counter dislodgment forces for the incising action are not so efficient as, for example, the balancing forces of the occlusal surfaces of the bicuspids and molars used in the chewing process. This makes the first step of eating food, the incising action, the most difficult of all three masticating actions.

The chewing and pulverizing of the bolus of the food are less difficult than biting or tearing food, but the coordination of the many muscles of mastication which produce the hinge and sliding movement of the mandible during eating requires some experience.

Actually, the easiest and least complex step in the eating process is that of swallowing. Deglutition, with the exception of the initial propelling of the bolus back to the pharynx, is an involuntary action.

Therefore, although the logical sequence of eating food is first biting, second chewing and third swallowing, it is much easier for the new denture patient to master this complex of masticatory movements in the reverse order, namely, swallowing first, chewing second, and biting last. Consequently, food of the consistency that will require only swallowing, such as liquids, should be prescribed for the first day or two after insertion of the denture. The use of soft foods can be advocated for the next few days and a firm or regular diet by the end of the week. Regardless of the consistency, the diet can be made varied, balanced and adequate, as will be shown in the following dietary suggestions.

DIET FOR THE FIRST DAY AFTER DENTURE INSERTION

The new denture wearer on his first postinsertion day might choose from the following foods, which are essentially liquid and arranged according to the 4 Food Groups:

Milk group: Fluid milk may be taken in any form

Meat group: For the first day or so eggs will be the food of choice; they may be taken in eggnogs; puréed meats or meat broths or soups may also be eaten

Vegetable-fruit group: Juices

Bread-cereal group: Gruels cooked in either milk or water

A sample menu using the liquid foods should include a glass of milk at least once a day, tea, coffee or caffein-free coffee, as desired.

Breakfast: Orange juice, strained oatmeal gruel with milk

Lunch: Cream of tomato soup, junket

Dinner: Strained cream of chicken soup, vanilla ice cream

Between meals: Eggnog, ginger ale with ice cream, cocoa or other milk drink or even the new low calorie liquid nutritionally complete mixtures

DIET FOR THE SECOND AND THIRD DAYS AFTER DENTURE INSERTION

For the second and third postinsertion days the denture patient may use soft foods which require a minimum of chewing.

Milk group: Fluid milk, cottage cheese

Meat group: Chopped beef, ground liver, tender chicken or fish in a cream sauce or even children's junior food preparations; eggs may be scrambled or soft cooked; dried peas may be used in a thick strained soup

Vegetable-fruit group: In addition to fruit and vegetable juices, tender cooked fruits and vegetables may be used (skins and seeds must be removed), such as the following: tender asparagus tips, cooked carrots, tender green beans, cooked or canned peaches, pears and apricots

Bread-
cereal group: Cooked cereals such as Cream of Wheat or Wheatena, milk toast and softened bread may be used; boiled rice, spaghetti, macaroni or noodles

These foods from the daily food guide can be readily arranged into three meals if the following simple steps are followed:

1. Bread or some breadstuff or cereal at each meal and some meals with both bread and cereal.

2. Milk as a beverage or in foods at two of the meals.

3. At least one serving from the vegetable-fruit group at each meal, citrus fruit usually at breakfast; the three or more other servings could be divided between noon and evening meals.

4. One of the servings from the meat group at the evening meal and the other at the noon meal or breakfast, or divided between them. We do need a serving from either the milk group or the meat group at each of the three meals; often we choose to have a serving of both.

A sample menu using soft foods would include butter or margarine with the meals, a glass of milk at least once a day, tea or coffee as desired.

Breakfast: Prune juice, soft-cooked egg, toast
Lunch: Cream of asparagus soup, cottage cheese and peaches, bread and gingerbread
Dinner: Meat loaf with tomato sauce, creamed potatoes, green peas, bread, tapioca pudding

DIETS FOR LATER PERIODS FOLLOWING INSERTION OF DENTURE

By the fourth day, or as soon as all the sore spots have healed, firm foods may be eaten in addition to those mentioned in the soft diet. In most instances, these foods should be cut into small pieces before eating.

A sample menu using firm foods would include butter or margarine with the meals, a glass of milk at least once a day, tea or coffee as desired.

Breakfast: Orange juice, scrambled eggs, toast
Lunch: Swiss steak, mashed potatoes, broccoli, bread, chocolate pudding
Dinner: Welsh rarebit, crisp bacon strips, carrot-apple salad, ice cream

In general, it has been found that raw vegetables and sandwiches are the foods least preferred by denture wearers (Yurkstas and Emerson, 1964). In fact, raw vegetables require more force during mastication to prepare for swallowing than most other foods (Yurkstas and Curby, 1953). Therefore, if the individual denture patient is able to manage sandwiches and salads, the ultimate in denture success and patient achievement has been realized.

SUMMARY

The best possible general advice is that our daily diets should include meat, milk, vegetables and fruits, and bread and cereals. For the older patient we suggest an emphasis on good quality protein foods, generous selection of vegetables and fruits and somewhat less on fats, starches and sugars to avoid an excess of calories. For the individual geriatric new denture wearer we might add that each diet prescription should be based on an analysis and evaluation of his individual food habits (the "why" of the diet) and actual food intake (the "what" of the diet). Furthermore, the physical nature of the diet should be consistent with the patient's experience and ability to swallow, chew and bite with his dental prosthesis.

REFERENCES

Beeuwkes, A. M.: Studying the food habits of the elderly. J. Amer. Diet. Assn., *37*:215, 1960.

Gordon, H.: Food for old people. South African Med. J., *38*:82, 1964.

Harris, R. S.: Cultural, geographical and technological influences on diet. J. Amer. Dent. Assn., *63*:465, 1961.

Lyons, J. S., and Trulson, M. F.: Food practices of older people living at home. J. Gerontol., *11*:66, 1956.

Ohlson, M. A., et al.: Symposium on geriatric nutrition intakes and retentions of nitrogen, calcium and phosphorus by 136 women between 30 and 85 years of age. Fed. Proc., *11*:775, 1952.

Saxton, J. A.: Nutrition and growth and their influence on longevity in rats. Biol. Symposia, *11*:177, 1945.

Sherman, H. C., and Campbell, H. L.: Further experiments on the influence of food upon longevity. J. Nutr., *2*:416, 1930.

Taylor, R. C., and Doku, C. H.: Dental survey of healthy older persons. J. Amer. Dent. Assn., *67*:63, 1963.

Watkins, D. M.: New findings in nutrition of older people. Amer. J. Public Health, *55*:548, 1965.

Yurkstas, A., and Curby, W. A.: Force analysis of prosthetic appliances during function. J. Prosth. Dent., *3*:83, 1953.

Yurkstas, A., and Emerson, W. H.: Dietary detections of persons with natural and artificial teeth. J. Prosth. Dent., *14*:695, 1964.

Appendices

APPENDIX I. Nutritive Values of Foods in Household Units

Foods largely in forms ready to eat

Dots appearing on the Table in place of numerical values indicate there may be a measurable amount of a constituent present, but no evidence is available to verify its presence.

The basic selection of foods for this table was made largely from government publications.

The table is reprinted from *Nutrition in Action*, by Ethel Austin Martin, copyright © 1965, 1963, by Holt, Rinehart and Winston, Inc. Used with the generous permission of the author and publisher. All rights reserved.

Food	Weight oz	Approximate Measure and Description	Calories	Protein g	Fat g	Carbohydrate g	Calcium mg	Iron mg	Vit. A Value IU	Thiamine mg	Riboflavin mg	Asc. acid mg
Almonds, shelled	0.5	12 nuts or 2 tbsp	90	3	8	3	35	0.7	0	0.04	0.14	tr
Apple, baked	4.6	1 med. apple, 2½ in. dia	120	tr	tr	30	8	0.4	50	0.04	0.02	3
Apple, raw	5.3	1 med. apple, 2½ in. dia	70	tr	tr	18	8	0.4	50	0.04	0.02	3
Apple, brown Betty	4.0	½ cup	175	2	4	35	21	0.7	135	0.07	0.05	tr
Apple pie (see Pies)												
Applejuice (sweet cider)	4.4	½ cup, fresh or can'd	60	tr	0	17	8	0.6	45	0.03	0.04	1
Applesauce	4.4	½ cup, sweetened, can'd	90	tr	tr	25	5	0.5	40	0.03	0.02	2
Apricots, canned	4.3	½ cup or 4 med. halves, 2 tbsp juice, sirup pack	105	1	tr	27	13	0.4	2130	0.02	0.03	5
Apricots, raw	1.3	1 med apricot	20	tr	tr	5	6	0.2	963	0.01	0.01	3
Apricots, dried, cooked	3.8	½ cup (scant) or 8 halves, 2 tbsp juice, sweetened	135	2	tr	34	26	1.5	2287	tr	0.04	3
Apricot whip	3.3	½ cup made with whipped nonfat dry milk	130	4	tr	30	64	1.2	2342	0.02	0.12	4
Asparagus, green, cooked	3.1	½ cup or 9 green stalks 4½ in. long	20	2	tr	3	17	0.9	910	0.12	0.15	20
Avocado, raw	3.8	½ avocado, 3⅓ x 4¼ in. peeled and pitted	185	2	18	6	11	0.6	310	0.12	0.21	15
Bacon, broiled or fried	0.6	2 sl., cooked crisp	95	5	8	1	2	0.5	0	0.08	0.05	...
Bacon, Canadian, cooked	1.5	3 sl., cooked crisp	100	18	12	tr	13	2.2	0	0.62	0.12	0
Banana, raw	5.3	1 med banana, 6 x 1½ in.	85	1	tr	23	8	0.7	190	0.05	0.06	10
Bavarian cream	3.5	½ cup (orange)	210	2	10	30	27	0.1	627	0.10	0.05	54
Bean sprouts, raw	1.6	½ cup, Mung sprouts	10	2	tr	2	13	0.4	5	0.03	0.04	7
Beans, white, dry, can'd	6.9	¾ cup, with pork and tomato or molasses	250	12	5	41	129	3.3	105	0.10	0.08	4
Beans, white, dry, can'd	6.9	¾ cup, without pork and with tomato or molasses	240	12	1	45	137	3.9	105	0.10	0.08	4
Beans, Lima, dry, cooked	5.0	¾ cup	195	12	1	36	42	4.2	tr	0.20	0.09	tr
Beans, Lima, green, cooked	2.8	½ cup	75	4	1	15	23	1.4	230	0.11	0.07	12
Beans, pinto, dry, raw	3.5	½ cup (Mexican red beans)	350	23	1	64	160	6.9	0	0.65	0.24	2

Food		Measure	g									
Beans, red, dry, can'd	6.7	¾ cup	175	11	1	32	56	3.5	0	0.10	0.10	tr
Beans, snap, green, cooked	2.2	½ cup	15	1	tr	3	23	0.5	415	0.05	0.06	9
Beef, corned, can'd	3.0	3 sl., 3 x 2 x ¼ in.	180	22	10	0	17	3.7	20	0.01	0.20	…
Beef, corned, hash, can'd	4.5	¾ cup	180	18	8	9	33	1.7	15	0.03	0.17	…
Beef, dried or chipped	2.0	4 thin sl., 4 x 5 in.	115	19	4	0	11	2.9	…	0.04	0.18	…
Beef, hamburger, cooked	3.0	1 patty, 3 in. dia (market ground)	245	21	17	0	9	2.7	30	0.07	0.02	…
Beef, heart, braised	3.0	2 round sl., 2½ in. dia, ½ in. thick	160	26	5	1	14	5.9	30	0.23	1.05	3
Beef, liver (see Liver)												
Beef, loaf (see Meat loaf)												
Beef, pot roast, cooked	3.0	1 piece, 4 x 3¾ x ½ in.	245	23	16	0	10	2.9	30	0.04	0.18	…
Beef, potpie, bkd.	7.9	1 indiv. pie, 4¼ in. dia	460	18	28	32	20	2.5	2830	0.07	0.14	tr
Beef, roast, oven cooked	3.0	2 sl., 6 x 3¼ x ⅛ in.	255	22	18	0	9	2.8	30	0.06	0.16	0
Beef, steak, broiled	3.0	1 piece, 3½ x 2 x ¾ in., no bone	375	19	32	0	9	2.6	60	0.06	0.16	0
Beef Stroganoff, cooked	4.6	½ cup	250	17	18	6	41	2.6	395	0.12	0.29	2
Beef tongue, boiled	3.0	7 sl., 2¼ x 2¼ x ⅛ in.	205	18	14	tr	7	2.5	…	0.04	0.26	6
Beets, cooked	2.9	½ cup, diced	35	1	tr	8	18	0.6	15	0.02	0.04	0
Beverages												
Beer	8.4	1 glass, 4 percent alcohol*	115	1	tr	11	10	tr	0	tr	0.06	0
Eggnog	4.0	½ cup (holiday type)*	225	3	11	12	56	0.4	536	0.03	0.12	…
Ginger ale	7.0	⅞ cup, carbonated soft drink	70	…	…	18	…	…	…	…	…	…
Kola-type	6.0	¾ cup, carbonated soft drink	80	…	…	21	…	…	…	…	…	…
Biscuits, baking powder	1.3	1 biscuit, 2½ in. dia, (enrch'd flour)	130	3	4	18	61	0.7	tr	0.09	0.09	tr
Blackberries, raw	2.5	½ cup	40	1	1	10	23	0.7	145	0.03	0.03	15
Blueberries, raw	2.5	½ cup	45	1	1	11	11	0.7	70	0.02	0.04	10
Bluefish, cooked	3.0	1 piece, 3½ x 2 x ½ in.	135	22	4	0	25	0.6	40	0.09	0.08	…
Bologna (see Sausage)												
Bouillon cubes	0.14	1 cube, ⅝ in.	2	tr	tr	0	…	…	…	…	0.07	0
Branflakes	1.0	¾ cup (40% bran), added thiamine	85	3	1	22	17	1.1	0	0.13	0.07	0

APPENDIX I—Continued

Food	Weight oz	Approximate Measure and Description	Calories	Protein g	Fat g	Carbohydrate g	Calcium mg	Iron mg	Vit. A Value IU	Thiamine mg	Riboflavin mg	Asc. acid mg
Bread, Boston brown	1.7	1 sl., 3 in. dia, ½ in. thick, (degermed cornmeal)	100	3	1	22	43	0.9	0	0.05	0.03	0
Bread, cracked wheat	0.8	1 sl., 3¾ x 3¾ x ⅓ in.	60	2	1	12	20	0.3	tr	0.03	0.02	tr
Bread, French or Vienna	0.7	1 sl., 3¼ x 2 x 1 in. (enrch'd flour)	60	2	1	11	9	0.4	tr	0.06	0.04	tr
Bread, Italian	0.7	1 sl., 3¼ x 2 x 1 in. (enrch'd flour)	55	2	tr	11	3	0.4	0	0.06	0.04	0
Bread, pumpernickel	1.2	1 sl., 4¾ x 3½ x ⅜ in. (dark rye flour)	85	3	0	19	30	0.8	0	0.08	0.05	0
Bread, raisin	0.8	1 sl., 3¾ x 3¾ x ⅓ in.	60	2	1	12	16	0.3	tr	0.01	0.02	tr
Bread, rye, light	0.8	1 sl., 2¾ x 2¼ x ½ in. (American type: ⅓ rye, ⅔ wheat)	55	2	tr	12	17	0.4	0	0.04	0.02	0
Bread, white	0.8	1 sl., 3¾ x 3¾ x ⅓ in. (enrch'd flour, 3-4% nonfat dry milk)	60	2	1	12	19	0.6	tr	0.06	0.05	tr
Bread, white	0.8	1 sl., 3¾ x 3¾ x ⅓ in. (unenrch'd flour, 3-4% nonfat dry milk)	60	2	1	12	19	0.2	tr	0.02	0.02	tr
Bread, white, toasted	0.7	1 sl., 3¾ x 3¾ x ⅓ in. (enrch'd flour, 3-4% nonfat dry milk)	60	2	1	12	19	0.6	tr	0.05	0.05	tr
Bread, whole-wheat	0.8	1 sl., 3¾ x 3¾ x ⅓ in. (graham or entire wheat bread)	55	2	1	11	23	0.5	tr	0.06	0.05	tr
Bread crumbs	0.8	¼ cup, dry	85	3	1	16	27	0.8	tr	0.05	0.07	tr
Broccoli, cooked	2.6	½ cup	20	3	tr	4	98	1.0	2550	0.05	0.11	56
Brussels sprouts, cooked	2.3	½ cup or 5 med sprouts	30	3	1	5	22	0.9	260	0.03	0.08	30
Bun (see Rolls)												
Butter	0.5	1 tbsp or 1 pat, ½ in. thick	100	tr	11	tr	3	tr	460	...	...	0

Buttermilk, cultured from skim milk (see Milk, skim)

Food	Measure											
Cabbage, cooked	½ cup, cooked briefly, little water	3.0	20	1	tr	5	39	0.4	75	0.04	0.04	27
Cabbage, raw	½ cup, finely shredded	1.8	10	1	tr	3	23	0.3	40	0.03	0.03	25
Cabbage, Chinese, raw	½ cup, 1 in. pieces	1.8	8	1	tr	1	22	0.5	130	0.02	0.02	16
Cake, angel food	2 in. sector or 1/12 of cake 8 in. dia without icing	1.4	110	3	tr	23	2	0.1	0	tr	0.05	tr
Cake, chocolate, layer	2 in. sector or 1/16 of cake 10 in. dia, fudge icing	4.2	420	5	14	70	118	0.5	140†	0.03	0.10	tr
Cake, fruit, dark	1 piece, 2 x 2 x ½ in.	1.1	105	2	4	17	29	0.8	50†	0.04	0.04	tr
Cake, pl., cupcake	1 cupcake, 2¾ in. dia without icing	1.4	130	3	3	23	62	0.2	50†	0.01	0.03	tr
Cake, pl., cupcake	1 cupcake, 2¾ in. dia with icing	1.8	160	3	3	31	58	0.2	50†	0.01	0.04	tr
Cake, pl., layer	2 in. sector or 1/16 of cake 10 in. dia with icing	3.5	320	5	6	62	117	0.4	90†	0.02	0.07	tr
Cake, pl., loaf	1 piece, 3 x 2 x 1½ in. without icing	1.9	180	4	5	31	85	0.2	70†	0.02	0.05	tr
Cake, pound	1 sl., 2¾ x 3 x ⅝ in.	1.1	130	2	7	15	16	0.5	100†	0.04	0.05	tr
Cake, sponge	2 in. sector or 1/12 of cake, 8 in. dia	1.4	115	3	2	22	11	0.6	210†	0.02	0.06	tr
Candy, caramels	4 small	1.0	120	1	3	22	36	0.7	50	0.01	0.04	tr
Candy, chocolate bar, pl. milk chocolate, sweet	1 bar, 3¾ x 1½ x ¼ in.	1.0	145	2	9	16	61	0.3	40	0.03	0.11	0
Candy, chocolate bar, almond	1 bar, 5⅓ x 1⅞ x ⅓ in.	1.75	265	4	19	25	102	1.4	70	0.07	0.25	0
Candy, chocolate creams	2 pieces, 1¼ in. dia (base) ⅝ in. thick	1.0	110	1	4	20	...	...	...	...	...	0
Candy, chocolate fudge	1 piece, 1¼ x 1¼ x 1 in.	1.0	115	tr	3	23	14	0.1	60	tr	0.02	tr
Candy, hard	6 pieces, 1 in. dia ¼ in. thick	1.0	110	0	0	28	0	0	0	0	0	0
Candy, peanut brittle	1 piece, 3¼ x 2½ x ¼ in.	1.0	125	2	4	21	11	0.6	10	0.03	0.01	0
Cantaloup, raw	½ melon, 5 in. dia	13.5	40	1	tr	9	33	0.8	6950 (orange flesh)	0.09	0.07	63

APPENDIX I—Continued

Food	Weight oz	Approximate Measure and Description	Calories	Protein g	Fat g	Carbohydrate g	Calcium mg	Iron mg	Vit. A Value IU	Thiamine mg	Riboflavin mg	Asc. acid mg
Carrots, cooked	2.6	½ cup, diced	20	1	1	5	19	0.5	9065	0.04	0.04	3
Carrots, raw	1.9	½ cup, grated	20	1	tr	5	22	0.5	6600	0.03	0.03	4
Carrots, raw	1.8	1 carrot, 5½ in. long or 25 thin strips	20	1	tr	5	20	0.4	6000	0.03	0.03	3
Catsup, tomato (see Tomato)												
Cauliflower, cooked	2.1	½ cup, flower buds	15	2	tr	3	13	0.7	55	0.04	0.05	17
Celery, raw	1.8	½ cup, diced	10	1	tr	2	25	0.3	0	0.03	0.02	4
Celery, raw	1.4	1 stalk, large outer, 8 in. long	5	1	tr	1	20	0.2	0	0.02	0.02	3
Cheese, blue mold	1.0	¾ in. sector or 3 tbsp (Roquefort type)	105	6	9	tr	122	0.2	350	0.01	0.17	0
Cheese, cheddar (American)	1.0	1 cube, 1⅛ in.	115	7	9	1	221	0.3	380	0.01	0.15	0
Cheese, cheddar (American)	0.25	1 tbsp, grated	30	2	2	tr	55	0.1	90	0.02	0.03	0
Cheese food, cheddar	1.0	2 round sl., 1⅝ in. dia, ¼ in. thick or 2 tbsp	95	6	7	2	163	0.2	300	0.01	0.17	0
Cheese, cottage	2.0	¼ cup creamed cottage cheese (made from skim milk)	60	8	2	2	50	0.2	100	0.02	0.16	0
Cheese, cottage	1.0	2 tbsp uncreamed cottage cheese (made from skim milk)	25	5	tr	1	26	0.1	tr	0.01	0.08	0
Cheese, cream	0.5	1 tbsp	55	1	6	tr	9	tr	230	tr	0.04	0
Cheese, Swiss	1.0	1 sl., 7 x 4 x ⅛ in.	105	7	8	1	271	0.3	320	0.01	0.06	0
Cheese sauce	2.1	¼ cup	110	5	9	4	156	0.1	337	0.02	0.14	1
Cheese soufflé	2.8	¾ cup	200	10	16	7	210	1.0	826	0.08	0.23	1
Cheesecake	5.7	1/10 of cake, 9 in. dia	400	15	23	35	128	0.8	958	0.08	0.33	1
Cherries, domestic, raw	4.0	1 cup, sour, sweet, hybrid	65	1	1	15	19	0.4	650	0.05	0.06	9
Cherries, West Indian, raw	0.4	2 med cherries (Acerola)	3	...		1	1	...	...		0.01	100
Chick peas, dry, raw	3.7	½ cup (garbanzos)	380	22	5	64	97	7.5	tr	0.58	0.19	2
Chicken, broiled	3.0	¼ small broiler, flesh and skin	185	23	9	0	10	1.4	260	0.04	0.15	...

Food	(oz)	Measure										
Chicken, can'd	3.0	1/3 cup, boned meat	170	25	7	0	12	1.5	160	0.03	0.14	⋯
Chicken, creamed‡	3.5	1/2 cup	222	20	12	6	84	1.1	445	0.04	0.20	1
Chicken, fried	3.3	1/2 breast, with bone	215	24	12	⋯	10	1.1	60	0.03	0.06	⋯
Chicken, fried	4.3	1 leg (thigh and drumstick)	245	27	15	⋯	13	1.8	220	0.05	0.18	⋯
Chicken pie (see Poultry potpie)												
Chile con carne, can'd	6.6	3/4 cup, made with beans	250	14	11	23	74	3.2	113	0.06	0.15	⋯
Chile con carne, can'd	6.7	3/4 cup, made without beans	385	20	29	11	73	2.7	285	0.04	0.23	⋯
Chili powder	0.5	1 tbsp (hot red peppers, ground)	50	2	1	9	20	1.2	11520	0.03	0.20	2
Chili sauce	0.6	1 tbsp (mainly tomatoes)	15	tr	tr	4	2	0.1	320	0.02	0.01	2
Chocolate, bitter	1.0	1 square baking chocolate	145	2	15	8	28	1.2	20	0.01	0.06	0
Chocolate candy (see Candy)												
Chocolate-flavored milk	8.8	1 cup chocolate milk drink	190	8	6	27	270	0.4	210	0.09	0.41	2
Chocolate morsels	0.5	30 morsels or 1 1/2 tbsp	80	1	4	10	5	0.3	tr	tr	tr	0
Chocolate sirup	1.4	2 tbsp	80	tr	tr	22	6	0.6	⋯	⋯	⋯	0
Chop suey, cooked	4.3	3/4 cup	325	19	20	16	43	2.9	85	0.11	0.13	17
Clams, can'd	3.0	1/2 scant cup or 3 med clams	45	7	1	2	74	5.4	70	0.04	0.08	⋯
Cocoa, beverage	6.4	3/4 cup, made with milk	175	7	8	20	215	0.7	293	0.07	0.34	2
Cocoa, dry	0.25	1 tbsp, powder	21	1	1	3	9	0.8	tr	0.01	0.03	0
Coconut, dried	0.6	1/4 cup, shredded, sweetened	86	1	6	8	3	0.4	0	0.01	0.01	0
Coconut, fresh	0.8	1/4 cup, shredded	83	1	8	3	4	0.4	⋯	0.02	0.01	1
Codfish, dried	1.8	1/2 cup	190	41	2	0	25	1.8	0	0.04	0.23	0
Coffee cake	2.8	1 piece, frosted, 3 x 3 x 1 1/4 in.	260	4	11	37	25	1.0	477	0.12	0.13	0
Cole slaw	2.1	1/2 cup	50	1	4	5	24	0.3	40	0.03	0.03	25
Collards (see Greens)												
Cookies												
Brownies	0.9	1 piece, 1 7/8 x 1 7/8 x 5/8 in.	145	2	9	17	12	0.5	231	0.03	0.04	⋯
Chocolate chip	0.4	1 cookie, 2 1/4 in. dia	60	1	3	7	4	0.2	81	0.01	0.01	⋯
Coconut bar chews	0.4	1 cookie, 3 x 7/8 x 1/3 in.	55	tr	2	9	7	0.3	76	0.01	0.01	0
Oatmeal cookies	0.4	1 cookie, 2 1/8 in. dia (raisins and nuts)	65	1	4	6	5	0.3	18	0.04	0.02	tr

APPENDIX I—Continued

Food	Weight oz	Approximate Measure and Description	Calories	Protein g	Fat g	Carbohydrate g	Calcium mg	Iron mg	Vit. A Value IU	Thiamine mg	Riboflavin mg	Asc. acid mg
Plain and assorted	0.9	1 cookie, 3 in. dia	110	2	3	19	6	0.2	0	0.01	0.01	0
Sugar cookies	0.3	1 cookie, 2½ in. dia	40	1	2	6	2	0.1	64	0.02	0.01	0
Corn, ears, cooked	4.9	1 ear sweetcorn, 5 in. long	65	2	1	16	4	0.5	300 (yellow corn)	0.09	0.08	6
Corn, sweet, can'd	4.5	½ cup, solids and liquid	85	3	1	21	5	0.7	260 (yellow corn)	0.04	0.07	7
Corned beef (see Beef)												
Corned beef hash (see Beef)												
Corn bread or muffin	1.7	1 muffin, 2¾ in. dia (enrch'd, degermed cornmeal)	155	4	5	22	79	0.9	240 (yellow corn)	0.10	0.15	tr
Cornflakes	1.0	1⅓ cup (added thiamine, niacin, and iron)	110	2	tr	24	3	0.5	0	0.12	0.03	0
Corn grits, cooked	5.6	⅔ cup, white (enrch'd degermed)	80	2	tr	18	1	0.5	tr	0.07	0.05	0
Corn meal, dry	5.0	1 cup, white or yellow (enrch'd, degermed)	525	11	2	114	9	4.2	430 (yellow corn)	0.64	0.38	0
Cow peas (see Peas)												
Crabmeat, cooked	3.0	½ cup flakes	90	14	2	1	38	0.8	...	0.04	0.05	...
Crackers, graham	0.3	1 cracker, 2½ in. square	30	1	1	5	2	0.2	...	0.02	0.01	...
Crackers, saltines	0.3	2 saltines, 2 in. square	35	1	1	6	2	0.1	0	tr	tr	0
Crackers, soda	0.2	1 cracker, 2½ in. square	25	1	1	4	1	tr	0	tr	tr	0
Cracker meal	0.4	1 tbsp	45	1	1	7	2	0.1	0	0.01	tr	0
Cranberry sauce, cooked	2.4	¼ cup, sweetened	140	tr	tr	36	6	0.2	20	0.02	0.02	1
Cream, half and half	0.5	1 tbsp (milk and cream)	20	tr	2	1	16	0	70	0	0.02	tr
Cream, heavy, whipping	0.5	1 tbsp unwhipped	55	tr	6	tr	10	0	240	0	0.02	tr
Cream, light coffee	0.5	1 tbsp	35	tr	3	1	15	0	130	0	0.02	tr
Cucumbers, raw	1.8	6 sl., pared, ⅛ in. thick	5	tr	tr	1	5	0.2	0	0.02	0.02	4
Custard, bkd.	4.3	½ cup	140	7	7	14	140	0.5	435	0.05	0.24	1
Dates, fresh or dried	1.6	¼ cup or 8 pitted dates	125	1	tr	34	26	1.4	25	0.04	0.04	0
Dessert topping, whipped	0.4	2 tbsp (low calorie, with nonfat dry milk)	17	1	...	3	29	...	1	0.01	0.04	1

Food		Measure	Weight (g)									
Doughnuts	1.1	1 doughnut, cake type	135	2	7	17	23	0.4	40	0.05	0.04	0
Egg, raw, boiled, poached	1.8	1 whole egg	80	6	6	tr	27	1.1	590	0.05	0.15	0
Egg white, raw	1.2	1 egg white	15	4	tr	tr	3	tr	0	tr	0.09	0
Egg yolk, raw	0.6	1 egg yolk	60	3	5	tr	24	0.9	580	0.04	0.07	0
Egg, creamed	4.0	½ cup (1 egg in ¼ cup white sauce)	190	9	14	7	103	1.2	928	0.07	0.25	tr
Egg, fried	1.9	1 egg, cooked in 1 tsp fat	115	6	10	tr	28	1.1	590	0.05	0.15	0
Egg, scrambled	2.2	1 egg, with milk and fat	110	7	8	1	51	1.1	690	0.05	0.18	0
Escarole, raw	2.0	3 leaves	10	1	tr	2	45	1.0	1700	0.04	0.07	6
Farina, cooked	5.6	⅔ cup (enrch'd with iron, thiamine, riboflavin, niacin, calcium)	70	2	tr	15	21	0.5	0	0.07	0.05	0
Fats, cooking, vegetable	0.4	1 tbsp solid fat	110	0	12	0	0	0	0	0	0	0
Figs, dried	0.7	1 large fig, 2 x 1 in.	60	1	tr	15	40	0.7	20	0.02	0.02	0
Fish (see various kinds of fish)												
Fish, creamed‡	4.8	½ cup (tuna, salmon, other)	220	20	13	8	81	0.9	385	0.05	0.18	tr
Fishsticks, breaded, cooked	4.0	5 sticks, 3¾ x 1 x ½ in.	200	19	10	8	13	0.5	...	0.05	0.08	...
Frankfurter, cooked	1.8	1 frankfurter (hot dog)	155	6	14	1	3	0.8	...	0.08	0.10	...
French toast, fried	2.8	1 sl. (enrch'd bread)	180	6	12	14	78	1.0	568	0.09	0.17	tr
Fruit balls, raw	0.4	1 ball, 1 in. dia (dried apricots, dates, nuts)	45	1	1	8	10	0.4	285	0.02	0.02	tr
Fruit cocktail, can'd	4.5	½ cup, solids and liquid	100	1	1	25	12	0.5	180	0.02	0.02	3
Gelatin, dry pl.	0.4	1 tbsp	35	9	tr	0	0	0	0	0	0	0
Gelatin dessert, pl.	4.2	½ cup, ready to eat	80	2	tr	18	0	0	0	0	0	0
Gelatin dessert, with fruit	4.2	½ cup, ready to eat	85	2	tr	21	7	0.4	135	0.04	0.03	4
Gingerbread	1.9	1 piece, 2 x 2 x 2 in.	180	2	7	28	63	1.4	50	0.02	0.05	tr
Grapefruit, can'd	4.4	½ cup or 4 sections with 4 tbsp juice, sirup pack	80	1	tr	22	16	0.4	10	0.04	0.02	38
Grapefruit, white, raw	10.0	½ med, 4¼ in. dia	50	1	tr	14	21	0.5	10	0.05	0.02	50
Grapefruit juice, can'd	4.3	½ cup, unsweetened	50	1	tr	12	10	0.5	10	0.04	0.02	42
Grapes, American type, raw	5.4	1 cup or 1 med bunch (slip skin)	70	1	1	16	13	0.4	100	0.05	0.03	4

APPENDIX I—Continued

Food	Weight oz	Approximate Measure and Description	Calories	Protein g	Fat g	Carbohydrate g	Calcium mg	Iron mg	Vit. A Value IU	Thiamine mg	Riboflavin mg	Asc. acid mg
Grapes, European type, raw	5.6	1 cup or 40 grapes	100	1	tr	26	18	0.6	150	0.08	0.04	7
Grapejuice, can'd	4.4	½ cup, sweetened	80	1	tr	21	14	0.4	....	0.05	0.03	tr
Greens, collards, cooked	3.3	½ cup	40	4	1	7	237	1.5	7250	0.08	0.23	42
Greens, dandelion, cooked	3.2	½ cup	40	3	1	8	169	2.8	13655	0.12	0.11	15
Greens, kale, cooked	1.9	½ cup	20	2	1	4	124	1.2	4610	0.04	0.13	28
Greens, mustard, cooked	2.5	½ cup	15	2	tr	3	154	2.1	5025	0.04	0.13	32
Greens, spinach, cooked	3.2	½ cup	20	3	1	3	112 (not usable)	1.8	10600	0.07	0.18	27
Greens, turnip, cooked	2.6	½ cup	20	2	1	4	188	1.8	7685	0.05	0.30	44
Guavas, raw	2.8	1 guava	50	1	tr	12	21	0.5	180	0.05	0.03	212
Haddock, fried	3.0	1 fillet, 4 x 2½ x ½ in.	135	16	5	6	15	0.5	50	0.03	0.08	...
Ham, smoked, cooked	3.0	2 sl., 5½ x 3¾ x ⅛ in.	290	18	24	1	8	2.2	0	0.39	0.15	...
Ham, luncheon meat, can'd	2.0	4 tbsp (spiced or unspiced)	165	8	14	1	5	1.2	0	0.18	0.12	...
Ham, luncheon meat, cooked	2.0	1 sl., 6¼ x 3¾ x ⅛ in.	170	13	13	0	5	1.5	0	0.57	0.15	...
Hamburger (see Beef)												
Honey, strained	0.7	1 tbsp	60	tr	0	17	1	0.2	0	tr	0.01	1
Ice, orange	4.7	½ cup	140	0	0	35	5	0.1	45	0.04	0.01	20
Ice cream, pl.	2.2	1 container, 3½ fluid oz, factory packed	130	2	8	13	76	0.1	320	0.03	0.12	1
Ice cream, pl., brick	2.5	1 sl., or cut (⅛ brick), factory packed	145	3	9	15	87	0.1	370	0.03	0.13	1
Ice cream, vanilla	3.5	1/6 qt scoop	200	4	12	21	117	0.12	720	0.04	0.34	0
Ice cream soda, chocolate	9.0	1 fountain-size serv.	255	3	8	46	75	0.7	297	0.03	0.13	1
Ice milk	3.3	½ cup	140	5	5	21	146	0.1	195	0.05	0.21	1
Jams, marmalades, preserves	0.7	1 tbsp	55	tr	tr	14	2	0.1	tr	tr	tr	1
Jellies	0.7	1 tbsp	50	0	0	13	2	0.1	tr	tr	tr	1

Kale (see Greens)

Food		Measure										
Lamb chop, broiled	4.8	1 shoulder chop, 5 x 3½ x ½ in., fat trimmed with bone	405	25	33	0	10	3.1	...	0.14	0.25	...
Lamb, leg, roasted	3.0	2 sl, 3 x 3¼ x ⅛ in. lean and fat, no bone	235	22	16	0	9	2.8	...	0.13	0.23	...
Lard	0.5	1 tbsp	135	0	14	0	0	0	0	0	0	0
Lemon juice, fresh	0.5	1 tbsp	5	tr	tr	1	1	tr	tr	tr	tr	7
Lemonade	8.7	1 cup, ready to serve, (from concentrate, frozen, sweetened)	110	tr	tr	28	2	0.1	10	0.01	0.01	17
Lentils, dry, cooked	3.5	½ cup	120	9	tr	22	12	2.5	200	0.20	0.09	0
Lettuce, headed, raw	16.0	1 hd., compact, 4¾ in. dia	70	5	1	13	100	2.3	2470	0.20	0.38	35
Lettuce, headed, raw	1.8	2 large leaves or 4 small	5	1	tr	1	11	0.2	270	0.02	0.04	4
Liver, beef, fried	2.0	1 sl, 3½ x 3 x ½ in.	120	13	4	6	5	4.4	30330	0.15	2.25	18
Liver, calf, fried	2.6	1 sl, 5 x 2 x ⅓ in. (dredged in flour)	230	15	15	4	5	9.0	19130	0.18	2.65	30
Liver, chicken, fried	3.0	3 med. livers (dredged in flour)	235	20	15	5	15	6.4	27370	0.19	2.11	17
Liver, pork, fried	2.5	1 sl, 3¾ x 1¾ x ½ in. (dredged in flour)	225	17	15	3	8	15.3	12070	0.34	2.53	19
Macaroni, cooked	3.7	¾ cup (enrch'd)	115	4	1	24	8	1.0	0	0.14	0.08	0
Macaroni and cheese, bk'd	5.8	¾ cup (macaroni enrch'd)	350	14	19	33	296	1.5	728	0.17	0.35	tr
Mackerel, can'd	3.0	⅗ cup, solids and liquid	155	18	9	0	221	1.9	20	0.02	0.28	...
Mangoes, raw	7.0	1 med. mango	90	1	...	23	12	0.3	8380	0.08	0.07	55
Margarine	0.5	1 tbsp or 1 pat ½ in. thick (fortified with vit. A)	100	tr	11	tr	3	tr	460	...	...	0
Marshmallows	0.3	1¼ in. dia	25	tr	0	6	0	0	0	0	0	0
Meat and bean stew, cooked	8.6	1 cup (Mexican dish)	345	17	16	34	90	4.6	1177	0.37	0.22	59
Meatloaf, beef, bak'd	2.7	1 sl, 3¾ x 2¼ x ¾ in.	240	19	17	3	34	2.9	138	0.10	0.21	tr
Milk, dry skim (nonfat)	0.7	¼ cup powder	73	7	tr	11	260	0.1	5	0.07	0.36	2
Milk, evaporated, can'd	4.4	½ cup, undiluted, unsweetened	170	9	10	12	318	0.2	410	0.05	0.42	2

APPENDIX I—Continued

Food	Weight oz	Approximate Measure and Description	Calories	Protein g	Fat g	Carbo-hydrate g	Cal-cium mg	Iron mg	Vit. A Value IU	Thia-mine mg	Ribo-flavin mg	Asc. acid mg
Milk, fluid, skim	8.6	1 cup (same as buttermilk)	90	9	tr	13	298	0.1	10	0.10	0.44	2
Milk, fluid, whole	8.5	1 cup	165	9	10	12	285	0.1	390	0.08	0.42	2
Milk, goat's, fluid, whole	8.5	1 cup	165	8	10	11	3.5	0.2	390	0.10	0.27	2
Milk, malted, pl., beverage	11.8	1 fountain-size glass	350	16	15	40	455	1.0	838	0.21	0.70	3
Milk, malted, pl., dry	0.4	1 tbsp (8 tablets)	40	1	1	6	28	0.2	92	0.03	0.05	0
Milkshake, chocolate	12.0	1 fountain-size glass	420	11	18	58	363	0.9	687	0.12	0.55	4
Molasses, cane, blackstrap	0.7	1 tbsp third extraction	45	...	...	11	116	2.3	...	0.02	0.04	...
Molasses, cane, light	0.7	1 tbsp first extraction	50	...	...	13	33	0.9	...	0.01	0.01	...
Muffins, pl.	1.7	1 muffin 2¾ in. dia (en-rch'd white flour)	135	4	5	19	74	0.7	60	0.08	0.11	tr
Mushrooms, can'd	4.3	½ cup, solids and liquid	15	2	tr	5	9	1.0	0	0.02	0.30	...
Noodles, egg, cooked	4.2	¾ cup, enrch'd	150	5	2	28	12	1.1	45	0.17	0.11	0
Nuts, cashew, roasted	1.2	¼ cup	195	6	16	9	13	1.3	...	0.12	0.12	...
Nuts, peanuts, roasted	0.5	20 kernels, or about 2 tbsp	90	4	7	3	11	0.3	0	0.05	0.02	0
Nuts, pecan halves	0.5	10 med. halves	100	1	11	2	11	0.4	20	0.13	0.02	tr
Nuts, walnut halves	0.5	8 med. halves	100	2	10	2	15	0.5	5	0.05	0.02	tr
Oatmeal or rolled oats, cooked	5.5	⅔ cup, regular or quick cooking	100	3	2	17	14	1.1	0	0.15	0.03	0
Oils, salad or cooking	0.5	1 tbsp	125	0	14	0	0	0	...	0	0	0
Okra, cooked	1.5	4 pods, 3 x ⅝ in.	15	1	tr	3	35	0.3	315	0.03	0.03	9
Olives, green, pickled	0.8	4 extra large	22	tr	2	tr	16	0.3	57	tr	...	...
Olives, ripe	0.8	4 extra large	30	tr	3	1	15	0.3	13	tr	tr	...
Onions, cooked	3.7	½ cup or 5 onions 1¼ in. dia	40	1	tr	9	34	0.5	55	0.02	0.03	7
Onions, raw	3.9	1 onion, 2½ in. dia	50	2	tr	11	35	0.6	60	0.04	0.04	10
Onions, raw	0.4	1 tbsp chopped	5	tr	0	1	3	0	tr	tr	tr	1
Orange, raw	7.4	1 orange, 3 in. dia, med.	70	1	tr	18	63	0.3	290	0.12	0.03	66
Orange juice, can'd	4.4	½ cup or 1 small glass (unsweetened)	60	1	tr	14	13	0.5	250	0.09	0.03	50

Food		Measure	Wt.									
Orange juice, fresh	4.3	½ cup or 1 small glass	60	1	tr	13	13	0.3	250	0.11	0.03	46
Orange juice, frozen	4.3	½ cup or 1 small glass (diluted ready to serve)	55	1	tr	14	11	0.1	250	0.11	0.02	56
Oyster stew	8.1	1 cup with 3–4 oysters	200	11	12	11	269	3.3	640	0.12	0.40	...
Oysters, raw	4.2	½ cup or 6–10 oysters	80	10	2	4	113	6.6	370	0.15	0.20	...
Pancakes, wheat	0.9	1 griddlecake, 4 in. dia (enrch'd flour)	60	2	2	8	34	0.3	30	0.05	0.06	tr
Papayas, raw	3.2	½ cup, in ½ in. cubes	35	1	tr	9	18	0.3	1595	0.04	0.04	51
Parsley, raw	0.12	1 tbsp chopped	1	tr	tr	tr	7	0.2	290	tr	0.01	7
Parsnips, cooked	2.7	½ cup	50	1	1	11	44	0.6	0	0.05	0.08	10
Peaches, can'd	4.1	2 med. halves with 2 tbsp juice, sirup pack	90	tr	tr	24	5	0.4	500	0.01	0.03	3
Peaches, raw	2.9	½ cup sl.	30	1	tr	8	8	0.4	1115 (yellow flesh)	0.02	0.04	6
Peaches, raw	4.0	1 med. 2½ in. dia	35	1	tr	10	9	0.5	1320 (yellow flesh)	0.02	0.05	7
Peanut butter	0.6	1 tbsp	90	4	8	3	12	0.4	0	0.02	0.02	0
Peanuts (see Nuts)												
Pears, can'd	4.1	2 med. halves with 2 tbsp juice, sirup pack	90	tr	tr	23	6	0.2	tr	0.01	0.02	2
Pears, raw	6.4	1 pear, 3 x 2½ in. dia	100	1	1	25	13	0.5	30	0.04	0.07	7
Peas, chick (see Chick peas, garbanzos)												
Peas, cowpeas, dry, cooked	4.3	½ cup (blackeye peas or frijoles)	95	7	1	17	21	1.6	10	0.21	0.06	tr
Peas, green, cooked	2.8	½ cup	55	4	1	10	18	1.5	575	0.20	0.11	12
Peas, pigeon, dry, raw	3.5	6 tbsp (gandules)	310	22	2	50	140	4.0	169	0.45	0.34	0
Peas, split, dry, cooked	4.4	½ cup	145	10	1	26	14	2.1	60	0.18	0.11	tr
Peppers, green, sweet, raw	0.7	¼ cup, diced	5	tr	tr	1	2	0.1	87	0.02	0.02	26
Peppers, green, stuffed, cooked	4.0	1 med. pepper (meat stuffing)	200	12	14	12	31	1.9	637	0.09	0.14	64
Peppers, red, sweet, raw	2.1	1 med. pod	20	1	tr	4	8	0.4	2670	0.05	0.05	122
Perch, ocean, fried	3.0	4 x 3 x ½ in.	195	16	11	6	14	1.3	50	0.09	0.10	...
Persimmon, raw	4.4	1 persimmon, 2½ in. dia (Japanese)	75	1	tr	20	6	0.4	2740	0.03	0.02	11

APPENDIX I—Continued

Food	Weight oz	Approximate Measure and Description	Calories	Protein g	Fat g	Carbohydrate g	Calcium mg	Iron mg	Vit. A Value IU	Thiamine mg	Riboflavin mg	Asc. acid mg
Pickle relish	0.4	1 tbsp	15	tr	tr	3	2	0.2	10	0	tr	1
Pickles, cucumber, bread and butter	1.5	6 sl, ¼ x 1½ in. dia	30	tr	tr	7	13	0.8	80	0.01	0.02	4
Pickles, cucumber, dill	4.7	1 large pickle, 4 x 1¾ in. dia	15	1	tr	3	34	1.6	420	tr	0.09	8
Pickles, cucumber, sweet	0.7	1 pickle, 2¾ x ¾ in. dia	20	tr	tr	5	3	0.3	20	0	tr	1
Pie, apple	4.7	4-in. sector or 1/7 of pie 9 in. dia	330	3	13	53	9	0.5	220	0.04	0.02	1
Pie, cherry	4.7	4-in. sector or 1/7 of pie 9 in. dia	340	3	13	55	14	0.5	520	0.04	0.02	2
Pie, custard	4.6	4-in. sector or 1/7 of pie 9 in. dia	265	7	11	34	162	1.6	290	0.07	0.21	0
Pie, lemon meringue	4.2	4-in. sector or 1/7 of pie 9 in. dia	300	4	12	45	24	0.6	210	0.04	0.10	1
Pie, mince	4.7	4-in. sector or 1/7 of pie 9 in. dia	340	3	9	62	22	3.0	10	0.09	0.05	1
Pie, pumpkin	4.6	4-in. sector or 1/7 of pie 9 in. dia	265	5	12	34	70	1.0	2480	0.04	0.15	0
Pimientos, can'd	1.3	1 med. pod (sweet pepper)	10	tr	tr	2	3	0.6	870	0.01	0.02	36
Pineapple, can'd	4.3	1 large sl, 2 small sl, 2 tbsp juice, sirup pack	95	tr	tr	26	35	0.7	100	0.09	0.02	11
Pineapple, raw	2.5	½ cup, diced, unsweetened	35	1	tr	10	11	0.2	90	0.06	0.02	17
Pineapple juice, can'd	4.4	½ cup or 1 small glass	60	1	tr	16	19	0.6	100	0.07	0.02	11
Pizza pie (cheese)	2.6	5½ in. sector or 1/8 of pie 14 in. dia	180	8	6	23	157	0.7	570	0.03	0.09	8
Plantain, green, raw	3.5	1 baking banana, 6 in. long	135	1	...	32	8	.08	380	0.07	0.04	28
Plums, can'd	4.3	½ cup or 3 plums with 2 tbsp juice, sirup pack	90	tr	tr	25	10	1.3	280	0.03	0.03	1
Plums, raw	2.1	1 plum, 2 in. dia	30	tr	tr	7	10	0.3	200	0.04	0.02	3
Popcorn, popped, buttered	0.5	1 cup	90	2	5	11	2	0.4	153	0.05	0.02	0

Food		Weight										
Pork, chop, cooked	3.4	1 chop, with bone	260	16	21	0	8	2.2	0	0.63	0.18	...
Pork, roast, cooked	3.0	2 sl, 5 x 4 x ⅛ in.	310	21	24	0	9	2.7	0	0.78	0.22	...
Potato, white, bk'd	3.4	1 med. 2½ in. dia	90	3	tr	21	9	0.7	tr	0.10	0.04	20
Potatoes, French fried	2.0	10 pieces, 3 x ½ x ¼ in.	155	2	7	20	9	0.7	tr	0.06	0.04	8
Potatoes, mashed	3.4	½ cup (milk added)	70	2	1	15	24	0.5	25	0.09	0.06	9
Potatoes, mashed	3.4	½ cup (milk and butter added)	115	2	6	14	23	0.5	235	0.08	0.05	8
Potato chips	0.7	10 med. chips, 2 in. dia	110	1	7	10	6	0.4	tr	0.04	0.02	2
Poultry potpie	7.9	1 individual pie, 4¼ in. dia (chicken or turkey)	485	17	28	39	41	1.6	1860	0.07	0.14	tr
Pretzels	0.2	5 small sticks	20	tr	tr	4	1	0	0	tr	tr	0
Prune juice, can'd	4.2	½ cup or small glass	85	1	tr	23	17	4.9	...	tr	0.02	2
Prunes, dried, cooked	3.7	5 med. prunes with 2 tbsp juice sweetened	160	1	tr	42	21	1.5	733	0.03	0.06	1
Pudding, chocolate	4.6	½ cup (chocolate blancmange)	190	6	8	26	158	0.9	211	0.06	0.27	1
Pudding, cornstarch	4.3	½ cup (plain blancmange)	140	5	5	20	145	0.1	195	0.04	0.20	1
Pudding, rice with raisins	4.8	½ cup (old-fashioned)	300	8	8	52	243	0.8	313	0.10	0.35	3
Pudding, tapioca	2.6	½ cup	140	5	5	12	104	0.4	327	0.04	0.19	1
Radishes, raw	1.4	4 small, without tops	10	tr	tr	2	15	0.4	10	0.01	0.01	10
Raisins, dried	0.4	1 tbsp	30	tr	tr	8	6	0.4	2	0.01	0.01	tr
Raspberries, red, raw	2.2	½ cup	35	1	1	9	14	0.6	80	0.02	0.05	16
Rhubarb, cooked	4.8	½ cup (sugar added)	190	1	tr	49	56	0.6	35	0.01	...	9
							(not usable)					
Rice, parboiled, cooked	4.6	¾ cup	150	3	tr	34	11	0.4	0	0.08	0.02	0
Rice, puffed	0.5	1 cup (added thiamine, niacin, and iron)	55	1	tr	12	2	0.3	...	0.06	0.01	0
Rice flakes	1.1	1 cup (added thiamine, niacin, and iron)	115	2	tr	26	9	0.5	...	0.11	0.01	0
Roll, barbecue bun	1.3	1 bun, 3½ in. dia (enrch'd)	115	3	2	20	28	0.7	tr	0.11	0.07	tr
Roll, hard, white	1.8	1 round roll	160	5	2	31	24	0.4	tr	0.03	0.05	tr
Roll, soft, white	1.3	1 roll (enrch'd)	115	3	2	20	28	0.7	tr	0.11	0.07	tr
Roll, sweet, pan	1.5	1 roll	135	4	4	21	37	0.3	30	0.03	0.06	0
Rutabagas, cooked	2.7	½ cup	25	1	tr	6	43	0.3	270	0.04	0.06	18

APPENDIX I—Continued

Food	Weight oz	Approximate Measure and Description	Calories	Protein g	Fat g	Carbohydrate g	Calcium mg	Iron mg	Vit. A Value IU	Thiamine mg	Riboflavin mg	Asc. acid mg
Salads§												
Salad, chicken	4.4	½ cup, with mayonnaise	280	25	19	1	20	1.7	200	0.04	0.15	1
Salad, egg	4.5	½ cup, with mayonnaise	190	6	18	1	35	1.3	630	0.06	0.16	1
Salad, fresh fruit	4.4	½ cup with French dressing (orange, apple, banana, grapes)	130	...	6	21	25	0.6	154	0.06	0.05	22
Salad, jellied, vegetable	4.3	½ cup, no dressing	70	3	...	16	14	0.2	25	0.03	0.02	20
Salad, lettuce	4.6	¼ solid hd., with French dressing	80	1	6	5	28	0.7	618	0.05	0.10	9
Salad, potato	4.9	½ cup with mayonnaise	185	2	12	17	21	0.8	40	0.11	0.05	17
Salad, tomato aspic	4.2	½ cup, no dressing	45	5	0	7	12	0.5	1441	0.07	0.05	22
Salad, tuna fish	3.6	½ cup, with mayonnaise	250	21	18	1	14	1.2	98	0.04	0.09	1
Salad dressing, blue cheese	0.6	1 tbsp	90	1	10	1	11	tr	30	tr	0.02	tr
Salad dressing, boiled	0.6	1 tbsp (homemade)	30	1	2	3	15	0.1	80	0.01	0.03	tr
Salad dressing, commercial	0.5	1 tbsp (plain, mayonnaise type)	60	tr	6	2	2	tr	30	tr	tr	0
Salad dressing, cottage cheese	0.9	2 tbsp (low calorie, no oil, nonfat dry milk)	17	2	0	2	31	0	18	0.01	0.06	1
Salad dressing, French	0.5	1 tbsp	60	tr	6	2	3	0.1	0	0	0	0
Salad dressing, mayonnaise	0.5	1 tbsp	110	tr	12	tr	2	0.1	40	tr	tr	0
Salad dressing, Thousand Island	0.5	1 tbsp	75	tr	8	1	2	0.1	60	tr	tr	2
Salmon, boiled or bk'd	4.2	1 steak, 4 x 3 x ½ in.	200	34	7	tr	...	1.4	...	0.12	0.33	...
Salmon, can'd	3.0	½ cup, pink salmon	120	17	5	0	159 (includes bones)	0.7	60 (pink)	0.03	0.16	...
Salmon loaf	4.0	½ cup or 1 sl, 4 x 1¼ x 1¼ in.	235	29	10	5	43	1.8	332	0.08	0.20	2
Sandwiches‖												
Sardines, can'd	2.0	5 pieces, 3 x 1 x ¼ in. (can'd in oil)	120	15	6	1	245	1.7	127	0.01	0.12	...
Sauce, chocolate	1.4	2 tbsp	75	1	4	9	32	0.2	87	0.01	0.05	...

Food		Measure										
Sauce, custard	1.1	2 tbsp (low calorie, with nonfat dry milk)	45	2	1	7	56	0.17	89	0.02	0.09	...
Sauce, hard	0.6	1 tbsp	90	...	6	11	1	0	231	0	0	0
Sauce, hollandaise (mock)	0.9	2 tbsp	75	2	7	3	36	0.2	353	0.02	0.06	1
Sauce, lemon	1.0	2 tbsp	40	0	1	8	...	...	34	...	...	2
Sauerkraut, can'd	2.6	½ cup, drained solids	15	1	tr	4	27	0.4	30	0.03	0.05	12
Sausage, bologna	2.0	3 sl., 3¼ in. dia 1/10 in. thick	170	7	16	1	4	1.0	...	0.09	0.12	...
Sausage, frankfurters (see Frankfurters)												
Sausage, liverwurst	2.0	3 sl., 2½ in. dia, ¼ in. thick	150	10	12	1	5	3.1	3260	0.10	0.63	0
Sausage, pork, bulk, can'd	2.0	3 small patties, 2 in. dia, ¼ in. thick	170	9	15	0	5	1.3	0	0.12	0.14	...
Sausage, salami	1.1	1 sl., 4½ in. dia, ⅛ in. thick	135	7	11	0	5	1.1	0	0.16	0.15	0
Sausage, Vienna, can'd	2.0	3 sausages, 2 in. long, ¾ in. dia	125	9	9	0	5	1.4	0	0.06	0.07	0
Shad, bk'd	3.0	1 piece, 4 x 3 x ½ in.	170	20	10	0	20	0.5	20	0.11	0.22	...
Sherbet	3.4	½ cup, factory packed	120	2	tr	29	48	0.05	48	0.02	0.08	0
Shrimp, can'd	3.0	½ cup, meat only	110	23	1	...	98	2.6	50	0.01	0.03	...
Sirup	0.7	1 tbsp, table blends	55	0	0	15	9	0.8	0	0	tr	0
Soup, bean, can'd	8.8	1 cup, ready-to-serve	190	8	5	30	95	2.8	...	0.10	0.10	0
Soup, chicken noodle, can'd	6.7	4/5 cup or ⅓ can, ready-to-serve	70	2	4	6	16	0.2	...	...	...	
Soup, clam chowder, can'd	8.9	1 cup, ready-to-serve	85	5	2	12	36	3.6	...	...	...	
Soup, consomme, broth, can'd	8.4	1 cup, ready-to-serve	10	2	...	0	2	1.0	0	0	0.05	
Soup, cream, vegetable, can'd	8.9	1 cup, ready-to-serve (asparagus, celery, or mushroom)	200	7	12	18	217	0.5	200	0.05	0.20	
Soup, minestrone	8.5	1 cup (homemade)	130	6	5	16	59	2.5	5177	0.13	0.15	31
Soup, tomato, can'd	8.6	1 cup, ready-to-serve	90	2	2	18	24	1.0	1230	0.02	0.10	10
Soup, vegetable, can'd	8.8	1 cup, ready-to-serve	80	4	2	14	32	0.8	...	0.05	0.08	8
Spaghetti, cooked	3.7	¾ cup, enrich'd	115	4	1	24	8	1.0	0	0.14	0.08	0
Spaghetti with meat sauce	6.6	¾ cup	215	10	8	26	19	1.5	518	0.05	0.08	10

APPENDIX I—*Continued*

Food	Weight oz	Approximate Measure and Description	Calories	Protein g	Fat g	Carbohydrate g	Calcium mg	Iron mg	Vit. A Value IU	Thiamine mg	Riboflavin mg	Asc. acid mg
Spaghetti with tomato sauce	6.6	¾ cup (with cheese)	160	5	4	27	34	0.8	623	0.05	0.06	11
Spinach (see Greens)												
Squash, summer, cooked	3.7	½ cup, diced	20	1	tr	4	16	0.4	275	0.04	0.08	12
Squash, winter, bk'd	3.6	½ cup, mashed	50	2	1	12	25	0.8	6345	0.05	0.16	7
Stew, beef and vegetable	6.2	¾ cup	140	11	8	11	23	2.1	1898	0.10	0.14	11
Strawberries, raw	2.6	½ cup, capped	30	1	1	7	16	0.8	45	0.02	0.05	44
Sugar, brown	0.5	1 tbsp (dark brown, firmly packed)	50	0	0	13	10	0.4	0	0	0	0
Sugar, granulated	0.4	1 tbsp (beet or cane)	50	0	0	12	...	...	0	0	0	0
Sugar, powdered	0.3	1 tbsp	30	0	0	8	...	...	0	0	0	0
Sugar, loaf	0.3	1 loaf (domino) 1⅛ x ¾ x ⅜ in.	25	0	0	7	...	...	0	0	0	0
Sweet potatoes, bk'd	3.9	1 med. potato, 5 x 2 in., peeled	155	2	1	36	44	1.0	8970	0.10	0.07	24
Sweet potatoes, candied	3.1	½ potato, 3½ x 2¼ in.	150	1	3	30	33	0.8	5515	0.05	0.04	9
Tangerine, raw	4.0	1 med., 2½ in. dia	40	1	tr	10	34	0.3	360	0.05	0.01	26
Toast, melba	0.2	1 sl., 3¾ x 1¾ in.	20	1	tr	3.9	5	0.1	0	0.01	0.01	0
Tomato, raw	5.3	1 med., 2 x 2½ in. dia	30	2	tr	6	16	0.9	1640	0.08	0.06	35
Tomatoes, can'd	4.2	½ cup	25	1	tr	5	14	0.8	1270	0.07	0.04	20
Tomato catsup	0.6	1 tbsp	15	tr	tr	4	2	0.1	320	0.02	0.01	2
Tomato juice, can'd	4.2	½ cup or 1 small glass	25	1	tr	5	9	0.5	1270	0.06	0.04	19
Tongue (see Beef)												
Tortillas	0.7	1 tortilla, 5 in. dia	50	1	1	10	22	0.4	40 (yellow corn)	0.04	0.01	...
Tuna, can'd	3.0	½ cup, drained solids	170	25	7	0	7	1.2	70	0.04	0.10	...
Tuna salad (see Salad)												
Turnip greens (see Greens)												
Turnips, cooked	2.7	½ cup	20	1	tr	5	31	0.4	tr	0.03	0.05	14
Veal, roast, cooked	3.0	2 sl., 3 x 2½ x ¼ in. (medium fat)	305	23	14	0	10	2.9	...	0.11	0.26	...
Veal cutlet, broiled	3.0	1 med. cutlet, 3¾ x 3 x ½ in. (without bone)	185	23	9	0	9	2.7	...	0.06	0.21	...
Veal cutlet, (breaded), cooked	4.8	2 sl., 2½ x 2½ x ¾ in. (wiener schnitzel)	315	26	21	5	37	4.2	295	0.22	0.41	...
Vinegar	0.5	1 tbsp	2	0	...	1	1	1.0	...	...	...	...

in. with enrch'd flour

Watermelon, raw	32.4	1 wedge, 4 x 8 in. (with rind)	120	2	1	29	30	0.9	2530	0.20	0.22	26
Welsh rarebit	4.4	½ cup	330	19	26	6	534	0.7	1118	0.04	0.40	....
Wheat flakes	1.0	1 cup	100	3	tr	23	13	1.2	0	0.16	0.05	0
Wheat flour, white	3.9	1 cup, enrch'd flour	400	12	1	84	18	3.2	0	0.48	0.29	0
Wheat flour, white	3.9	1 cup, unenrch'd flour	400	12	1	84	18	0.9	0	0.07	0.05	0
Wheat flour, whole-wheat	4.2	1 cup	400	16	2	85	49	4.0	0	0.66	0.14	0
Wheat germ	0.3	2 tbsp	30	2	1	4	7	0.7	0	0.17	0.07	0
Wheat, rolled, cooked	5.5	⅔ cup	120	3	1	27	13	1.1	0	0.11	0.04	0
Wheat, shredded	1.0	1 large biscuit, 4 x 2¼ in.	100	3	1	23	13	1.0	0	0.06	0.03	0
White sauce (medium)	2.3	¼ cup	110	3	8	6	76	0.1	338	0.02	0.11	tr
Yeast, brewer's, dry	0.3	1 tbsp	25	3	tr	3	17	1.4	tr	1.25	0.34	tr
Yoghurt	8.6	1 cup, from partially skimmed milk	120	8	4	13	295	0.1	170	0.09	0.43	2

* See page 286 for calorie values of additional alcohol beverages.

† Vitamin A value varies with the kind of fat used in making cakes. Butter or fortified margarine used in cakes listed in this Appendix.

‡ One-half cup of a creamed dish calls for ¼ cup white sauce and about ⅓ cup of any one of a wide variety of meats, vegetables, or other foods which may be combined. (See Egg, creamed and Fish, creamed.)

§ Only a few common types of salad are listed here. The calorie and nutritive values of any combination of foods in a salad are easily estimated. The ½ cup servings of salad in this table, using mayonnaise, were made with ⅓ cup of the main ingredient, such as chicken, plus 2 tbsp of a crisp vegetable, such as celery, plus 1 tbsp of mayonnaise. One tbsp of French dressing was used in the fruit and lettuce salads. The total calorie yield of a salad may vary greatly, depending on the kind and the amount of salad dressing added.

|| Sandwiches, as such, are not included in this list, but the "makings" are available for any type desired. For example, a simple sandwich might consist of 2 slices of enriched bread, 2 tsp butter, 1 slice salami sausage, 1 tsp mayonnaise, plus 2 leaves of lettuce. The sum of the calories and of the nutrients gives the total contribution of the sandwich. For salad sandwiches, the procedure is essentially the same. Chicken, egg, or tuna salads, in the measures given in Appendix G, provide the amount of filling needed for a sandwich made with a large bun or 2 slices of bread.

ALCOHOLIC BEVERAGES

Beverage	Amount	Number of Calories
Beer (4 percent alcohol)	8-oz glass	115
Eggnog (holiday variety made with whiskey and rum)	½ cup	225
Whiskey, Gin, Rum		
100 proof	1 jigger (1½ oz)	125
90 proof	1 jigger (1½ oz)	110
86 proof	1 jigger (1½ oz)	105
80 proof	1 jigger (1½ oz)	100
70 proof	1 jigger (1½ oz)	85
Wines		
Table wines (such as Chablis, claret, Rhine wine and sauterne)	1 wine glass, about 3 oz	70–90
Sweet or dessert wines (such as muscatel, port, sherry or Tokay)	1 wine glass, about 3 oz	120–160

All values, except for Eggnog, from Food and Your Weight, Home and Garden Bulletin, No. 74. Washington, D.C. U.S. Department of Agriculture, 1960.

APPENDIX II.   Foods and Mixed Dishes Classified According to Food Group and Amounts Commonly Considered as One Serving

(Unless specified each serving = 1 unit of the food group)

Foods and Mixed Dishes	Amount Commonly Considered as One Serving	Food Group
apple	1 med. size, 3–4 oz.	veg.-fruit
apple juice	½ cup	veg.-fruit
apricots	3–4 oz., 2–3 med.	veg.-fruit
asparagus	½ cup (4 oz.)	veg.-fruit
avocado	½ cup (4 oz.)	veg.-fruit
bacon	2 slices	fat
bananas	1 med.	veg.-fruit
beans (dry)	½ cup, cooked	meat
beans (fresh) green or wax	½ cup, cooked (4 oz.)	veg.-fruit
beef	2–3 oz., cooked hamburger	meat
beet greens	½ cup (4 oz.)	veg.-fruit
beets	½ cup (4 oz.)	veg.-fruit
berries	½ cup (4 oz.)	veg.-fruit
biscuits (baking powder)	1 medium, 2″ in diameter	bread-cereal
blanc mange	½ cup	milk (½)
		sugar—3 tsp.
bread, corn	1 piece—2″ square	bread-cereal
bread, all varieties	1 slice (1 oz.)	bread-cereal
broccoli	½ cup (4 oz.)	veg.-fruit
brussel sprouts	½ cup (4 oz.)	veg.-fruit
butter	1 tsp.	fat
buttermilk	1 cup	milk
cabbage	½ cup (4 oz.)	veg.-fruit
cantaloupe	¼ medium melon	veg.-fruit
carrots	½ cup (4 oz.)	veg.-fruit
cauliflower	½ cup (4 oz.)	veg.-fruit
cereals, cooked (oatmeal, corn meal, Cream of Wheat, etc.)	½ cup (1 oz.)	bread-cereal
cereals, ready-to-eat, flaked or puffed	¾ cup–1 cup (1 oz.)	bread-cereal
celery	½ cup (4 oz.)	veg.-fruit
chard	½ cup (4 oz.)	veg.-fruit
cheese, cheddar, American, Swiss	1 oz.	milk
cheese, cream	2 tbsp.	fat
cheese, soft type, cottage	½ cup	milk (⅓)
cheese bits	½ cup (10 to 20 crackers)	bread-cereal
cherries	3–4 oz., 15 large	veg.-fruit
chicken	½ breast or 1 leg and thigh (4 oz.)	meat
chickory	½ cup (4 oz.)	veg.-fruit
clams	3–4 oz., cooked (½) cup	meat
cocoa, made with milk	1 cup	milk
collards	½ cup (4 oz.)	veg.-fruit
corn	½ cup (4 oz.)	veg.-fruit
	1 ear—5 inches	

APPENDIX II—*Continued*

Foods and Mixed Dishes	*Amount Commonly Considered as One Serving*	*Food Group*
crackers, round, thin	6 crackers	bread-cereal
saltines	3 crackers	bread-cereal
graham	3 crackers	bread-cereal
oyster	24 crackers	bread-cereal
cress	½ cup (4 oz.)	veg.-fruit
cucumbers	½ cup (4 oz.)	veg.-fruit
custard pudding (½ cup)	½ cup	milk (½) sugar—3 tsp.
dandelion greens	½ cup (4 oz.)	veg.-fruit
duck	2–3 oz., cooked	meat
egg, in any form	2	meat
eggplant	½ cup (4 oz.)	veg.-fruit
English muffins	1 muffin	bread-cereal
escarole	½ cup (4 oz.)	veg.-fruit
figs (fresh)	3 small	veg.-fruit
fish—cod, haddock, bass, mackerel, flounder, halibut	3–4 oz., cooked	meat
fish chowder	1⅓ cup	meat (½) milk veg.-fruit (½)
grapefruit	½ medium	veg.-fruit
grapes	3–4 oz., 22 Tokay 60 green, seedless	veg.-fruit
greens, all kinds cooked	½ cup	veg.-fruit
grits	½ cup (1 oz.)	bread-cereal
guava	3 oz.	veg.-fruit
heart	2–3 oz., cooked	meat
ice cream	½ cup	milk (¼)
kale	½ cup (4 oz.)	veg.-fruit
kidney	2–3 oz., cooked	meat
lamb	2 rib chops, ½ inch thick	meat
lentils, dried	½ cup, cooked	meat
lettuce	½ cup (4 oz.)	veg.-fruit
liver	2–3 oz., cooked	meat
lobster	2–3 oz., cooked	meat
macaroni	½ cup	bread-cereal
macaroni and cheese	1 cup (8 oz.)	bread-cereal milk
mango	3–4 oz.	veg.-fruit
margarine	1 tsp.	fat
mayonnaise	1 tbsp.	fat
meat loaf	3–4 oz.	meat bread-cereal (¼)
meat stew	1 cup (8 oz.)	veg.-fruit meat
meat, lean,—beef, lamb, pork, veal	3–4 oz.	meat
melons, honeydew	¼ melon; ½ cup diced	veg.-fruit
milk (fresh, diluted, evaporated, reconstructed or dried)	½ cup	milk

APPENDIX II—*Continued*

Foods and Mixed Dishes	*Amount Commonly Considered as One Serving*	*Food Group*
muffins	1 med.	bread-cereal
mushrooms	½ cup	veg.-fruit
nectarines	1 medium	veg.-fruit
noodles	½ cup	bread-cereal
nuts	2 tbsp., ½ oz.	meat (¼)
okra	½ cup (4 oz.)	veg.-fruit
olives	10–12	fat
onions	½ cup (4 oz.)	veg.-fruit
oranges	1 medium, 3–4 oz.	veg.-fruit
oysters	6–8 medium	meat
pancakes	1, 4 inch pancake	bread-cereal
papaya	3–4 oz. (½ cup)	veg.-fruit
parsnips	½ cup (4 oz.)	veg.-fruit
peaches	1 medium, 3–4 oz.	veg.-fruit
peanut butter	2 tbsp.	meat (½)
pears	1 medium, 3–4 oz.	veg.-fruit
peas, fresh or canned	½ cup, cooked	veg.-fruit
peas, dried	½ cup, cooked	meat
peppers	½ cup (4 oz.)	veg.-fruit
pineapple	3–4 oz., ½ cup diced	veg.-fruit
plums	½ cup (1 medium)	veg.-fruit
popcorn	¾–1 cup	bread-cereal
pork	1 chop, 1 inch thick	meat
pies, fruit—2 crusts (apple, berry, peach, cherry, etc.)	⅙ of a pie	bread-cereal veg.-fruit sugar, 2 tbsp.
pies, cream—1 crust (custard, squash)	⅙ of a pie	bread-cereal (½) milk (½) sugar, 2 tbsp.
pie, lemon meringue 1 crust	⅙ of a pie	bread-cereal (½) sugar, 3 tbsp.
popovers	1 popover	bread-cereal
potato chips	8–10 pieces	bread-cereal
potatoes	1 med. size, 3–4 oz.	veg.-fruit
pretzel sticks	½ cup (10 to 20 crackers)	bread-cereal
prunes	4 medium	veg.-fruit
rabbit	2–3 oz., cooked	meat
radishes	½ cup (4 oz.)	veg.-fruit
rice	½ cup	bread-cereal
rolls, plain	1 roll, medium, Parker House or Cloverleaf	bread-cereal
rutabaga	½ cup (4 oz.)	veg.-fruit
Ry-Krisp	4 crackers	bread-cereal
salmon	2–3 oz., cooked	meat
sandwiches	2 slices bread	bread-cereal (2)
	Filling:	
	2 oz. meat, fish, chicken, egg or peanut butter	meat (½)
	1 slice cheese	milk
	lettuce, tomato	veg.-fruit

APPENDIX II—*Continued*

Foods and Mixed Dishes	*Amount Commonly Considered as One Serving*	*Food Group*
sardines	2–3 oz.	meat
sauerkraut	½ cup	veg.-fruit
sausage (bologna, frankforts, liverwurst, etc.)	2–3 oz., 3 slices 1 large or 2 small frankforts	meat
shredded wheat	¾–1 cup, 1 oz.	bread-cereal
shrimp	2–3 oz., cooked	meat
soup, vegetable	1 cup	veg.-fruit
soup, cream of tomato, asparagus, corn	1 cup	milk (½)
soup, clear, chicken or beef bouillon	1 cup	meat (½)
soup, noodle, rice or barley	1 cup	bread-cereal (½)
spaghetti	½ cup	bread-cereal
spaghetti (Italian style) with meat sauce	1 cup spaghetti ½ cup meat sauce	bread-cereal (2) meat, (½)
spinach	½ cup (4 oz.)	veg.-fruit
squash	½ cup (4 oz.)	veg.-fruit
strawberries	3–4 oz., 1 cup	veg.-fruit
swordfish	2–3 oz., cooked	meat
tangerines	3–4 oz., 1 medium	veg.-fruit
tapioca pudding	½ cup	milk (½) sugar, 3 tsp.
tomatoes	1 med. size, 3–4 oz.	veg.-fruit
tortillas	1 medium	bread-cereal
tuna	2–3 oz., cooked	meat
turkey	2–3 oz., cooked	meat
turnips	½ cup (4 oz.)	veg.-fruit
turnip greens	½ cup (4 oz.)	veg.-fruit
veal	2–3 oz., cooked	meat
venison	2–3 oz., cooked	meat
waffles	½ medium	bread-cereal
watermelon	$\frac{1}{16}$ of a melon, ½ cup diced	veg.-fruit
white sauce, for creamed chicken, meat, fish, or vegetables	½ cup	milk (½)
yams	1 medium	veg.-fruit
yoghurt	1 cup	milk

APPENDIX III.   Some Representative Vitamin Preparations from a Few Different Drug Houses

	Product	Mfg.	B1 mg.	B2 mg.	Niac. mg.	Pant. mg.	B6 mg.	C mg.	A units	D units	B12 mcg.	Folacin mg.	Misc.
Vitamins B complex and C in Maintenance Doses	B-compules with Vitamin C	Abbott	3	3	20	3	0.1	30					Liver fraction 330 mg.
	Betacebrin Syrup	Lilly (in pints)	1	1	8			40					Yeast 250 mg. per 5 cc.
	Beta-Cevalin Comp.	Lilly	5	5	25	12.5	2.5	75			1		Liver 455 mg.
	Cebefortis Tab.	Upjohn	5	5	50	25	1.5	150			2	0.5	
	Novo-Basic	E. R. Squibb & Sons	5	5	50	10	1	150			2	15	
	Totabex Caps	C. D. Smith	3	2	15	2	0.25	30					
	Vitilkon Syrup	Upjohn (in 4 & 12 oz.) per teaspoon	2	3	30			100					
Vitamins B complex and C in Therapeutic Doses	Allbee with Vitamin C	A. H. Robins Co., Inc.	15	10	50	10		250					
	A.S.F. Caps	Roerig	10	10	100	20	2	300			4	1.5	Menadione (Kanalogue) 2 mg.
	Convalets Filmtabs	Abbott	10	10	100	20	2	300			4	1.5	

	E. R. Squibb & Sons											
Novogran	E. R. Squibb & Sons	10	10	100	20	2	300			4	1.5	Menadione (Kanalogue) 2 mg.
Panalins-T	Mead Johnson	10	10	100	20	2	300			4	1.5	
Probec-Tabs (formerly Stuart Therapeutic B complex, C)	Stuart	10	10	150	10	5	150			5 mcg. 50% USP crystalline 50% B_{12} concentrate		
Provite B with C	Ives-Cameron	25	12.5	100	5	1.5	100			2		
Stress Caps	Lederle	10	10	100	20	2	300			4	1.5	Menadione (Kanalogue) 2 mg.
Surgimin-T	Walker	10	10	100	20	2	300			4	1.5	
Thera-combex Kaps	Parke, Davis	25	15	100	10	1	150			5	2.5	
Vio-Bec Caps	Rowell	25	25	100	40	10	250					
Mixed Vitamin Caps	Mead-Johnson	3	4	25			100	10,000	1,000			
Multi-Vi Caps	White	1.5	2	10			37.5	5,000	500			
Pan-Concemin	Merrell	3	2	20	2	1	50	5,000	800			
Unicaps	Upjohn	2.5	2.5	20	5	0.5	50	5,000	500	2	0.25	
Vi-Zo-8	Pitman-Moore	1.5	2	5	1		30	5,000	800			E-6.2 mg.

Group: Vitamins B complex and C in Therapeutic Doses

Group: Vitamins A, D, B Complex and C in Maintenance Doses

APPENDIX III—Continued

	Product	Mfg.	B₁ mg.	B₂ mg.	Niac. mg.	Pant. mg.	B₆ mg.	C mg.	A units	D units	B₁₂ mcg.	Folacin mg.	Misc.
Vitamins A, D, B Complex and C in Maintenance Doses	Paladac Multi-V	Parke, Davis (3 therapeutic doses per 4 cc.)	3	3	20	5	1	50	5,000	1,000	5		
	Procebrin Multi-Vitamin Drops	Lilly (no therapeutic doses) contains per 0.6 cc.	2	1	16	3	1	120	6,000	1,600			
Vitamins A, D, B Complex and C in Therapeutic Doses	Alin-Diem Caps	McNiel	5	10	50			150	25,000	1,000			
	Bio Formula	Walker	25	10	150			150	25,000	2,500			
	Magnicaps	Plessner	10	5	150			150	25,000	1,000			
	Theragran (also liquid, same contents)	E. R. Squibb & Sons	10	10	150			150	25,000	1,000			
	Therapeutic Formula E	Ives-Cameron	10	10	150	10		150	25,000	1,000			E-3.4 mg.
	Thera-Vita	Warner	10	10	100	10	1	150	12,500	1,250			

There are many other preparations listed in Edwin P. Jordan, ed.: Modern Drug Encyclopedia and Therapeutic Index. 7th ed. New York, Drug Publications, Inc., 1958.

APPENDIX IV.   Equivalent Weights and Measures

EQUIVALENT WEIGHTS AND MEASURES

COMPARATIVE VALUES OF WEIGHT AND VOLUME OF WATER

1 liter	=	1 kilo. = 2.2 lbs.
1 fluid ounce	=	30 Gm. = 1.04 ozs.
1 pint	=	473 Gm. = 1.04 lbs.
1 quart	=	.946 kilo. = 2.1 lbs.

TABLE OF COMMON MEASURES AND METRIC EQUIVALENTS

1 tsp.	=	5 cc.
1 tbsp.	=	14 cc. (approx. 15 Gm.)
1 cup	=	225 cc. (approx. 240 Gm.)

(*See* Table of Measures and Approximate Weights, p. 407)

COMPARATIVE TEMPERATURES

Boiling water, sea level	100	212
Body temperature	37	98.6
Tropical temperature	30	89
Room temperature, average	20	70
Freezing	0	32

TABLE OF MEASURES AND APPROXIMATE WEIGHTS

3 teaspoons	1 tbsp.
16 tablespoons	1 cup
½ cup	1 gill
2 cups	1 pt.
4 cups	1 qt.
2 pints	1 qt.
4 quarts	1 gal.
1 tablespoon butter	½ oz.
*1 tablespoon liquid	½ oz.
1 tablespoon flour	¼ oz.
1 tablespoon sugar	⅜ oz.
*1 cup liquid	8 ozs.
1 cup flour	4½ ozs.
1 cup butter	8 ozs.
1 cup sugar	10 ozs.

* Water or milk.

APPENDIX IV—*Continued*

EQUIVALENT WEIGHTS AND MEASURES

WEIGHT EQUIVALENTS

	Milligram	Gram	Kilogram	Grain	Ounce	Pound
1 microgram (mcg.)	.001	.000001				
1 milligram (mg.)	1.	.001		.0154		
1 gram (Gm.)	1,000.	1.	.001	15.4	.035	.0022
1 kilogram (Kg.)	1,000,000.	1,000.	1.	15,400.	35.2	2.2
1 grain (gr.)	64.8	.065		1.		
1 ounce (oz.)		28.3		437.5	1.	.063
1 pound (lb.)		453.6	.454		16.0	1.

VOLUME EQUIVALENTS

	Cubic Millimeter	Cubic Centimeter	Liter	Fluid Ounce	Pint	Quart
1 cubic millimeter (cu. mm.)	1.	.001				
1 cubic centimeter (cc.)	1,000.	1.	.001			
1 liter (L.)	1,000,000.	1,000.	1.	33.8	2.1	1.05
1 fluid ounce		30.(29.57)	.03	1.		
1 pint (pt.)		473.	.473	16.	1.	
1 quart (qt.)		946.	.946	32.	2.	1.

LINEAR EQUIVALENTS

	Millimeter	Centimeter	Meter	Inch	Foot	Yard
1 millimeter (mm.)	1.	.1	.001	.039	.00325	.0011
1 centimeter (cm.)	10.	1.		.39	.0325	.011
1 meter (M.)	1,000.	100.	1.	39.37	3.25	1.08
1 inch (in.)	25.4	2.54	.025	1.	.083	.028
1 foot (ft.)	304.8	30.48	.305	12.	1.	.33
1 yard (yd.)	914.4	91.44	.914	36.	3.	1.

As given in Cooper, L. F., Barber, E. M., Mitchell, H. S., and Rymbergen, H. J.: Nutrition in Health and Disease. Philadelphia, J. B. Lippincott Company.

APPENDIX V. Average Height and Weight Tables for Children

HEIGHT-WEIGHT TABLES FOR GIRLS *

JUVENILE AND ADOLESCENT AGES

HEIGHT IN INCHES	5 Yrs	6 Yrs	7 Yrs	8 Yrs	9 Yrs	10 Yrs	11 Yrs	12 Yrs	13 Yrs	14 Yrs	15 Yrs	16 Yrs	17 Yrs	18 Yrs
38	33	33												
39	34	34												
40	36	36	36											
41	37	37	37											
42	39	39	39											
43	41	41	41	41										
44	42	42	42	42										
45	45	45	45	45	45									
46	47	47	47	48	48									
47	49	50	50	50	50	50								
48		52	52	52	52	53	53							
49			54	55	55	56	56							
50			56	57	58	59	61	62						
51			59	60	61	61	63	65						
52			63	64	64	64	65	67						
53			66	67	67	68	68	69	71					
54				69	70	70	71	71	73					
55				72	74	74	74	75	77	78				
56					76	78	78	79	81	83				
57					80	82	82	82	84	88	92			
58						84	86	86	88	93	96	101		
59						87	90	90	92	96	100	103	104	
60						91	95	95	97	101	105	108	109	111
61							99	100	101	105	108	112	113	116
62							104	105	106	109	113	115	117	118
63								110	110	112	116	117	119	120
64								114	115	117	119	120	122	123
65								118	120	121	122	123	125	126
66									124	124	125	128	129	130
67									128	130	131	133	133	135
68									131	133	135	136	138	138
69										135	137	138	140	142
70										136	138	140	142	144
71										138	140	142	144	145

* Prepared by Bird T. Baldwin, Ph.D., and Thomas D. Wood, M.D. Published originally by American Child Health Association.

APPENDIX V—*Continued*

HEIGHT-WEIGHT TABLES FOR BOYS *

JUVENILE AND ADOLESCENT AGES

HEIGHT IN INCHES	5 Yrs	6 Yrs	7 Yrs	8 Yrs	9 Yrs	10 Yrs	11 Yrs	12 Yrs	13 Yrs	14 Yrs	15 Yrs	16 Yrs	17 Yrs	18 Yrs	19 Yrs
38	34	34													
39	35	35													
40	36	36													
41	38	38	38												
42	39	39	39	39											
43	41	41	41												
44	44	44	44	44											
45	46	46	46	46	46										
46	47	48	48	48	48										
47	49	50	50	50	50	50									
48		52	53	53	53	53									
49		55	55	55	55	55	55								
50		57	58	58	58	58	58	58							
51			61	61	61	61	61	61							
52			63	64	64	64	64	64	64						
53			66	67	67	67	67	68	68						
54				70	70	70	70	71	71	72					
55				72	72	73	73	74	74	74					
56				75	76	77	77	77	78	78	80				
57					79	80	81	81	82	83	83				
58					83	84	84	85	85	86	87				
59						87	88	89	89	90	90	90			
60						91	92	92	93	94	95	96			
61							95	96	97	99	100	103	106		
62							100	101	102	103	104	107	111	116	
63							105	106	107	108	110	113	118	123	127
64								109	111	113	115	117	121	126	130
65								114	117	118	120	122	127	131	134
66									119	122	125	128	132	136	139
67									124	128	130	134	136	139	142
68										134	134	137	141	143	147
69										137	139	143	146	149	152
70										143	144	145	148	151	155
71										148	150	151	152	154	159
72											153	155	156	158	163
73											157	160	162	164	167
74											160	164	168	170	171

* Prepared by Bird T. Baldwin, Ph.D., and Thomas D. Wood, M.D. Published originally by American Child Health Association.

APPENDIX VI. Desirable Weights for Men and Women

Weight in Pounds According to Frame (in Indoor Clothing)*

DESIRABLE WEIGHTS FOR MEN AGED 25 AND OVER					DESIRABLE WEIGHTS FOR WOMEN AGED 25 AND OVER				
Height with Shoes, 1″ Heels		Small Frame	Medium Frame	Large Frame	Height with Shoes, 2″ Heels		Small Frame	Medium Frame	Large Frame
Feet	Inches				Feet	Inches			
5	2	112–120	118–129	126–141	4	10	92–98	96–107	104–119
5	3	115–123	121–133	129–144	4	11	94–101	98–110	106–122
5	4	118–126	124–136	132–148	5	0	96–104	101–113	109–125
5	5	121–129	127–139	135–152	5	1	99–107	104–116	112–128
5	6	124–133	130–143	138–156	5	2	102–110	107–119	115–131
5	7	128–137	134–147	142–161	5	3	105–113	110–122	118–134
5	8	132–141	138–152	147–166	5	4	108–116	113–126	121–138
5	9	136–145	142–156	151–170	5	5	111–119	116–130	125–142
5	10	140–150	146–160	155–174	5	6	114–123	120–135	129–146
5	11	144–154	150–165	159–179	5	7	118–127	124–139	133–150
6	0	148–158	154–170	164–184	5	8	122–131	128–143	137–154
6	1	152–162	158–175	168–189	5	9	126–135	132–147	141–158
6	2	156–167	162–180	173–194	5	10	130–140	136–151	145–163
6	3	160–171	167–185	178–199	5	11	134–144	140–155	149–168
6	4	164–175	172–190	182–204	6	0	138–148	144–159	153–173

* For nude weight, deduct 5 to 7 pounds (male) and 2 to 4 pounds (female).

Prepared by Metropolitan Life Insurance Company, 1960. Derived primarily from data of the Build and Blood Pressure Study, 1959.

APPENDIX VII.   Standards of Identity for Vitamin and Mineral Supplements as Suggested by the Food and Drug Administration in 1966

Multivitamin Supplement (per daily dose)

Vitamin A	1250 to 5000 U.S.P. units
Vitamin D	200 to 400 U.S.P. units
Ascorbic acid	25 to 75 mg.
Thiamine	0.35 to 1.4 mg.
Riboflavin	0.50 to 2.0 mg.
Niacin	5.5 to 22 mg.

Multimineral Supplement

Calcium	350 to 1400 mg.
Iron	5 to 20 mg.

Multivitamin and Mineral Supplement
To the above formulation, these optimal vitamins and minerals may be added

Vitamin E	10 to 30 I.U.
Vitamin B_6	0.50 to 2.0 mg.
Folic acid	0.03 to 0.10 mg.
Pantothenic acid	2.5 to 10 mg.
Vitamin B_{12}	2 to 5 mcg.
Phosphorus	350 to 1400 mg.
Magnesium	75 to 300 mg.
Copper	0.5 to 2 mg.
Iodine	0.04 to 0.15 mg.

APPENDIX VIII. Food and Drug Administration 1966 Regulations Applying to Foods

*I. Rules for Fortifying Foods as Suggested by
the Food and Drug Administration in 1966*

1. Fortification of cereal is authorized when a 25 per cent or more loss in potency results from processing the natural grain.

2. Fortified fruit or vegetable juices and drinks should contain 30 to 50 mg. of vitamin C per 4 oz. serving.

3. Fortified milk should contain 100 U.S.P. units of vitamin D per 8 oz. serving.

II. Rules for Labeling in General

Labels must not suggest that:

1. Foods are effective or adequate for treatment or prevention of disease because of their added vitamins and minerals.

2. A diet of ordinary food does not supply adequate amounts of vitamins and minerals.

3. A significant segment of the population of the United States is suffering from dietary deficiency of vitamins or minerals.

III. Rules for Labeling Low Calorie Foods

1. Low calorie foods to be labeled as such must contain only 15 calories per serving or provide only 30 calories in the average amount eaten daily.

2. Labels on artificially sweetened products must contain a comparison of the number of calories in a serving of the product made with an equivalent amount of natural sugar.

3. If the food calories are not reduced by at least 50 per cent by the use of artificial sweeteners, then these sweeteners should not be used.

(1 gm. of saccharin = 300 gm. of sugar in sweetness)

(1 gm. of cyclamic acid or sodium or calcium cyclamate = 30 gm. of sugar in sweetness)

Index